Advances in Experimental Medicine and Biology

Volume 1278

Advances in Experimental Medicine and Biology provides a platform for scientific contributions in the main disciplines of the biomedicine and the life sciences. This series publishes thematic volumes on contemporary research in the areas of microbiology, immunology, neurosciences, biochemistry, biomedical engineering, genetics, physiology, and cancer research. Covering emerging topics and techniques in basic and clinical science, it brings together clinicians and researchers from various fields.

Advances in Experimental Medicine and Biology has been publishing exceptional works in the field for over 40 years, and is indexed in SCOPUS, Medline (PubMed), Journal Citation Reports/Science Edition, Science Citation Index Expanded (SciSearch, Web of Science), EMBASE, BIOSIS, Reaxys, EMBiology, the Chemical Abstracts Service (CAS), and Pathway Studio.

2019 Impact Factor: 2.450 5 Year Impact Factor: 2.324

More information about this series at http://www.springer.com/series/5584

Song-Guo Zheng
Editor

T Regulatory Cells in Human Health and Diseases

Springer

Editor
Song-Guo Zheng
Department of Internal Medicine
Ohio State University College of Medicine and Wexner Medical Center
Columbus, OH, USA

ISSN 0065-2598 ISSN 2214-8019 (electronic)
Advances in Experimental Medicine and Biology
ISBN 978-981-15-6406-2 ISBN 978-981-15-6407-9 (eBook)
https://doi.org/10.1007/978-981-15-6407-9

This Springer imprint is published by the registered company Springer Nature Singapore Pte Ltd.
The registered company address is: 152 Beach Road, #21-01/04 Gateway East, Singapore 189721, Singapore

Contents

1 **Regulatory T Cells: Concept, Classification, Phenotype, and Biological Characteristics** 1
Yang Du, Qiannan Fang, and Song-Guo Zheng

2 **A Structure-Guided Delineation of FOXP3 Regulation Mechanism in IPEX** 33
Tengyun Ma, Xiaomin Song, Ciriaco A. Piccirillo, Guoping Deng, and Mark I. Greene

3 **Regulation of Treg Functions by the Ubiquitin Pathway** . . 47
Elena Montauti and Deyu Fang

4 **Ubiquitin-Dependent Regulation of Treg Function and Plasticity** 63
Yi Dong and Fan Pan

5 **Metabolic Choice Tunes Foxp3+ Regulatory T Cell Function** 81
Xiaoxia Wang, Hao Cheng, Yige Shen, and Bin Li

6 **Treg Cells and Epigenetic Regulation** 95
Joseph A. Bellanti and Dongmei Li

7 **TCR/ITK Signaling in Type 1 Regulatory T cells** 115
Michael C. McGee, Avery August, and Weishan Huang

8 **Adipose Tissue T Regulatory Cells: Implications for Health and Disease** 125
David Bradley, Dharti Shantaram, Alan Smith, and Willa A. Hsueh

9 **Intestinal Regulatory T Cells** 141
Vanessa R. Figliuolo da Paz, Deepa R. Jamwal, and Pawel R. Kiela

10 **The Association of Gut Microbiota and Treg Dysfunction in Autoimmune Diseases** 191
Yuying Liu, Dat Q. Tran, John William Lindsey, and Jon Marc Rhoads

11 Tregs in Autoimmune Uveitis 205
Zhaohao Huang, Wenli Li, and Wenru Su

12 Understanding and Targeting Human Cancer Regulatory T Cells to Improve Therapy 229
H. Ryan Kolb, Nicholas Borcherding, and Weizhou Zhang

13 TNF–TNFR2 Signal Plays a Decisive Role in the Activation of $CD4^+Foxp3^+$ Regulatory T Cells: Implications in the Treatment of Autoimmune Diseases and Cancer 257
Md Sahidul Islam, Yang Yang, and Xin Chen

14 The Pursuit of Regulatory T Cells in the Induction of Transplant Tolerance 273
Preston R. Arnold and Xian C. Li

15 Regulatory T Cells for the Induction of Transplantation Tolerance 289
Weitao Que and Xiao-Kang Li

Contributors

Preston R. Arnold Immunobiology and Transplant Science Center and Department of Surgery, Houston Methodist Hospital, Texas Medical Center, Houston, TX, USA
Department of Surgery, Weill Cornell Medical College of Cornell University, New York, NY, USA

Avery August Department of Microbiology and Immunology, College of Veterinary Medicine, Cornell University, Ithaca, NY, USA

Joseph A. Bellanti Department of Pediatrics, Georgetown University Medical Center, Washington, DC, USA
Department of Microbiology-Immunology, Georgetown University Medical Center, Washington, DC, USA
International Center for Interdisciplinary Studies of Immunology (ICISI), Georgetown University Medical Center, Washington, DC, USA

Nicholas Borcherding Department of Pathology and Medical Science Training Program, University of Iowa, Iowa City, IA, USA

David Bradley Diabetes and Metabolism Research Center, Division of Endocrinology, Diabetes and Metabolism, Department of Internal Medicine, The Ohio State University, Columbus, OH, USA
The Ohio State University Wexner Medical Center, Columbus, OH, USA

Xin Chen State Key Laboratory of Quality Research in Chinese Medicine, Institute of Chinese Medical Sciences, University of Macau, Macau SAR, China

Hao Cheng Bio-X Institutes, Shanghai Jiao Tong University, Shanghai, China

Guoping Deng Department of Immunology, School of Basic Medical Sciences, Peking University, Beijing, China

Yi Dong Department of Cell Biology, Johns Hopkins University School of Medicine, Baltimore, MD, USA

Yang Du Department of Pathology and Physiopathology, Guilin Medical University, Guilin, Guangxi, China

Laboratory of Tumor Immunology and Microenvironmental Regulation, Guilin Medical University, Guilin, Guangxi, China

Deyu Fang Department of Pathology, Northwestern University Feinberg School of Medicine, Chicago, IL, USA

Qiannan Fang Department of Clinical Immunology, The Third Affiliated Hospital of Sun Yat-sen University, Guangzhou, China

Vanessa R. Figliuolo da Paz Department of Pediatrics, University of Arizona, Tucson, AZ, USA

Mark I. Greene Department of Pathology and Laboratory Medicine, Perelman School of Medicine, University of Pennsylvania, Philadelphia, PA, USA

Willa A. Hsueh Diabetes and Metabolism Research Center, Division of Endocrinology, Diabetes and Metabolism, Department of Internal Medicine, The Ohio State University, Columbus, OH, USA

Weishan Huang Department of Pathobiological Sciences, School of Veterinary Medicine, Louisiana State University, Baton Rouge, LA, USA
Department of Microbiology and Immunology, College of Veterinary Medicine, Cornell University, Ithaca, NY, USA

Zhaohao Huang State Key Laboratory of Ophthalmology, Zhongshan Ophthalmic Center, Sun Yat-sen University, Guangzhou, China

Md Sahidul Islam State Key Laboratory of Quality Research in Chinese Medicine, Institute of Chinese Medical Sciences, University of Macau, Macau SAR, China

Deepa R. Jamwal Department of Pediatrics, University of Arizona, Tucson, AZ, USA

Pawel R. Kiela Department of Pediatrics, University of Arizona, Tucson, AZ, USA
Department of Immunobiology, University of Arizona, Tucson, AZ, USA

H. Ryan Kolb Department of Pathology, Immunology and Laboratory Medicine, University of Florida, Gainesville, FL, USA

Bin Li Shanghai Institute of Immunology, Department of Immunology and Microbiology, Shanghai Jiao Tong University School of Medicine, Shanghai Jiao Tong University, Shanghai, China

Dongmei Li Department of Microbiology-Immunology, Georgetown University Medical Center, Washington, DC, USA
International Center for Interdisciplinary Studies of Immunology (ICISI), Georgetown University Medical Center, Washington, DC, USA

Wenli Li State Key Laboratory of Ophthalmology, Zhongshan Ophthalmic Center, Sun Yat-sen University, Guangzhou, China

Xian C. Li Immunobiology and Transplant Science Center and Department of Surgery, Houston Methodist Hospital, Texas Medical Center, Houston, TX, USA
Department of Surgery, Weill Cornell Medical College of Cornell University, New York, NY, USA

Xiao-Kang Li Division of Transplantation Immunology, National Research Institute for Child Health and Development, Tokyo, Japan

John William Lindsey Department of Neurology, Division of Gastroenterology, The University of Texas Health Science Center at Houston McGovern Medical School, Houston, TX, USA

Yuying Liu Department of Pediatrics, Division of Gastroenterology, The University of Texas Health Science Center at Houston McGovern Medical School, Houston, TX, USA

Tengyun Ma Department of Immunology, School of Basic Medical Sciences, Peking University, Beijing, China

Michael C. McGee Department of Pathobiological Sciences, School of Veterinary Medicine, Louisiana State University, Baton Rouge, LA, USA

Elena Montauti Department of Pathology, Northwestern University Feinberg School of Medicine, Chicago, IL, USA

Fan Pan Center for Cancer Immunology Research, Shenzhen Institutes of Advanced Technology, Chinese Academy of Sciences, Shenzhen, Guangdong, China

Ciriaco A. Piccirillo Department of Microbiology and Immunology, McGill University, Montreal, QC, Canada
Centre of Excellence in Translational Immunology (CETI), Research Institute of the McGill University Health Centre, Montreal, QC, Canada

Weitao Que Division of Transplantation Immunology, National Research Institute for Child Health and Development, Tokyo, Japan

Jon Marc Rhoads Department of Pediatrics, Division of Gastroenterology, The University of Texas Health Science Center at Houston McGovern Medical School, Houston, TX, USA

Dharti Shantaram Diabetes and Metabolism Research Center, Division of Endocrinology, Diabetes and Metabolism, Department of Internal Medicine, The Ohio State University, Columbus, OH, USA

Yige Shen Shanghai Jiao Tong University School of Medicine, Shanghai Jiao Tong University, Shanghai, China

Alan Smith Diabetes and Metabolism Research Center, Division of Endocrinology, Diabetes and Metabolism, Department of Internal Medicine, The Ohio State University, Columbus, OH, USA

Xiaomin Song State Key Laboratory of Molecular Biology, CAS Center for Excellence in Molecular Cell Science, Institute of Biochemistry and Cell Biology, Chinese Academy of Sciences, Shanghai, China

Wenru Su State Key Laboratory of Ophthalmology, Zhongshan Ophthalmic Center, Sun Yat-sen University, Guangzhou, China

Dat Q. Tran Department of Pediatrics, Division of Gastroenterology, The University of Texas Health Science Center at Houston McGovern Medical School, Houston, TX, USA

Xiaoxia Wang Shanghai Institute of Immunology, Department of Immunology and Microbiology, Shanghai Jiao Tong University School of Medicine, Shanghai Jiao Tong University, Shanghai, China

Yang Yang State Key Laboratory of Quality Research in Chinese Medicine, Institute of Chinese Medical Sciences, University of Macau, Macau SAR, China

Weizhou Zhang Department of Pathology, Immunology and Laboratory Medicine, University of Florida, Gainesville, FL, USA

Song-Guo Zheng Department of Internal Medicine, Ohio State University College of Medicine and Wexner Medical Center, Columbus, OH, USA

Regulatory T Cells: Concept, Classification, Phenotype, and Biological Characteristics

1

Yang Du, Qiannan Fang, and Song-Guo Zheng

Abstract

Regulatory T cells (Treg) play an indispensable role in maintaining the body's immune nonresponse to self-antigens and suppressing the body's unwarranted and potentially harmful immune responses. Their absence, reduction, dysfunction, transformation, and instability can lead to numerous autoimmune diseases. There are several distinct subtypes of the Treg cells, although they share certain biological characteristics and have unique phenotypes with different regulatory functions, as well as mechanistic abilities. In this book chapter, we introduce the latest advances in Treg cell subtypes pertaining to classification, phenotype, biological characteristics, and mechanisms. We also highlight the relationship between Treg cells and various diseases, including autoimmune, infectious, as well as tumors and organ transplants.

Keywords

Treg cells · Immune response · Classification · Immunological diseases · Cell therapy

1.1 Introduction

Five decades ago, a subset of T cells with immunosuppressive properties was described by Dr. Richard Gershon and colleagues at Yale University (Gershon and Kondo 1970, 1971). However, the existence of this suppressor T cell was questioned due to lack of reliable markers. However, two decades later, Sakaguchi and his colleagues demonstrated that a CD4+ subpopulation that constitutively expresses CD25 (IL-2 receptor alpha chain) in mouse thymus showed inhibitory activity. This discovery led scientists to revisit the study of suppressor or regulatory T cells (Treg) (Sakaguchi et al. 1995). These thymus-derived naturally occurring CD4+CD25+ (tTreg) cells migrate to the periphery 3 days after birth and constitute 5–10% of the peripheral CD4+ T cells in normal naïve mice. Moreover, these cells have critical functional abilities since the removal of these CD4+CD25+ T cells causes a discrete autoinflammatory clinical phenotype: gastritis, thyroiditis, and type 1 diabetes (T1D). These manifestations were similar to which was

Y. Du
Department of Pathology and Physiopathology, Guilin Medical University, Guilin, Guangxi, China

Laboratory of Tumor Immunology and Microenvironmental Regulation, Guilin Medical University, Guilin, Guangxi, China

Q. Fang
Department of Clinical Immunology, The Third Affiliated Hospital of Sun Yat-Sen University, Guangzhou, China

S.-G. Zheng (✉)
Department of Internal Medicine, Ohio State University College of Medicine and Wexner Medical Center, Columbus, OH, USA

S.-G. Zheng (ed.), *T Regulatory Cells in Human Health and Diseases*, Advances in Experimental Medicine and Biology 1278, https://doi.org/10.1007/978-981-15-6407-9_1

seen in the neonatal thymectomy (NTx) mice. For a limited time after thymectomy, adoptively transferring normal CD4+CD25+ T cells can prevent the development of autoimmunity in these mice (Kojima and Prehn 1981; Sakaguchi et al. 1995; Itoh et al. 1999). Thus, CD4+CD25+ was reestablished as a Treg population. However, since CD25 is also an activation marker for CD4 + cells, its expression on CD4+ cells cannot be viewed as a specific marker for Treg cells.

A decade later, a milestone observation was made; describing the expression of the transcription factor Foxp3 (mice) or Foxp3 (human) in CD25+CD4+ in rodents and humans, respectively, can specifically define the Treg population (Fontenot et al. 2003; Khattri et al. 2003; Hori et al. 2003). Functional deletions and mutations in the human Foxp3 gene can cause severe autoinflammatory disease. The clinical syndrome resulting from this deletion was described much earlier and was called immune dysregulation, polyendocrinopathy, enteropathy, and X-linked (IPEX) syndrome (Powell et al. 1982). Thus, Foxp3 is a specific marker for these cells, and CD4+CD25+Foxp3+ cell population is designated regulatory T cells. Indeed, Treg cells, which naturally express Foxp3 in the nucleus and CD25 on the cell surface, are present in normal individuals and actively participate in suppressing aberrant or unwarranted immune responses against self, microorganisms, and bacteria (Martín-Orozco et al. 2017; Grant et al. 2015). Additionally, recent studies have established that the Treg cell lineage requires not only the transcription factor Foxp3 but also the establishment of Treg cell-specific CpG hypomethylation pattern (Ohkura et al. 2012).

Much progress has been made in demonstrating the role of Treg cells in the pathogenesis and development of many autoimmune diseases (Yang et al. 2019a, b). Furthermore, the role of Tregs in the tumor microenvironment is being closely examined since their numbers are increased in tumors (Kashima et al. 2020; Solis-Castillo et al. 2020). Whether these Treg cells belong to tTreg or they are induced locally is not clear (Horwitz et al. 2008; Lin et al. 2013). In addition to tTreg, induced Treg in the periphery (pTreg) or those in ex vivo with TGF-β and IL-2 (iTreg) represents a new development in the history of Treg cells (Zheng et al. 2002, 2004, 2007). In both immune responses to self-antigens and to tumor antigens, Foxp3+ Treg plays an inhibitory role. While the former, inhibition of immune response to self-antigens, is a beneficial; the latter is deleterious. Consistent with this concept is the observation that large numbers of Treg cells infiltrating a tumor tissue are often associated with poor prognosis (Tanaka and Sakaguchi 2017). More researches are needed to delineate in more details such as the role and mechanism of Treg cells in various immune responses in autoimmune diseases and in tumors. Enhancing the suppressive role of Treg cells in immune responses in autoimmunity and dampening those responses in malignancy are likely to become novel treatments for these diseases.

1.2 Treg and the Immune Response

Regulatory T cells (Tregs) are necessary for the maintenance of immune self-tolerance and homeostasis (Josefowicz et al. 2012). Treg cells have unique surface expression profiles, including CD4, CD25, CD62L, and specific CD45 isoforms. With the discovery of the specific transcription factor Foxp3 in Treg cell population, the concept of a regulatory immune cell has changed from a rare CD4+ T-cell subtype to an important immune homeostatic regulator (Hori et al. 2003). Although it is formally named as a regulatory T cell, its role mostly relates to immune suppression. Treg cells inhibit various immune responses, thus a defect or reduction of Treg number and/or function; or other Treg biological changes have been observed in various autoimmune diseases, such as RA (Prakken et al. 2013), SLE (Scheinecker et al. 2010), and IPEX

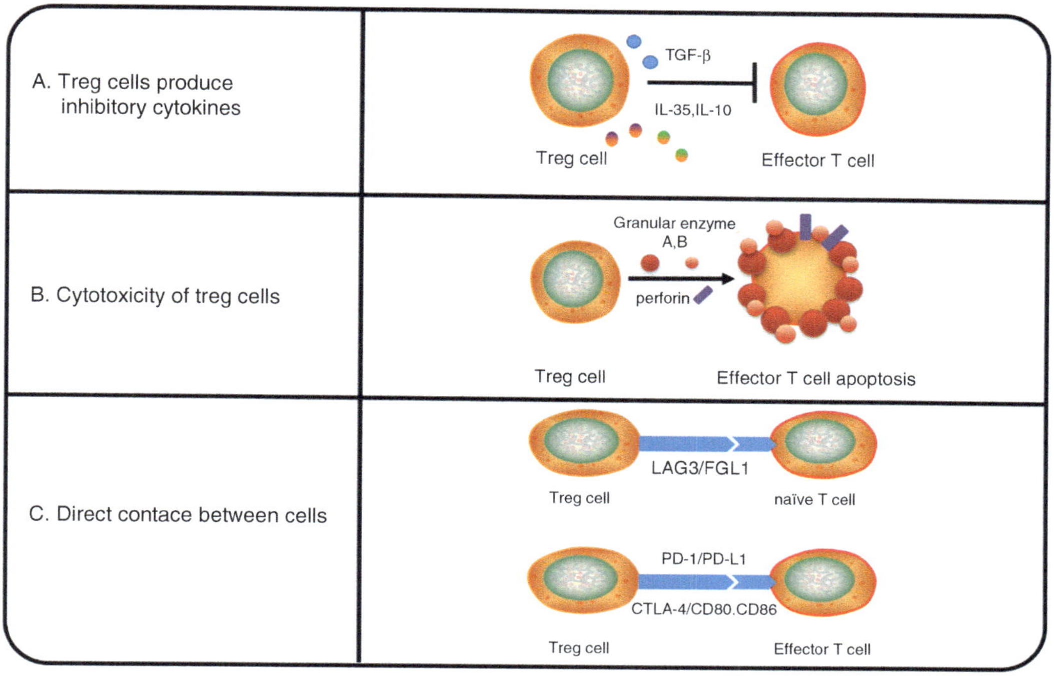

Fig. 1.1 Schematic diagram of the immunosuppressive mechanism of Treg

syndrome (Horino et al. 2014). Moreover, the changes in Treg in these various disorders are not incidental but have been shown to contribute to their development (Scheinecker et al. 2019). Conversely, an increase in Treg cells inhibits abnormal immune responses to autoantigens as well as antitumor immune responses. Large numbers of Treg cells infiltrating tumor tissue are often associated with poor prognosis (Tanaka and Sakaguchi 2017). In contrast, an increase in Tregs in the peripheral blood and in the graft microenvironment is considered important for inducing graft tolerance (Bahmani et al. 2018; Graca et al. 2002; Lee et al. 2005). Furthermore, an overabundance of Treg cells impairs the immune response in patients with infection (Costa et al. 2013). In general, Treg cells are part of the adaptive immune system playing a key role in maintaining homeostasis by exerting their immunosuppressive effects in normal and disease state (Mohr et al. 2019). In addition, we also describe various possible immunosuppressive mechanisms of Treg cells (Fig. 1.1).

1.3 Classification of Treg Cells

It is well known that there are many subpopulations of the Treg cell family (Fig. 1.2), including CD4+Foxp3+ cells, "Tr1" cells producing interleukin 10 (IL-10), CD8+ T suppressor cells, natural killer T cells, CD4−CD8− T cells, and γδ T cells. In this chapter, we focus on the first three types of Treg cells (Tang and Bluestone 2008; Zhou et al. 2011).

1.4 CD4+Foxp3+ Treg Cells

Among various Treg populations, CD4+Foxp3+ Treg cells are a key member of the Treg networks and display many important functional characteristics. Foxp3+ Treg cells are often classified as either "natural" or "induced" subsets. Natural tTregs are long-lived and develop in the thymus, while induced Tregs are thought to arise in the periphery after stimulation (pTreg). In

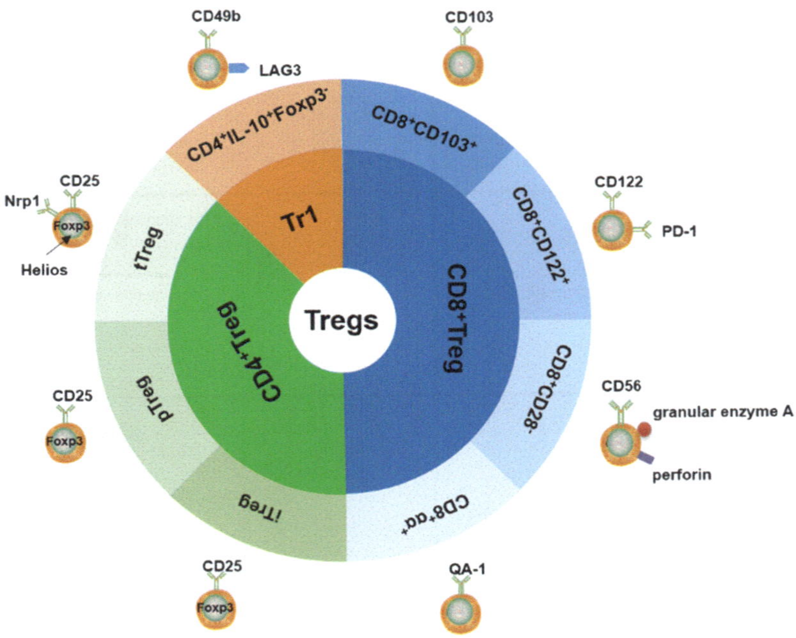

Fig. 1.2 Classification of Treg cells. The figure shows the classification of different subtypes of Treg cells and the characteristic markers of each type of Treg cells. Treg cells can be divided into CD4+ and CD8+ Treg cells. These two types of Treg cells can be further classified as various subpopulations as shown in the figure. In addition, in CD4+ Treg cells, another type of Tregs called Tr1 was identified. Unlike traditional CD4+Foxp3+ Treg, Tr1 cells were Foxp3− with high expression of IL-10

addition, TGF-β and IL-2 are crucial cytokines to induce Treg ex vivo (iTreg) (Zheng et al. 2002, 2007). These cell populations all express CD25 and Foxp3; however, they also have some differences that will be discussed later in this chapter.

Foxp3 is an important marker for the development and function of tTreg cells. It is expressed in mostly CD4+CD25+ T cells and a small number of CD4+CD25− T cells. However, CD25 is a surface marker that effectively identifies live Treg cells while Foxp3 is an intranuclear transcription factor that require permeabilization of the cell membrane; therefore, it cannot be used as a marker in live cells (Fontenot et al. 2003). Initially, the three research teams found that Foxp3 was expressed in a large and stable amount in mouse CD25+CD4+ Treg cells, but not in naïve CD25−CD4+ T cells or activated CD4+ T cells (Fontenot et al. 2003; Khattri et al. 2003; Hori et al. 2003). These findings support the critical role of Foxp3 in the differentiation of Treg cells. Foxp3-mutant scurfy mice spontaneously die at 3 weeks of age of a systemic autoinflammatory diseases (Godfrey et al. 1991). In humans, the loss of function or mutation of Foxp3 also causes severe autoinflammatory disease, like IPEX. These are due to the deficiency or abnormal function of CD25+CD4+ natural tTreg (Powell et al. 1982; Bacchetta et al. 2018). All these observations suggest that Foxp3 is a functional molecule of Treg cells. Nonetheless, it has recently been reported that Foxp3 expression alone is not sufficient for Treg lineage commitment. It has been shown that demethylation of a Treg-specific demethylation region (TSDR) in the Foxp3 promoter plays a crucial role in the maintenance of tTreg lineage and that this demethylation of TSDR is associated with a stable tTreg cell phenotype (Ohkura et al. 2012).

tTregs are produced by immature Treg precursors of heat-stable antigen $(HAS)^{hi}$ CD4 single-positive (SP) stage when Foxp3 is induced and tTreg lineage commitment is established (Lee and Hsieh 2009). In the presence of IL-2 and TGF-β, peripherally derived Treg (pTreg) cells differentiate from naïve T cells in peripheral sites. Those induced by TGF-β and IL-2 in vitro are called induced Treg (iTreg) cells (Fig. 1.3). pTreg and iTreg share some similarities;

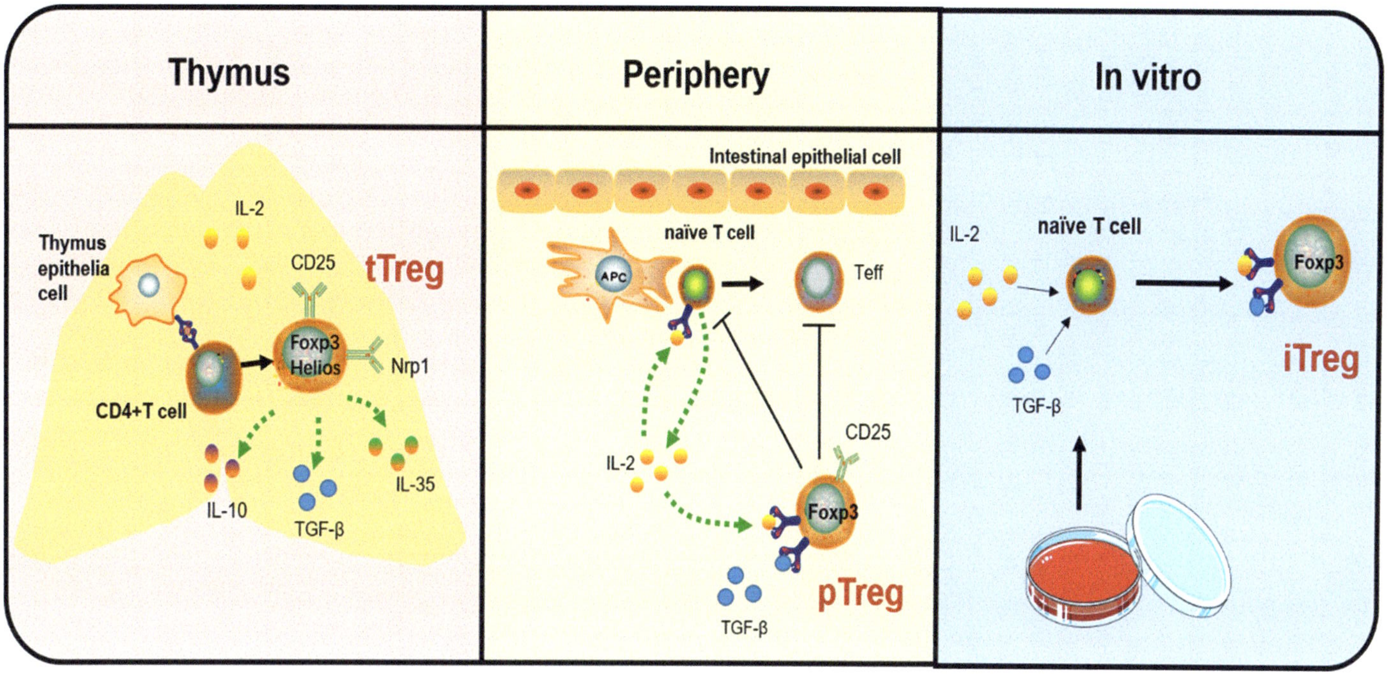

Fig. 1.3 Development and differentiation of Treg cells. The development and differentiation of Treg cells from thymus, peripheral, and in vitro are shown. Thymus-derived Tregs (tTreg) originates from the thymocytes in response to self-antigen stimulation during T cell development in vivo. In addition, induced Tregs develop in the periphery (pTreg), and induced Tregs (iTreg) differentiate from naïve T cells by means of T-cell receptor (TCR) stimulation through exposure to the IL-2 and TGF-β ex vivo

however, they also have some differences that will be discussed later. Human CD4+Foxp3+ T cells are composed of three subsets with different phenotypes and functions: CD45RA(+) Foxp3 (lo) inactive Treg cells (rTreg cells), CD45RA (−) Foxp3 (hi) activated Treg cells (aTreg cells), and cytokine-secreting CD45RA(−) Foxp3 (lo) non-inhibitory T cells (Miyara et al. 2009). It has been shown that the first two populations had inhibitory effects in vitro. The terminally differentiated aTreg cells die rapidly, and the rTreg cells proliferate and become aTreg cells in vitro and in vivo (Miyara et al. 2009).

Although CD4+Foxp3+ Treg population has attracted widespread attention for their role in maintaining immune homeostasis, studies have also found that CD4+ Tr1 and CD8+ Treg showed the immunomodulatory functions as well. The number and/or function of these cells are impaired in several autoimmune diseases and autoimmune experimental animal models (Roncarolo et al. 2018; Pellegrino et al. 2019), suggesting that immunotherapy targeting these cells can also improve autoimmune status management (Dinesh et al. 2010; Huang et al. 2017, 2018).

1.5 Tr1 Cells

Unlike other CD4+ T-cell subsets, Tr1 cells are a Treg cell type characterized by the expression of CD49b and LAG3 (Gagliani et al. 2013), as well as the lack or transient expression of the transcription factor Foxp3 and a significant expression of IL-10 (Vieira et al. 2004; Groux et al. 1997). Since their discovery, Tr1 cells have been shown to play an important role in maintaining immune homeostasis and preventing T-cell-mediated diseases (Vieira et al. 2004). Tr1 cells are made in the periphery after antigen exposure and under tolerogenic conditions. In vivo and in vitro studies have confirmed their inhibitory effects in mouse and human microenvironment. Specifically, Tr1 cells prevented and downregulated aberrant immune responses to pathogenic and nonpathogenic antigens and was also associated with long-term tolerance in humans (Bacchetta et al. 1994; Gianfrani et al. 2006; Serafini et al. 2009; Roncarolo et al. 2014; Globinska et al. 2018). Immunosuppression observed in some infectious diseases is associated with a higher frequency of Tr1 cells (Chang 2007; Koch et al. 2015).

Tr1 cells do not constitutionally express Foxp3 (Vieira et al. 2004), but once activated, they can rapidly upregulate Foxp3 (Levings et al. 2005; Brun et al. 2009, 2011), but this upregulation is not comparable to tTreg or iTreg. In addition to secreting large amounts of IL-10, Tr1 cells also secrete TGF-β, IL-5, GM-CSF, and IFN-γ and small amounts of IL-4, IL-17, and IL-2 (Bacchetta et al. 1990, 1994; Groux et al. 1997). A key aspect of Tr1 cell-mediated regulation is that these cells need to be activated by their TCR to show their regulatory activity, which reflects their antigen specificity. After activation, Tr1 cells secrete IL-10 and TGF-β, which directly and indirectly inhibit T-cell response (Roncarolo et al. 2014). Generally, IL-10 limits the magnitude of the immune response and inhibits the response of T effector cells by downregulating the expression of MHC II glycoproteins (de Waal Malefyt et al. 1991), co-stimulatory molecules, and pro-inflammatory cytokines that are produced by APC (Gregori and Roncarolo, 2018; Roncarolo et al. 2014).

The differentiation and function of Tr1 cells depend on the presence of IL-10, an effective immunosuppressive cytokine even though it has pleotropic effects. Tr1 cells can inhibit the response of effector cells in an IL-10-dependent manner by interacting with CTLA-4 and PD-1 (Roncarolo et al. 2014; Gregori et al. 2012) or by killing pro-inflammatory cells directly with a granzymes (Gagliani et al. 2013; Huber et al. 2011). Tr1 cells also express extracellular enzymes CD39 and CD73, which produce adenosine through enzymatic hydrolysis of extracellular ATP and disrupt the metabolic state of effector T cells (Mandapathil et al. 2010; Mascanfroni et al. 2015; Su et al. 2019). Moreover, the role of synergistic inhibitory receptors (such as LAG-3, TIGIT, and TIM3) in the regulation of Tr1 cells is under investigation. However, it was observed that injection of antibodies against

LAG-3 reversed the state of immune tolerance induced by Tr1 cells, and it was predicted that these co-inhibitory receptors may contribute to the suppressive function of Tr1 cells (Jofra et al. 2018). Interestingly, IL-27 also plays a key role in Tr1 cell induction in mice and provides an alternative mechanism for the generation of these cells. As such, it was demonstrated that the development of Tr1 cells in intestinal associated lymphoid tissues is independent of IL-10 (Maynard et al. 2007). Short-term activation of mouse T cells in the presence of IL-27 leads to the production of Tr1 cells both in vitro and in vivo (Awasthi et al. 2007; Batten et al. 2008; Fitzgerald et al. 2007; Iwasaki et al. 2013; Pot et al. 2009; Stumhofer et al. 2007).

IL-27 promotes IL-10 production in mouse CD4+ T cells by activating the STAT1 and STAT3 pathways (Awasthi et al. 2007; Fitzgerald et al. 2007; Stumhofer et al. 2007). However, the role of IL-27 in inducing human Tr1 cells remains unclear. Recent research has established that the functional development of Tr1 cells requires TCR/ITK signaling through the Ras/IRF4 pathway (Huang et al. 2017).

Like Foxp3+ Treg cells, Foxp3-IL10+ Tr1 cells also have a therapeutic potential in the treatment of inflammatory diseases. The overactivation of NLRP3 inflammasome is the basis of several common chronic inflammatory diseases (Robbins et al. 2014; Agostini et al. 2004). It was speculated that Foxp3+ Treg and/or Tr1 cells may play a role in regulating inflammasome activities. This was examined and only Tr1 cells can regulate this pathway (Yao et al. 2015). More studies are needed to compare the functional difference between Foxp3+ Treg and Tr1 cells.

1.6 CD8+ Treg Cells

CD8+ Treg subset is a subgroup of CD8+ T cells with inhibitory potential first discovered in 1972 (Gershon et al. 1972). However, the lack of unique markers to identify these inhibitory T cells led to long hiatus in this field of study. Interest in these studies began to re-emerge in the 1990s (Noble et al. 1998). In 2007, the function of these cells in the context of viral infections and tumorigenesis was highlighted (Guillonneau et al. 2007). Like CD4+ Treg cells, they also can be divided into multiple subtypes, including CD8+CD103+, CD8+CD122+, CD8+CD28−, and CD8+ Qa-1-restricted CD8αα+TCRαβ+ Treg cell populations.

1.6.1 CD8+CD103+ Treg Cells

CD8+CD103+ T cells are considered to be suppressive CD8+ T cells because they have been shown to control immune responses both in vivo and in vitro (Ma et al. 2015; Lerret et al. 2012). However, the exact properties of the CD8+CD103+ regulatory T cells, as well as their suppressive function and related mechanism in allogeneic transplantation, remain elusive.

CD103 (αE7β integrin) is an E-cadherin receptor that is present on the surface of CD8+ T lymphocytes in thymus, small intestine, bronchoalveolar fluid, and allografts in mice and humans (Pauls et al. 2001; Rihs et al. 1996; Sarnacki et al. 1992; Liu et al. 2014; Zhong et al. 2018; Zhang et al. 2019). CD103 has been described as the first affinity ligand that mediates cell adhesion and migration and has the function of directing T lymphocytes to epithelial cells that express E-cadherin (Cepek et al. 1994; Karecla et al. 1995). In humans, CD103 expression is limited to 1% of circulating memory T cells (Picker et al. 1990). In mice, CD103 is expressed in 40–60% of peripheral CD8+ T cells (Hadley et al. 1997), mainly in those showing a naïve phenotype (Wang et al. 2004).

The relationship between CD103 and CD8+ T cells has been established in studies of renal transplant rejection and intestinal graft versus host disease (El-Asady et al. 2005). It was found that CD103 expression was significantly upregulated in a TGF-β-dependent manner on CD8+ T cells entering kidney transplantation sites or intestinal epithelial cells. This demonstrated that CD103 characterizes CD8+ effector T cells (Wang et al. 2004; Hadley et al. 2001; El-Asady et al. 2005). However, subsequent research found that these cells have regulatory activity. In these studies, alloreactive CD8+CD103+ T cells are produced by the stimulation using alloantigen but not anti-CD3/CD28 antibodies (Uss et al. 2006). CD8+CD103+ Tregs are mostly CD28+; they also lacked CD25, Foxp3, CTLA-4, LAG-3, and GITR expression. Moreover, the presence of TGF-β resulted in increased expression of CD103 (Hadley et al. 1997).

CD8+CD103+ Treg cells inhibit T-cell proliferation in mixed lymphocyte cultures in a manner that depends on cell-to-cell contact (Koch et al. 2008). Even though they also secreted IL-10 and TGF-β, this inhibitory effect is not mediated through these cytokines. This makes them to be different from Tr1 cells. However, in other related studies, it was demonstrated that IFN-γ binding to CD8 Tregs is required to mediate its TGF-β-based suppression (Myers et al. 2005). Meanwhile, they identified a murine peptide-specific CD8+ T regulatory cell population that inhibited the response of CD4+ T cells. However, TGF-β expression in situ tightly associates with CD103 expression (Robertson et al. 2001). Chronic ileitis in TNF-driven model as well as liver transplantation provides compelling evidence that CD8+CD103+ Treg cells that secreted TGF-β have a protective effect (Ho et al. 2008; Lu et al. 2009).

Recent studies have provided multiple pathways for the induction of antigen-specific CD8+ Treg with inhibitory functions. The latest research has successfully established a method using TGF-β plus RAPA in addition to CD3/CD28 and IL-2 stimulation to effectively induce human CD8+ Tregs in vitro. These newly induced hCD8+ Tregs express stable and high levels of Foxp3, CD103, and PD-1 but have no IL-17A secretion, and survive in vivo after adoptive transfer in a CIA animal model (Sun et al. 2019). Moreover, these cells have a strong inhibitory ability and can alleviate collagen-induced arthritis. Our group has reported that CD8+CD103+ iTreg cells can be induced ex vivo in the presence of TGF-β and IL-2 (Liu et al. 2014; Zhong et al. 2018; Zhang et al. 2019). Interestingly, the functional activity of these Treg populations does not require Foxp3 expression (Liu et al. 2014; Zhong et al. 2018). Moreover, we further found that CD39 expression participated in the functional activity of these Treg population (Zhang et al. 2019). Further studies are warranted to determine the differences among these Treg populations.

1.6.2 CD8+CD122+ Treg Cells

CD8+CD122+ Treg cells are another natural CD8 + Treg subtype, which has been shown to have inhibitory capacity in transplantation and autoimmunity (Suciu-Foca et al. 2003; Rifa'i et al. 2004). Previous studies have shown that CD8 +CD122+ Treg cells can control immune homeostasis, inhibit traditional T-cell responses (Rifa'i et al. 2004; Endharti et al. 2005; Chen et al. 2008; Shi et al. 2008;Molloy et al. 2011;Endharti et al. 2011), and regulate autoimmune responses (Kim et al. 2011; Mangalam et al. 2012), including experiments autoimmune encephalomyelitis (Mangalam et al. 2012; Lee et al. 2008; Seifert et al. 2017), Graves' disease (Saitoh et al. 2007), and colitis (Endharti et al. 2011). Earlier studies described CD8+CD122+ T cells as antigen-specific memory T cells (Zhang et al. 1998; Ku et al. 2000; Judge et al. 2002). CD122 expression was initially described in nonregulatory memory lymphocytes (Zhang et al. 1998; Judge et al. 2002). Subsequently, it was reported that central memory CD8+CD122+ T cells ($CD44^{high}CD62L^{high}$) also play a role in regulating T-cell homeostasis and act as regulatory T cells (Rifa'i et al. 2004). A series of studies in recent years have further shown that CD8+CD122+ T cells indeed inhibit conventional T-cell responses (Rifa'i et al. 2004; Endharti et al. 2005; Chen et al. 2008; Shi et al. 2008; Molloy et al. 2011; Endharti et al. 2011; Wang et al. 2010) and control autoimmune diseases (Kim et al. 2011; Mangalam et al. 2012). Moreover, some studies have found that memory CD8+CD122+ T cells and bystander central memory CD8+ T cells also belong to Treg that inhibit mouse allograft rejection (Wan et al. 2008; Dai et al. 2010). Furthermore, other studies have suggested that central memory CD8

+ T cells can modulate the acceptance of allogeneic lungs (Krupnick et al. 2014).

CD122 is β subunit of IL-2 receptor, while CD25 is an α subunit of the same receptor on T cells (Sakaguchi et al. 1995). Murine CD8+CD122+ Tregs express CD122 (IL-2Rβ) and CXCR3 but do not express CD25, while their CD4+CD25+ counterpart does not express CD122. CD8+CD122+ Tregs are also $CD44^{high}$, $CD62L^{high}$ CCR7+, and most CD127− (Suzuki et al. 2008; Dai et al. 2010). However, CD8+CD122+ Tregs are Foxp3 negative (Dai et al. 2010), suggesting that they are a different subset of induced CD8+Foxp3+ Treg cells (Lerret et al. 2012).

In a comparison study, investigators demonstrated that CD8+CD122+ Tregs were more effective than CD4+Foxp3+ Tregs in inhibiting allograft rejection (Dai et al. 2014). In fact, PD-1 is a key marker in identifying whether CD8+CD122+ T cells are Treg cells. The CD8+CD122+PD-1+ group in the CD8+CD122+ population is mainly responsible for CD8+CD122+ Treg-mediated inhibition (Dai et al. 2010). We previously also reported that PD1 and TNFRII expression contributes to CD8+ Treg functional capacity (Horwitz et al. 2013; Jacob et al. 2009; Yang et al. 2018, 2019a, b). Nonetheless, PD1 is also a marker of T-cell exhaustion, contributing to functional decline of these cells (Simon and Labarriere 2017). The expression of PD1 in Treg function and vitality needs more in-depth studies.

The effects of combined CD8+CD122+PD-1+ Treg and conventional co-stimulatory blockade on allograft rejection and allogeneic immunity were examined recently. A synergistic response was demonstrated with CD8+CD122+PD-1+ Treg and CD40/CD154, but no synergism was seen with B7/CD28 blockade on prolonging the survival time of skin allografts in wild-type mice (Liu et al. 2019). The B7/CD28 co-stimulatory block, but not CD40/CD154, had a negative impact on the in vivo expansion of adoptively transferred Treg and its IL-10 production in vitro (Liu et al. 2019). The increased effects of CD8+CD122+PD-1+ Treg cells on the survival rate of allografts depend to a considerable extent on the expression of IL-10 by Treg cells (Dai et al. 2010; Liu et al. 2017). CD8+CD122+ Tregs may regulate the immune response by producing IL-10, TGF-β1 and IFNγ, but the exact mechanism of their inhibitory effect is still unclear.

1.6.3 CD8+CD28− Cells

CD8+CD28− T cells are also classified as natural CD8+ Treg population. Among different CD8+ Treg subgroups, non-antigen-specific CD8+CD28− Tregs have been associated with a variety of clinical conditions such as pregnancy, cancer, organ transplants, and infectious diseases. Different reports showed that CD8+CD28−Tregs inhibit T-cell proliferation in an IL-10-dependent or TGF-β-dependent manner (Fenoglio et al. 2008; Miller et al. 2010).

CD8+ T cells play a key role in the recognition and clearance of intracellular pathogen-infected cells (Nagata and Koide 2010) and anti-tumor response (Mempel and Bauer 2009). Binding of CD8+ T-cell surface receptor TCR and MHC-I-binding antigen expressed on the surface of professional antigen-presenting cells (pAPC) led to the activation of CD8+ T cells (Strioga et al. 2011). However, the optimal activation of CD8+ T cells cannot be maintained by TCR stimulation alone, and a second co-stimulation signal is required for the complete activation and survival of these cells (Boesteanu and Katsikis 2009). The interaction between CD28 molecule on T lymphocytes and CD86 and CD80 molecules expressed on the surface of pAPC provides a co-stimulating signal that has been well described (Strioga et al. 2011). When a sufficient signal is delivered to naive CD8+ T cells, they will proliferate and differentiate into two different cell types. One is cytotoxic T lymphocytes (CTLS), which undergo apoptosis upon maturation and effector function. The other is CD8+ memory T cells, which are both central and effector cells (Zheng et al. 2006). Their continuous presence in the circulation is important for controlling another potential exposure to the same antigen in a faster and more efficient manner (Kaech et al. 2002). Chronic antigenic stimulation leads to repeated activation cycles, which progressively loses CD28 molecule expression with each cycle of activation. This leads to the accumulation of "highly antigenic experienced" T cells with the CD8+CD28− phenotype, characterized by extremely short telomeres (Vallejo 2005). There is a close relationship between CD28 and telomerase denaturation. Telomerase activity is essential for cell proliferation, production of cytokines and chemokines, and antiviral activity. However, lack of CD28 leads to loss of the ability to activate telomerase activity in the cells. Maintaining the presence of CD28 through in vitro gene transduction slows down the rate of "immune aging" and improves the efficiency of the immune system (Cohen et al. 2013). Telomeres build on the ends of chromosomes and ensure their stability. Unprotected ends of chromosomes are at high risk of degradation, which leads to loss of genetic information and cell death (Kim Sh et al. 2002). Studies have found an association between a shortened telomere length in peripheral blood cells and autoimmune diseases, such as SLE (Haque et al. 2013; Honda et al. 2001), rheumatoid arthritis (Colmegna et al. 2008), systemic sclerosis (SSC) (Artlett et al. 1996), ANCA-related vasculitis (AAV) (Vogt et al. 2003), psoriasis, and atopic dermatitis (Wu et al. 2000). It is believed now that one of the main causes of abnormal immune response is abnormal telomeres in autoimmunity (Montoya-Ortiz 2013). Moreover, the loss of CD28 has been observed to be associated with increased surface expression of CD57 molecules. CD8+CD28− (CD8+CD57+) T cells are known to be antigen-specific, terminally differentiated, but are also known as functional memory or effector T cells that undergo multiple cell division cycles. These cells are characterized by reduced or even loss of telomerase five activity and low levels of expression of genes involved in cell cycle regulation. CD8+CD28− (CD8+CD57+) T cells are generally limited in their ability to proliferate after stimulation and are thought to have reached a state of "replicative senescence" or "clonal failure" (Strioga et al. 2011; Focosi et al. 2010).

The relationship between CD8+CD28− (CD8+CD57+) lymphocytes and apoptotic sensitivity is controversial. Some researchers (Borthwick et al. 2000; Wood et al. 2009) reported that these cells are highly sensitive to activation-induced apoptosis because they have observed an increased expression of Fas and caspase-3 and a decreased expression of anti-apoptotic molecules such as survivin or heat shock protein 27 (HSP27). Others suggested that CD8+CD28− (CD8+CD57+) T lymphocytes are highly resistant to apoptosis and thus accumulate gradually throughout life (Spaulding et al. 1999; Effros 2011).

Most autoimmune diseases are associated with an increase in CD8+CD28− (CD8+CD57+) T cells, which exhibit high cytotoxic activity, and their presence may correlate with more severe manifestations of the disease. Such changes in CD8+CD57+ population has been observed in multiple sclerosis (Mikulkova et al. 2010), type 1 diabetes (Mikulkova et al. 2010), Graves' disease (Sun et al. 2008), and rheumatoid arthritis (Wang et al. 1997). Moreover, regulatory properties have been attributed to lymphocytes with a CD8+CD28− phenotype. Further analysis confirmed the expression of Foxp3 in these cells (Frisullo et al. 2010; Manavalan et al. 2004). However, this finding was not confirmed by other groups (Korecka-Polak et al. 2011; Scotto et al. 2004). In addition to the lack of Foxp3 expression, characteristic markers of cytotoxic cells, such as granzyme A or perforin, were detected on the surface of CD8+CD28− population (Baeten et al. 2006).

1.6.4 CD8+ Qa-1-Restricted CD8αα+TCRαβ+ Regulatory T Cell

Major histocompatibility complex (MHC) class Ib molecules consist of rodent QA-1, QA-2, H2-M3, and CD1d and human leukocyte antigen (HLA)-E, HLA-G, and CD1(Braud et al. 1999). Qa-1-restricted CD8αα+TCRαβ+ T cells have been recognized as another CD8+ Treg subset (referred to as CD8αα+ Treg) that recognizes an antigenic determinant in the conserved CDR2 region of the TCRβ.82 chain (Smith et al. 2010). CD8αα+ Tregs can control experimental autoimmune encephalomyelitis (EAE), a prototype for multiple sclerosis. They control EAE by inducing apoptosis in activated and pathogenic CD4+ cells after recognition of the TCR–peptide/Qa-1 complex on their cell surface (Smith et al. 2009, 2010). There were data showing that a unique phenotype of CD8αα+TCRαβ+ Treg cells is enriched in a number of molecules expressed by NK cells, members of TNF-superfamily as well as in negative signaling molecules, including CD200 (Fanchiang et al. 2012). In general, peripheral class Ib-responsive CD8αα+TCRαβ+ T cells represent a unique type of regulatory T cells that are different from class Ia MHC-restricted conventional T cells. These findings have important significance for understanding the regulatory mechanism mediated by the CD8+ Treg cell population, and much work needs to be done to clarify their relationship to other CD8+ Treg populations.

1.7 Similarity and Differences Between tTreg and iTreg Cells

As we discussed before, CD4+Foxp3+ Tregs can be classified into three subtypes, thymus-derived Treg (tTreg) originates from the thymocytes in response to self-antigen stimulation during T-cell development in vivo, induced Treg in the periphery (pTreg) and induced Treg (iTreg) from naïve T cells by means of T-cell receptor (TCR) stimulation through exposure to the IL-2 and TGF-β ex vivo (iTreg) (Zheng et al. 2002; Sakaguchi 2004). The different Treg subtypes share significant similarities, such as their high levels of IL-2 receptor alpha chain (CD25) expression and dependence on the forkhead box P3 transcription factor (Foxp3). As pTregs are developed in vivo, it is hard to distinguish them from tTregs, and Foxp3+CD4+ cells in the periphery actually could be a mixture population of both tTreg and pTreg cells. We will focus here on similarities and differences between tTreg and iTreg subpopulations.

First, developmentally, iTreg requires TGF-β signaling since lack of this signal fails to induce iTreg (Zheng et al. 2002, 2004; Lu et al. 2010). However, tTreg development is independent upon TGF-β signaling (Piccirillo et al. 2002; Jordan et al. 2001). Second, CTLA-4 has an essential role in the generation of Foxp3 and acquisition of suppressor activity by naïve cells activated with TGF-β in vitro while it is not necessary for the development of tTreg cells in the thymus (Zheng et al. 2006; Liang et al. 2005). Third, phenotypically, both tTreg and iTreg have negative immune regulation function in vitro and in vivo through the secretion of inhibitory cytokines, such as TGF-β, IL-10, IL-35, and cAMP, expression of ectoenzymes CD39 and CD73 to degrade extracellular ATP, and both express similar levels of Foxp3. In contrast, phenotypic distinguishable features include tTregs overexpress Helios (a member of the Ikaros family of transcription factors) and Nrp1(a type 1 transmembrane protein), which are the two immunosuppressive protein molecules, while iTregs frequently express less of those proteins (Atif et al. 2020; Thornton et al. 2010; Weiss et al. 2012). However, there is controversy surrounding these observations, and others suggested that iTregs also express similar levels of Helios and Nrp1 (Akimova et al. 2011; Verhagen and Wraith 2010; Szurek et al. 2015). Therefore, there is a need to identify specific markers that reliably distinguish these two populations (Wing et al. 2019). Fourth, the two Treg populations have functional differences in inflammation and high salt diet. tTreg population tends to convert from regulatory cells to effector cells with loss of immune regulation ability in inflammatory and high salt microenvironment (Wu et al. 2013a, b; Kleinewietfeld et al. 2013; Sakaguchi et al. 2013). However, we and others have demonstrated that iTreg sustained their function even under inflammation and high salt conditions (Zheng et al. 2008; O'Connor et al. 2010; Zhou et al. 2010; Kong et al. 2012a, b), demonstrating potential advantages of iTreg versus tTreg in treating these diseases. Interestingly, *all-trans* retinoic acid could overcome tTreg instability in the presence of inflammatory milieu in mice and humans

Table 1.1 Similarity and differences between tTreg and iTreg cells

		Thymus-derived (tTreg)	Induced in vitro (iTreg)
Similarities	High levels of CD25 and Foxp3 secreting TGF-β, IL-10, IL-35 expression CD39 and CD73 have suppressive activities in vitro and in vivo		
Differences	**Development** TGF-β CLTA-4	Not Required Not dependent	Required Dependent
	Phenotype Helios/Nrp1	Overexpression	Less expression
	Stability (inflammatory and high salt environment)	Not stable (convert to effector cells)	Stable
	Function of B cells	Killing ability	Cytokine secretion
	Antigen-specificity	Polyclonal	Antigen-specific
	Epigenetic	Hypomethylation	Hypermethylation
	Treatment on established CIA	Not effective	Effective

(Zhou et al. 2010; Lu et al. 2014; Luo et al. 2019). Fifth, iTreg and tTreg cells have a unique mechanism in targeting B cells. It has been previously reported that tTregs suppressed B-cell function mainly through killing target cells (Zhao et al. 2006; Iikuni et al. 2009), while iTregs suppressed B cells via their cytokine secretion rather than killing (Xu et al. 2016). The latter mechanism may be more optimal. Sixth, tTregs are polyclonal while iTregs can be developed into antigen-specific iTregs (Zheng et al. 2006; Kong et al. 2012a, b). In addition, tTreg cells are rare cell population while iTreg cells can be manipulated in a large scale, demonstrating their translational value. Seventh, there are also some similarities and differences between the two Treg subpopulations in the epigenetic mechanisms such as specific DNA methylation, histone modification which show the important role for the differentiation and stabilization of cells. tTreg cells exhibit the CpG hypomethylation pattern of the Treg cell representative regions such as Foxp3, Tnfrsf18, Ctla4, Ikzf4, and Il2ra even after anti-CD3/CD28 stimulation. The methylation status of those regions is stably high in iTreg except Il2ra which is also gradually demethylated with TCR stimulation. Interestingly, histone modifications are less specific for tTreg because of the similar histone such as H3K4me3 and H3K27me3 modification in the Treg associated genes of tTreg and iTreg cells (Ohkura et al. 2012). The combination of hypomethylation establishment and Foxp3 expression are essential for regulating and stabilizing the expression of the molecules required for tTreg cell development and function. In fact, many studies demonstrated that iTregs are superior to tTreg in treating the established autoimmune and inflammatory diseases (Kong et al. 2012a, b; Su et al. 2012). The studies are needed to delineate the significance of these Treg populations. The difference between these two Treg populations were compared in detail in Table 1.1.

1.8 Treg in Diseases

1.8.1 Treg in Autoimmune Diseases

Autoimmune diseases are characterized by a breakdown in immune tolerance. Regulatory T cells regulate peripheral immune tolerance and homeostasis and play an important role in the development of autoimmune diseases. Functional defects and reduced numbers of Treg cells have been observed in a variety of autoimmune diseases (Scheinecker et al. 2019) including type 1 diabetes, multiple sclerosis, systemic lupus erythematosus (SLE), myasthenia gravis, and rheumatoid arthritis (RA) (Chen et al. 2013; Goschl et al. 2019). Patients with CD25 (IL-2R) deficiency suffer from autoimmune phenomena, lymphadenopathy, and persistent viral infections similar to the mentioned IPEX syndrome. However, in spite of the presence of Foxp3+CD4+ T cells, impaired production of IL-10 was demonstrated following T-cell activation (Malek 2008). RA is a systemic autoimmune disease that causes chronic inflammation and tissue destruction of joints (Zou et al. 2018). The role of Treg cells in pathogenesis is not fully understood, as changes in the number of Treg cells in synovium and peripheral blood have been described in the past few years, but these observations have not been uniform (Han et al. 2008; Cao et al. 2004; Jiao et al. 2007; Sempere-Ortells et al. 2009; Kawashiri et al. 2011; Niu et al. 2012; Samson et al. 2012; Lina et al. 2011; Yang et al. 2019a, b). In addition to the quantitative defects, there have been reports of functional defects in patients with RA (Rapetti et al. 2015; Flores-Borja et al. 2008; Nadkarni et al. 2007). Anti-TNF and anti-IL-6 therapies can restore the balance of Treg and Th17 cells in RA patients and affect Treg function (Samson et al. 2012, Nadkarni et al. 2007; Ehrenstein et al. 2004). Over the past two decades, improved anti-rheumatic drugs (bDMARD) have revolutionized the treatment of RA patients. Among them, TNF blockade, IL-6R blockade, and inhibition of T-cell activation through targeting co-stimulation signals are second-line drugs for treating patients with refractory RA with methotrexate (Smolen et al. 2016). The latest data showed that anti-TNF therapy in RA patients increases the frequency of TNFR2+ Treg cells, and the lack of TNFR2 leads to an increase in the methylation of Foxp3 TSDR, suggesting that TNFR2 expression has an effect on Treg function and stability in RA patients (Santinon et al. 2019). This is consistent with our recent reports that TNFRII favors Tregs while TNFRI promotes T effector cells (Yang et al. 2018; 2019a, b). IL-6 is another driver of inflammation in RA patients and is involved in the plasticity of Treg cells (Xu et al. 2007; Zheng et al. 2008; Luo and Zheng 2016). After treatment with tocilizumab, the number of Treg cells in RA patients increased in patients with good clinical response (Thiolat et al. 2014; Kikuchi et al. 2015).

Similarly, in the study of systemic lupus erythematosus (SLE), it was found that helper T cells (Th17) can induce tissue inflammation and immune responses. In contrast, Treg cells can mediate immune tolerance and inhibit inflammatory responses. The imbalance of Th17 cells and Treg cells affects the occurrence and pathogenesis of SLE (Ma et al. 2010; Szmyrka-Kaczmarek et al. 2014). Therefore, inhibiting the differentiation of Th17 cells and promoting the differentiation of Treg cells can help restore the balance of Th17/Treg cells and provide a potential new therapeutic approach in SLE. Low doses of IL-2 can increase the number of Treg cells, maintain immune homeostasis, and temporarily relieve clinical symptoms in patients with SLE (Goudy et al. 2013; Hartemann et al. 2013; Saadoun et al. 2011; He et al. 2016; von Spee-Mayer et al. 2016; Ye et al. 2018). However, most of these studies focus on short-term effects (von Spee-Mayer et al. 2016) and are not feasible long-term treatment (Humrich et al. 2015). The decrease in Treg and the imbalance of Th17/Treg cells are related to the occurrence and development of refractory SLE. It was proposed that the use of low-dose IL-2 combined with rapamycin can restore the Treg number and the balance of Th17/Treg cells. This method can induce immune tolerance, promote clinical response, reduce the dosage of prednisone or cytotoxic drugs, reduce drug side effects, and collectively benefit patients (Zhao et al. 2019).

Ankylosing spondylitis (AS) is an inflammatory autoimmune disease related to HLA-B27. Similar to RA, Treg cells also have functional deficits in AS (Guo et al. 2016). Recent studies have found that treatment with anti-TNF may affect the Th17/Treg ratio (Dulic et al. 2018; Liao et al. 2015). While most have reported that

Treg cells are involved in the pathogenesis of AS (Cao et al. 2004; Wang et al. 2015; Ye et al. 2016), there are few reports that are inconsistent with that concept. Several studies have reported a significant decrease in circulating Treg cells in AS patients compared to healthy controls (Wu et al. 2011; Zhao et al. 2011). Other studies have found no change in the percentage of circulating Treg cells in patients with AS (Guo et al. 2016; Cao et al. 2004; Wang et al. 2015; Ye et al. 2016).

Systemic sclerosis (SSc) is a connective tissue disease characterized by immune dysfunction, microvascular damage, and fibrosis of the skin and various organs (Frantz et al. 2018). A growing body of evidence indicates that T-cell proliferation and cytokine secretion play a major role in SSc (Meloni et al. 2009; Kalogerou et al. 2005). There is also controversy regarding the role of T regs in SSC with some reports showing impaired function while others demonstrating normal function (Liu et al. 2013a, b; Radstake et al. 2009; Papp et al. 2011; Fenoglio et al. 2011; Mathian et al. 2012; Klein et al. 2011). Studies have shown that Treg cells in the blood of patients with SSc have a normal phenotype and do not produce T effector cytokines. In contrast, skin Treg cells affected by SSc produce large amounts of IL-4 and IL-13 (MacDonald et al. 2015). Mathian et al. found that in SSc patients, activated and resting Treg cells were not functionally matched. The number of activated Treg cells decreased early in the disease (Mathian et al. 2012). Some authors have found that Tregs are significantly upregulated in all SSc phenotypes (Radstake et al. 2009; Giovannetti et al. 2010), especially in active and severe diseases (Slobodin et al. 2010). These inconsistencies could be explained by disease stages, new onset and treated patients, and methods used to define Treg cells. However, most studies reported that their functional capacity is reduced or T effector cells have gained the increased resistant ability to Treg suppression in disease status. Thus, the increase in the number of Treg cells and the restoration of their functions are important ways to induce immune tolerance and treat autoimmune diseases. Although Treg cells have heterogeneity and stability issues in clinical application, this treatment strategy is an important direction in the prevention and treatment of patients with autoimmune diseases.

1.8.2 Tregs in Infectious Diseases

In the field of infectious diseases, immune checkpoints limit protective immunity in many chronic infections, but therapies based on immune checkpoint suppression have not been well developed. Clearly, most pathogens have developed mechanisms to evade and suppress the host's protective immune response. Although Treg cells must control the activation of T-effector cells to prevent autoimmunity, it is also known that enhanced activation of Treg cells may lead to suppression of host immunity against microorganisms (viruses, bacteria, protozoa, fungi, and worms), resulting in reduced antimicrobial immunity and the persistence of pathogens (Belkaid et al. 2002, 2006). Many animal models of bacterial infection (e.g., *Listeria monocytogenes*, *Salmonella enteritis*, and *Mycobacterium tuberculosis*) exhibit the proliferation of Foxp3+ Treg. The inhibitory function of Treg cells can lead to increased bacterial load and invasion of systemic tissues (Johanns et al. 2010; Scott-Browne et al. 2007; Rowe et al. 2011). In addition, higher Treg frequency was associated with increased titer of hepatitis C virus RNA and dengue virus (Cabrera et al. 2004; Lühn et al. 2007). Paradoxically, Tregs play an early protective role in local infection in animal models of herpes simplex virus 2 and west Nile virus (Lanteri et al. 2009; Lund et al. 2008).

In the early stages of HIV infection, Tregs were found to control viral replication in target CD4+ T cells (Moreno-Fernandez et al. 2011). Moreover, Tregs may play an important and beneficial role in preventing the vigorous inflammatory response in the course of infection by parasites such as *Pneumocystis carinii* (Hori et al. 2002) and *Schistosoma mansoni* (Layland et al. 2013). Similarly, Tregs protect the host from parasites such as plasmodium and toxoplasma as well as the fungus *Candida albicans* (Haque et al. 2010; Oldenhove et al. 2009; Pandiyan et al.

2011). These complicated roles of Treg cells in acute and chronic microbial infections require a delicate balance between Foxp3+ Treg and effector T cells to provide an effective immune response against pathogens without inducing destructive autoimmunity.

Immunosuppressive cytokines IL-10 and TGF-β have established roles in suppressing anti-pathogenic effector T-cell responses (Mills 2004), but there is also growing evidence that immune checkpoints and Treg cells play a role. The use of immune checkpoint inhibitors may help reverse chronic infections, especially the immunosuppressive state in parasitic infections, and has great potential to promote an immune response that cures parasites. However, the greatest potential may lie in combining immune checkpoint blockades with therapeutic vaccination. Indeed, one study has demonstrated that vaccine-induced immune responses can be enhanced by reducing Treg cells or blocking the production or function of immunosuppressive cytokines (Jarnicki et al. 2008; Moore et al. 2005). Although blocking Treg cells or immune checkpoints is unlikely to be used to enhance the efficacy of routine vaccination, it does have great potential to enhance the efficacy of therapeutic vaccines against many chronic infections such as malaria, tuberculosis, and HIV.

1.8.3 Treg in Cancer

The association between Treg and cancers has been widely studied for decades. Many studies have shown that Tregs infiltrate into various types of cancers, such as breast, lung, liver, and gastrointestinal tract (Liyanage et al. 2002; Anna et al. 2002; Lars et al. 2005; Fumiko Ichihara et al. 2003). Additionally, the immunosuppressive function of Treg cells in patients with cancers is also significantly increased compared with healthy controls (Ju et al. 2009). Furthermore, Tregs account for 10–50% of the CD4+ T cells in tumors sites and, therefore, contribute substantially to tumor development by impairing the activation, survival, and expansion of anti-tumor T cells. The high levels of Treg and decreased ratios of tumor-infiltrating CD8+ T cells to Foxp3 + Treg have been demonstrated to correlate with the poor prognosis (Sato et al. 2005; Bates et al. 2006). Investigations on tumor specimens are crucial to understanding the phenotypes and origins of tumor-infiltrating Tregs which is more complicated than what we expected. It is still unclear whether the large amounts of Treg in tumor tissues have migrated from the periphery, amplified tTreg, or iTreg differentiated from naïve or effector CD4+ T cells in the tumor microenvironment. It is noted that the increase in intra-tumoral expression of chemokines such as CCL17, CCL22, and CCL28 may facilitate the recruitment of Treg cells (Speiser et al. 2016). With the rapid development of single cell sequencing technologies and the adoption of novel deeper immunophenotyping, a more comprehensive understanding of Treg heterogeneity is within reach. This is critical to safely and effectively exploit their anti-tumor immune response and to enhance their targeting therapeutic potential. Considering the crucial role of Treg in promoting tumorigenesis, cancer immunotherapy targeting Treg got more attention and achieved great progress. Therefore, the functional molecules related to Treg such as CTLA-4, GITR, PD-1, OX-40, and LAG3 are potential candidates that can be used for Treg depletion or functional modulation (Tanaka and Sakaguchi 2017). Checkpoint blockade therapy such as the clinical use of anti-CTLA-4 antibody predominantly affected Treg cells, thereby enhancing the anti-tumor immune response (Simpson et al. 2013). Additionally, agonistic interference such as anti-GITR antibody could abrogate Treg suppressive function because Tregs express elevated level of GITR and upregulate it upon activation (Shimizu et al. 2002). In general, Tregs have the opposite effect in cancers and autoimmune diseases. Thus, an effective anti-tumor response such as systemic Tregs depletion is likely to result in autoimmunity. Therefore, selective targeting specific subpopulations of Treg cells such as effector Treg in tumor tissues may be a more effective approach to promote tumor immunity without causing serious autoimmune reaction.

Table 1.2 Relationship between Treg cells and diseases

Disease	Circulating Treg frequency (compared to normal people)	Frequency of Tregs at focus	Function of Tregs	Changes in cytokines associated with Treg cells	Treg types most relevant to disease activity	Treg-related treatments
Autoimmune disease						
RA	↓	↑	↓	TNF-α IL-6 IL-1 ↑	$CD4^+CD25^{hi}CD127^{low/-}$ $Foxp3^+Helios^+Treg$	1. Treg adoptive transfer 2. Pharmacological -based boosting Tregs 3. Foxp3(+)Treg-induction in vivo 4. "Tregitope" therapy
SLE	↓	–	↓	IL-6 IL-17 IL-23 ↑ TGF-β IL-2 ↓	$Helios^+Foxp3^+Treg$	
AS	↓	Unclear	↓	TNF-α IL-17 IL-6 IL-23 ↑	$CD4^+CD25^+Foxp3^+$ $CD127^-Treg$	
SSc	↓	Contradictory	↓	IL-4 IL-13 ↑ IL-10 ↓	$CD4^+CD25^+Foxp3^+$ Treg	
Type 1 diabetes	Unaltered	–	↓	IL-2 IL-10 ↓	$CD4^+CD25^+Foxp3^+$ Treg	
Infection disease	↑	↑	↑	IL-2 TGF-β ↑	$Foxp3^+$ Treg	Blocking Treg cells
Cancer	↑	↑	↑	IL-2 TGF-β ↑	$Foxp3^+$ Treg	1. Anti-CTLA-4 antibody 2. Anti-GITR antibody
Organ transplantation	↓	↓	↓	IL-6 IL-17 ↑ TGF-β IL-2 ↓	$Foxp3^+$ Treg	Treg adoptive transfer

1.8.4 Treg in Organ Transplantation

The achievement of graft tolerance and prevention of graft-versus-host disease (GvHD) have always been the important aims in the field of organ transplantation (Atif et al. 2020). Research has been focusing on preventing the chronic allograft dysfunction and reducing the side effects of long-term immunosuppression (Issa et al. 2013). In this regard, Treg within the peripheral circulating blood and graft microenvironment contribute to induce immune tolerance through employing multiple regulatory mechanisms (Liu et al. 2013a, b; Ferrer et al. 2014). Tregs limit graft damage by migrating to the organ and then retreating to the draining lymph nodes to maintain tolerance. Additionally, Tregs constitutively express cytotoxic T-lymphocyte antigen-4 (CTLA-4), which can interact with the costimulatory ligands CD80, CD86 of APCs, thereby inhibiting the activation of T cells (Tang and Vincenti 2017). Importantly, adoptive transfer of Treg has been shown to prevent transplant rejection and GVHD in mouse models which spur the development of experimental Treg therapies, particularly with antigen-specific Treg cells (Wu et al. 2013a, b; Brunstein et al. 2016; Liao et al. 2017; Gu et al. 2014). Currently, several Treg phase I clinical trials have shown encouraging safety and efficacy (Brunstein et al. 2011; Satoru et al. 2016). One trial was designed to identify the function of ex vivo expanded recipient polyclonal Treg in inducing immune tolerance in kidney transplant. The expanded Tregs amplified circulating Treg levels in a sustained manner without infusion-related side effects, infectious, or rejection events up to 2 years posttransplant (Mathew et al. 2018). Another trial was to determine the safety and feasibility of expanding polyclonal Treg ex vivo in kidney transplant recipients with subclinical graft inflammation noted on 6-month surveillance biopsy (Chandran et al. 2017). Treg infusion was safe and well tolerated without the impaired stability. In addition, compared with polyclonal Treg, alloantigen-specific Tregs have superior effects which potentially led to more targeted suppression, lower dose, and better safety (Esensten et al. 2018). There are still many challenges regarding Treg therapies in organ transplantation that require more basic and clinical studies. The above diseases are summarized in Table 1.2.

1.9 Conclusion

Treg cells play a vital role in regulating immune tolerance and balance, preventing autoimmunity, and have distinct roles in various disease environments. Restoring the frequency and function of Treg cells favor the therapy of autoimmune diseases and maintenance of allo-graft organ transplantation. Conversely, they probably are detrimental in patients with cancers and infection. Thus, they are a double-edged sword and selective manipulation of their frequency and function may be therapeutic in various immunological diseases. Indeed, it has been demonstrated to be effective in a series of clinical trials. However, it is still necessary to further explore the molecular mechanisms of how Treg cells regulate immune response. Treg surface molecular markers are still needed to be further identified since this will help to isolate live Treg cells in the clinical setting. In addition, the difference in phenotype and biological characteristics of various Treg subpopulations require to be further analyzed and compared. The epigenetic modification and mechanisms remain an in-depth study as well. Overall, understanding in-depth the role, mechanisms, and approaches in operating Treg cells helps to develop an innovative strategy for the treatment of many relevant diseases.

References

Agostini L, Martinon F, Burns K, McDermott MF, Hawkins PN, Tschopp J (2004) NALP3 forms an IL-1beta-processing inflammasome with increased activity in Muckle-Wells autoinflammatory disorder. Immunity 20(3):319–325. https://doi.org/10.1016/s1074-7613(04)00046-9

Akimova T, Beier UH, Wang L, Levine MH, Hancock WW (2011) Helios expression is a marker of T cell activation and proliferation. PLoS One 6(8):e24226. https://doi.org/10.1371/journal.pone.0024226

Anna Maria Wolf DW et al (2002) Increase of regulatory T Cells in the peripheral blood of cancer patients. Clin Cancer Res 9:606–612

Artlett CM, Black CM, Briggs DC, Stevens CO, Welsh KI (1996) Telomere reduction in scleroderma patients: a possible cause for chromosomal instability. Br J Rheumatol 35(8):732–737. https://doi.org/10.1093/rheumatology/35.8.732

Atif M, Conti F, Gorochov G, Oo YH, Miyara M (2020) Regulatory T cells in solid organ transplantation. Clin Transl Immunol 9(2):e01099. https://doi.org/10.1002/cti2.1099

Awasthi A, Carrier Y, Peron JP, Bettelli E, Kamanaka M, Flavell RA, Kuchroo VK, Oukka M, Weiner HL (2007) A dominant function for interleukin 27 in generating interleukin 10-producing anti-inflammatory T cells. Nat Immunol 8(12):1380–1389. https://doi.org/10.1038/ni1541

Bacchetta R, de Waal Malefijt R, Yssel H, Abrams J, de Vries JE, Spits H, Roncarolo MG (1990) Host-reactive CD4+ and CD8+ T cell clones isolated from a human chimera produce IL-5, IL-2, IFN-gamma and granulocyte/macrophage-colony-stimulating factor but not IL-4. J Immunol 144(3):902–908

Bacchetta R, Bigler M, Touraine JL, Parkman R, Tovo PA, Abrams J, de Waal Malefyt R, de Vries JE, Roncarolo MG (1994) High levels of interleukin 10 production in vivo are associated with tolerance in SCID patients transplanted with HLA mismatched hematopoietic stem cells. J Exp Med 179(2):493–502. https://doi.org/10.1084/jem.179.2.493

Bacchetta R, Barzaghi F, Roncarolo MG (2018) From IPEX syndrome to FOXP3 mutation: a lesson on immune dysregulation. Ann N Y Acad Sci 1417(1):5–22. https://doi.org/10.1111/nyas.13011

Bahmani B, Uehara M, Jiang L, Ordikhani F, Banouni N, Ichimura T, Solhjou Z, Furtmüller GJ, Brandacher G, Alvarez D, von Andrian UH, Uchimura K, Xu Q, Vohra I, Yilmam OA, Haik Y, Azzi J, Kasinath V, Bromberg JS, McGrath MM, Abdi R (2018) Targeted delivery of immune therapeutics to lymph nodes prolongs cardiac allograft survival. J Clin Invest 128(11):4770–4786. https://doi.org/10.1172/jci120923

Bates GJ, Fox SB, Han C, Leek RD, Garcia JF, Harris AL, Banham AH (2006) Quantification of regulatory T cells enables the identification of high-risk breast cancer patients and those at risk of late relapse. J Clin Oncol 24(34):5373–5380. https://doi.org/10.1200/jco.2006.05.9584

Batten M, Kljavin NM, Li J, Walter MJ, de Sauvage FJ, Ghilardi N (2008) Cutting edge: IL-27 is a potent inducer of IL-10 but not FoxP3 in murine T cells. J Immunol 180(5):2752–2756. https://doi.org/10.4049/jimmunol.180.5.2752

Baeten D et al (2006) Phenotypically and functionally distinct CD8+ lymphocyte populations in long-term drug-free tolerance and chronic rejection in human kidney graft recipients. J Am Soc Nephrol 17(1):294–304

Belkaid Y, Piccirillo CA, Mendez S, Shevach EM, Sacks DL (2002) CD4+CD25+ regulatory T cells control Leishmania major persistence and immunity. Nature 420(6915):502–507. https://doi.org/10.1038/nature01152

Belkaid Y, Blank RB, Suffia I (2006) Natural regulatory T cells and parasites: a common quest for host homeostasis. Immunol Rev 212:287–300. https://doi.org/10.1111/j.0105-2896.2006.00409.x

Boesteanu AC, Katsikis PD (2009) Memory T cells need CD28 costimulation to remember. Semin Immunol 21(2):69–77. https://doi.org/10.1016/j.smim.2009.02.005

Borthwick NJ, Lowdell M, Salmon M, Akbar AN (2000) Loss of CD28 expression on CD8(+) T cells is induced by IL-2 receptor gamma chain signalling cytokines and type I IFN, and increases susceptibility to activation-induced apoptosis. Int Immunol 12(7):1005–1013. https://doi.org/10.1093/intimm/12.7.1005

Braud VM, Allan DS, McMichael AJ (1999) Functions of nonclassical MHC and non-MHC-encoded class I molecules. Curr Opin Immunol 11(1):100–108

Brun V, Bastian H, Neveu V, Foussat A (2009) Clinical grade production of IL-10 producing regulatory Tr1 lymphocytes for cell therapy of chronic inflammatory diseases. Int Immunopharmacol 9(5):609–613. https://doi.org/10.1016/j.intimp.2009.01.032

Brun V, Neveu V, Pers YM, Fabre S, Quatannens B, Bastian H, Clerget-Chossat N, Jorgensen C, Foussat A (2011) Isolation of functional autologous collagen-II specific IL-10 producing Tr1 cell clones from rheumatoid arthritis blood. Int Immunopharmacol 11(8):1074–1078. https://doi.org/10.1016/j.intimp.2011.03.001

Brunstein CG, Miller JS, Cao Q, McKenna DH, Hippen KL, Curtsinger J, Defor T, Levine BL, June CH, Rubinstein P, McGlave PB, Blazar BR, Wagner JE (2011) Infusion of ex vivo expanded T regulatory cells in adults transplanted with umbilical cord blood: safety profile and detection kinetics. Blood 117(3):1061–1070. https://doi.org/10.1182/blood-2010-07-293795

Brunstein CG, Miller JS, McKenna DH, Hippen KL, DeFor TE, Sumstad D, Curtsinger J, Verneris MR, MacMillan ML, Levine BL, Riley JL, June CH, Le C, Weisdorf DJ, McGlave PB, Blazar BR, Wagner JE (2016) Umbilical cord blood-derived T regulatory cells to prevent GVHD: kinetics, toxicity profile, and clinical effect. Blood 127(8):1044–1051. https://doi.org/10.1182/blood-2015-06-653667

Cabrera R, Tu Z, Xu Y, Firpi RJ, Rosen HR, Liu C, Nelson DR (2004) An immunomodulatory role for CD4(+)CD25(+) regulatory T lymphocytes in hepatitis C virus infection. Hepatology 40(5):1062–1071. https://doi.org/10.1002/hep.20454

Cao D, van Vollenhoven R, Klareskog L, Trollmo C, Malmstrom V (2004) CD25brightCD4+ regulatory T cells are enriched in inflamed joints of patients with chronic rheumatic disease. Arthritis Res Ther 6(4): R335–R346. https://doi.org/10.1186/ar1192

Cepek KL, Shaw SK, Parker CM, Russell GJ, Morrow JS, Rimm DL, Brenner MB (1994) Adhesion between epithelial cells and T lymphocytes mediated by E-cadherin and the alpha E beta 7 integrin. Nature 372(6502):190–193. https://doi.org/10.1038/372190a0

Chandran S, Tang Q, Sarwal M, Laszik ZG, Putnam AL, Lee K, Leung J, Nguyen V, Sigdel T, Tavares EC, Yang JYC, Hellerstein M, Fitch M, Bluestone JA, Vincenti F (2017) Polyclonal regulatory T cell therapy for control of inflammation in kidney transplants. Am J Transplant 17(11):2945–2954. https://doi.org/10.1111/ajt.14415

Chang KM (2007) Regulatory T cells in hepatitis C virus infection. Hepatol Res 37(Suppl 3):S327–S330. https://doi.org/10.1111/j.1872-034X.2007.00220.x

Chen X, Priatel JJ, Chow MT, Teh HS (2008) Preferential development of CD4 and CD8 T regulatory cells in RasGRP1-deficient mice. J Immunol 180 (9):5973–5982. https://doi.org/10.4049/jimmunol.180.9.5973

Chen M, Su W, Lin X, Guo Z, Wang J, Zhang Q, Brand D, Ryffel B, Huang J, Liu Z, He X, Le AD, Zheng SG (2013) Adoptive transfer of human gingiva-derived mesenchymal stem cells ameliorates collagen-induced arthritis via suppression of Th1 and Th17 cells and enhancement of regulatory T cell differentiation. Arthritis Rheum 65(5):1181–1193. https://doi.org/10.1002/art.37894

Cohen S, Janicki-Deverts D, Turner RB, Casselbrant ML, Li-Korotky HS, Epel ES, Doyle WJ (2013) Association between telomere length and experimentally induced upper respiratory viral infection in healthy adults. JAMA 309(7):699–705. https://doi.org/10.1001/jama.2013.613

Colmegna I, Diaz-Borjon A, Fujii H, Schaefer L, Goronzy JJ, Weyand CM (2008) Defective proliferative capacity and accelerated telomeric loss of hematopoietic progenitor cells in rheumatoid arthritis. Arthritis Rheum 58(4):990–1000. https://doi.org/10.1002/art.23287

Costa DL, Guimaraes LH, Cardoso TM, Queiroz A, Lago E, Roselino AM, Bacellar O, Carvalho EM, Silva JS (2013) Characterization of regulatory T cell (Treg) function in patients infected with Leishmania braziliensis. Hum Immunol 74(12):1491–1500. https://doi.org/10.1016/j.humimm.2013.08.269

Dai H, Wan N, Zhang S, Moore Y, Wan F, Dai Z (2010) Cutting edge: programmed death-1 defines CD8+CD122+ T cells as regulatory versus memory T cells. J Immunol 185(2):803–807. https://doi.org/10.4049/jimmunol.1000661

Dai Z, Zhang S, Xie Q, Wu S, Su J, Li S, Xu Y, Li XC (2014) Natural CD8+CD122+ T cells are more potent in suppression of allograft rejection than CD4+CD25+ regulatory T cells. Am J Transplant 14(1):39–48. https://doi.org/10.1111/ajt.12515

de Waal Malefyt R, Haanen J, Spits H, Roncarolo MG, te Velde A, Figdor C, Johnson K, Kastelein R, Yssel H, de Vries JE (1991) Interleukin 10 (IL-10) and viral IL-10 strongly reduce antigen-specific human T cell proliferation by diminishing the antigen-presenting capacity of monocytes via downregulation of class II major histocompatibility complex expression. J Exp Med 174(4):915–924. https://doi.org/10.1084/jem.174.4.915

Dinesh RK, Skaggs BJ, La Cava A, Hahn BH, Singh RP (2010) CD8+ Tregs in lupus, autoimmunity, and beyond. Autoimmun Rev 9(8):560–568. https://doi.org/10.1016/j.autrev.2010.03.006

Dulic S, Vasarhelyi Z, Bajnok A, Szalay B, Toldi G, Kovacs L, Balog A (2018) The impact of anti-TNF therapy on CD4+ and CD8+ cell subsets in ankylosing spondylitis. Pathobiology 85(3):201–210. https://doi.org/10.1159/000484250

Effros RB (2011) Telomere/telomerase dynamics within the human immune system: effect of chronic infection and stress. Exp Gerontol 46(2–3):135–140. https://doi.org/10.1016/j.exger.2010.08.027

Ehrenstein MR, Evans JG, Singh A, Moore S, Warnes G, Isenberg DA, Mauri C (2004) Compromised function of regulatory T cells in rheumatoid arthritis and reversal by anti-TNFalpha therapy. J Exp Med 200 (3):277–285. https://doi.org/10.1084/jem.20040165

El-Asady R, Yuan R, Liu K, Wang D, Gress RE, Lucas PJ, Drachenberg CB, Hadley GA (2005) TGF-{beta}-dependent CD103 expression by CD8(+) T cells promotes selective destruction of the host intestinal epithelium during graft-versus-host disease. J Exp Med 201(10):1647–1657. https://doi.org/10.1084/jem.20041044

Endharti AT, Rifa IM, Shi Z, Fukuoka Y, Nakahara Y, Kawamoto Y, Takeda K, Isobe K, Suzuki H (2005) Cutting edge: CD8+CD122+ regulatory T cells produce IL-10 to suppress IFN-gamma production and proliferation of CD8+ T cells. J Immunol 175 (11):7093–7097. https://doi.org/10.4049/jimmunol.175.11.7093

Endharti AT, Okuno Y, Shi Z, Misawa N, Toyokuni S, Ito M, Isobe K, Suzuki H (2011) CD8+CD122+ regulatory T cells (Tregs) and CD4+ Tregs cooperatively prevent and cure CD4+ cell-induced colitis. J Immunol 186(1):41–52. https://doi.org/10.4049/jimmunol.1000800

Esensten JH, Muller YD, Bluestone JA, Tang Q (2018) Regulatory T-cell therapy for autoimmune and autoinflammatory diseases: the next frontier. J Allergy Clin Immunol 142(6):1710–1718. https://doi.org/10.1016/j.jaci.2018.10.015

Fanchiang SS et al (2012) Global expression profiling of peripheral Qa-1-restricted CD8alphaalpha +TCRalphabeta+ regulatory T cells reveals innate-like features: implications for immune-regulatory repertoire. Hum Immunol 73(3):214–222

Fenoglio D, Ferrera F, Fravega M, Balestra P, Battaglia F, Proietti M, Andrei C, Olive D, Antonio LC, Indiveri F,

Filaci G (2008) Advancements on phenotypic and functional characterization of non-antigen-specific CD8+CD28- regulatory T cells. Hum Immunol 69 (11):745–750. https://doi.org/10.1016/j.humimm.2008.08.282

Fenoglio D, Battaglia F, Parodi A, Stringara S, Negrini S, Panico N, Rizzi M, Kalli F, Conteduca G, Ghio M, De Palma R, Indiveri F, Filaci G (2011) Alteration of Th17 and Treg cell subpopulations co-exist in patients affected with systemic sclerosis. Clin Immunol 139 (3):249–257. https://doi.org/10.1016/j.clim.2011.01.013

Ferrer IR, Hester J, Bushell A, Wood KJ (2014) Induction of transplantation tolerance through regulatory cells: from mice to men. Immunol Rev 258(1):102–116. https://doi.org/10.1111/imr.12158

Fitzgerald DC, Zhang GX, El-Behi M, Fonseca-Kelly Z, Li H, Yu S, Saris CJ, Gran B, Ciric B, Rostami A (2007) Suppression of autoimmune inflammation of the central nervous system by interleukin 10 secreted by interleukin 27-stimulated T cells. Nat Immunol 8 (12):1372–1379. https://doi.org/10.1038/ni1540

Flores-Borja F, Jury EC, Mauri C, Ehrenstein MR (2008) Defects in CTLA-4 are associated with abnormal regulatory T cell function in rheumatoid arthritis. Proc Natl Acad Sci U S A 105(49):19396–19401. https://doi.org/10.1073/pnas.0806855105

Focosi D, Bestagno M, Burrone O, Petrini M (2010) CD57 + T lymphocytes and functional immune deficiency. J Leukoc Biol 87(1):107–116. https://doi.org/10.1189/jlb.0809566

Fontenot JD, Gavin MA, Rudensky AY (2003) Foxp3 programs the development and function of CD4 +CD25+ regulatory T cells. Nat Immunol 4 (4):330–336. https://doi.org/10.1038/ni904

Frantz C, Auffray C, Avouac J, Allanore Y (2018) Regulatory T cells in systemic sclerosis. Front Immunol 9:2356. https://doi.org/10.3389/fimmu.2018.02356

Frisullo G et al (2010) CD8(+)Foxp3(+) T cells in peripheral blood of relapsing-remitting multiple sclerosis patients. Hum Immunol 71(5):437–441

Gagliani N, Magnani CF, Huber S, Gianolini ME, Pala M, Licona-Limon P, Guo B, Herbert DR, Bulfone A, Trentini F, Di Serio C, Bacchetta R, Andreani M, Brockmann L, Gregori S, Flavell RA, Roncarolo MG (2013) Coexpression of CD49b and LAG-3 identifies human and mouse T regulatory type 1 cells. Nat Med 19(6):739–746. https://doi.org/10.1038/nm.3179

Gershon RK, Kondo K (1970) Cell interactions in the induction of tolerance: the role of thymic lymphocytes. Immunology 18(5):723–737

Gershon RK, Kondo K (1971) Infectious immunological tolerance. Immunology 21(6):903–914

Gershon RK, Cohen P, Hencin R, Liebhaber SA (1972) Suppressor T cells. J Immunol 108(3):586–590

Gianfrani C, Levings MK, Sartirana C, Mazzarella G, Barba G, Zanzi D, Camarca A, Iaquinto G, Giardullo N, Auricchio S, Troncone R, Roncarolo MG (2006) Gliadin-specific type 1 regulatory T cells from the intestinal mucosa of treated celiac patients inhibit pathogenic T cells. J Immunol 177 (6):4178–4186. https://doi.org/10.4049/jimmunol.177.6.4178

Giovannetti A, Rosato E, Renzi C, Maselli A, Gambardella L, Giammarioli AM, Palange P, Paoletti P, Pisarri S, Salsano F, Malorni W, Pierdominici M (2010) Analyses of T cell phenotype and function reveal an altered T cell homeostasis in systemic sclerosis. Correlations with disease severity and phenotypes. Clin Immunol 137(1):122–133. https://doi.org/10.1016/j.clim.2010.06.004

Globinska A, Boonpiyathad T, Satitsuksanoa P, Kleuskens M, van de Veen W, Sokolowska M, Akdis M (2018) Mechanisms of allergen-specific immunotherapy: diverse mechanisms of immune tolerance to allergens. Ann Allergy Asthma Immunol 121 (3):306–312. https://doi.org/10.1016/j.anai.2018.06.026

Godfrey VL, Wilkinson JE, Rinchik EM, Russell LB (1991) Fatal lymphoreticular disease in the scurfy (sf) mouse requires T cells that mature in a sf thymic environment: potential model for thymic education. Proc Natl Acad Sci U S A 88(13):5528–5532. https://doi.org/10.1073/pnas.88.13.5528

Goschl L, Scheinecker C, Bonelli M (2019) Treg cells in autoimmunity: from identification to Treg-based therapies. Semin Immunopathol 41(3):301–314. https://doi.org/10.1007/s00281-019-00741-8

Goudy K, Aydin D, Barzaghi F, Gambineri E, Vignoli M, Ciullini Mannurita S, Doglioni C, Ponzoni M, Cicalese MP, Assanelli A, Tommasini A, Brigida I, Dellepiane RM, Martino S, Olek S, Aiuti A, Ciceri F, Roncarolo MG, Bacchetta R (2013) Human IL2RA null mutation mediates immunodeficiency with lymphoproliferation and autoimmunity. Clin Immunol 146(3):248–261. https://doi.org/10.1016/j.clim.2013.01.004

Graca L, Cobbold SP, Waldmann H (2002) Identification of regulatory T cells in tolerated allografts. J Exp Med 195(12):1641–1646. https://doi.org/10.1084/jem.20012097

Grant CR, Liberal R, Mieli-Vergani G, Vergani D, Longhi MS (2015) Regulatory T-cells in autoimmune diseases: challenges, controversies and--yet--unanswered questions. Autoimmun Rev 14(2):105–116. https://doi.org/10.1016/j.autrev.2014.10.012

Gregori S, Roncarolo MG (2018) Engineered T regulatory type 1 cells for clinical application. Front Immunol 9:233. https://doi.org/10.3389/fimmu.2018.00233

Gregori S, Goudy KS, Roncarolo MG (2012) The cellular and molecular mechanisms of immunosuppression by human type 1 regulatory T cells. Front Immunol 3:30. https://doi.org/10.3389/fimmu.2012.00030

Groux H, O'Garra A, Bigler M, Rouleau M, Antonenko S, de Vries JE, Roncarolo MG (1997) A CD4+ T-cell subset inhibits antigen-specific T-cell responses and prevents colitis. Nature 389(6652):737–742. https://doi.org/10.1038/39614

Gu J, Lu L, Chen M, Xu L, Lan Q, Li Q, Liu Z, Chen G, Wang P, Wang X, Brand D, Olsen N, Zheng SG (2014) TGF-β-induced CD4+Foxp3+ T cells attenuate acute graft-versus-host disease by suppressing expansion and killing of effector CD8+ cells. J Immunol 193 (7):3388–3397. https://doi.org/10.4049/jimmunol.1400207

Guillonneau C, Hill M, Hubert FX, Chiffoleau E, Herve C, Li XL, Heslan M, Usal C, Tesson L, Menoret S, Saoudi A, Le Mauff B, Josien R, Cuturi MC, Anegon I (2007) CD40Ig treatment results in allograft acceptance mediated by CD8CD45RC T cells, IFN-gamma, and indoleamine 2,3-dioxygenase. J Clin Invest 117 (4):1096–1106. https://doi.org/10.1172/jci28801

Guo H, Zheng M, Zhang K, Yang F, Zhang X, Han Q, Chen ZN, Zhu P (2016) Functional defects in CD4(+) CD25(high) FoxP3(+) regulatory cells in ankylosing spondylitis. Sci Rep 6:37559. https://doi.org/10.1038/srep37559

Hadley GA, Bartlett ST, Via CS, Rostapshova EA, Moainie S (1997) The epithelial cell-specific integrin, CD103 (alpha E integrin), defines a novel subset of alloreactive CD8+ CTL. J Immunol 159(8):3748–3756

Hadley GA, Charandee C, Weir MR, Wang D, Bartlett ST, Drachenberg CB (2001) CD103+ CTL accumulate within the graft epithelium during clinical renal allograft rejection. Transplantation 72(9):1548–1555. https://doi.org/10.1097/00007890-200111150-00013

Han GM, O'Neil-Andersen NJ, Zurier RB, Lawrence DA (2008) CD4+CD25high T cell numbers are enriched in the peripheral blood of patients with rheumatoid arthritis. Cell Immunol 253(1–2):92–101. https://doi.org/10.1016/j.cellimm.2008.05.007

Haque A, Best SE, Amante FH, Mustafah S, Desbarrieres L, de Labastida F, Sparwasser T, Hill GR, Engwerda CR (2010) CD4+ natural regulatory T cells prevent experimental cerebral malaria via CTLA-4 when expanded in vivo. PLoS Pathog 6(12):e1001221. https://doi.org/10.1371/journal.ppat.1001221

Haque S, Rakieh C, Marriage F, Ho P, Gorodkin R, Teh LS, Snowden N, Day PJ, Bruce IN (2013) Shortened telomere length in patients with systemic lupus erythematosus. Arthritis Rheum 65(5):1319–1323. https://doi.org/10.1002/art.37895

Hartemann A, Bensimon G, Payan CA, Jacqueminet S, Bourron O, Nicolas N, Fonfrede M, Rosenzwajg M, Bernard C, Klatzmann D (2013) Low-dose interleukin 2 in patients with type 1 diabetes: a phase 1/2 randomised, double-blind, placebo-controlled trial. Lancet Diabetes Endocrinol 1(4):295–305. https://doi.org/10.1016/s2213-8587(13)70113-x

He J, Zhang X, Wei Y, Sun X, Chen Y, Deng J, Jin Y, Gan Y, Hu X, Jia R, Xu C, Hou Z, Leong YA, Zhu L, Feng J, An Y, Jia Y, Li C, Liu X, Ye H, Ren L, Li R, Yao H, Li Y, Chen S, Zhang X, Su Y, Guo J, Shen N, Morand EF, Yu D, Li Z (2016) Low-dose interleukin-2 treatment selectively modulates CD4(+) T cell subsets in patients with systemic lupus erythematosus. Nat Med 22(9):991–993. https://doi.org/10.1038/nm.4148

Ho J, Kurtz CC, Naganuma M, Ernst PB, Cominelli F, Rivera-Nieves J (2008) A CD8+/CD103high T cell subset regulates TNF-mediated chronic murine ileitis. J Immunol 180(4):2573–2580. https://doi.org/10.4049/jimmunol.180.4.2573

Honda M, Mengesha E, Albano S, Nichols WS, Wallace DJ, Metzger A, Klinenberg JR, Linker-Israeli M (2001) Telomere shortening and decreased replicative potential, contrasted by continued proliferation of telomerase-positive CD8+CD28(lo) T cells in patients with systemic lupus erythematosus. Clin Immunol 99(2):211–221. https://doi.org/10.1006/clim.2001.5023

Hori S, Carvalho TL, Demengeot J (2002) CD25+CD4+ regulatory T cells suppress CD4+ T cell-mediated pulmonary hyperinflammation driven by Pneumocystis carinii in immunodeficient mice. Eur J Immunol 32 (5):1282–1291. https://doi.org/10.1002/1521-4141(200205)32:5<1282:Aid-immu1282>3.0.Co;2-#

Hori S, Nomura T, Sakaguchi S (2003) Control of regulatory T cell development by the transcription factor Foxp3. Science 299(5609):1057–1061. https://doi.org/10.1126/science.1079490

Horino S, Sasahara Y, Sato M, Niizuma H, Kumaki S, Abukawa D, Sato A, Imaizumi M, Kanegane H, Kamachi Y, Sasaki S, Terui K, Ito E, Kobayashi I, Ariga T, Tsuchiya S, Kure S (2014) Selective expansion of donor-derived regulatory T cells after allogeneic bone marrow transplantation in a patient with IPEX syndrome. Pediatr Transplant 18(1):E25–E30. https://doi.org/10.1111/petr.12184

Horwitz DA, Zheng SG, Gray JD (2008) Natural and TGF-beta-induced Foxp3(+) CD4(+) CD25(+) regulatory T cells are not mirror images of each other. Trends Immunol 29(9):429–435. https://doi.org/10.1016/j.it.2008.06.005

Horwitz DA, Pan S, Ou JN, Wang J, Chen M, Gray JD, Zheng SG (2013) Therapeutic polyclonal human CD8 + CD25+ Fox3+ TNFR2+ PD-L1+ regulatory cells induced ex-vivo. Clin Immunol 149(3):450–463. https://doi.org/10.1016/j.clim.2013.08.007

Huang W, Solouki S, Koylass N, Zheng SG, August A (2017) ITK signalling via the Ras/IRF4 pathway regulates the development and function of Tr1 cells. Nat Commun 8:15871. https://doi.org/10.1038/ncomms15871

Huang W, Solouki S, Carter C, Zheng SG, August A (2018) Beyond type 1 regulatory T cells: co-expression of LAG3 and CD49b in IL-10-producing T cell lineages. Front Immunol 9:2625. https://doi.org/10.3389/fimmu.2018.02625

Huber S, Gagliani N, Esplugues E, O'Connor W Jr, Huber FJ, Chaudhry A, Kamanaka M, Kobayashi Y, Booth CJ, Rudensky AY, Roncarolo MG, Battaglia M, Flavell RA (2011) Th17 cells express interleukin-10 receptor and are controlled by Foxp3(−) and Foxp3+ regulatory CD4+ T cells in an interleukin-10-

dependent manner. Immunity 34(4):554–565. https://doi.org/10.1016/j.immuni.2011.01.020
Humrich JY, von Spee-Mayer C, Siegert E, Alexander T, Hiepe F, Radbruch A, Burmester GR, Riemekasten G (2015) Rapid induction of clinical remission by low-dose interleukin-2 in a patient with refractory SLE. Ann Rheum Dis 74(4):791–792. https://doi.org/10.1136/annrheumdis-2014-206506
Ichihara F, Kono K, Takahashi A, Kawaida H, Sugai H, Fujii H (2003) Increased populations of regulatory T cells in peripheral blood and tumor-infiltrating lymphocytes in patients with gastric and esophageal cancers. Clin Cancer Res 9:4404–4408
Iikuni N, Lourenço EV, Hahn BH, La Cava A (2009) Cutting edge: regulatory T cells directly suppress B cells in systemic lupus erythematosus. J Immunol 183 (3):1518–1522. doi:https://doi.org/10.4049/jimmunol.0901163
Issa F, Robb RJ, Wood KJ (2013) The where and when of T cell regulation in transplantation. Trends Immunol 34(3):107–113. https://doi.org/10.1016/j.it.2012.11.003
Itoh M, Takahashi T, Sakaguchi N, Kuniyasu Y, Shimizu J, Otsuka F, Sakaguchi S (1999) Thymus and autoimmunity: production of CD25+CD4+ naturally anergic and suppressive T cells as a key function of the thymus in maintaining immunologic self-tolerance. J Immunol 162(9):5317–5326
Iwasaki Y, Fujio K, Okamura T, Yanai A, Sumitomo S, Shoda H, Tamura T, Yoshida H, Charnay P, Yamamoto K (2013) Egr-2 transcription factor is required for Blimp-1-mediated IL-10 production in IL-27-stimulated CD4+ T cells. Eur J Immunol 43 (4):1063–1073. https://doi.org/10.1002/eji.201242942
Jacob N, Yang H, Pricop L, Liu Y, Gao X, Zheng SG, Wang J, Gao HX, Putterman C, Koss MN, Stohl W, Jacob CO (2009) Accelerated pathological and clinical nephritis in systemic lupus erythematosus-prone New Zealand Mixed 2328 mice doubly deficient in TNF receptor 1 and TNF receptor 2 via a Th17-associated pathway. J Immunol 182(4):2532–2541. https://doi.org/10.4049/jimmunol.0802948
Jarnicki AG, Conroy H, Brereton C, Donnelly G, Toomey D, Walsh K, Sweeney C, Leavy O, Fletcher J, Lavelle EC, Dunne P, Mills KH (2008) Attenuating regulatory T cell induction by TLR agonists through inhibition of p38 MAPK signaling in dendritic cells enhances their efficacy as vaccine adjuvants and cancer immunotherapeutics. J Immunol 180(6):3797–3806. https://doi.org/10.4049/jimmunol.180.6.3797
Jiao Z, Wang W, Jia R, Li J, You H, Chen L, Wang Y (2007) Accumulation of FoxP3-expressing CD4+CD25+ T cells with distinct chemokine receptors in synovial fluid of patients with active rheumatoid arthritis. Scand J Rheumatol 36(6):428–433. https://doi.org/10.1080/03009740701482800
Jofra T, Di Fonte R, Galvani G, Kuka M, Iannacone M, Battaglia M, Fousteri G (2018) Tr1 cell immunotherapy promotes transplant tolerance via de novo Tr1 cell induction in mice and is safe and effective during acute viral infection. Eur J Immunol 48 (8):1389–1399. https://doi.org/10.1002/eji.201747316
Johanns TM, Ertelt JM, Rowe JH, Way SS (2010) Regulatory T cell suppressive potency dictates the balance between bacterial proliferation and clearance during persistent Salmonella infection. PLoS Pathog 6(8): e1001043. https://doi.org/10.1371/journal.ppat.1001043
Jordan MS, Boesteanu A, Reed AJ, Petrone AL, Holenbeck AE, Lerman MA, Naji A, Caton AJ (2001) Thymic selection of CD4+CD25+ regulatory T cells induced by an agonist self-peptide. Nat Immunol 2(4):301–306. https://doi.org/10.1038/86302
Josefowicz SZ, Lu LF, Rudensky AY (2012) Regulatory T cells: mechanisms of differentiation and function. Annu Rev Immunol 30:531–564. https://doi.org/10.1146/annurev.immunol.25.022106.141623
Ju S, Qiu H, Zhou X, Zhu B, Lv X, Huang X, Li J, Zhang Y, Liu L, Ge Y, Johnson DE, Ju S, Shu Y (2009) CD13+CD4+CD25hi regulatory T cells exhibit higher suppressive function and increase with tumor stage in non-small cell lung cancer patients. Cell Cycle 8(16):2578–2585. https://doi.org/10.4161/cc.8.16.9302
Judge AD, Zhang X, Fujii H, Surh CD, Sprent J (2002) Interleukin 15 controls both proliferation and survival of a subset of memory-phenotype CD8(+) T cells. J Exp Med 196(7):935–946. https://doi.org/10.1084/jem.20020772
Kaech SM, Wherry EJ, Ahmed R (2002) Effector and memory T-cell differentiation: implications for vaccine development. Nat Rev Immunol 2(4):251–262. https://doi.org/10.1038/nri778
Kalogerou A, Gelou E, Mountantonakis S, Settas L, Zafiriou E, Sakkas L (2005) Early T cell activation in the skin from patients with systemic sclerosis. Ann Rheum Dis 64(8):1233–1235. https://doi.org/10.1136/ard.2004.027094
Karecla PI, Bowden SJ, Green SJ, Kilshaw PJ (1995) Recognition of E-cadherin on epithelial cells by the mucosal T cell integrin alpha M290 beta 7 (alpha E beta 7). Eur J Immunol 25(3):852–856. https://doi.org/10.1002/eji.1830250333
Kashima Y, Nishii N, Tachinami H, Furusawa E, Nagai S, Harada H, Azuma M (2020) Orthotopic tongue squamous cell carcinoma (SCC) model exhibiting a different tumor-infiltrating T-cell status with margin-restricted CD8(+) T cells and regulatory T cell-dominance, compared to skin SCC. Biochem Biophys Res Commun 526(1):218–224. https://doi.org/10.1016/j.bbrc.2020.03.022
Kawashiri SY, Kawakami A, Okada A, Koga T, Tamai M, Yamasaki S, Nakamura H, Origuchi T, Ida H, Eguchi K (2011) CD4+CD25(high)CD127(low/-) Treg cell frequency from peripheral blood correlates with disease activity in patients with rheumatoid arthritis. J

Rheumatol 38(12):2517–2521. https://doi.org/10.3899/jrheum.110283

Khattri R, Cox T, Yasayko SA, Ramsdell F (2003) An essential role for scurfin in CD4+CD25+ T regulatory cells. Nat Immunol 4(4):337–342. https://doi.org/10.1038/ni909

Kikuchi J, Hashizume M, Kaneko Y, Yoshimoto K, Nishina N, Takeuchi T (2015) Peripheral blood CD4 (+) CD25(+) CD127(low) regulatory T cells are significantly increased by tocilizumab treatment in patients with rheumatoid arthritis: increase in regulatory T cells correlates with clinical response. Arthritis Res Ther 17:10. https://doi.org/10.1186/s13075-015-0526-4

Kim Sh SH, Kaminker P, Campisi J (2002) Telomeres, aging and cancer: in search of a happy ending. Oncogene 21(4):503–511. https://doi.org/10.1038/sj.onc.1205077

Kim HJ, Wang X, Radfar S, Sproule TJ, Roopenian DC, Cantor H (2011) CD8+ T regulatory cells express the Ly49 Class I MHC receptor and are defective in autoimmune prone B6-Yaa mice. Proc Natl Acad Sci U S A 108(5):2010–2015. https://doi.org/10.1073/pnas.1018974108

Klein S, Kretz CC, Ruland V, Stumpf C, Haust M, Hartschuh W, Hartmann M, Enk A, Suri-Payer E, Oberle N, Krammer PH, Kuhn A (2011) Reduction of regulatory T cells in skin lesions but not in peripheral blood of patients with systemic scleroderma. Ann Rheum Dis 70(8):1475–1481. https://doi.org/10.1136/ard.2009.116525

Kleinewietfeld M, Manzel A, Titze J, Kvakan H, Yosef N, Linker RA, Muller DN, Hafler DA (2013) Sodium chloride drives autoimmune disease by the induction of pathogenic TH17 cells. Nature 496(7446):518–522. https://doi.org/10.1038/nature11868

Koch SD, Uss E, van Lier RA, ten Berge IJ (2008) Alloantigen-induced regulatory CD8+CD103+ T cells. Hum Immunol 69(11):737–744. https://doi.org/10.1016/j.humimm.2008.08.281

Koch K, Koch N, Sandaradura de Silva U, Jung N, Schulze zur Wiesch J, Fatkenheuer G, Hartmann P, Romerio F, Lehmann C (2015) Increased frequency of CD49b/LAG-3(+) type 1 regulatory T cells in HIV-infected individuals. AIDS Res Hum Retrovir 31 (12):1238–1246. https://doi.org/10.1089/aid.2014.0356

Kojima A, Prehn RT (1981) Genetic susceptibility to post-thymectomy autoimmune diseases in mice. Immunogenetics 14(1–2):15–27. https://doi.org/10.1007/bf00344296

Kong N, Lan Q, Chen M, Wang J, Shi W, Horwitz DA, Quesniaux V, Ryffel B, Liu Z, Brand D, Zou H, Zheng SG (2012a) Antigen-specific transforming growth factor β-induced Treg cells, but not natural Treg cells, ameliorate autoimmune arthritis in mice by shifting the Th17/Treg cell balance from Th17 predominance to Treg cell predominance. Arthritis Rheum 64 (8):2548–2558. https://doi.org/10.1002/art.34513

Kong N, Lan Q, Su W, Chen M, Wang J, Yang Z, Park R, Dagliyan G, Conti PS, Brand D, Liu Z, Stohl W, Zou H, Zheng SG (2012b) Induced T regulatory cells suppress osteoclastogenesis and bone erosion in collagen-induced arthritis better than natural T regulatory cells. Ann Rheum Dis 71(9):1567–1572. https://doi.org/10.1136/annrheumdis-2011-201052

Korecka-Polak A et al (2011) Human peripheral blood CD8+ CD28- T cells of renal allograft recipients do not express FOXP3 protein. Transplant Proc 43 (8):2917–2921

Krupnick AS, Lin X, Li W, Higashikubo R, Zinselmeyer BH, Hartzler H, Toth K, Ritter JH, Berezin MY, Wang ST, Miller MJ, Gelman AE, Kreisel D (2014) Central memory CD8+ T lymphocytes mediate lung allograft acceptance. J Clin Invest 124(3):1130–1143. https://doi.org/10.1172/jci71359

Ku CC et al (2000) Control of homeostasis of CD8+ memory T cells by opposing cytokines. Science 288 (5466):675–678

Lanteri MC, O'Brien KM, Purtha WE, Cameron MJ, Lund JM, Owen RE, Heitman JW, Custer B, Hirschkorn DF, Tobler LH, Kiely N, Prince HE, Ndhlovu LC, Nixon DF, Kamel HT, Kelvin DJ, Busch MP, Rudensky AY, Diamond MS, Norris PJ (2009) Tregs control the development of symptomatic West Nile virus infection in humans and mice. J Clin Invest 119(11):3266–3277. https://doi.org/10.1172/jci39387

Lars A, Ormandy TH, Wedemeyer H, Manns MP, Greten TF, Korangy F (2005) Increased populations of regulatory T cells in peripheral blood of patients with hepatocellular carcinoma. Cancer Res 65:6

Layland LE, Straubinger K, Ritter M, Loffredo-Verde E, Garn H, Sparwasser T, Prazeres da Costa C (2013) Schistosoma mansoni-mediated suppression of allergic airway inflammation requires patency and Foxp3+ Treg cells. PLoS Negl Trop Dis 7(8):e2379. https://doi.org/10.1371/journal.pntd.0002379

Lee HM, Hsieh CS (2009) Rare development of Foxp3+ thymocytes in the CD4+CD8+ subset. J Immunol 183 (4):2261–2266. https://doi.org/10.4049/jimmunol.0901304

Lee I, Wang L, Wells AD, Dorf ME, Ozkaynak E, Hancock WW (2005) Recruitment of Foxp3+ T regulatory cells mediating allograft tolerance depends on the CCR4 chemokine receptor. J Exp Med 201 (7):1037–1044. https://doi.org/10.1084/jem.20041709

Lee YH, Ishida Y, Rifa'i M, Shi Z, Isobe K, Suzuki H (2008) Essential role of CD8+CD122+ regulatory T cells in the recovery from experimental autoimmune encephalomyelitis. J Immunol 180(2):825–832. https://doi.org/10.4049/jimmunol.180.2.825

Lerret NM, Houlihan JL, Kheradmand T, Pothoven KL, Zhang ZJ, Luo X (2012) Donor-specific CD8+ Foxp3+ T cells protect skin allografts and facilitate induction of conventional CD4+ Foxp3+ regulatory T cells. Am J Transplant 12(9):2335–2347. https://doi.org/10.1111/j.1600-6143.2012.04120.x

Levings MK, Gregori S, Tresoldi E, Cazzaniga S, Bonini C, Roncarolo MG (2005) Differentiation of Tr1 cells by immature dendritic cells requires IL-10 but not CD25+CD4+ Tr cells. Blood 105 (3):1162–1169. https://doi.org/10.1182/blood-2004-03-1211
Liang S, Alard P, Zhao Y, Parnell S, Clark SL, Kosiewicz MM (2005) Conversion of CD4+ CD25- cells into CD4+ CD25+ regulatory T cells in vivo requires B7 costimulation, but not the thymus. J Exp Med 201 (1):127–137. https://doi.org/10.1084/jem.20041201
Liao HT, Lin YF, Tsai CY, Chou CT (2015) Regulatory T cells in ankylosing spondylitis and the response after adalimumab treatment. Joint Bone Spine 82 (6):423–427. https://doi.org/10.1016/j.jbspin.2015.03.003
Liao T, Xue Y, Zhao D, Li S, Liu M, Chen J, Brand DD, Zheng H, Zhang Y, Zheng SG, Sun Q (2017) In vivo attenuation of antibody-mediated acute renal allograft rejection by ex vivo TGF-β-induced CD4(+)Foxp3(+) regulatory T cells. Front Immunol 8:1334. https://doi.org/10.3389/fimmu.2017.01334
Lin X, Chen M, Liu Y, Guo Z, He X, Brand D, Zheng SG (2013) Advances in distinguishing natural from induced Foxp3(+) regulatory T cells. Int J Clin Exp Pathol 6(2):116–123
Lina C, Conghua W, Nan L, Ping Z (2011) Combined treatment of etanercept and MTX reverses Th1/Th2, Th17/Treg imbalance in patients with rheumatoid arthritis. J Clin Immunol 31(4):596–605. https://doi.org/10.1007/s10875-011-9542-6
Liu X, Gao N, Li M, Xu D, Hou Y, Wang Q, Zhang G, Sun Q, Zhang H, Zeng X (2013a) Elevated levels of CD4(+)CD25(+)FoxP3(+) T cells in systemic sclerosis patients contribute to the secretion of IL-17 and immunosuppression dysfunction. PLoS One 8(6):e64531. https://doi.org/10.1371/journal.pone.0064531
Liu Z, Fan H, Jiang S (2013b) CD4(+) T-cell subsets in transplantation. Immunol Rev 252(1):183–191. https://doi.org/10.1111/imr.12038
Liu Y, Lan Q, Lu L, Chen M, Xia Z, Ma J, Wang J, Fan H, Shen Y, Ryffel B, Brand D, Quismorio F, Liu Z, Horwitz DA, Xu A, Zheng SG (2014) Phenotypic and functional characteristic of a newly identified CD8+ Foxp3- CD103+ regulatory T cells. J Mol Cell Biol 6(1):81–92. https://doi.org/10.1093/jmcb/mjt026
Liu H, Wang Y, Zeng Q, Zeng YQ, Liang CL, Qiu F, Nie H, Dai Z (2017) Suppression of allograft rejection by CD8+CD122+PD-1+ Tregs is dictated by their Fas ligand-initiated killing of effector T cells versus Fas-mediated own apoptosis. Oncotarget 8 (15):24187–24195. https://doi.org/10.18632/oncotarget.15551
Liu H, Qiu F, Wang Y, Zeng Q, Liu C, Chen Y, Liang CL, Zhang Q, Han L, Dai Z (2019) CD8+CD122+PD-1+ Tregs synergize with costimulatory blockade of CD40/CD154, but not B7/CD28, to prolong murine allograft survival. Front Immunol 10:306. https://doi.org/10.3389/fimmu.2019.00306
Liyanage UK, Moore TT, Joo HG, Tanaka Y, Herrmann V, Doherty G, Drebin JA, Strasberg SM, Eberlein TJ, Goedegebuure PS, Linehan DC (2002) Prevalence of regulatory T cells is increased in peripheral blood and tumor microenvironment of patients with pancreas or breast adenocarcinoma. J Immunol 169(5):2756–2761. https://doi.org/10.4049/jimmunol.169.5.2756
Lu L, Yu Y, Li G, Pu L, Zhang F, Zheng S, Wang X (2009) CD8(+) CD103(+) regulatory T cells in spontaneous tolerance of liver allografts. Int Immunopharmacol 9(5):546–548. https://doi.org/10.1016/j.intimp.2009.01.021
Lu L, Wang J, Zhang F, Chai Y, Brand D, Wang X, Horwitz DA, Shi W, Zheng SG (2010) Role of SMAD and non-SMAD signals in the development of Th17 and regulatory T cells. J Immunol 184 (8):4295–4306. https://doi.org/10.4049/jimmunol.0903418
Lu L, Lan Q, Li Z, Zhou X, Gu J, Li Q, Wang J, Chen M, Liu Y, Shen Y, Brand DD, Ryffel B, Horwitz DA, Quismorio FP, Liu Z, Li B, Olsen NJ, Zheng SG (2014) Critical role of all-trans retinoic acid in stabilizing human natural regulatory T cells under inflammatory conditions. Proc Natl Acad Sci U S A 111(33):E3432–E3440
Lühn K, Simmons CP, Moran E, Dung NT, Chau TN, Quyen NT, Thao le TT, Van Ngoc T, Dung NM, Wills B, Farrar J, McMichael AJ, Dong T, Rowland-Jones S (2007) Increased frequencies of CD4+ CD25 (high) regulatory T cells in acute dengue infection. J Exp Med 204(5):979–985. https://doi.org/10.1084/jem.20061381
Lund JM, Hsing L, Pham TT, Rudensky AY (2008) Coordination of early protective immunity to viral infection by regulatory T cells. Science 320(5880):1220–1224. https://doi.org/10.1126/science.1155209
Luo Y, Zheng SG (2016) Hall of fame among pro-inflammatory cytokines: interleukin-6 gene and its transcriptional regulation mechanisms. Front Immunol 7:604. https://doi.org/10.3389/fimmu.2016.00604
Luo Y, Xue Y, Wang J, Dang J, Fang Q, Huang G, Olsen N, Zheng SG (2019) Negligible effect of sodium chloride on the development and function of TGF-beta-induced CD4(+) Foxp3(+) regulatory T cells. Cell Rep 26(7):1869–1879 e1863. https://doi.org/10.1016/j.celrep.2019.01.066
Ma J, Yu J, Tao X, Cai L, Wang J, Zheng SG (2010) The imbalance between regulatory and IL-17-secreting CD4+ T cells in lupus patients. Clin Rheumatol 29 (11):1251–1258. https://doi.org/10.1007/s10067-010-1510-7
Ma L, Simpson E, Li J, Xuan M, Xu M, Baker L, Shi Y, Yougbare I, Wang X, Zhu G, Chen P, Prud'homme GJ, Lazarus AH, Freedman J, Ni H (2015) CD8+ T cells are predominantly protective and required for effective steroid therapy in murine models of immune

thrombocytopenia. Blood 126(2):247–256. https://doi.org/10.1182/blood-2015-03-635417
MacDonald KG, Dawson NA, Huang Q, Dunne JV, Levings MK, Broady R (2015) Regulatory T cells produce profibrotic cytokines in the skin of patients with systemic sclerosis. J Allergy Clin Immunol 135 (4):946–9e9. https://doi.org/10.1016/j.jaci.2014.12.1932
Malek TR (2008) The biology of interleukin-2. Annu Rev Immunol 26:453–479. https://doi.org/10.1146/annurev.immunol.26.021607.090357
Manavalan JS et al (2004) Alloantigen specific CD8 +CD28- FOXP3+ T suppressor cells induce ILT3+ ILT4+ tolerogenic endothelial cells, inhibiting alloreactivity. Int Immunol 16(8):1055–1068
Mandapathil M, Szczepanski MJ, Szajnik M, Ren J, Jackson EK, Johnson JT, Gorelik E, Lang S, Whiteside TL (2010) Adenosine and prostaglandin E2 cooperate in the suppression of immune responses mediated by adaptive regulatory T cells. J Biol Chem 285 (36):27571–27580. https://doi.org/10.1074/jbc.M110.127100
Mangalam AK, Luckey D, Giri S, Smart M, Pease LR, Rodriguez M, David CS (2012) Two discreet subsets of CD8 T cells modulate PLP (91-110) induced experimental autoimmune encephalomyelitis in HLA-DR3 transgenic mice. J Autoimmun 38(4):344–353. https://doi.org/10.1016/j.jaut.2012.02.004
Martín-Orozco E, Norte-Muñoz M, Martínez-García J (2017) Regulatory T cells in allergy and asthma. Front Pediatr 5:117. https://doi.org/10.3389/fped.2017.00117
Mascanfroni ID, Takenaka MC, Yeste A, Patel B, Wu Y, Kenison JE, Siddiqui S, Basso AS, Otterbein LE, Pardoll DM, Pan F, Priel A, Clish CB, Robson SC, Quintana FJ (2015) Metabolic control of type 1 regulatory T cell differentiation by AHR and HIF1-alpha. Nat Med 21(6):638–646. https://doi.org/10.1038/nm.3868
Mathew JM, Voss JH, LeFever A, Konieczna I, Stratton C, He J, Huang X, Gallon L, Skaro A, Ansari MJ, Leventhal JR (2018) A phase I clinical trial with ex vivo expanded recipient regulatory T cells in living donor kidney transplants. Sci Rep 8(1):7428. https://doi.org/10.1038/s41598-018-25574-7
Mathian A, Parizot C, Dorgham K, Trad S, Arnaud L, Larsen M, Miyara M, Hié M, Piette JC, Frances C, Yssel H, Amoura Z, Gorochov G (2012) Activated and resting regulatory T cell exhaustion concurs with high levels of interleukin-22 expression in systemic sclerosis lesions. Ann Rheum Dis 71(7):1227–1234. https://doi.org/10.1136/annrheumdis-2011-200709
Maynard CL, Harrington LE, Janowski KM, Oliver JR, Zindl CL, Rudensky AY, Weaver CT (2007) Regulatory T cells expressing interleukin 10 develop from Foxp3+ and Foxp3- precursor cells in the absence of interleukin 10. Nat Immunol 8(9):931–941. https://doi.org/10.1038/ni1504
Meloni F, Solari N, Cavagna L, Morosini M, Montecucco CM, Fietta AM (2009) Frequency of Th1, Th2 and Th17 producing T lymphocytes in bronchoalveolar lavage of patients with systemic sclerosis. Clin Exp Rheumatol 27(5):765–772
Mempel TR, Bauer CA (2009) Intravital imaging of CD8+ T cell function in cancer. Clin Exp Metastasis 26 (4):311–327. https://doi.org/10.1007/s10585-008-9196-9
Mikulkova Z, Praksova P, Stourac P, Bednarik J, Strajtova L, Pacasova R, Belobradkova J, Dite P, Michalek J (2010) Numerical defects in CD8+CD28- T-suppressor lymphocyte population in patients with type 1 diabetes mellitus and multiple sclerosis. Cell Immunol 262(2):75–79. https://doi.org/10.1016/j.cellimm.2010.02.002
Miller WP, Srinivasan S, Panoskaltsis-Mortari A, Singh K, Sen S, Hamby K, Deane T, Stempora L, Beus J, Turner A, Wheeler C, Anderson DC, Sharma P, Garcia A, Strobert E, Elder E, Crocker I, Crenshaw T, Penedo MC, Ward T, Song M, Horan J, Larsen CP, Blazar BR, Kean LS (2010) GVHD after haploidentical transplantation: a novel, MHC-defined rhesus macaque model identifies CD28- CD8+ T cells as a reservoir of breakthrough T-cell proliferation during costimulation blockade and sirolimus-based immunosuppression. Blood 116 (24):5403–5418. https://doi.org/10.1182/blood-2010-06-289272
Mills KH (2004) Regulatory T cells: friend or foe in immunity to infection? Nat Rev Immunol 4 (11):841–855. https://doi.org/10.1038/nri1485
Miyara M, Yoshioka Y, Kitoh A, Shima T, Wing K, Niwa A, Parizot C, Taflin C, Heike T, Valeyre D, Mathian A, Nakahata T, Yamaguchi T, Nomura T, Ono M, Amoura Z, Gorochov G, Sakaguchi S (2009) Functional delineation and differentiation dynamics of human CD4+ T cells expressing the FoxP3 transcription factor. Immunity 30(6):899–911. https://doi.org/10.1016/j.immuni.2009.03.019
Mohr A, Atif M, Balderas R, Gorochov G, Miyara M (2019) The role of FOXP3(+) regulatory T cells in human autoimmune and inflammatory diseases. Clin Exp Immunol 197(1):24–35. https://doi.org/10.1111/cei.13288
Molloy MJ, Zhang W, Usherwood EJ (2011) Suppressive CD8+ T cells arise in the absence of CD4 help and compromise control of persistent virus. J Immunol 186 (11):6218–6226. https://doi.org/10.4049/jimmunol.1003812
Montoya-Ortiz G (2013) Immunosenescence, aging, and systemic lupus erythematous. Autoimmune Dis 2013:267078. https://doi.org/10.1155/2013/267078
Moore AC, Gallimore A, Draper SJ, Watkins KR, Gilbert SC, Hill AV (2005) Anti-CD25 antibody enhancement of vaccine-induced immunogenicity: increased durable cellular immunity with reduced immunodominance. J Immunol 175(11):7264–7273. https://doi.org/10.4049/jimmunol.175.11.7264

Moreno-Fernandez ME, Rueda CM, Rusie LK, Chougnet CA (2011) Regulatory T cells control HIV replication in activated T cells through a cAMP-dependent mechanism. Blood 117(20):5372–5380. https://doi.org/10.1182/blood-2010-12-323162

Myers L, Croft M, Kwon BS, Mittler RS, Vella AT (2005) Peptide-specific CD8 T regulatory cells use IFN-gamma to elaborate TGF-beta-based suppression. J Immunol 174(12):7625–7632. https://doi.org/10.4049/jimmunol.174.12.7625

Nadkarni S, Mauri C, Ehrenstein MR (2007) Anti-TNF-alpha therapy induces a distinct regulatory T cell population in patients with rheumatoid arthritis via TGF-beta. J Exp Med 204(1):33–39. https://doi.org/10.1084/jem.20061531

Nagata T, Koide Y (2010) Induction of specific CD8 T cells against intracellular bacteria by CD8 T-cell-oriented immunization approaches. J Biomed Biotechnol 2010:764542. https://doi.org/10.1155/2010/764542

Niu Q, Cai B, Huang ZC, Shi YY, Wang LL (2012) Disturbed Th17/Treg balance in patients with rheumatoid arthritis. Rheumatol Int 32(9):2731–2736. https://doi.org/10.1007/s00296-011-1984-x

Noble A, Pestano GA, Cantor H (1998) Suppression of immune responses by CD8 cells. I. Superantigen-activated CD8 cells induce unidirectional Fas-mediated apoptosis of antigen-activated CD4 cells. J Immunol 160(2):559–565

O'Connor RA, Leech MD, Suffner J, Hämmerling GJ, Anderton SM (2010) Myelin-reactive, TGF-β-induced regulatory T cells can be programmed to develop Th1-like effector function but remain less proinflammatory than myelin-reactive Th1 effectors and can suppress pathogenic T cell clonal expansion in vivo. J Immunol 185(12):7235–7243. https://doi.org/10.4049/jimmunol.1001551

Ohkura N, Hamaguchi M, Morikawa H, Sugimura K, Tanaka A, Ito Y, Osaki M, Tanaka Y, Yamashita R, Nakano N, Huehn J, Fehling HJ, Sparwasser T, Nakai K, Sakaguchi S (2012) T cell receptor stimulation-induced epigenetic changes and Foxp3 expression are independent and complementary events required for Treg cell development. Immunity 37(5):785–799. https://doi.org/10.1016/j.immuni.2012.09.010

Oldenhove G, Bouladoux N, Wohlfert EA, Hall JA, Chou D, Dos Santos L, O'Brien S, Blank R, Lamb E, Natarajan S, Kastenmayer R, Hunter C, Grigg ME, Belkaid Y (2009) Decrease of Foxp3+ Treg cell number and acquisition of effector cell phenotype during lethal infection. Immunity 31(5):772–786. https://doi.org/10.1016/j.immuni.2009.10.001

Pandiyan P, Conti HR, Zheng L, Peterson AC, Mathern DR, Hernández-Santos N, Edgerton M, Gaffen SL, Lenardo MJ (2011) CD4(+) CD25(+) Foxp3(+) regulatory T cells promote Th17 cells in vitro and enhance host resistance in mouse Candida albicans Th17 cell infection model. Immunity 34(3):422–434. https://doi.org/10.1016/j.immuni.2011.03.002

Papp G, Horvath IF, Barath S, Gyimesi E, Sipka S, Szodoray P, Zeher M (2011) Altered T-cell and regulatory cell repertoire in patients with diffuse cutaneous systemic sclerosis. Scand J Rheumatol 40(3):205–210. https://doi.org/10.3109/03009742.2010.528021

Pauls K, Schon M, Kubitza RC, Homey B, Wiesenborn A, Lehmann P, Ruzicka T, Parker CM, Schon MP (2001) Role of integrin alphaE(CD103)beta7 for tissue-specific epidermal localization of CD8+ T lymphocytes. J Invest Dermatol 117(3):569–575. https://doi.org/10.1046/j.0022-202x.2001.01481.x

Pellegrino M, Crino A, Rosado MM, Fierabracci A (2019) Identification and functional characterization of CD8+ T regulatory cells in type 1 diabetes patients. PLoS One 14(1):e0210839. https://doi.org/10.1371/journal.pone.0210839

Piccirillo CA, Letterio JJ, Thornton AM, McHugh RS, Mamura M, Mizuhara H, Shevach EM (2002) CD4(+) CD25(+) regulatory T cells can mediate suppressor function in the absence of transforming growth factor beta1 production and responsiveness. J Exp Med 196(2):237–246. https://doi.org/10.1084/jem.20020590

Picker LJ, Terstappen LW, Rott LS, Streeter PR, Stein H, Butcher EC (1990) Differential expression of homing-associated adhesion molecules by T cell subsets in man. J Immunol 145(10):3247–3255

Pot C, Jin H, Awasthi A, Liu SM, Lai CY, Madan R, Sharpe AH, Karp CL, Miaw SC, Ho IC, Kuchroo VK (2009) Cutting edge: IL-27 induces the transcription factor c-Maf, cytokine IL-21, and the costimulatory receptor ICOS that coordinately act together to promote differentiation of IL-10-producing Tr1 cells. J Immunol 183(2):797–801. https://doi.org/10.4049/jimmunol.0901233

Powell BR, Buist NR, Stenzel P (1982) An X-linked syndrome of diarrhea, polyendocrinopathy, and fatal infection in infancy. J Pediatr 100(5):731–737. https://doi.org/10.1016/s0022-3476(82)80573-8

Prakken B, Ellen W, van Wijk F (2013) Editorial: quality or quantity? Unraveling the role of Treg cells in rheumatoid arthritis. Arthritis Rheum 65(3):552–554. https://doi.org/10.1002/art.37831

Radstake TR, van Bon L, Broen J, Wenink M, Santegoets K, Deng Y, Hussaini A, Simms R, Cruikshank WW, Lafyatis R (2009) Increased frequency and compromised function of T regulatory cells in systemic sclerosis (SSc) is related to a diminished CD69 and TGFbeta expression. PLoS One 4(6):e5981. https://doi.org/10.1371/journal.pone.0005981

Rapetti L, Chavele KM, Evans CM, Ehrenstein MR (2015) B cell resistance to Fas-mediated apoptosis contributes to their ineffective control by regulatory T cells in rheumatoid arthritis. Ann Rheum Dis 74(1):294–302. https://doi.org/10.1136/annrheumdis-2013-204049

Rifa'i M, Kawamoto Y, Nakashima I, Suzuki H (2004) Essential roles of CD8+CD122+ regulatory T cells in the maintenance of T cell homeostasis. J Exp Med 200(9):1123–1134. https://doi.org/10.1084/jem.20040395

Rihs S, Walker C, Virchow JC Jr, Boer C, Kroegel C, Giri SN, Braun RK (1996) Differential expression of alpha E beta 7 integrins on bronchoalveolar lavage T lymphocyte subsets: regulation by alpha 4 beta 1-integrin crosslinking and TGF-beta. Am J Respir Cell Mol Biol 15(5):600–610. https://doi.org/10.1165/ajrcmb.15.5.8918367

Robbins GR, Wen H, Ting JP (2014) Inflammasomes and metabolic disorders: old genes in modern diseases. Mol Cell 54(2):297–308. https://doi.org/10.1016/j.molcel.2014.03.029

Robertson H, Wong WK, Talbot D, Burt AD, Kirby JA (2001) Tubulitis after renal transplantation: demonstration of an association between CD103+ T cells, transforming growth factor beta1 expression and rejection grade. Transplantation 71(2):306–313. https://doi.org/10.1097/00007890-200101270-00024

Roncarolo MG, Gregori S, Bacchetta R, Battaglia M (2014) Tr1 cells and the counter-regulation of immunity: natural mechanisms and therapeutic applications. Curr Top Microbiol Immunol 380:39–68. https://doi.org/10.1007/978-3-662-43492-5_3

Roncarolo MG, Gregori S, Bacchetta R, Battaglia M, Gagliani N (2018) The biology of T regulatory type 1 cells and their therapeutic application in immune-mediated diseases. Immunity 49(6):1004–1019. https://doi.org/10.1016/j.immuni.2018.12.001

Rowe JH, Ertelt JM, Aguilera MN, Farrar MA, Way SS (2011) Foxp3(+) regulatory T cell expansion required for sustaining pregnancy compromises host defense against prenatal bacterial pathogens. Cell Host Microbe 10(1):54–64. https://doi.org/10.1016/j.chom.2011.06.005

Saadoun D, Rosenzwajg M, Joly F, Six A, Carrat F, Thibault V, Sene D, Cacoub P, Klatzmann D (2011) Regulatory T-cell responses to low-dose interleukin-2 in HCV-induced vasculitis. N Engl J Med 365 (22):2067–2077. https://doi.org/10.1056/NEJMoa1105143

Saitoh O, Abiru N, Nakahara M, Nagayama Y (2007) CD8+CD122+ T cells, a newly identified regulatory T subset, negatively regulate Graves' hyperthyroidism in a murine model. Endocrinology 148 (12):6040–6046. https://doi.org/10.1210/en.2007-0300

Sakaguchi S (2004) Naturally arising CD4+ regulatory t cells for immunologic self-tolerance and negative control of immune responses. Annu Rev Immunol 22:531–562. https://doi.org/10.1146/annurev.immunol.21.120601.141122

Sakaguchi S, Sakaguchi N, Asano M, Itoh M, Toda M (1995) Immunologic self-tolerance maintained by activated T cells expressing IL-2 receptor alpha-chains (CD25). Breakdown of a single mechanism of self-tolerance causes various autoimmune diseases. J Immunol 155(3):1151–1164

Sakaguchi S, Vignali DA, Rudensky AY, Niec RE, Waldmann H (2013) The plasticity and stability of regulatory T cells. Nat Rev Immunol 13(6):461–467. https://doi.org/10.1038/nri3464

Samson M, Audia S, Janikashvili N, Ciudad M, Trad M, Fraszczak J, Ornetti P, Maillefert JF, Miossec P, Bonnotte B (2012) Brief report: inhibition of interleukin-6 function corrects Th17/Treg cell imbalance in patients with rheumatoid arthritis. Arthritis Rheum 64(8):2499–2503. https://doi.org/10.1002/art.34477

Santinon F, Batignes M, Mebrek ML, Biton J, Clavel G, Herve R, Lemeiter D, Breckler M, Busato F, Tost J, Ziol M, Boissier MC, Decker P, Semerano L, Bessis N (2019) Involvement of tumor necrosis factor receptor type II in FoxP3 stability and as a marker of Treg cells specifically expanded by anti-tumor necrosis factor treatments in rheumatoid arthritis. Arthritis Rheumatol 72(4):576–587. https://doi.org/10.1002/art.41134

Sarnacki S, Begue B, Buc H, Le Deist F, Cerf-Bensussan N (1992) Enhancement of CD3-induced activation of human intestinal intraepithelial lymphocytes by stimulation of the beta 7-containing integrin defined by HML-1 monoclonal antibody. Eur J Immunol 22 (11):2887–2892. https://doi.org/10.1002/eji.1830221120

Sato E, Olson SH, Ahn J, Bundy B, Nishikawa H, Qian F, Jungbluth AA, Frosina D, Gnjatic S, Ambrosone C, Kepner J, Odunsi T, Ritter G, Lele S, Chen YT, Ohtani H, Old LJ, Odunsi K (2005) Intraepithelial CD8+ tumor-infiltrating lymphocytes and a high CD8+/regulatory T cell ratio are associated with favorable prognosis in ovarian cancer. Proc Natl Acad Sci U S A 102(51):18538–18543. https://doi.org/10.1073/pnas.0509182102

Satoru Todo KY, Goto R, Zaitsu M, Nagatsu A, Oura T, Masaaki W, Takeshi Aoyagi TS, Shimamura T, Kamiyama T, Sato N, Sugita J, Kanako Hatanaka HB, Habu S, Demetris AJ, Ko O (2016) A pilot study of operational tolerance with a regulatory T-Cell-Based cell therapy in living donor liver transplantation. Hepatology 64:2

Scheinecker C, Bonelli M, Smolen JS (2010) Pathogenetic aspects of systemic lupus erythematosus with an emphasis on regulatory T cells. J Autoimmun 35 (3):269–275. https://doi.org/10.1016/j.jaut.2010.06.018

Scheinecker C, Goschl L, Bonelli M (2019) Treg cells in health and autoimmune diseases: new insights from single cell analysis. J Autoimmun 110:102376. https://doi.org/10.1016/j.jaut.2019.102376

Scotto L et al (2004) Overlap between molecular markers expressed by naturally occurring CD4+CD25+ regulatory T cells and antigen specific CD4+CD25+ and CD8+CD28- T suppressor cells. Hum Immunol 65 (11):1297–1306

Scott-Browne JP, Shafiani S, Tucker-Heard G, Ishida-Tsubota K, Fontenot JD, Rudensky AY, Bevan MJ, Urdahl KB (2007) Expansion and function of Foxp3-expressing T regulatory cells during tuberculosis. J

Exp Med 204(9):2159–2169. https://doi.org/10.1084/jem.20062105

Seifert HA, Benedek G, Nguyen H, Kent G, Vandenbark AA, Offner H (2017) Estrogen protects both sexes against EAE by promoting common regulatory cell subtypes independent of endogenous estrogen. Metab Brain Dis 32(5):1747–1754. https://doi.org/10.1007/s11011-017-0063-8

Sempere-Ortells JM, Perez-Garcia V, Marin-Alberca G, Peris-Pertusa A, Benito JM, Marco FM, Zubcoff JJ, Navarro-Blasco FJ (2009) Quantification and phenotype of regulatory T cells in rheumatoid arthritis according to disease activity score-28. Autoimmunity 42(8):636–645. https://doi.org/10.3109/08916930903061491

Serafini G, Andreani M, Testi M, Battarra M, Bontadini A, Biral E, Fleischhauer K, Marktel S, Lucarelli G, Roncarolo MG, Bacchetta R (2009) Type 1 regulatory T cells are associated with persistent split erythroid/lymphoid chimerism after allogeneic hematopoietic stem cell transplantation for thalassemia. Haematologica 94(10):1415–1426. https://doi.org/10.3324/haematol.2008.003129

Shi Z, Rifa'i M, Lee YH, Shiku H, Isobe K, Suzuki H (2008) Importance of CD80/CD86-CD28 interactions in the recognition of target cells by CD8+CD122+ regulatory T cells. Immunology 124(1):121–128. https://doi.org/10.1111/j.1365-2567.2007.02747.x

Shimizu J, Yamazaki S, Takahashi T, Ishida Y, Sakaguchi S (2002) Stimulation of CD25+CD4+ regulatory T cells through GITR breaks immunological self-tolerance. Nat Immunol 3(2):135–142. https://doi.org/10.1038/ni759

Simon S, Labarriere N (2017) PD-1 expression on tumor-specific T cells: friend or foe for immunotherapy? Onco Targets Ther 7(1):e1364828. https://doi.org/10.1080/2162402x.2017.1364828

Simpson TR, Li F, Montalvo-Ortiz W, Sepulveda MA, Bergerhoff K, Arce F, Roddie C, Henry JY, Yagita H, Wolchok JD, Peggs KS, Ravetch JV, Allison JP, Quezada SA (2013) Fc-dependent depletion of tumor-infiltrating regulatory T cells co-defines the efficacy of anti-CTLA-4 therapy against melanoma. J Exp Med 210(9):1695–1710. https://doi.org/10.1084/jem.20130579

Slobodin G, Ahmad MS, Rosner I, Peri R, Rozenbaum M, Kessel A, Toubi E, Odeh M (2010) Regulatory T cells (CD4(+) CD25(bright)FoxP3(+)) expansion in systemic sclerosis correlates with disease activity and severity. Cell Immunol 261(2):77–80. https://doi.org/10.1016/j.cellimm.2009.12.009

Smith TR et al (2009) Dendritic cells use endocytic pathway for cross-priming class Ib MHC-restricted CD8alphaalpha+TCRalphabeta+ T cells with regulatory properties. J Immunol 182(11):6959–6968

Smith TR et al (2010) CD8alpha+ dendritic cells prime TCR-peptide-reactive regulatory CD4+FOXP3- T cells. Eur J Immunol 40(7):1906–1915

Smolen JS, Breedveld FC, Burmester GR, Bykerk V, Dougados M, Emery P, Kvien TK, Navarro-Compan MV, Oliver S, Schoels M, Scholte-Voshaar M, Stamm T, Stoffer M, Takeuchi T, Aletaha D, Andreu JL, Aringer M, Bergman M, Betteridge N, Bijlsma H, Burkhardt H, Cardiel M, Combe B, Durez P, Fonseca JE, Gibofsky A, Gomez-Reino JJ, Graninger W, Hannonen P, Haraoui B, Kouloumas M, Landewe R, Martin-Mola E, Nash P, Ostergaard M, Ostor A, Richards P, Sokka-Isler T, Thorne C, Tzioufas AG, van Vollenhoven R, de Wit M, van der Heijde D (2016) Treating rheumatoid arthritis to target: 2014 update of the recommendations of an international task force. Ann Rheum Dis 75(1):3–15. https://doi.org/10.1136/annrheumdis-2015-207524

Solis-Castillo LA, Garcia-Romo GS, Diaz-Rodriguez A, Reyes-Hernandez D, Tellez-Rivera E, Rosales-Garcia VH, Mendez-Cruz AR, Jimenez-Flores JR, Villafana-Vazquez VH, Pedroza-Gonzalez A (2020) Tumor-infiltrating regulatory T cells, CD8/Treg ratio, and cancer stem cells are correlated with lymph node metastasis in patients with early breast cancer. Breast Cancer 27(5):837–849. https://doi.org/10.1007/s12282-020-01079-y

Spaulding C, Guo W, Effros RB (1999) Resistance to apoptosis in human CD8+ T cells that reach replicative senescence after multiple rounds of antigen-specific proliferation. Exp Gerontol 34(5):633–644. https://doi.org/10.1016/s0531-5565(99)00033-9

Speiser DE, Ho PC, Verdeil G (2016) Regulatory circuits of T cell function in cancer. Nat Rev Immunol 16 (10):599–611. https://doi.org/10.1038/nri.2016.80

Strioga M, Pasukoniene V, Characiejus D (2011) CD8+ CD28- and CD8+ CD57+ T cells and their role in health and disease. Immunology 134(1):17–32. https://doi.org/10.1111/j.1365-2567.2011.03470.x

Stumhofer JS, Silver JS, Laurence A, Porrett PM, Harris TH, Turka LA, Ernst M, Saris CJ, O'Shea JJ, Hunter CA (2007) Interleukins 27 and 6 induce STAT3-mediated T cell production of interleukin 10. Nat Immunol 8(12):1363–1371. https://doi.org/10.1038/ni1537

Su W, Fan H, Chen M, Wang J, Brand D, He X, Quesniaux V, Ryffel B, Zhu L, Liang D, Zheng SG (2012) Induced CD4+ forkhead box protein-positive T cells inhibit mast cell function and established contact hypersensitivity through TGF-β1. J Allergy Clin Immunol 130(2):444–452.e447. https://doi.org/10.1016/j.jaci.2012.05.011

Su W, Chen X, Zhu W, Yu J, Li W, Li Y, Li Z, Olsen N, Liang D, Zheng SG (2019) The cAMP-adenosine feedback loop maintains the suppressive function of regulatory T cells. J Immunol 203(6):1436–1446. https://doi.org/10.4049/jimmunol.1801306

Suciu-Foca N, Manavalan JS, Cortesini R (2003) Generation and function of antigen-specific suppressor and regulatory T cells. Transpl Immunol 11 (3–4):235–244. https://doi.org/10.1016/s0966-3274(03)00052-2

Sun Z, Zhong W, Lu X, Shi B, Zhu Y, Chen L, Zhang G, Zhang X (2008) Association of Graves' disease and prevalence of circulating IFN-gamma-producing CD28(-) T cells. J Clin Immunol 28(5):464–472. https://doi.org/10.1007/s10875-008-9213-4

Sun J, Yang Y, Huo X, Zhu B, Li Z, Jiang X, Xie R, Gao L, Sun Y, Fan H, Zhu Y, Yang J (2019) Efficient therapeutic function and mechanisms of human polyclonal CD8(+)CD103(+)Foxp3(+) regulatory T cells on collagen-induced arthritis in mice. J Immunol Res 2019:8575407. https://doi.org/10.1155/2019/8575407

Suzuki H, Shi Z, Okuno Y, Isobe K (2008) Are CD8+CD122+ cells regulatory T cells or memory T cells? Hum Immunol 69(11):751–754. https://doi.org/10.1016/j.humimm.2008.08.285

Szmyrka-Kaczmarek M, Kosmaczewska A, Ciszak L, Szteblich A, Wiland P (2014) Peripheral blood Th17/Treg imbalance in patients with low-active systemic lupus erythematosus. Postepy Higi Med Dos (Online) 68:893–898. https://doi.org/10.5604/17322693.1111127

Szurek E, Cebula A, Wojciech L, Pietrzak M, Rempala G, Kisielow P, Ignatowicz L (2015) Differences in expression level of helios and neuropilin-1 do not distinguish thymus-derived from extrathymically-induced CD4+Foxp3+ regulatory T cells. PLoS One 10(10):e0141161. https://doi.org/10.1371/journal.pone.0141161

Tanaka A, Sakaguchi S (2017) Regulatory T cells in cancer immunotherapy. Cell Res 27(1):109–118. https://doi.org/10.1038/cr.2016.151

Tang Q, Bluestone JA (2008) The Foxp3+ regulatory T cell: a jack of all trades, master of regulation. Nat Immunol 9(3):239–244. https://doi.org/10.1038/ni1572

Tang Q, Vincenti F (2017) Transplant trials with Tregs: perils and promises. J Clin Invest 127(7):2505–2512. https://doi.org/10.1172/JCI90598

Thiolat A, Semerano L, Pers YM, Biton J, Lemeiter D, Portales P, Quentin J, Jorgensen C, Decker P, Boissier MC, Louis-Plence P, Bessis N (2014) Interleukin-6 receptor blockade enhances CD39+ regulatory T cell development in rheumatoid arthritis and in experimental arthritis. Arthritis Rheumatol 66(2):273–283. https://doi.org/10.1002/art.38246

Thornton AM, Korty PE, Tran DQ, Wohlfert EA, Murray PE, Belkaid Y, Shevach EM (2010) Expression of Helios, an Ikaros transcription factor family member, differentiates thymic-derived from peripherally induced Foxp3+ T regulatory cells. J Immunol 184(7):3433–3441. https://doi.org/10.4049/jimmunol.0904028

Todo S, Yamashita K, Goto R, Zaitsu M, Nagatsu A, Oura T, Watanabe M, Aoyagi T, Suzuki T, Shimamura T, Kamiyama T, Sato N, Sugita J, Hatanaka K, Bashuda H, Habu S, Demetris AJ, Okumura K (2016) A pilot study of operational tolerance with a regulatory T-cell-based cell therapy in living donor liver transplantation. Hepatology 64:2. https://doi.org/10.1002/hep.28459/suppinfo

Uss E, Rowshani AT, Hooibrink B, Lardy NM, van Lier RA, ten Berge IJ (2006) CD103 is a marker for alloantigen-induced regulatory CD8+ T cells. J Immunol 177(5):2775–2783. https://doi.org/10.4049/jimmunol.177.5.2775

Vallejo AN (2005) CD28 extinction in human T cells: altered functions and the program of T-cell senescence. Immunol Rev 205:158–169. https://doi.org/10.1111/j.0105-2896.2005.00256.x

Verhagen J, Wraith DC (2010) Comment on "Expression of Helios, an Ikaros transcription factor family member, differentiates thymic-derived from peripherally induced Foxp3+ T regulatory cells". J Immunol 185(12):7129.; ; author reply 7130. https://doi.org/10.4049/jimmunol.1090105

Vieira PL, Christensen JR, Minaee S, O'Neill EJ, Barrat FJ, Boonstra A, Barthlott T, Stockinger B, Wraith DC, O'Garra A (2004) IL-10-secreting regulatory T cells do not express Foxp3 but have comparable regulatory function to naturally occurring CD4+CD25+ regulatory T cells. J Immunol 172(10):5986–5993. https://doi.org/10.4049/jimmunol.172.10.5986

Vogt S, Iking-Konert C, Hug F, Andrassy K, Hänsch GM (2003) Shortening of telomeres: evidence for replicative senescence of T cells derived from patients with Wegener's granulomatosis. Kidney Int 63(6):2144–2151. https://doi.org/10.1046/j.1523-1755.2003.0003.x

von Spee-Mayer C, Siegert E, Abdirama D, Rose A, Klaus A, Alexander T, Enghard P, Sawitzki B, Hiepe F, Radbruch A, Burmester GR, Riemekasten G, Humrich JY (2016) Low-dose interleukin-2 selectively corrects regulatory T cell defects in patients with systemic lupus erythematosus. Ann Rheum Dis 75(7):1407–1415. https://doi.org/10.1136/annrheumdis-2015-207776

Wan N, Dai H, Wang T, Moore Y, Zheng XX, Dai Z (2008) Bystander central memory but not effector memory CD8+ T cells suppress allograft rejection. J Immunol 180(1):113–121. https://doi.org/10.4049/jimmunol.180.1.113

Wang EC, Lawson TM, Vedhara K, Moss PA, Lehner PJ, Borysiewicz LK (1997) CD8high+ (CD57+) T cells in patients with rheumatoid arthritis. Arthritis Rheum 40(2):237–248. https://doi.org/10.1002/art.1780400208

Wang D, Yuan R, Feng Y, El-Asady R, Farber DL, Gress RE, Lucas PJ, Hadley GA (2004) Regulation of CD103 expression by CD8+ T cells responding to renal allografts. J Immunol 172(1):214–221. https://doi.org/10.4049/jimmunol.172.1.214

Wang LX, Li Y, Yang G, Pang PY, Haley D, Walker EB, Urba WJ, Hu HM (2010) CD122+CD8+ Treg suppress vaccine-induced antitumor immune responses in lymphodepleted mice. Eur J Immunol 40(5):1375–1385. https://doi.org/10.1002/eji.200839210

Wang C, Liao Q, Hu Y, Zhong D (2015) T lymphocyte subset imbalances in patients contribute to ankylosing spondylitis. Exp Ther Med 9(1):250–256. https://doi.org/10.3892/etm.2014.2046

Weiss JM, Bilate AM, Gobert M, Ding Y, Curotto de Lafaille MA, Parkhurst CN, Xiong H, Dolpady J, Frey AB, Ruocco MG, Yang Y, Floess S, Huehn J, Oh S, Li MO, Niec RE, Rudensky AY, Dustin ML, Littman DR, Lafaille JJ (2012) Neuropilin 1 is expressed on thymus-derived natural regulatory T cells, but not mucosa-generated induced Foxp3+ T reg cells. J Exp Med 209(10):1723–1742., , s1721. https://doi.org/10.1084/jem.20120914

Wing JB, Tanaka A, Sakaguchi S (2019) Human FOXP3 (+) regulatory T cell heterogeneity and function in autoimmunity and cancer. Immunity 50(2):302–316. https://doi.org/10.1016/j.immuni.2019.01.020

Wood KL, Twigg HL 3rd, Doseff AI (2009) Dysregulation of CD8+ lymphocyte apoptosis, chronic disease, and immune regulation. Front Biosci (Landmark edition) 14:3771–3781. https://doi.org/10.2741/3487

Wu K, Higashi N, Hansen ER, Lund M, Bang K, Thestrup-Pedersen K (2000) Telomerase activity is increased and telomere length shortened in T cells from blood of patients with atopic dermatitis and psoriasis. J Immunol 165(8):4742–4747. https://doi.org/10.4049/jimmunol.165.8.4742

Wu Y, Ren M, Yang R, Liang X, Ma Y, Tang Y, Huang L, Ye J, Chen K, Wang P, Shen H (2011) Reduced immunomodulation potential of bone marrow-derived mesenchymal stem cells induced CCR4+CCR6+ Th/Treg cell subset imbalance in ankylosing spondylitis. Arthritis Res Ther 13(1):R29. https://doi.org/10.1186/ar3257

Wu C, Yosef N, Thalhamer T, Zhu C, Xiao S, Kishi Y, Regev A, Kuchroo VK (2013a) Induction of pathogenic TH17 cells by inducible salt-sensing kinase SGK1. Nature 496(7446):513–517. https://doi.org/10.1038/nature11984

Wu DC, Hester J, Nadig SN, Zhang W, Trzonkowski P, Gray D, Hughes S, Johnson P, Wood KJ (2013b) Ex vivo expanded human regulatory T cells can prolong survival of a human islet allograft in a humanized mouse model. Transplantation 96(8):707–716. https://doi.org/10.1097/TP.0b013e31829fa271

Xu L, Kitani A, Fuss I, Strober W (2007) Cutting edge: regulatory T cells induce CD4+CD25-Foxp3- T cells or are self-induced to become Th17 cells in the absence of exogenous TGF-beta. J Immunol 178 (11):6725–6729. https://doi.org/10.4049/jimmunol.178.11.6725

Xu A, Liu Y, Chen W, Wang J, Xue Y, Huang F, Rong L, Lin J, Liu D, Yan M, Li QZ, Li B, Song J, Olsen N, Zheng SG (2016) TGF-β-induced regulatory T cells directly suppress B cell responses through a noncytotoxic mechanism. J Immunol 196 (9):3631–3641. https://doi.org/10.4049/jimmunol.1501740

Yang S, Wang J, Brand DD, Zheng SG (2018) Role of TNF-TNF receptor 2 signal in regulatory T cells and its therapeutic implications. Front Immunol 9:784. https://doi.org/10.3389/fimmu.2018.00784

Yang M, Liu Y, Mo B, Xue Y, Ye C, Jiang Y, Bi X, Liu M, Wu Y, Wang J, Olsen N, Pan Y, Zheng SG (2019a) Helios but not CD226, TIGIT and Foxp3 is a potential marker for CD4(+) Treg cells in patients with rheumatoid arthritis. Cell Physiol Biochem 52 (5):1178–1192. https://doi.org/10.33594/000000080

Yang S, Xie C, Chen Y, Wang J, Chen X, Lu Z, June RR, Zheng SG (2019b) Differential roles of TNFalpha-TNFR1 and TNFalpha-TNFR2 in the differentiation and function of CD4(+) Foxp3(+) induced Treg cells in vitro and in vivo periphery in autoimmune diseases. Cell Death Dis 10(1):27. https://doi.org/10.1038/s41419-018-1266-6

Yao Y, Vent-Schmidt J, McGeough MD, Wong M, Hoffman HM, Steiner TS, Levings MK (2015) Tr1 Cells, but not Foxp3+ regulatory T cells, suppress NLRP3 inflammasome activation via an IL-10-dependent mechanism. J Immunol 195(2):488–497. https://doi.org/10.4049/jimmunol.1403225

Ye L, Goodall JC, Zhang L, Putintseva EV, Lam B, Jiang L, Liu W, Yin J, Lin L, Li T, Wu X, Yeo G, Shugay M, Chudakov DM, Gaston H, Xu H (2016) TCR usage, gene expression and function of two distinct FOXP3(+)Treg subsets within CD4(+)CD25 (hi) T cells identified by expression of CD39 and CD45RO. Immunol Cell Biol 94(3):293–305. https://doi.org/10.1038/icb.2015.90

Ye C, Brand D, Zheng SG (2018) Targeting IL-2: an unexpected effect in treating immunological diseases. Signal Transduct Target Ther 3:2. https://doi.org/10.1038/s41392-017-0002-5

Zhang X, Sun S, Hwang I, Tough DF, Sprent J (1998) Potent and selective stimulation of memory-phenotype CD8+ T cells in vivo by IL-15. Immunity 8(5):591–599. https://doi.org/10.1016/s1074-7613(00)80564-6

Zhang X, Ouyang X, Xu Z, Chen J, Huang Q, Liu Y, Xu T, Wang J, Olsen N, Xu A, Zheng SG (2019) CD8 +CD103+ iTregs inhibit chronic graft-versus-host disease with lupus nephritis by the increased expression of CD39. Mol Ther 27(11):1963–1973. https://doi.org/10.1016/j.ymthe.2019.07.014

Zhao DM, Thornton AM, DiPaolo RJ, Shevach EM (2006) Activated CD4+CD25+ T cells selectively kill B lymphocytes. Blood 107(10):3925–3932. https://doi.org/10.1182/blood-2005-11-4502

Zhao SS, Hu JW, Wang J, Lou XJ, Zhou LL (2011) Inverse correlation between CD4+ CD25high CD127low/- regulatory T-cells and serum immunoglobulin A in patients with new-onset ankylosing spondylitis. J Int Med Res 39(5):1968–1974. https://doi.org/10.1177/147323001103900543

Zhao C, Chu Y, Liang Z, Zhang B, Wang X, Jing X, Hao M, Wang Y, An J, Zhang X, Sun L, Chen J (2019) Low dose of IL-2 combined with rapamycin restores and maintains the long-term balance of Th17/

Treg cells in refractory SLE patients. BMC Immunol 20(1):32. https://doi.org/10.1186/s12865-019-0305-0

Zheng SG, Gray JD, Ohtsuka K, Yamagiwa S, Horwitz DA (2002) Generation ex vivo of TGF-beta-producing regulatory T cells from CD4+CD25- precursors. J Immunol 169(8):4183–4189. https://doi.org/10.4049/jimmunol.169.8.4183

Zheng SG, Wang JH, Gray JD, Soucier H, Horwitz DA (2004) Natural and induced CD4+CD25+ cells educate CD4+CD25- cells to develop suppressive activity: the role of IL-2, TGF-beta, and IL-10. J Immunol 172 (9):5213–5221. https://doi.org/10.4049/jimmunol.172.9.5213

Zheng SG, Meng L, Wang JH, Watanabe M, Barr ML, Cramer DV, Gray JD, Horwitz DA (2006) Transfer of regulatory T cells generated ex vivo modifies graft rejection through induction of tolerogenic CD4+CD25+ cells in the recipient. Int Immunol 18 (2):279–289. https://doi.org/10.1093/intimm/dxh368

Zheng SG, Wang J, Wang P, Gray JD, Horwitz DA (2007) IL-2 is essential for TGF-beta to convert naive CD4+CD25- cells to CD25+Foxp3+ regulatory T cells and for expansion of these cells. J Immunol 178 (4):2018–2027. https://doi.org/10.4049/jimmunol.178.4.2018

Zheng SG, Wang J, Horwitz DA (2008) Cutting edge: Foxp3+CD4+CD25+ regulatory T cells induced by IL-2 and TGF-beta are resistant to Th17 conversion by IL-6. J Immunol 180(11):7112–7116. https://doi.org/10.4049/jimmunol.180.11.7112

Zhong H, Liu Y, Xu Z, Liang P, Yang H, Zhang X, Zhao J, Chen J, Fu S, Tang Y, Lv J, Wang J, Olsen N, Xu A, Zheng SG (2018) TGF-beta-induced CD8(+)CD103(+) regulatory T cells show potent therapeutic effect on chronic graft-versus-host disease lupus by suppressing B cells. Front Immunol 9:35. https://doi.org/10.3389/fimmu.2018.00035

Zhou X, Kong N, Wang J, Fan H, Zou H, Horwitz D, Brand D, Liu Z, Zheng SG (2010) Cutting edge: all-trans retinoic acid sustains the stability and function of natural regulatory T cells in an inflammatory milieu. J Immunol 185(5):2675–2679. https://doi.org/10.4049/jimmunol.1000598

Zhou X, Kong N, Zou H, Brand D, Li X, Liu Z, Zheng SG (2011) Therapeutic potential of TGF-beta-induced CD4(+) Foxp3(+) regulatory T cells in autoimmune diseases. Autoimmunity 44(1):43–50. https://doi.org/10.3109/08916931003782163

Zou Y, Xu S, Xiao Y, Qiu Q, Shi M, Wang J, Liang L, Zhan Z, Yang X, Olsen N, Zheng SG, Xu H (2018) Long noncoding RNA LERFS negatively regulates rheumatoid synovial aggression and proliferation. J Clin Invest 128(10):4510–4524. https://doi.org/10.1172/jci97965

A Structure-Guided Delineation of FOXP3 Regulation Mechanism in IPEX

2

Tengyun Ma, Xiaomin Song, Ciriaco A. Piccirillo, Guoping Deng, and Mark I. Greene

Abstract

The FOXP3 transcription factor acts as a master regulator in the development and function of regulatory T cells (Tregs). Insufficient expression or mutation of FOXP3 gene impairs Treg abundancy and function and causes fatal autoimmune lymphoproliferative diseases in mice and humans. The available crystal structures of FOXP3 protein fragments provide insights into understanding details of the FOXP3 work mechanism in Tregs. This chapter consists of four sections. First, we introduce some features of Treg cells indispensable for the establishment of immune tolerance; second, we describe the critical roles of FOXP3 in Treg development and function; third, we summarize the current available crystal structures of FOXP3 functional domains and related pathogenic mutations in autoimmune diseases; finally, we discuss the potential functional and pathological relevance of FOXP3 protein structure modulation, partner interaction, and posttranslation modification based on the clinical significance in IPEX disease. The information presented in this chapter will help to consider therapeutic strategies to enhance FOXP3 activity and Treg function in the settings of autoimmune disease. Targeting Treg suppression based on FOXP3 structure and interactions hold great promises for the therapy of autoimmune diseases.

Keywords

Regulatory T cells · FOXP3 · Crystal structure · Protein interactions · Posttranslational modifications · IPEX

2.1 Treg Cells Play a Central Role in Peripheral Immune Tolerance

The immune system mounts an immune response against invading foreign pathogens while maintains tolerance to self-tissues. The concept

T. Ma · G. Deng (✉)
Department of Immunology, School of Basic Medical Sciences, Peking University, Beijing, China
e-mail: gdeng@pku.edu.cn

X. Song
State Key Laboratory of Molecular Biology, CAS Center for Excellence in Molecular Cell Science, Institute of Biochemistry and Cell Biology, Chinese Academy of Sciences, Shanghai, China

C. A. Piccirillo
Department of Microbiology and Immunology, McGill University, Montreal, QC, Canada

Centre of Excellence in Translational Immunology (CETI), Research Institute of the McGill University Health Centre, Montreal, QC, Canada

M. I. Greene
Department of Pathology and Laboratory Medicine, Perelman School of Medicine, University of Pennsylvania, Philadelphia, PA, USA

S.-G. Zheng (ed.), *T Regulatory Cells in Human Health and Diseases*, Advances in Experimental Medicine and Biology 1278, https://doi.org/10.1007/978-981-15-6407-9_2

of T-cell-mediated immune tolerance, i.e., suppressor T cells, has been described since early 1970s (Gershon and Kondo 1971). There are two main mechanisms to establish immune tolerance. In the central immune tolerance, self-reactive T cells are deleted in the thymus by clone selection; however, some autoreactive cells escape deletion. There are peripheral immune tolerance processes. Some escaping autoreactive T cells from thymus are actively inhibited by various immune suppressors. $CD4^{+}CD25^{+}$ regulatory T cells (Tregs) highly express CD25, the alpha-chain molecule of IL-2 receptor. Tregs possess immune suppressive capacity and are actively engaged in the peripheral immune tolerance (Sakaguchi et al. 1995). Reconstitution of Tregs prevented autoimmune developments caused by the transfer of effector T cells into BALB/c athymic nude mice; furthermore, these Treg cells limited immune responses to allogeneic skin graft. These studies demonstrated that Treg cells can suppress immune response toward both self and non-self-antigens (Sakaguchi et al. 1995).

Treg cells are categorized into at least two main subsets according to their origins (Horwitz et al. 2008). The majority of Treg cells develop and mature in the thymus requiring stimulation of self-antigens and co-stimulation from CD28, i.e., thymus-derived Treg cells (Tai et al. 2005; Lio and Hsieh 2008); other Treg cells are generated in the periphery, i.e., peripheral Treg cells, by conversion from conventional T cells upon exposure to certain extracellular stimulation such as cytokine TGF-β, foreign agonist peptide, and commensal bacteria (Atarashi et al. 2011; Kretschmer et al. 2005; Zheng et al. 2002, 2004). Treg cells develop in thymus via a two-step process whereby Treg cell precursors upregulate IL-2R (CD25) expression upon TCR activation, and then Foxp3 expression is induced concomitant with the acquisition of IL-2 cytokine (Lio and Hsieh 2008). IL-2 signal is also crucial for the differentiation of induced Treg cells (Zheng et al. 2007a; Davidson et al. 2007). Helios, an Ikaros transcription factor family member, is preferentially expressed in thymocytes and considered as a marker to distinguish thymus-derived Treg cells and peripheral Treg cells under physiological conditions (Thornton et al. 2010). Helios is required for the stability of effector Treg cells (Kim et al. 2015; Nakagawa et al. 2016). In patients with rheumatoid arthritis, Helios seems to be better than Foxp3 to identify Treg cells (Yang et al. 2019). A surface marker Neuropilin-1 is selectively expressed on naturally thymus-derived Treg cells (Yadav et al. 2012; Weiss et al. 2012). The signaling of neuropilin-1–semaphorin-4a axis is required for the stability and function of regulatory T cells (Delgoffe et al. 2013). In addition, thymus-derived Treg cells display unique genomic imprints such as DNA hypomethylation and modification of Treg lineage-specific genes in comparison with that of periphery-derived Treg cells. This epigenetic feature confers thymus-derived Tregs to be more stable during cell passage than periphery-derived Treg cells (Lal et al. 2009; Ohkura et al. 2012; Feng et al. 2014).

The molecular mechanisms underlying the immune suppression of Treg cells are incompletely understood yet. There are many excellent review articles summarize the mechanisms of Treg suppression (Vignali 2012; Schmidt et al. 2012; Sakaguchi et al. 2009; Miyara and Sakaguchi 2007; Shevach 2009). Treg cells derived in the periphery also function in suppression of T-cell immune response. Tregs can directly suppress immune responses by cell-to-cell contact through expression of surface inhibitory receptors, such as CTLA-4, PD-1, GITR, and LAG-3. Inducible deletion of CTLA-4 in adult mice promoted peripheral expansion of Treg cells and enhanced the protection from EAE disease (Klocke et al. 2016; Paterson et al. 2015). Although lack of CTLA-4 does not affect the thymus-derived Treg cells (Salomon et al. 2000), it indeed reduces the development of induced Treg cells (Zheng et al. 2006). Treg cells produce immune suppressive cytokines including IL-10, TGF-beta, and IL-35, or compete for consumption of the IL-2 cytokine by expression of CD25 to restrain fitness of effector T cells.

Treg cells also mediate immune suppression by the expression of CD39 and CD73 to generate pericellular adenosine from extracellular nucleotides (Deaglio et al. 2007; Su et al. 2019b). Of note, Treg suppression is driven by TCR engagement with cognate antigens, although the dominant bystander suppression of Treg cells occurs in an antigen-non-specific manner (Legoux et al. 2015). Adoptive transfer of antigen-specific rather than polyclonal expanded Treg cells can effectively prevent the development of autoimmune diseases in mice and humans (Bluestone et al. 2015; Jaeckel et al. 2005; Kong et al. 2012). In addition, Treg cells may convey suppressive activity to nearby Tconv helper cells via a cell–cell contact manner which has been suggested as a phenomenon named infectious tolerance (Jonuleit et al. 2002; Gershon and Kondo 1971).

2.2 Foxp3 Is the Dominant Regulator of Treg Cell Lineage Commitment and Maintenance

The role of FOXP3 transcription factor involved in immune tolerance was initially revealed from the studies of *Scurfy* mice and human IPEX syndrome. *Scurfy* mice arise spontaneously with profound X-linked recessive autoimmune diseases including external dermatitis and internal lymphadenopathy, splenomegaly, and hepatomegaly (Ramsdell and Ziegler 2014). Genetic analysis of *Scurfy* mice found that a mutation in *Foxp3* transcriptional factor is responsible for the autoimmune phenotypes mechanically (Brunkow et al. 2001). In addition, human immune dysregulation, polyendocrinopathy, enteropathy, and X-linked syndrome (IPEX) contains multiple mutations in the forkhead/winged-helix domain of human FOXP3, which disrupts the interaction between FOXP3 and DNA (Wildin et al. 2001). We summarize mutation types and mutation location of FOXP3 functional domains and gene locus according to the available published data (Fig. 2.1a–c). In addition to the site mutations, other mutation types including unstable mRNA, splicing, nonsense mutation, missense, frameshift and deletion, as well as complex mutation of FOXP3 have been reported from IPEX patients. Many of the mutations which occur in the FKH and Leucine zipper FOXP3 domains identify the importance of these two functional domains in regulating FOXP3 transcriptional activity. A few mutations of other genes including CTLA-4, IL2RA, and STAT5 may also be causative of autoimmune symptoms of IPEX (Fig. 2.1d). These findings demonstrate that FOXP3 acts as a dominant regulator in establishing the immune tolerance and preventing the progression of autoimmune diseases.

Foxp3 is predominantly expressed in mouse $CD4^+CD25^+$ Treg cells and acts as the lineage-specific transcriptional factor for their development and function in a cell-intrinsic manner (Fontenot et al. 2003). Dysfunction or mutation of Foxp3 led to lethal autoimmune phenotypes in mice accompanied with loss of Treg abundancy in vivo, as described above as *Scurfy* phenotypes (Khattri et al. 2003). Ectopic expression of Foxp3 confers conventional $CD4^+$ T-cell regulatory capacity to inhibit autoimmune diseases in vivo (Hori et al. 2003). In addition, Foxp3-transgenic mice in which *Foxp3* gene is overexpressed possessed fewer peripheral $CD4^+$ T cells. These $CD4^+$ T cells are compromised in their functional capability as shown in reduced proliferative responses and IL-2 production upon antigen stimulation in comparison with the normal counterparts (Kasprowicz et al. 2003; Khattri et al. 2001). These observations identify a negative role of Foxp3 in T-cell activation by attenuating TCR signaling (Carson and Ziegler 2007).

Human FOXP3 is exclusively expressed in $CD4^+CD25^+$ T cells and is correlated with immune suppression as seen in mice. $CD4^+CD25^-$ T cells can also transiently elevate FOXP3 expression upon TCR stimulation (Walker et al. 2003; Jonuleit et al. 2001). Transient expression of FOXP3 in human $CD4^+CD25^-$ T cells leads to hyporesponsiveness of activated T cells, but is not sufficient to induce a full regulatory T-cell activity (Wang et al. 2007). The human FOXP3 protein possesses two isoforms: one is full-length of gene expression which represents the ortholog to mouse Foxp3, and the other one is a truncated isoform

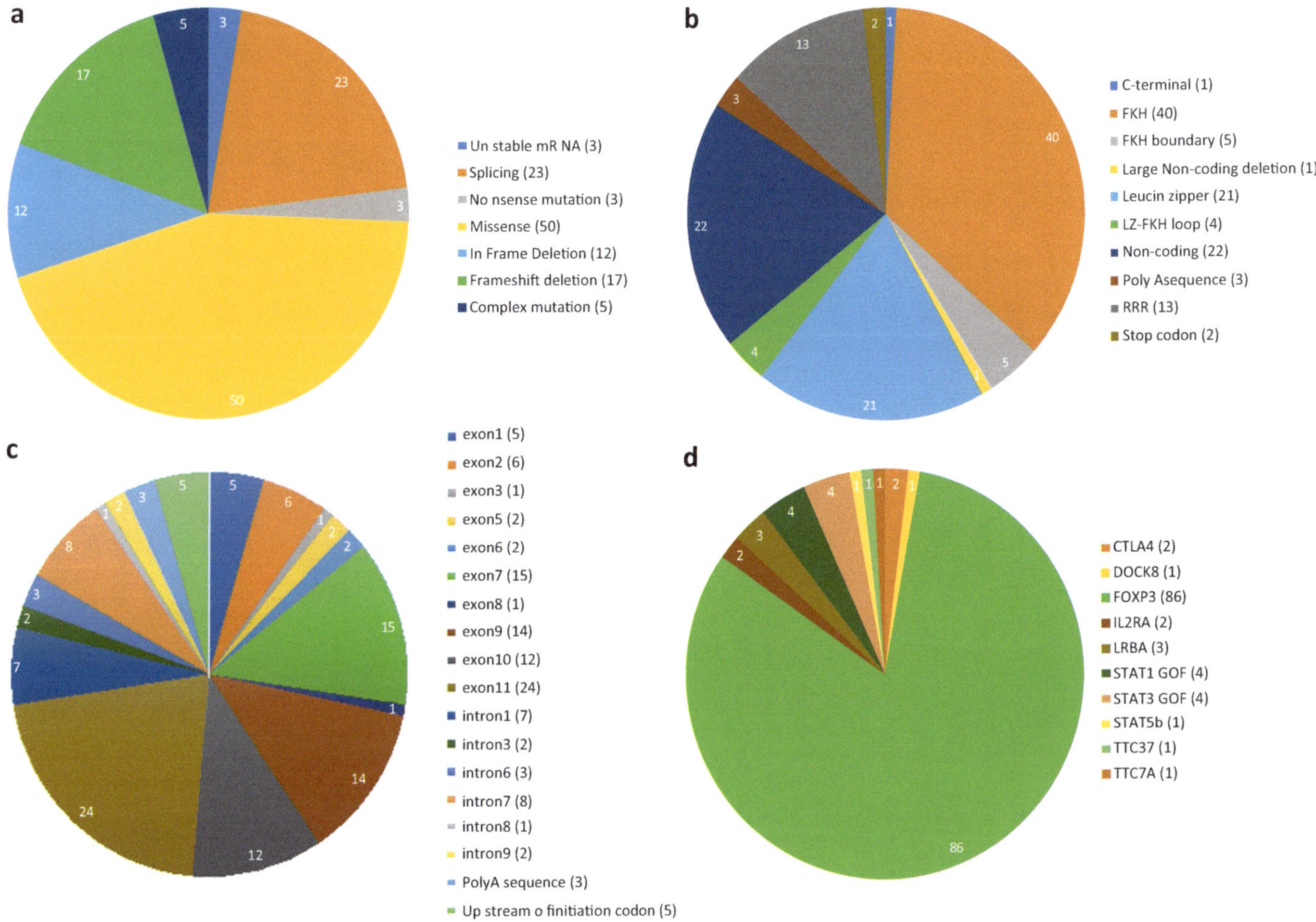

Fig. 2.1 (**a–c**) A summary of known mutation types, mutation locations within functional domain and gene locus of FOXP3 described in IPEX patients; (**d**) known mutant genes responsible for in IPEX patients. The statistics data used in this figure are derived and adapted from several published papers (Gambineri et al. 2018; Agakidis et al. 2019; Kadakia et al. 2019; Griswold et al. 2018; Luo et al. 2018; Magg et al. 2018; Lin et al. 2018; Frith et al. 2019; Louie et al. 2017; Duclaux-Loras et al. 2018; d'Hennezel et al. 2012)

(FOXP3Δ2) encoded from an mRNA lacking exon 2; however, ectopic expression of FOXP3 alone or together with FOXP3Δ2 is not sufficient to induce a complete Treg suppressive activity (Allan et al. 2005), which suggests that other factors or mechanisms are required for the development of complete Treg cell lineage. In fact, an epigenome-based definition of Treg cells, prior to Foxp3 expression, is proposed to understand the functional stability, plasticity, and heterogeneity of Treg lineage (Morikawa and Sakaguchi 2014; Morikawa et al. 2014). Satb1 is thought to be responsible for the activation of Treg cell-specific super-enhancers (Treg-SEs) and the subsequent expression of Treg signature genes including Foxp3 in Treg precursor cells during thymocyte development (Kitagawa et al. 2017).

Treg cells co-express Foxp3 with other transcriptional factors of effector CD4 T cells for more timely and efficient control of immune responses. For instance, dermal and large intestinal lamina propria Treg cells express high levels of GATA3 upon activation, GATA3 interacts and promotes Foxp3 expression, and *GATA3*-deficient Treg cells showed impaired suppression capability, failed to accumulate at inflamed sites, and produced effector cytokine IL-17a (Wohlfert et al. 2011; Wang et al. 2011). $Foxp3^{+}ROR\gamma t^{+}$ T cells represent a stable Treg lineage with enhanced suppressive capability in T-cell-mediated intestinal inflammation as compared with $Foxp3^{+}ROR\gamma t^{-}$ Tregs (Yang et al. 2016). Th1-polarizing infection increased specialized $T\text{-}bet^{+}$ Treg cells that potently inhibit Th1-type immune responses (Koch et al. 2012), intestinal Treg cells prevalently co-express T-bet and RORγt, and acute ablation of $T\text{-}bet^{+}$ Treg cells resulted in Th1 autoimmunity (Levine et al. 2017). These findings indicate that phenotypic heterogeneity and functional plasticity of Treg cells are critical features of immune tolerance.

In summary, Foxp3 is the lineage-specific transcriptional factor of Treg cells, Foxp3 together with other factors as well as genome epigenetic modification program the development of stable Treg cells.

2.3 Structural Features of Foxp3 Protein

The forkhead box (FOX) proteins are a superfamily of transcription factors which is composed of 50 FOX protein members and 19 subfamilies classified by their sequence homologies. Forkhead transcription factors are involved in oncogenesis including aspects of tumor initiation, maintenance, and drug resistance (Lam et al. 2013). The Foxp3 protein mainly localizes in nucleus of Treg cells (Lopes et al. 2006) and binds to ~700 genes and miRNA involved in the regulation of many facets of Treg cell biology (Marson et al. 2007; Zheng et al. 2007b).

The Foxp3 protein contains three main functional domains: (1) the N-terminal domain is a proline-rich domain involved in transcriptional activation and repression; (2) the middle localized zinc finger leucine-zipper domain involved in its dimerization or association with other factors; (3) and the conserved C-terminal forkhead domain is responsible for DNA binding (Deng et al. 2020; Lopes et al. 2006). Three subdomains, i.e., the C-terminal 12 amino acids, an immediate N-terminal to the forkhead domain, and the first 51 amino acids within the N-terminal, contribute to the nuclear transport of murine Foxp3 (Hancock and Ozkaynak 2009).

The proline-rich Foxp3 N-terminal domain regulates its transcriptional activity through interacting with many factors, including Foxp1, AML1/Runx1, NFAT, TIP60, HDAC7, and so on (as reviewed otherwhere (Zhou et al. 2008; Deng et al. 2012)). Two independent studies of transgenic mice with $Foxp3^{gfp}$ revealed that a GFP insertion at Foxp3 N-terminal impaired Treg development and stability by blocking the interaction of Foxp3 with Eos, Tip60, HDAC7, and HIF-1α. These $Foxp3^{gfp}$ mice were predisposed to autoimmune diabetes (Darce et al. 2012; Bettini et al. 2012). These observations suggest that a proper stereo conformation and interactions of Foxp3 protein are totally required for the normal function of Foxp3 in Treg cells.

The crystal structures of the FOXP3 domains facilitate insight into the mechanism by which

FOXP3 regulates the development and function of Treg cells. Bandukwala and collogues have solved a high-resolution crystal structure of a ternary complex which contains the NFAT1 DNA-binding domain and the FOXP3 forkhead domain bound to DNA (Bandukwala et al. 2011). In this structure, the forkhead domain of FOXP3 forms a domain-swapped dimer to bridge two molecules of DNA in an antiparallel orientation (Fig. 2.2a). The highly conserved hydrophobic aromatic residues, including Tyr364, Trp366, Phe367, Phe371, Phe373, Phe374, and Trp381 of forkhead domain, form the interface of domain swapping. Two IPEX mutations F371C and F373A localized in this interface could impair the stability of the domain-swapped dimer, suggesting that domain-swapping is functionally relevant to the pathogenies of IPEX (Fig. 2.2b). Of note, disruption of the domain-swap interface of FOXP3 does not impair DNA binding as assumed, but attenuates FOXP3-mediated suppressor functions. These findings indicated that FOXP3 regulates gene expression through the assembly of higher-order stereo transcription complexes by the domain-swapped FOXP3 dimer (Bandukwala et al. 2011). This idea was verified by a further study from the same research group which demonstrated that the domain-swapped dimer allows FOXP3 to coordinate the expression of genes through bridging two distal regions of DNA and reorganizing the genome architecture (Chen et al. 2015). The IPEX mutation M370I in the FKH domain-swap interface of FOXP3 generates Th2-like Treg cells expressing GATA3 and Th2 cytokines, resulting in unrestrained Th2 immune response (Van Gool et al. 2019). These findings demonstrate that FOXP3 orchestrates chromosomal interaction dynamics during regulating Treg development and function.

The mutant FOXP3 leucine-zipper domain (ΔE201) as well as a truncated mutant forkhead can lead to the IPEX dysregulation syndrome (Chatila et al. 2000). Mechanistically, the mutant FOXP3 leucine-zipper impairs FOXP3 homo/heterodimerization and compromises Treg suppressive function (Chae et al. 2006). FOXP3-ZL also mediates the heterodimerization with FOXP1, which regulates Foxp3 chromatin binding and coordinates Treg function (Wang et al. 2003; Konopacki et al. 2019).

This mechanism was explained by the structurally characterized mouse Foxp3 zinc finger and leucine-zipper region (mFOXP3-ZL) (Song et al. 2012). The crystal structure of mFOXP3-ZL homodimer features in a two-stranded anti-parallel α-helical coiled coil with a twofold symmetry (Fig. 2.3a). Unlike typical coiled coil structures, FOXP3 dimer association is flexible and dynamic with half residues in the coiled coil core packing being polar ones. Stabilization of the mFOXP3-ZL coiled coil is influenced by electrostatic interactions such as E242-K249 and E248-Q234 inter-subunit hydrogen bonds and salt bridges. The importance of K249 (which is equivalent to K250 of human FOXP3) in stabilizing the coiled coil structure explains a functional and pathological relevance in IPEX disease cases (Wildin et al. 2002). Lysine residues are candidates for acetylation modification. We discuss how acetylation could affect the conformation of Foxp3 structure and dimerization to fine-tune Treg function in the next section.

2.4 Foxp3 Activity Is Modified by Protein Posttranslational Modifications and Interactions

The roles of Foxp3 posttranslational modifications in regulating Treg function have been reviewed elsewhere (van Loosdregt and Coffer 2014; Deng et al. 2019). The Foxp3 protein has multiple types of posttranslational modifications, such as phosphorylation, acetylation, ubiquitylation, methylation, and so on. These modifications reflect environmental and intrinsic signals in regulating Foxp3 transcriptional activity and functional property. Accumulating evidence demonstrates that phosphorylation, acetylation, and ubiquitylation are largely involved in regulating Foxp3 protein stabilization by preventing its proteasome-dependent degradation (Morawski et al. 2013; Deng et al. 2015; van Loosdregt et al. 2010; Chen et al. 2013).

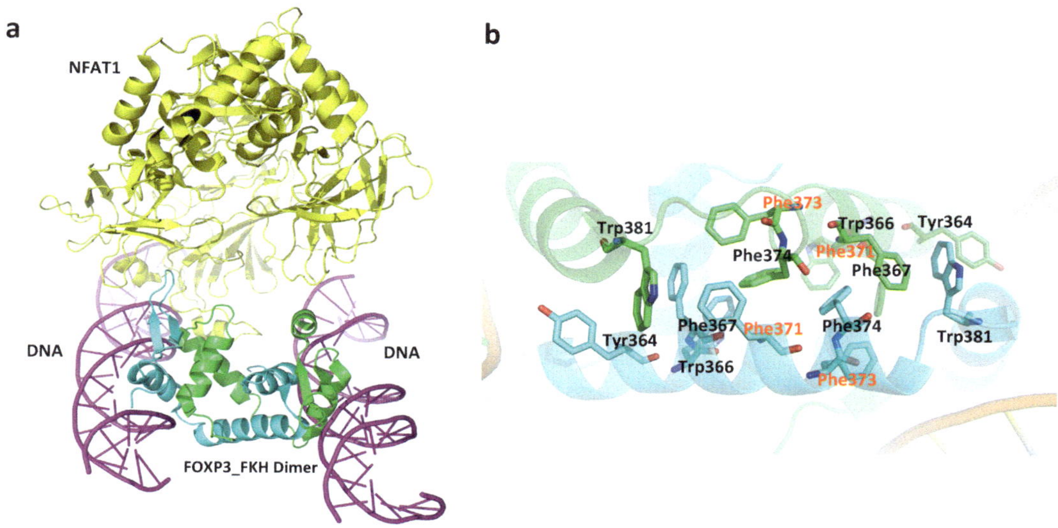

Fig. 2.2 (**a**) The crystal structure of the complex of the NFAT1 DNA-binding domain (yellow), the FOXP3 FKH domain dimer (blue and cyan) and DNA molecules (purple); (**b**) the interface of domain swapping is formed by hydrophobic aromatic residues, including Tyr364, Trp366, Phe367, Phe371, Phe373, Phe374, and Trp381 of FOXP3 forkhead domain. Two IPEX mutations F371C and F373A (red highlighted) localized in this interface could impair the stability of the domain-swapped dimer and cause autoimmune diseases. The image is based on the crystal structure of FoxP3 forkhead domain complexed with DNA and NFAT (PDB code 3QRF) using Pymol (Bandukwala et al. 2011)

Foxp3 posttranslational modifications can fine-tune transcriptional activity by adjusting its protein conformation. Protein acetylation modification is one of the most extensively studied modifications of Foxp3. TIP60 and p300 are two well-known histone acetyltransferases (HATs) involved in Foxp3 acetylation and Treg function regulation, and multiple acetylated sites of Foxp3 have been identified (Xiao et al. 2014; Li et al. 2007). Acetylation acts as a competitive event on the lysine residue with ubiquitination and enhances Treg function by limiting Foxp3 protein from ubiquitination-dependent degradation (van Loosdregt et al. 2010; Kwon et al. 2012). However, some IPEX patients have similar frequency of Treg population as the same-aged healthy controls but still develop autoimmune disorders (Brusko et al. 2005, 2007). As mentioned above, deletions of FOXP3 ΔK250 and ΔE251 (equivalent to mouse Foxp3 K249 and E250) accounted for the pathogenesis of IPEX patients (Chatila et al. 2000; Wildin et al. 2002). Deletions of these key residues may break the E242-K249 inter-subunit salt bridge or Q234-E248 inter-subunit hydrogen bond and consequently disrupt the FOXP3 homodimerization as described in coiled coil crystal structure of FOXP3-ZL domain (Song et al. 2012). Indeed, acetylation of K249 and K251 will neutralize the positive charges of these residues and destabilize the FOXP3 homodimerization in a regulated manner (Fig. 2.3b, c), which proposes a model of how Foxp3 posttranslational modifications in modulating Foxp3 activity (Song et al. 2012).

In addition to the posttranslational modifications, the interactions of Foxp3 with other factors also mediate Foxp3 activity. Foxp3 forms multiprotein complexes of 400–800 kDa with ~361 partner proteins (Rudra et al. 2012). As discussed above, disruption of the interaction of Foxp3 with Eos, Tip60, HDAC7, and HIF-1α by inserting a GFP tag at Foxp3 N-terminal compromised Foxp3 activity and Treg function (Darce et al. 2012; Bettini et al. 2012). Downregulated expression of TIP60 is observed in rheumatoid arthritis patients with impaired

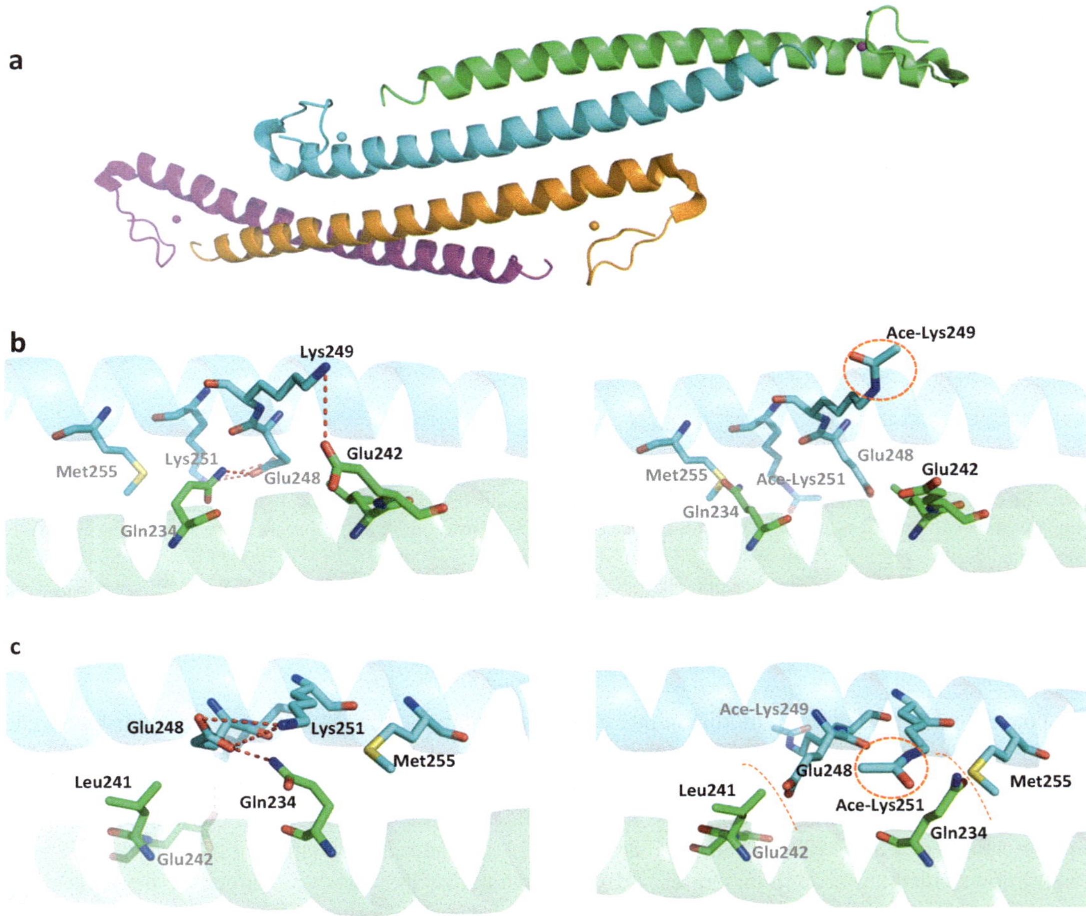

Fig. 2.3 (**a**) Coiled coil structure of Foxp3 ZL (zinc finger and leucine-zipper) domain which is modified as described in context; (**b**, **c**) electrostatic interactions such as E242-K249 and E248-Q234 inter-subunit hydrogen bonds and salt bridges enhance the stabilization of Foxp3-ZL coiled coil (left), acetylation of K249 and K251 (red circle) neutralize the positive charges of these residues and destabilize the FOXP3 homodimerization (right). The scheme is generated based on the crystal structure of FoxP3 ZL domain (PDB code 3QRF) using Pymol (Song et al. 2012)

Foxp3 activity and Treg function (Su et al. 2019a). The FOXP3-A384T mutation occurred in IPEX patient impairs the interaction of FOXP3 with TIP60 and decreases Treg suppressive function. Pharmacologic enhancement of FOXP3–TIP60 interactions by an allosteric modifier could rescue the loss of Treg function (Bin Dhuban et al. 2017). The allosteric modification of TIP60–FOXP3 interaction offers a new strategy to modulate $FOXP3^{+}$ Treg cell function specifically without affecting effector T-cell responses pharmacologically.

2.5 Conclusion

In this chapter, we discussed FOXP3 regulatory features and mechanisms in Treg cells based on the available structure information of FOXP3 functional domains. The high-resolution crystal structures of FOXP3 FHK and Leucine-Zipper domains give us a deep insight into delineating the work model of FOXP3 in Treg cells. In addition, posttranslational modifications and partner interactions fine-tune the FOXP3 protein stereo conformation and structure, which modulates FOXP3 activity and Treg suppression. Allosteric modification of FOXP3 interaction by drugs offers a new strategy to mediate Treg cell function without affecting effector T-cell responses.

Acknowledgments *Funding*: We gratefully acknowledge the funding support from Peking University Health Science Center (BMU2018YJ011) and 2019 Clinical-Medicine Plus-X Young Scholars Project.
Disclosures: The authors declare that there is no conflict of interest regarding the publication of this article.

References

Agakidis C, Agakidou E, Sarafidis K, Papoulidis I, Xinias I, Farmaki E (2019) Immune dysregulation, polyendocrinopathy, enteropathy, X-linked syndrome associated with a novel mutation of FOXP3 gene. Front Pediatr 7:20. https://doi.org/10.3389/fped.2019.00020

Allan SE, Passerini L, Bacchetta R, Crellin N, Dai M, Orban PC, Ziegler SF, Roncarolo MG, Levings MK (2005) The role of 2 FOXP3 isoforms in the generation of human CD4+ Tregs. J Clin Invest 115(11):3276–3284. https://doi.org/10.1172/JCI24685

Atarashi K, Tanoue T, Shima T, Imaoka A, Kuwahara T, Momose Y, Cheng G, Yamasaki S, Saito T, Ohba Y, Taniguchi T, Takeda K, Hori S, Ivanov II, Umesaki Y, Itoh K, Honda K (2011) Induction of colonic regulatory T cells by indigenous Clostridium species. Science 331(6015):337–341. https://doi.org/10.1126/science.1198469

Bandukwala HS, Wu Y, Feuerer M, Chen Y, Barboza B, Ghosh S, Stroud JC, Benoist C, Mathis D, Rao A, Chen L (2011) Structure of a domain-swapped FOXP3 dimer on DNA and its function in regulatory T cells. Immunity 34(4):479–491. https://doi.org/10.1016/j.immuni.2011.02.017

Bettini ML, Pan F, Bettini M, Finkelstein D, Rehg JE, Floess S, Bell BD, Ziegler SF, Huehn J, Pardoll DM, Vignali DA (2012) Loss of epigenetic modification driven by the Foxp3 transcription factor leads to regulatory T cell insufficiency. Immunity 36(5):717–730. https://doi.org/10.1016/j.immuni.2012.03.020

Bin Dhuban K, d'Hennezel E, Nagai Y, Xiao Y, Shao S, Istomine R, Alvarez F, Ben-Shoshan M, Ochs H, Mazer B, Li B, Sekine C, Berezov A, Hancock W, Torgerson TR, Greene MI, Piccirillo CA (2017) Suppression by human FOXP3(+) regulatory T cells requires FOXP3-TIP60 interactions. Sci Immunol 2(12):eaai9297. https://doi.org/10.1126/sciimmunol.aai9297

Bluestone JA, Buckner JH, Fitch M, Gitelman SE, Gupta S, Hellerstein MK, Herold KC, Lares A, Lee MR, Li K, Liu W, Long SA, Masiello LM, Nguyen V, Putnam AL, Rieck M, Sayre PH, Tang Q (2015) Type 1 diabetes immunotherapy using polyclonal regulatory T cells. Sci Transl Med 7(315):315ra189. https://doi.org/10.1126/scitranslmed.aad4134

Brunkow ME, Jeffery EW, Hjerrild KA, Paeper B, Clark LB, Yasayko SA, Wilkinson JE, Galas D, Ziegler SF, Ramsdell F (2001) Disruption of a new forkhead/winged-helix protein, scurfin, results in the fatal lymphoproliferative disorder of the scurfy mouse. Nat Genet 27(1):68–73. https://doi.org/10.1038/83784

Brusko TM, Wasserfall CH, Clare-Salzler MJ, Schatz DA, Atkinson MA (2005) Functional defects and the influence of age on the frequency of CD4+ CD25+ T-cells in type 1 diabetes. Diabetes 54(5):1407–1414. https://doi.org/10.2337/diabetes.54.5.1407

Brusko T, Wasserfall C, McGrail K, Schatz R, Viener HL, Schatz D, Haller M, Rockell J, Gottlieb P, Clare-Salzler M, Atkinson M (2007) No alterations in the frequency of FOXP3+ regulatory T-cells in type 1 diabetes. Diabetes 56(3):604–612. https://doi.org/10.2337/db06-1248

Carson BD, Ziegler SF (2007) Impaired T cell receptor signaling in Foxp3+ CD4 T cells. Ann N Y Acad Sci 1103:167–178. https://doi.org/10.1196/annals.1394.022

Chae WJ, Henegariu O, Lee SK, Bothwell AL (2006) The mutant leucine-zipper domain impairs both dimerization and suppressive function of Foxp3 in T cells. Proc Natl Acad Sci U S A 103(25):9631–9636. https://doi.org/10.1073/pnas.0600225103

Chatila TA, Blaeser F, Ho N, Lederman HM, Voulgaropoulos C, Helms C, Bowcock AM (2000) JM2, encoding a fork head-related protein, is mutated in X-linked autoimmunity-allergic disregulation syndrome. J Clin Invest 106(12):R75–R81. https://doi.org/10.1172/JCI11679

Chen Z, Barbi J, Bu S, Yang HY, Li Z, Gao Y, Jinasena D, Fu J, Lin F, Chen C, Zhang J, Yu N, Li X, Shan Z, Nie J, Gao Z, Tian H, Li Y, Yao Z, Zheng Y, Park BV, Pan Z, Zhang J, Dang E, Li Z, Wang H, Luo W, Li L, Semenza GL, Zheng SG, Loser K, Tsun A, Greene MI, Pardoll DM, Pan F, Li B (2013) The ubiquitin ligase Stub1 negatively modulates regulatory T cell suppressive activity by promoting degradation of the transcription factor Foxp3. Immunity 39(2):272–285. https://doi.org/10.1016/j.immuni.2013.08.006

Chen Y, Chen C, Zhang Z, Liu CC, Johnson ME, Espinoza CA, Edsall LE, Ren B, Zhou XJ, Grant SF, Wells AD, Chen L (2015) DNA binding by FOXP3 domain-swapped dimer suggests mechanisms of long-range chromosomal interactions. Nucleic Acids Res 43 (2):1268–1282. https://doi.org/10.1093/nar/gku1373

d'Hennezel E, Bin Dhuban K, Torgerson T, Piccirillo CA (2012) The immunogenetics of immune dysregulation, polyendocrinopathy, enteropathy, X linked (IPEX) syndrome. J Med Genet 49(5):291–302. https://doi.org/10.1136/jmedgenet-2012-100759

Darce J, Rudra D, Li L, Nishio J, Cipolletta D, Rudensky AY, Mathis D, Benoist C (2012) An N-terminal mutation of the Foxp3 transcription factor alleviates arthritis but exacerbates diabetes. Immunity 36(5):731–741. https://doi.org/10.1016/j.immuni.2012.04.007

Davidson TS, DiPaolo RJ, Andersson J, Shevach EM (2007) Cutting edge: IL-2 is essential for TGF-beta-mediated induction of Foxp3+ T regulatory cells. J Immunol 178(7):4022–4026. https://doi.org/10.4049/jimmunol.178.7.4022

Deaglio S, Dwyer KM, Gao W, Friedman D, Usheva A, Erat A, Chen JF, Enjyoji K, Linden J, Oukka M, Kuchroo VK, Strom TB, Robson SC (2007) Adenosine generation catalyzed by CD39 and CD73 expressed on regulatory T cells mediates immune suppression. J Exp Med 204(6):1257–1265. https://doi.org/10.1084/jem.20062512

Delgoffe GM, Woo SR, Turnis ME, Gravano DM, Guy C, Overacre AE, Bettini ML, Vogel P, Finkelstein D, Bonnevier J, Workman CJ, Vignali DA (2013) Stability and function of regulatory T cells is maintained by a neuropilin-1-semaphorin-4a axis. Nature 501 (7466):252–256. https://doi.org/10.1038/nature12428

Deng G, Xiao Y, Zhou Z, Nagai Y, Zhang H, Li B, Greene MI (2012) Molecular and biological role of the FOXP3 N-terminal domain in immune regulation by T regulatory/suppressor cells. Exp Mol Pathol 93 (3):334–338. https://doi.org/10.1016/j.yexmp.2012.09.013

Deng G, Nagai Y, Xiao Y, Li Z, Dai S, Ohtani T, Banham A, Li B, Wu SL, Hancock W, Samanta A, Zhang H, Greene MI (2015) Pim-2 kinase influences regulatory T cell function and stability by mediating Foxp3 protein N-terminal phosphorylation. J Biol Chem 290(33):20211–20220. https://doi.org/10.1074/jbc.M115.638221

Deng G, Song X, Fujimoto S, Piccirillo CA, Nagai Y, Greene MI (2019) Foxp3 post-translational modifications and Treg suppressive activity. Front Immunol 10:2486. https://doi.org/10.3389/fimmu.2019.02486

Deng G, Song X, Greene MI (2020) FoxP3 in Treg cell biology: a molecular and structural perspective. Clin Exp Immunol 199(3):255–262. https://doi.org/10.1111/cei.13357

Duclaux-Loras R, Charbit-Henrion F, Neven B, Nowak J, Collardeau-Frachon S, Malcus C, Ray PF, Moshous D, Beltrand J, Goulet O, Cerf-Bensussan N, Lachaux A, Rieux-Laucat F, Ruemmele FM (2018) Clinical heterogeneity of immune dysregulation, polyendocrinopathy, enteropathy, X-linked syndrome: a French multicenter retrospective study. Clin Transl Gastroenterol 9(10):201. https://doi.org/10.1038/s41424-018-0064-x

Feng Y, Arvey A, Chinen T, van der Veeken J, Gasteiger G, Rudensky AY (2014) Control of the inheritance of regulatory T cell identity by a cis element in the Foxp3 locus. Cell 158(4):749–763. https://doi.org/10.1016/j.cell.2014.07.031

Fontenot JD, Gavin MA, Rudensky AY (2003) Foxp3 programs the development and function of CD4+CD25+ regulatory T cells. Nat Immunol 4 (4):330–336. https://doi.org/10.1038/ni904

Frith K, Joly AL, Ma CS, Tangye SG, Lohse Z, Seitz C, Verge CF, Andersson J, Gray P (2019) The FOXP3Delta2 isoform supports Treg cell development and protects against severe IPEX syndrome. J Allergy Clin Immunol 144(1):317–320 e318. https://doi.org/10.1016/j.jaci.2019.03.003

Gambineri E, Ciullini Mannurita S, Hagin D, Vignoli M, Anover-Sombke S, DeBoer S, Segundo GRS, Allenspach EJ, Favre C, Ochs HD, Torgerson TR (2018) Clinical, immunological, and molecular heterogeneity of 173 patients with the phenotype of immune dysregulation, polyendocrinopathy, enteropathy, X-linked (IPEX) syndrome. Front Immunol 9:2411. https://doi.org/10.3389/fimmu.2018.02411

Gershon RK, Kondo K (1971) Infectious immunological tolerance. Immunology 21(6):903–914

Griswold C, Durica AR, Dennis LG, Jewell AF (2018) Prenatal bowel findings in male siblings with a confirmed FOXP3 mutation. J Ultrasound Med 37 (4):1033–1037. https://doi.org/10.1002/jum.14428

Hancock WW, Ozkaynak E (2009) Three distinct domains contribute to nuclear transport of murine Foxp3. PLoS One 4(11):e7890. https://doi.org/10.1371/journal.pone.0007890

Hori S, Nomura T, Sakaguchi S (2003) Control of regulatory T cell development by the transcription factor Foxp3. Science 299(5609):1057–1061. https://doi.org/10.1126/science.1079490

Horwitz DA, Zheng SG, Gray JD (2008) Natural and TGF-beta-induced Foxp3(+)CD4(+) CD25(+) regulatory T cells are not mirror images of each other. Trends Immunol 29(9):429–435. https://doi.org/10.1016/j.it.2008.06.005

Jaeckel E, von Boehmer H, Manns MP (2005) Antigen-specific FoxP3-transduced T-cells can control established type 1 diabetes. Diabetes 54(2):306–310. https://doi.org/10.2337/diabetes.54.2.306

Jonuleit H, Schmitt E, Stassen M, Tuettenberg A, Knop J, Enk AH (2001) Identification and functional characterization of human CD4(+)CD25(+) T cells with regulatory properties isolated from peripheral blood. J Exp

Med 193(11):1285–1294. https://doi.org/10.1084/jem.193.11.1285
Jonuleit H, Schmitt E, Kakirman H, Stassen M, Knop J, Enk AH (2002) Infectious tolerance: human CD25(+) regulatory T cells convey suppressor activity to conventional CD4(+) T helper cells. J Exp Med 196 (2):255–260. https://doi.org/10.1084/jem.20020394
Kadakia S, Farnaes L, Dimmock D, Chowdhury S, Ding Y, Anderson EJ, Kingsmore S, Newfield RS (2019) Diagnosis and treatment of a boy with IPEX syndrome presenting with diabetes in early infancy. Clin Case Rep 7(11):2123–2127. https://doi.org/10.1002/ccr3.2438
Kasprowicz DJ, Smallwood PS, Tyznik AJ, Ziegler SF (2003) Scurfin (FoxP3) controls T-dependent immune responses in vivo through regulation of CD4+ T cell effector function. J Immunol 171(3):1216–1223. https://doi.org/10.4049/jimmunol.171.3.1216
Khattri R, Kasprowicz D, Cox T, Mortrud M, Appleby MW, Brunkow ME, Ziegler SF, Ramsdell F (2001) The amount of scurfin protein determines peripheral T cell number and responsiveness. J Immunol 167 (11):6312–6320. https://doi.org/10.4049/jimmunol.167.11.6312
Khattri R, Cox T, Yasayko SA, Ramsdell F (2003) An essential role for scurfin in CD4+CD25+ T regulatory cells. Nat Immunol 4(4):337–342. https://doi.org/10.1038/ni909
Kim HJ, Barnitz RA, Kreslavsky T, Brown FD, Moffett H, Lemieux ME, Kaygusuz Y, Meissner T, Holderried TA, Chan S, Kastner P, Haining WN, Cantor H (2015) Stable inhibitory activity of regulatory T cells requires the transcription factor Helios. Science 350 (6258):334–339. https://doi.org/10.1126/science.aad0616
Kitagawa Y, Ohkura N, Kidani Y, Vandenbon A, Hirota K, Kawakami R, Yasuda K, Motooka D, Nakamura S, Kondo M, Taniuchi I, Kohwi-Shigematsu T, Sakaguchi S (2017) Guidance of regulatory T cell development by Satb1-dependent super-enhancer establishment. Nat Immunol 18(2):173–183. https://doi.org/10.1038/ni.3646
Klocke K, Sakaguchi S, Holmdahl R, Wing K (2016) Induction of autoimmune disease by deletion of CTLA-4 in mice in adulthood. Proc Natl Acad Sci U S A 113(17):E2383–E2392. https://doi.org/10.1073/pnas.1603892113
Koch MA, Thomas KR, Perdue NR, Smigiel KS, Srivastava S, Campbell DJ (2012) T-bet(+) Treg cells undergo abortive Th1 cell differentiation due to impaired expression of IL-12 receptor beta2. Immunity 37(3):501–510. https://doi.org/10.1016/j.immuni.2012.05.031
Kong N, Lan Q, Chen M, Wang J, Shi W, Horwitz DA, Quesniaux V, Ryffel B, Liu Z, Brand D, Zou H, Zheng SG (2012) Antigen-specific transforming growth factor beta-induced Treg cells, but not natural Treg cells, ameliorate autoimmune arthritis in mice by shifting the Th17/Treg cell balance from Th17 predominance to Treg cell predominance. Arthritis Rheum 64 (8):2548–2558. https://doi.org/10.1002/art.34513
Konopacki C, Pritykin Y, Rubtsov Y, Leslie CS, Rudensky AY (2019) Transcription factor Foxp1 regulates Foxp3 chromatin binding and coordinates regulatory T cell function. Nat Immunol 20 (2):232–242. https://doi.org/10.1038/s41590-018-0291-z
Kretschmer K, Apostolou I, Hawiger D, Khazaie K, Nussenzweig MC, von Boehmer H (2005) Inducing and expanding regulatory T cell populations by foreign antigen. Nat Immunol 6(12):1219–1227. https://doi.org/10.1038/ni1265
Kwon HS, Lim HW, Wu J, Schnolzer M, Verdin E, Ott M (2012) Three novel acetylation sites in the Foxp3 transcription factor regulate the suppressive activity of regulatory T cells. J Immunol 188(6):2712–2721. https://doi.org/10.4049/jimmunol.1100903
Lal G, Zhang N, van der Touw W, Ding Y, Ju W, Bottinger EP, Reid SP, Levy DE, Bromberg JS (2009) Epigenetic regulation of Foxp3 expression in regulatory T cells by DNA methylation. J Immunol 182(1):259–273. https://doi.org/10.4049/jimmunol.182.1.259
Lam EW, Brosens JJ, Gomes AR, Koo CY (2013) Forkhead box proteins: tuning forks for transcriptional harmony. Nat Rev Cancer 13(7):482–495. https://doi.org/10.1038/nrc3539
Legoux FP, Lim JB, Cauley AW, Dikiy S, Ertelt J, Mariani TJ, Sparwasser T, Way SS, Moon JJ (2015) CD4+ T cell tolerance to tissue-restricted self antigens is mediated by antigen-specific regulatory T cells rather than deletion. Immunity 43(5):896–908. https://doi.org/10.1016/j.immuni.2015.10.011
Levine AG, Mendoza A, Hemmers S, Moltedo B, Niec RE, Schizas M, Hoyos BE, Putintseva EV, Chaudhry A, Dikiy S, Fujisawa S, Chudakov DM, Treuting PM, Rudensky AY (2017) Stability and function of regulatory T cells expressing the transcription factor T-bet. Nature 546(7658):421–425. https://doi.org/10.1038/nature22360
Li B, Samanta A, Song X, Iacono KT, Bembas K, Tao R, Basu S, Riley JL, Hancock WW, Shen Y, Saouaf SJ, Greene MI (2007) FOXP3 interactions with histone acetyltransferase and class II histone deacetylases are required for repression. Proc Natl Acad Sci U S A 104 (11):4571–4576. https://doi.org/10.1073/pnas.0700298104
Lin Y, Xu A, Zeng C, Cheng J, Li N, Niu H, Liu L, Li X (2018) Somatic and germline FOXP3 mosaicism in the mother of a boy with IPEX syndrome. Eur J Immunol 48(5):885–887. https://doi.org/10.1002/eji.201747445
Lio CW, Hsieh CS (2008) A two-step process for thymic regulatory T cell development. Immunity 28 (1):100–111. https://doi.org/10.1016/j.immuni.2007.11.021
Lopes JE, Torgerson TR, Schubert LA, Anover SD, Ocheltree EL, Ochs HD, Ziegler SF (2006) Analysis of FOXP3 reveals multiple domains required for its

function as a transcriptional repressor. J Immunol 177 (5):3133–3142. https://doi.org/10.4049/jimmunol.177.5.3133
Louie RJ, Tan QK, Gilner JB, Rogers RC, Younge N, Wechsler SB, McDonald MT, Gordon B, Saski CA, Jones JR, Chapman SJ, Stevenson RE, Sleasman JW, Friez MJ (2017) Novel pathogenic variants in FOXP3 in fetuses with echogenic bowel and skin desquamation identified by ultrasound. Am J Med Genet A 173 (5):1219–1225. https://doi.org/10.1002/ajmg.a.38144
Luo Y, Chen J, Fang Y, Lou J, Yu J (2018) A case of metaplastic atrophic gastritis in immune dysregulation, polyendocrinopathy, enteropathy, X-linked (IPEX) syndrome. BMC Pediatr 18(1):191. https://doi.org/10.1186/s12887-018-1169-9
Magg T, Wiebking V, Conca R, Krebs S, Arens S, Schmid I, Klein C, Albert MH, Hauck F (2018) IPEX due to an exon 7 skipping FOXP3 mutation with autoimmune diabetes mellitus cured by selective TReg cell engraftment. Clin Immunol 191:52–58. https://doi.org/10.1016/j.clim.2018.03.008
Marson A, Kretschmer K, Frampton GM, Jacobsen ES, Polansky JK, MacIsaac KD, Levine SS, Fraenkel E, von Boehmer H, Young RA (2007) Foxp3 occupancy and regulation of key target genes during T-cell stimulation. Nature 445(7130):931–935. https://doi.org/10.1038/nature05478
Miyara M, Sakaguchi S (2007) Natural regulatory T cells: mechanisms of suppression. Trends Mol Med 13 (3):108–116. https://doi.org/10.1016/j.molmed.2007.01.003
Morawski PA, Mehra P, Chen C, Bhatti T, Wells AD (2013) Foxp3 protein stability is regulated by cyclin-dependent kinase 2. J Biol Chem 288 (34):24494–24502. https://doi.org/10.1074/jbc.M113.467704
Morikawa H, Sakaguchi S (2014) Genetic and epigenetic basis of Treg cell development and function: from a FoxP3-centered view to an epigenome-defined view of natural Treg cells. Immunol Rev 259(1):192–205. https://doi.org/10.1111/imr.12174
Morikawa H, Ohkura N, Vandenbon A, Itoh M, Nagao-Sato S, Kawaji H, Lassmann T, Carninci P, Hayashizaki Y, Forrest AR, Standley DM, Date H, Sakaguchi S, FANTOM Consortium (2014) Differential roles of epigenetic changes and Foxp3 expression in regulatory T cell-specific transcriptional regulation. Proc Natl Acad Sci U S A 111(14):5289–5294. https://doi.org/10.1073/pnas.1312717110
Nakagawa H, Sido JM, Reyes EE, Kiers V, Cantor H, Kim HJ (2016) Instability of Helios-deficient Tregs is associated with conversion to a T-effector phenotype and enhanced antitumor immunity. Proc Natl Acad Sci U S A 113(22):6248–6253. https://doi.org/10.1073/pnas.1604765113
Ohkura N, Hamaguchi M, Morikawa H, Sugimura K, Tanaka A, Ito Y, Osaki M, Tanaka Y, Yamashita R, Nakano N, Huehn J, Fehling HJ, Sparwasser T, Nakai K, Sakaguchi S (2012) T cell receptor stimulation-induced epigenetic changes and Foxp3 expression are independent and complementary events required for Treg cell development. Immunity 37 (5):785–799. https://doi.org/10.1016/j.immuni.2012.09.010
Paterson AM, Lovitch SB, Sage PT, Juneja VR, Lee Y, Trombley JD, Arancibia-Carcamo CV, Sobel RA, Rudensky AY, Kuchroo VK, Freeman GJ, Sharpe AH (2015) Deletion of CTLA-4 on regulatory T cells during adulthood leads to resistance to autoimmunity. J Exp Med 212(10):1603–1621. https://doi.org/10.1084/jem.20141030
Ramsdell F, Ziegler SF (2014) FOXP3 and scurfy: how it all began. Nat Rev Immunol 14(5):343–349. https://doi.org/10.1038/nri3650
Rudra D, deRoos P, Chaudhry A, Niec RE, Arvey A, Samstein RM, Leslie C, Shaffer SA, Goodlett DR, Rudensky AY (2012) Transcription factor Foxp3 and its protein partners form a complex regulatory network. Nat Immunol 13(10):1010–1019. https://doi.org/10.1038/ni.2402
Sakaguchi S, Sakaguchi N, Asano M, Itoh M, Toda M (1995) Immunologic self-tolerance maintained by activated T cells expressing IL-2 receptor alpha-chains (CD25). Breakdown of a single mechanism of self-tolerance causes various autoimmune diseases. J Immunol 155(3):1151–1164
Sakaguchi S, Wing K, Onishi Y, Prieto-Martin P, Yamaguchi T (2009) Regulatory T cells: how do they suppress immune responses? Int Immunol 21 (10):1105–1111. https://doi.org/10.1093/intimm/dxp095
Salomon B, Lenschow DJ, Rhee L, Ashourian N, Singh B, Sharpe A, Bluestone JA (2000) B7/CD28 costimulation is essential for the homeostasis of the CD4+CD25+ immunoregulatory T cells that control autoimmune diabetes. Immunity 12(4):431–440. https://doi.org/10.1016/s1074-7613(00)80195-8
Schmidt A, Oberle N, Krammer PH (2012) Molecular mechanisms of treg-mediated T cell suppression. Front Immunol 3:51. https://doi.org/10.3389/fimmu.2012.00051
Shevach EM (2009) Mechanisms of foxp3+ T regulatory cell-mediated suppression. Immunity 30(5):636–645. https://doi.org/10.1016/j.immuni.2009.04.010
Song X, Li B, Xiao Y, Chen C, Wang Q, Liu Y, Berezov A, Xu C, Gao Y, Li Z, Wu SL, Cai Z, Zhang H, Karger BL, Hancock WW, Wells AD, Zhou Z, Greene MI (2012) Structural and biological features of FOXP3 dimerization relevant to regulatory T cell function. Cell Rep 1(6):665–675. https://doi.org/10.1016/j.celrep.2012.04.012
Su Q, Jing J, Li W, Ma J, Zhang X, Wang Z, Zhou Z, Dai L, Shao L (2019a) Impaired Tip60-mediated Foxp3 acetylation attenuates regulatory T cell development in rheumatoid arthritis. J Autoimmun 100:27–39. https://doi.org/10.1016/j.jaut.2019.02.007
Su W, Chen X, Zhu W, Yu J, Li W, Li Y, Li Z, Olsen N, Liang D, Zheng SG (2019b) The cAMP-adenosine

feedback loop maintains the suppressive function of regulatory T cells. J Immunol 203(6):1436–1446. https://doi.org/10.4049/jimmunol.1801306

Tai X, Cowan M, Feigenbaum L, Singer A (2005) CD28 costimulation of developing thymocytes induces Foxp3 expression and regulatory T cell differentiation independently of interleukin 2. Nat Immunol 6 (2):152–162. https://doi.org/10.1038/ni1160

Thornton AM, Korty PE, Tran DQ, Wohlfert EA, Murray PE, Belkaid Y, Shevach EM (2010) Expression of Helios, an Ikaros transcription factor family member, differentiates thymic-derived from peripherally induced Foxp3+ T regulatory cells. J Immunol 184 (7):3433–3441. https://doi.org/10.4049/jimmunol.0904028

Van Gool F, Nguyen MLT, Mumbach MR, Satpathy AT, Rosenthal WL, Giacometti S, Le DT, Liu W, Brusko TM, Anderson MS, Rudensky AY, Marson A, Chang HY, Bluestone JA (2019) A mutation in the transcription factor Foxp3 drives T helper 2 effector function in regulatory T cells. Immunity 50(2):362–377 e366. https://doi.org/10.1016/j.immuni.2018.12.016

van Loosdregt J, Coffer PJ (2014) Post-translational modification networks regulating FOXP3 function. Trends Immunol 35(8):368–378. https://doi.org/10.1016/j.it.2014.06.005

van Loosdregt J, Vercoulen Y, Guichelaar T, Gent YY, Beekman JM, van Beekum O, Brenkman AB, Hijnen DJ, Mutis T, Kalkhoven E, Prakken BJ, Coffer PJ (2010) Regulation of Treg functionality by acetylation-mediated Foxp3 protein stabilization. Blood 115(5):965–974. https://doi.org/10.1182/blood-2009-02-207118

Vignali DA (2012) Mechanisms of T(reg) suppression: still a long way to go. Front Immunol 3:191. https://doi.org/10.3389/fimmu.2012.00191

Walker MR, Kasprowicz DJ, Gersuk VH, Benard A, Van Landeghen M, Buckner JH, Ziegler SF (2003) Induction of FoxP3 and acquisition of T regulatory activity by stimulated human CD4+CD25- T cells. J Clin Invest 112(9):1437–1443. https://doi.org/10.1172/JCI19441

Wang B, Lin D, Li C, Tucker P (2003) Multiple domains define the expression and regulatory properties of Foxp1 forkhead transcriptional repressors. J Biol Chem 278(27):24259–24268. https://doi.org/10.1074/jbc.M207174200

Wang J, Ioan-Facsinay A, van der Voort EI, Huizinga TW, Toes RE (2007) Transient expression of FOXP3 in human activated nonregulatory CD4+ T cells. Eur J Immunol 37(1):129–138. https://doi.org/10.1002/eji.200636435

Wang Y, Su MA, Wan YY (2011) An essential role of the transcription factor GATA-3 for the function of regulatory T cells. Immunity 35(3):337–348. https://doi.org/10.1016/j.immuni.2011.08.012

Weiss JM, Bilate AM, Gobert M, Ding Y, Curotto de Lafaille MA, Parkhurst CN, Xiong H, Dolpady J, Frey AB, Ruocco MG, Yang Y, Floess S, Huehn J, Oh S, Li MO, Niec RE, Rudensky AY, Dustin ML, Littman DR, Lafaille JJ (2012) Neuropilin 1 is expressed on thymus-derived natural regulatory T cells, but not mucosa-generated induced Foxp3+ T reg cells. J Exp Med 209(10):1723–1742, S1721. https://doi.org/10.1084/jem.20120914

Wildin RS, Ramsdell F, Peake J, Faravelli F, Casanova JL, Buist N, Levy-Lahad E, Mazzella M, Goulet O, Perroni L, Bricarelli FD, Byrne G, McEuen M, Proll S, Appleby M, Brunkow ME (2001) X-linked neonatal diabetes mellitus, enteropathy and endocrinopathy syndrome is the human equivalent of mouse scurfy. Nat Genet 27(1):18–20. https://doi.org/10.1038/83707

Wildin RS, Smyk-Pearson S, Filipovich AH (2002) Clinical and molecular features of the immunodysregulation, polyendocrinopathy, enteropathy, X linked (IPEX) syndrome. J Med Genet 39(8):537–545. https://doi.org/10.1136/jmg.39.8.537

Wohlfert EA, Grainger JR, Bouladoux N, Konkel JE, Oldenhove G, Ribeiro CH, Hall JA, Yagi R, Naik S, Bhairavabhotla R, Paul WE, Bosselut R, Wei G, Zhao K, Oukka M, Zhu J, Belkaid Y (2011) GATA3 controls Foxp3(+) regulatory T cell fate during inflammation in mice. J Clin Invest 121(11):4503–4515. https://doi.org/10.1172/JCI57456

Xiao Y, Nagai Y, Deng G, Ohtani T, Zhu Z, Zhou Z, Zhang H, Ji MQ, Lough JW, Samanta A, Hancock WW, Greene MI (2014) Dynamic interactions between TIP60 and p300 regulate FOXP3 function through a structural switch defined by a single lysine on TIP60. Cell Rep 7(5):1471–1480. https://doi.org/10.1016/j.celrep.2014.04.021

Yadav M, Louvet C, Davini D, Gardner JM, Martinez-Llordella M, Bailey-Bucktrout S, Anthony BA, Sverdrup FM, Head R, Kuster DJ, Ruminski P, Weiss D, Von Schack D, Bluestone JA (2012) Neuropilin-1 distinguishes natural and inducible regulatory T cells among regulatory T cell subsets in vivo. J Exp Med 209(10):1713–1722. , S1711–1719. https://doi.org/10.1084/jem.20120822

Yang BH, Hagemann S, Mamareli P, Lauer U, Hoffmann U, Beckstette M, Fohse L, Prinz I, Pezoldt J, Suerbaum S, Sparwasser T, Hamann A, Floess S, Huehn J, Lochner M (2016) Foxp3(+) T cells expressing RORgammat represent a stable regulatory T-cell effector lineage with enhanced suppressive capacity during intestinal inflammation. Mucosal Immunol 9(2):444–457. https://doi.org/10.1038/mi.2015.74

Yang M, Liu Y, Mo B, Xue Y, Ye C, Jiang Y, Bi X, Liu M, Wu Y, Wang J, Olsen N, Pan Y, Zheng SG (2019) Helios but not CD226, TIGIT and Foxp3 is a potential marker for CD4(+) Treg cells in patients with rheumatoid arthritis. Cell Physiol Biochem 52 (5):1178–1192. https://doi.org/10.33594/000000080

Zheng SG, Gray JD, Ohtsuka K, Yamagiwa S, Horwitz DA (2002) Generation ex vivo of TGF-beta-producing regulatory T cells from CD4+CD25- precursors. J

Immunol 169(8):4183–4189. https://doi.org/10.4049/jimmunol.169.8.4183

Zheng SG, Wang JH, Gray JD, Soucier H, Horwitz DA (2004) Natural and induced CD4+CD25+ cells educate CD4+CD25- cells to develop suppressive activity: the role of IL-2, TGF-beta, and IL-10. J Immunol 172 (9):5213–5221. https://doi.org/10.4049/jimmunol.172.9.5213

Zheng SG, Wang JH, Stohl W, Kim KS, Gray JD, Horwitz DA (2006) TGF-beta requires CTLA-4 early after T cell activation to induce FoxP3 and generate adaptive CD4+CD25+ regulatory cells. J Immunol 176 (6):3321–3329. https://doi.org/10.4049/jimmunol.176.6.3321

Zheng SG, Wang J, Wang P, Gray JD, Horwitz DA (2007a) IL-2 is essential for TGF-beta to convert naive CD4+CD25- cells to CD25+Foxp3+ regulatory T cells and for expansion of these cells. J Immunol 178 (4):2018–2027. https://doi.org/10.4049/jimmunol.178.4.2018

Zheng Y, Josefowicz SZ, Kas A, Chu TT, Gavin MA, Rudensky AY (2007b) Genome-wide analysis of Foxp3 target genes in developing and mature regulatory T cells. Nature 445(7130):936–940. https://doi.org/10.1038/nature05563

Zhou Z, Song X, Li B, Greene MI (2008) FOXP3 and its partners: structural and biochemical insights into the regulation of FOXP3 activity. Immunol Res 42 (1–3):19–28. https://doi.org/10.1007/s12026-008-8029-x

Regulation of Treg Functions by the Ubiquitin Pathway

3

Elena Montauti and Deyu Fang

Abstract

Regulatory T (Tregs) cells, required to maintain immune homeostasis, have significant power in disease outcomes. Treg dysfunction, predominantly characterized by the loss of the master transcription factor FoxP3 and the acquisition of T_{eff}-like phenotypes, can promote autoimmunity as well as enhance antitumor immunity. As FoxP3 expression and stability are pinnacle for Treg suppressive functions, understanding the pathways that regulate FoxP3 is crucial to ascertain Treg-mediated therapies for autoimmune diseases and cancer. Mechanisms controlling FoxP3 expression and stability range from transcriptional to posttranslational, revealing multiple therapeutic opportunities. While many of the transcriptional pathways have been explored in detail, a recent surge in interest on the posttranslational mechanisms regulating FoxP3 has arisen. Particularly, the role of ubiquitination on Tregs both directly and indirectly involving FoxP3 has gained interest. Here, we summarize the current knowledge on ubiquitin-dependent, FoxP3-mediated control of Treg function as it pertains to human diseases.

Keywords

Treg · FoxP3 · Ubiquitin · E3 ligase · Deubiquitination

3.1 Introduction

The power of the human immune system, capable of defeating both internal and external threats, is kept in check to avoid host injury by immune tolerizing T-regulatory cells (Tregs), creating a delicate balance between immune activation and suppression crucial for host survival (Mahajan et al. 2006; Maloy et al. 2003; Sakaguchi et al. 1995; Singh et al. 2001). To maintain peripheral tolerance, Tregs implement a variety of mechanisms including but not limited to the production of inflammatory cytokines, expression of co-inhibitory molecules, and the scavenging of growth factors (Horwitz et al. 2008; Kong et al. 2012; Vignali et al. 2008). Problematically, tumor hijacking of Treg suppressive capabilities promotes tumor immune evasion and enhances tumor growth (Akimova et al. 2017; Angelin et al. 2017; Clark et al. 2007; De Simone et al. 2016; Mao et al. 2017; Plitas et al. 2016; Sawant et al. 2019; Woo et al. 2001). Additionally, the tumor residing Tregs have an altered

Deyu Fang is a Lead Contact.

E. Montauti · D. Fang (✉)
Department of Pathology, Northwestern University Feinberg School of Medicine, Chicago, IL, USA
e-mail: ElenaMontauti2015@u.northwestern.edu; FangD@Northwestern.edu

S.-G. Zheng (ed.), *T Regulatory Cells in Human Health and Diseases*, Advances in Experimental Medicine and Biology 1278, https://doi.org/10.1007/978-981-15-6407-9_3

transcriptional landscape, resulting in heightened suppressive functions and creating a major hurdle when targeting tumors for immunotherapy (Akimova et al. 2017; Angelin et al. 2017; De Simone et al. 2016; Plitas et al. 2016). Depletion of Tregs in these highly immunogenic cancers through Treg-specific markers is a promising strategy for the enhancement of tumor immunotherapy. Unfortunately, most Treg markers are nonspecific, such as CD25 and GITR, and clinical administration may affect activation of tumor-targeting T-effector (Teff) cells (Rech et al. 2012; Robb et al. 1981; Sakaguchi et al. 1995; Schaer et al. 2013). The most specific marker for Treg lineage is FoxP3, the master regulator of Treg immune regulatory functions (Fontenot et al. 2003). However, direct targeting of FoxP3 has proven difficult, as is a transcription factor. Therefore, a promising new strategy of tumor immunotherapy is through regulation of FoxP3 expression, resulting in the attenuation of Treg suppressive function.

Understanding the fundamental regulators of Foxp3 is critical, for the foundation of new potential Treg therapies capable of modulating the immune response during cancer, chronic infection, or autoimmune diseases. FoxP3 expression can be regulated at the transcriptional, posttranscriptional, and posttranslational levels, resulting in various degrees of suppression (Gao et al. 2015; Isomura et al. 2009; Kerdiles et al. 2010; Liston et al. 2008; Tai et al. 2005; Tang et al. 2003; Zheng et al. 2008; Zhou et al. 2008). These regulatory mechanisms have major consequences for the stability and overall function of Tregs. Particularly, a newly appreciated layer of FoxP3 regulation and Treg functional modulation is through ubiquitination. In the following sections, we will discuss ubiquitin-dependent pathways that modulate Treg generation, function, and stability, some of which directly regulate Foxp3 expression and activity.

3.2 The Ubiquitination Pathway

As one of the most conserved mechanisms in all biology, ubiquitination is one of the most intensely studied of the posttranslational modifications. A pathway with many end goals, ubiquitination refers to the sequential three-step cascade initiated by ubiquitin activation by ubiquitin-activating enzymes (E1s), followed by conjugation by ubiquitin-conjugating enzymes (E2s), and culminated with covalent ligation of ubiquitin to the target protein by ubiquitin ligases (E3s) (Ciechanover et al. 1982; Ciehanover et al. 1978). Whether it be protein degradation, transcription, activation, or cellular localization, the particular lysine residue by which the ubiquitin is attached, and the length of the ubiquitin chain, determines the fate of the target protein (Johnson et al. 1995; Kim et al. 2007; Kirisako et al. 2006; Peng et al. 2003; van Loosdregt et al. 2013; Xu et al. 2009). Ligation of polyubiquitin chains can occur at one of seven lysine residues on the ubiquitin protein: K6, K11, K27, K29, K33, K48, and K63. K48 and K63 modifications are the best understood. K63-ubiquitin chains result in the association of the target protein with proteins involved in DNA repair, endocytic trafficking, and signal transduction (Tenno et al. 2004). Contrarily, K48 polyubiquitination culminates in the degradation of target proteins by the 26S proteasome (Johnson et al. 1995). Furthermore, mono-ubiquitination on key histone lysine residues can regulate transcriptional accessibility.

Appropriately, the process of ubiquitination is dynamic and reversible. The highly regulated removal of ubiquitin, termed deubiquitination, has been implicated in numerous cellular functions, including cell cycle regulation (Hanna et al. 2003), proteasome-dependent degradation (Wilkinson 1997), and gene expression (Suresh et al. 2016). Deubiquitination is catalyzed by a group of cysteine proteases called deubiquitinases (DUBS) that encompass more than 100 proteins shown to regulate p53 activity, WNT, NF-κB pathways, EGFR, cell cycle progression, apoptosis, and response to DNA damage (Fraile et al. 2011; Wilkinson 1997). Clearly, ubiquitination mediates many diverse cellular processes, and their dysregulation has been specifically implicated in cancer and autoimmunity. In the following sections, we discuss how ubiquitination facilitates both immune activation and restraint, particularly through the ubiquitin-specific pathways that modulate Treg generation, stability, and function.

3.3 Regulation of Treg Function by E3 Ligases Directly Targeting FoxP3

As the primary function of Tregs is the maintenance of peripheral tolerance, FoxP3 stability is key to controlling the immune system (Lu et al. 2014). Therefore, stabilizing FoxP3 expression enhances control over the aberrant immune activation that occurs in autoimmune diseases (Fontenot et al. 2003; Mahajan et al. 2006; Zhou et al. 2010). As DUBs remove ubiquitin from target proteins, E3 ligases place them on. Therefore, targeting E3 ligases does downregulate the ubiquitination of FoxP3 that is an attractive therapeutic strategy for autoimmune diseases.

3.4 Stub 1 Potentiates Environmental Cues to Regulate Treg Function

STIP1 homology and U-Box containing protein 1 (Stub1) is a stress activated, U-box domain-type E3 ubiquitin ligase that was found to interact with FoxP3 through the chaperone molecule heat shock 70 kDa protein (HSP70) (Fig. 3.1) (Ballinger et al. 1999; Chen et al. 2013). Although not highly expressed in Treg cells at baseline, Stub1 was noted to be upregulated in vitro in the presence of proinflammatory cytokines and lipopolysaccharides, environmental conditions that downregulated FoxP3. This, along with HSP70-dependent Stub1 interaction with FoxP3, suggested it may be the driving force behind FoxP3 ubiquitination (Chen et al. 2013). To support this, retroviral overexpression of Stub1 in Tregs abolished their ability to control immune responses in vitro and in vivo and enhanced Treg expression of inflammatory cytokines. Conversely, silencing Stub1 augmented Treg function in vitro and in vivo (Chen et al. 2013). The relationship between Stub1 and FoxP3, and the resulting fate of Treg suppressive function, highlights the importance of environmental cues of mechanical pathways.

Although in vivo suppressive assays revealed a Stub1 importance on Treg suppressive function, the degree of which Stub1 induction reduces Treg functions under physiological conditions is still unknown. Particularly, the observation that a small pool of Tregs retains FoxP3 regardless of enhanced Stub1 expression suggests the eventual growth of a population of Tregs that are likely more stable and suppressive. The physiological implications of these more stable Tregs may likely put a damper on immune surveillance, possibly resulting in tumor colonization. Similarly, the level Stub1-specific loss of FoxP3 would push Tregs into long-lasting dysfunction, resulting in autoimmune diseases that needs to be explored before Stub1-specific therapies can be envisioned.

3.5 TRAF6 Controls FoxP3 Localization Via K63 Ubiquitination

The E3 ligase TNF receptor-associated factor 6 (TRAF6) was found to be an important signaling molecule involved in the priming of T-cell-mediated immune responses (Kobayashi et al. 2003). Apart from this, TRAF6 deficiency resulted in enhanced Th17 commitment as well as multi-organ autoimmunity and dermatitis, suggesting involvement in immune regulation (Akiyama et al. 2005; Cejas et al. 2010; Chiffoleau et al. 2003; Muto et al. 2013). At baseline, preferential expression of TRAF6 by multiple Treg subsets implicated TRAF6 as an important factor in Treg biology (Ni et al. 2019). Particularly, Muto et al. (2013) detailed that mice lacking TRAF6 specifically in Tregs showed elevated T_{eff}-cell activation, supporting the notion of insufficient immune control. Despite Tregs being elevated in frequency, TRAF6-null Tregs displayed an unstable phenotype in vivo and resulted in enhanced anti-tumor immunity

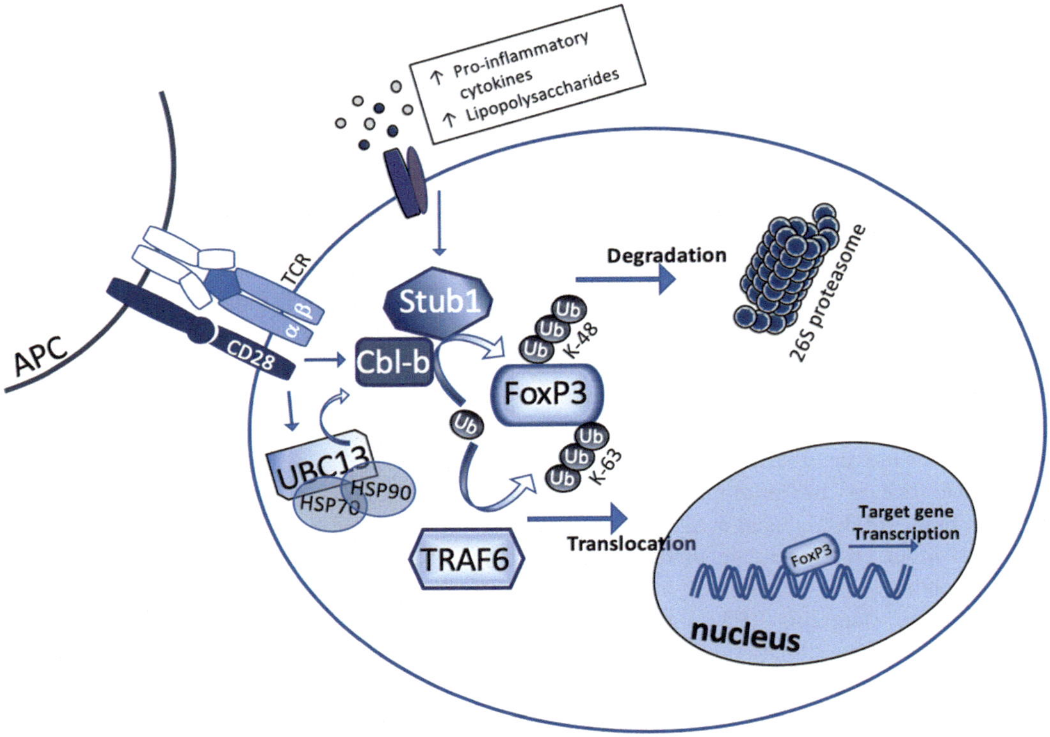

Fig. 3.1 Schematic showing direct E3 regulatory elements on FoxP3. Solid arrows indicate stimulatory effects. Inflammatory signals from the cell environment stabilize Stub1 levels to deubiquitinate FoxP3. Upon TCR/CDX28 co-stimulation, Cbl-b is recruited to Stub1 as aid. Similarly, Cbl-b is also recruited by the UBC13/HSP complex. These pathways lead to FoxP3 K48 ubiquitination and subsequent degradation of the proteasome. Uniquely, TRAF6 ubiquitinates FoxP3 via K63, resulting in FoxP3 translocation into the nucleus. These two pathways of ubiquitination are fundamentally important to Treg-suppressive function

(Muto et al. 2013; Ni et al. 2019). Interestingly, TRAF6 deficiency did not alter TGF-β driven iTreg differentiation, nor Treg suppressive capabilities in vitro suggesting that TRAF6 may be more important for determining the commitment to the Treg lineage (Cejas et al. 2010; Shimo et al. 2011). Indeed, TGF-β signal is crucial for Treg cell differentiation and function (Zheng et al. 2002, 2004). However, recent findings suggest an underappreciated ubiquitin-mediated pathway involving TRAF6 and FoxP3.

As stated in the introduction, there are various ubiquitin linkages, each with their unique downstream pathways. While FoxP3 can be ubiquitinated via the K48-linkage, ultimately leading to degradation, it was revealed to also be ubiquitinated through the K63-type ubiquitination via TRAF6 (Fig. 3.1) (Ni et al. 2019). This ubiquitination was largely through the zinc finger and leucine zipper domains on FoxP3 as well as the TRAF6 target residue K262. Tregs insensitive to FoxP3-specific K626 ubiquitination resulted in Treg dysfunction in vivo (Ni et al. 2019). Ultimately, this modification promoted FoxP3 translocation into the nucleus and gene regulatory functions (Wang et al. 2012). These data reveal an unprecedented role for TRAF6, confirming its importance in many immunological pathways, presenting TRAF6 as an enticing target for anti-tumor immunotherapy. However, targeting TRAF6 may prove counterproductive due to its involvement in T-cell priming and the NFκB pathways downstream of the TCR. Fortunately, TRAF6 seems dispensable for TCR-mediated activation of T cells as well as NFκB activation (Chiffoleau et al. 2003). Furthermore, T-cell-specific deletion of TRAF6 phenocopies the FoxP3-specific knockout, and global TRAF6 knockout resulted in enhanced leukocyte activation and inflammation (Muto et al. 2013; Ni et al. 2019). Therefore, TRAF6 seems to have a more pronounced role in Treg immune restraint rather than in T_{eff} immune activation, positioning TRAF6 as a likely target focused on breaking tolerance in tumor immunotherapy.

3.6 Cbl-b Induces FoxP3 Degradation Upon TCR/CD28 Stimulation

The initial ubiquitin-mediated influence on Tregs occurs during the thymic generation of $Foxp3^{+}$ Tregs influenced by ubiquitin-mediated regulation of NFκB signaling (Schmidt-Supprian et al. 2004). Signaling cascades initiated at the TCR, combined with co-stimulation, are central for proper Treg development. Upon TCR stimulation, downstream phosphorylation of IκB (creating PI3K) leads to its ubiquitination, allowing for key NFκB family members to translocate into the nucleus and aid in the transcriptional activation of the *Foxp3* gene (Gerondakis et al. 2012; Schmidt-Supprian et al. 2004).

Casitas-B-lineage lymphoma protein-b (Cbl-b) is an E3 ligase known to modulate T-cell activation through non-proteolytic tagging of PI3K, preventing its association with CD28 and the TCR. When TCR activation is coupled with CD28 co-stimulation, the ubiquitination of Cbl-b lowers the threshold required for T-cell activation by releasing PI3K. Without co-stimulation, or with coinhibitory signals from CTLA-4 engagement, Cbl-b levels stabilize and cling to PI3K downregulating TCR signaling. In fact, both co-stimulatory and co-inhibitory signals affect Treg cell development (Tang et al. 2003; Zheng et al. 2006). Cbl-b-mediated restraint of PI3K also counters AKT/mTOR activation, a pathway that negatively effects FoxP3 expression (Bachmaier et al. 2000; Chiang et al. 2000). Furthermore, FoxP3 levels in Cbl-$b^{-/-}$ mice can be restored with the administration of a PI3K inhibitor, defining Cbl-b as a key factor controlling FoxP3 levels (Harada et al. 2010).

Although Cbl-b facilitates FoxP3 expression through decreasing the T-cell activation threshold, more recent results have demonstrated that Cbl-b actually acts to downregulate FoxP3 posttranslationally (Zhao et al. 2015). Cbl-b acts in concert with Stub1 to sequentially induce FoxP3 degradation in tTregs through ubiquitination upon TCR/CD28 stimulation (Fig. 3.1). Cbl-b associates with the Stub1-FoxP3 complex through its UBA domain, and siRNA knockout of Cbl-b alone could stabilize FoxP3 protein levels and reduce FoxP3 polyubiquitination. Treatment of $Cd28^{-/-}$ mice with a proteasome inhibitor also completely rescues defective tTreg development in these mice, demonstrating the importance of Cbl-b function in tTreg development (Zhao et al. 2015). It is noteworthy to mention the contrasting roles played by Cbl-b in peripheral verses thymic FoxP3 induction—an important point in dissecting the dynamic role of Cbl-b in Tregs.

3.7 UBC13, a Critical E2 in Treg Functions

Another mechanism in place to check aberrant gene expression is through the K63-type E2 ubiquitin-conjugating enzyme UBC13 (also known as CHIP), known to be involved in TNFα signaling and NFκB activation (Fukushima et al. 2007). UBC13 is a member of a RING-like domain containing E3 ligases such as heat shock protein 70 (HSP70) and HSP90 functioning to facilitate the polyubiquitination of target proteins (Cyr et al. 2002). Deletion of UBC13 disrupts NFκB signaling, leading to a reduction in IL-10 and SOCS1 expression (Takahashi et al. 2011). SOCS1 is an important regulator for IFNγ and IL-17 production, and its altered expression could result in T_{eff}-like phenotypes. Supporting this, mice with a Treg-specific deletion of UBC13 demonstrate high levels of inflammation due to highly proliferative but non-functioning Tregs. These Tregs express high levels of both IFNγ and IL-17, validating the importance for UBC13 in proper gene regulation within Treg cells (Chang et al. 2012).

UBC13 acts directly on FoxP3 polyubiquitination through the recruitment of Cbl-b to the Ring-like complex. This results in synergistic enhancement of FoxP3 degradation by the proteasome (Fig. 3.1) (Zhao et al. 2015). Furthermore, UBC13 plays an important role in peripheral (p)Tregs during inflammatory conditions via its direct interaction with HSP70, promoting K48-linked FoxP3 ubiquitination and subsequent proteosomal degradation. Overexpression of UBC13 induced a T_{eff}-like Treg phenotype, compromising many Treg-associated gene expression (Zhang et al. 2016). Overall, UBC13 functions as an influencer in Treg stability.

3.8 Indirect FoxP3 Targeting Through Ubiquitination

Although targeting a direct E3 ligase of FoxP3 seems the most streamlined way to regulate Treg function, indirect FoxP3 regulators also reveal promising therapeutic targets.

3.9 RNF20 Decreases FoxP3 Transcription Through H2B Ubiquitination

Ring finger protein 20 (RNF20) was shown to be a negative regulator of FoxP3 in a CRISPR-based targeted loss of function screen, where knockdown resulted in increased FoxP3 expression (Cortez et al. 2020). RNF20 is a known E3 ligase, specifically functioning to ubiquitinate histones H2A and H2B. Interestingly, the same screen flagged another protein with the opposing effect, ubiquitin-specific peptidase 22 (see below). Therefore, it is possible that RNF20 and USP22 could have an epistatic relationship due to their reciprocal effects on histone ubiquitination, particularly at the FoxP3 locus. Indeed, Western blot analysis demonstrated that targeting RNF20 in USP22-deficient Tregs restored H2BK120ub levels. Consistent with this model, RNF20 and USP22 co-occupy regions along the FoxP3 locus. Furthermore, targeting Rnf20 tended to reduce H2BK120 ubiquitination (H2BK120ub) at sites along the FoxP3 locus that had increased H2BK120ub upon USP22 deletion, suggesting a reciprocal ubiquitin switch on the FoxP3 promoter. Finally, RNP-mediated knockout of RNF20 was able to rescue the impairment of FoxP3 transcription exhibited by the USP22-null Tregs. Ultimately, RNF20 and USP22 demonstrate a novel and enticing avenue to easily “switch” Treg-suppressive capabilities through FoxP3 transcription (Fig. 3.2). Harnessing the power of this regulatory mechanism could aid in the battle between anti-tumor immunity and the consequential auto-immunity that often follows.

3.10 Hrd1 Destabilizes FoxP3 Through an Inflammation-Induced ER Stress Response

It has been well established that inflammatory conditions destabilize Treg-suppressive capabilities, yet the role of the ER stress response had not been previously examined (Martinon et al. 2010; Xu et al. 2019). Best known for its ability to catalyze the degradation of misfolded or unfolded proteins through ER-associated degradation, Hrd1 is an E3 ubiquitin ligase that was first implicated in the immune system as a positive regulator for T cells. Specifically, genetic deletion resulted in substandard T-cell development, differentiation, and function (Xu et al. 2016). Of particular interest, deletion of Hrd1 enhances the inflammation-induced ER stress response in Tregs, further destabilizing FoxP3 levels and Treg suppressive function in vivo. Furthermore, Hrd1-null naive CD4 T cells are resistant to TGF-β-induced Treg polarization. Hrd1-mediated stabilization of Tregs seems to function through Hrd1 suppression of IRE1α kinase (Fig. 3.2), as Hrd-1-null Treg instability was rescued by the administration of IRE-α inhibitor (Xu et al. 2019). These results define a critical role in Hrd1 in suppressing ER stress–induced Treg depolarization under inflammation conditions and indicates another pathway with which the cellular environment is key to Treg function.

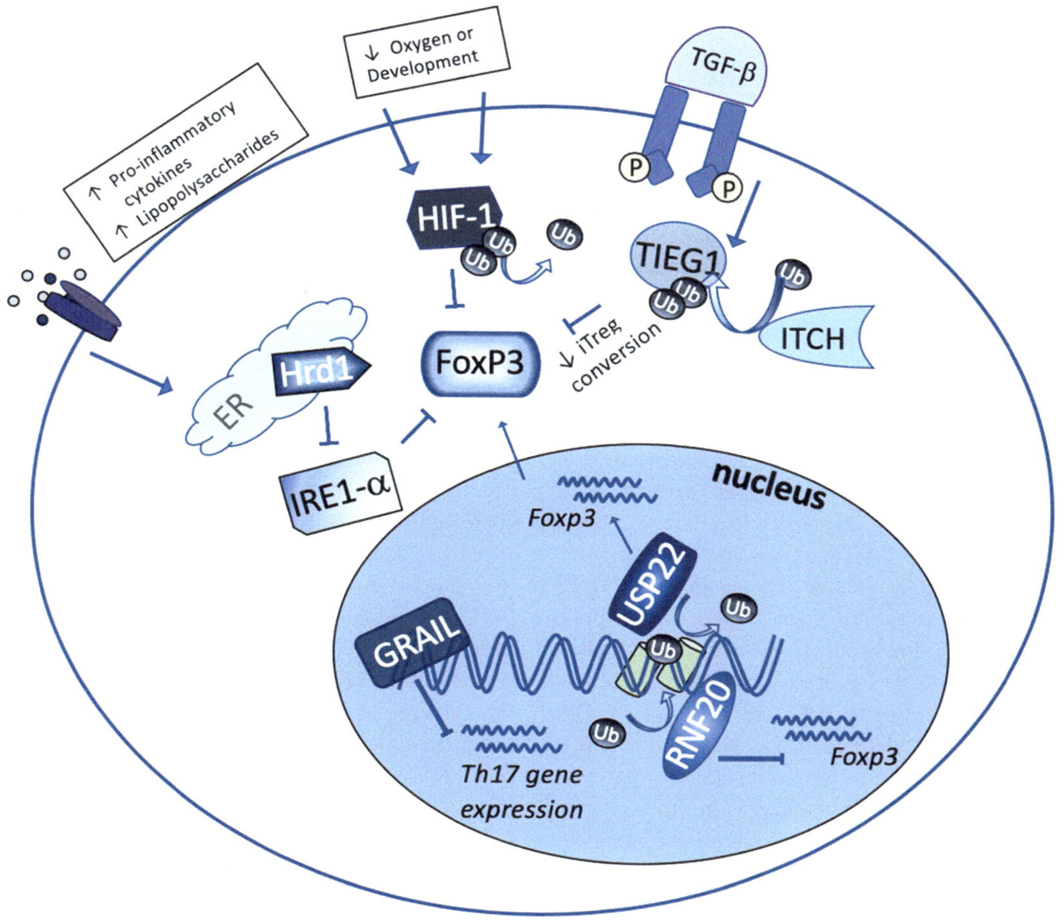

Fig. 3.2 Schematic showing E3 ubiquitin ligases indirectly regulating FoxP3. Solid arrows indicate stimulatory effects. T bars show inhibitory effects. Inflammatory signals induce ER stress to block the inhibitory effects of IRE1α on FoxP3. Hypoxia stabilizes HIF-1 proteins that naturally inhibit FoxP3 through a co-degradation mechanism. Similarly, HIF-1 is stabilized during development through a STAT3-dependent manner in order to keep a balance between Tregs and TH17 cells. The TGF-β pathway, known for its importance in FoxP3 induction and stabilization, can also act to stabilize ITCH, which blocks TIEG1 suppression on FoxP3 through ubiquitination. Transcriptionally, GRAIL maintains Treg suppressive functions by downregulating inappropriate gene expression, while USP22 and RNF20 act together to regulate FoxP3 transcription through histone H2B ubiquitination

3.11 HIF-1, an Important VHL in FoxP3 Stability

During differentiation, naïve CD4+ T cells acquire specialized functions in response to cytokine signals. While pro-inflammatory Th17 cells and suppressive Tregs are functionally opposite, they share common signals in their differentiation pathways. Both cell lineages require TGF-β for their generation: Tregs require just high levels of TGF-β, while Th17 require both TGF-β and STAT3 activating cytokines (Bettelli et al. 2006). Interestingly, both Tregs and Th17 express FoxP3 in the early stages of their lineages. Since FoxP3 is known to suppress RORγt-driven expression of TH17-associated genes, the timely removal of FoxP3 is necessary for the development of the Th17 lineage.

Hypoxia-inducible factor 1 (HIF-1), although known for its role under hypoxic conditions, plays a key role in this Treg/Th17 balance (Dang et al. 2011). During Th17 differentiation, HIF-1 is stabilized in a STAT3-dependent manner despite the presence of oxygen, suggesting HIF-1's importance in Th17 development (Guan et al. 2017). HIF-1 was shown to directly upregulate RORγt transcription as well as directly interact with RORγt to activate more Th17 signature genes. Correspondingly, deficiency of the alpha subunit of HIF-1 (HIF-1α) in $CD4^+$ T cells stunted *Th17* gene upregulation as well as reciprocal upregulation of FoxP3 protein level. Similarly, elevated levels of HIF-1 have a detrimental effect to FoxP3 expression and Treg stability (Fig. 3.2) (Hsiao et al. 2015). FoxP3 upregulation was not seen at the transcription level, suggesting a posttranslational regulation of FoxP3 by HIF-1 at the Treg/Th17 differentiation point (Dang et al. 2011; Shi et al. 2011). This regulation was dependent on HIF-1 interaction with FoxP3, and its own ability to be ubiquitinated. Interestingly, both oxygen-resistant HIF-1 mutants and mutations along the HIF-1 degradation pathway resulted in FoxP3 protein stability. Furthermore, inhibition of the proteasome resulted in the stability of both proteins, proposing a co-degradation method where FoxP3 is degraded by its association with HIF-1 along the HIF1 degradation pathway (Dang et al. 2011). Although the molecular mechanism that determines Hif-1/FoxP3 co-degradation is still not definitively known, it is clear that the duality of HIF-1 function along the Treg/Th17 lineage crossroads is highly important for proper immune function.

Contradicting the previous study, Hsiao et al. demonstrated that a Treg-restricted loss of Deltex (DTX1), a promoter of HIF-1 degradation, resulted in loss of suppressive function and FoxP3 levels in both in vivo models of airway inflammation and colitis (Hsiao et al. 2015). Importantly, simultaneous deletion of HIF-1 and DTX1 in Tregs largely corrected the deficiency seen in the DTX1 single knockouts (Hsiao et al. 2015). Therefore, under DTX1 deletion, FoxP3 protein loss was more pronounced even though HIF-1 degradation was decreased, challenging the model of co-degradation. However, DTX1 did not appear to drive the direct ubiquitination of HIF-1, therefore there could still be basal levels of HIF-1 degradation which reduce the FoxP3 protein pool. Furthermore, the precise details of DTX1 stabilizing effects on FoxP3 are still unknown, raising the possibility that DTX1 may have unknown consequences to the HIF-1-FoxP3 interaction and co-degradation. Regardless, both support that higher levels of HIF-1 result in FoxP3 degradation. The push of this regulatory pathway from pro- to anti-inflammatory through HIF-1 could enable the regulation of the immune system through the Treg/Th17 lineage balance. Importantly, the duration of hypoxia is crucial for HIF-1 stabilization, possibly halting of the co-degradation of FoxP3 (Facciabene et al. 2011). Since aberrant cellular proliferation within tumor sites results in a significant drop in oxygen availability, studying the importance of this pathway on Treg stability reveals in new avenues for tumor therapies.

3.12 ITCH Deficiency Results in Th2-Like Treg Cells

The HECT-domain-type E3 ligase ITCH, named for the incessant itching displayed by the mice lacking it, promotes the generation and maintenance of Treg cells. Murine ITCH global deletion results in diminished Treg ability to control inflammation, particularly that driven by Th2 T cells. A Treg-specific deletion of ITCH phenocopied the global knockout, and the mice showed signs of spontaneous autoimmunity (Jin et al. 2013). Furthermore, ITCH-deficient Tregs acquired Th2-like phenotypes, including heightened STAT6 activation and GATA-3 expression, suggesting ITCH as an important suppressor for Th2 associated gene expression (Jin et al. 2013).

Furthermore, as ITCH is also involved in the TGF-β pathway through the phosphorylation of Smad2 and the ubiquitination of TGF-β-inducible early gene 1 (*TIEG1*), the conversion into iTregs in the periphery TGF-β (Fig. 3.2) is of particular importance (Bai et al. 2004). Mice lacking either

ITCH or TIEG1 failed to suppress airway inflammation in vivo and hindered Treg differentiation in vitro (Venuprasad et al. 2008). These studies cement the importance of ITCH in Treg phenotypic stability.

3.13 GRAIL Restrains Inappropriate Gene Expression in Tregs

As its name states, the RING finger E3 ligase "gene related to anergy in lymphocytes" (GRAIL) is highly important for the induction of anergy in T cells (Anandasabapathy et al. 2003), maintaining immune tolerance. GRAIL-null T cells are hyperproliferative and less sensitive to immune-mediated suppression. Although GRAIL is naturally upregulated by Tregs, its deletion appears to have no effect on the population size in vivo and in skewing ability in vitro (MacKenzie et al. 2007). Despite having normal FoxP3 levels, GRAIL-null Tregs are less suppressive than their wild-type counterparts and exhibit higher levels of *Th17* gene expression. Particularly, the high level of IL-21 expressed by these cells are attributed to unchecked NFAT signaling due to the absence of GRAIL (Nurieva et al. 2010). Therefore, GRAIL can function to preserve peripheral tolerance by restraining inappropriate gene expression capable of undermining Treg suppressive function.

3.14 DUBs of FoxP3 in Regulating Treg Functions

Since understanding the fundamental regulators of Tregs is critical, scientists have set to identify novel regulators of FoxP3 in the hopes of finding potential therapeutic targets. Observations that posttranslational modifications of FoxP3, including ubiquitination, result in protein destabilization that has attracted a lot of recent attention (Dang et al. 2011; van Loosdregt et al. 2009, 2011). Since direct ubiquitination of Foxp3 can drive its degradation, it stands to reason that counteracting this process should preserve levels of this important regulatory hub of the Treg phenotype. Three novel DUBs of FoxP3 have been identified in the past 7 years, all of which belong to the ubiquitin-specific peptidase (USP) family.

3.15 USP22 Functions as Both a Transcriptional and Posttranslational Regulator of FoxP3

Ubiquitin-specific peptidase 22 (USP22) is a highly conserved member of the USP family, and is known for its importance in transcription through histone deubiquitination as part of the Spt-Ada-Gen5 Acetyl transferase (SAGA) complex (Zhang et al. 2008). In addition, USP22 has been implicated in cancer through its involvement in regulating genes involved in cell cycle progression, apoptosis, and development and has been identified as one of 11 "death from cancer" signature genes (Glinsky et al. 2005; Li et al. 2014; Lin et al. 2012, 2015). The silencing of USP22 in many cancer types results in the cessation of proliferation. These oncolytic functions are not shared by the other members of the USP family implicated in Treg function. Of interest, USP22 was recently flagged in the same pooled Crispr screen as RNF20, but as a positive regulator of FoxP3 (Cortez et al. 2020). Upon USP22 deletion in Treg cells both in vitro and in vivo mouse model ($USP22^{f/f}FoxP3^{YFP\text{-}Cre}$), both FoxP3 mean fluorescent intensity (MFI) and mRNA level decreased, resulting in a loss of Treg suppressive function in vitro. Further analysis by Chip-qPCR, and subsequent confirmation by Chip-seq, revealed USP22 transcriptional control over FoxP3 through deubiquitination of Histone H2B in the conserved non-coding sequence 1 (CNS1) region of the FoxP3 locus. Looking more broadly across the genome, the sites that USP22 functioned were also enriched for activating histone modifications, suggesting that USP22 binds to regions containing gene regulatory elements, including Treg super enhancers. Furthermore, USP22 dually functions to reverse FoxP3 polyubiquitination, preserving Treg suppressive function through both transcriptional and posttranslational mechanisms. Further validation of USP22 requirements on Treg suppressive capabilities was seen in both an in vivo model of colitis and an EAE model of multiple sclerosis, where the USP22-deficient Tregs were unable to maintain immune tolerance.

The functional duality of USP22, being important in both the tumor cell itself and the Tregs within the tumor microenvironment, signal USP22 as an attractive tumor immune therapeutic. Indeed, upon induction of four different tumor types (B16 melanoma, LLC1 Lewis Lung Carcinoma, MC38 colon carcinoma, and EG7 Lymphoma) USP22$^{f/f}$FoxP3$^{YFP-Cre}$ mice demonstrated an increased anti-tumor immune phenotype, resulting in enhanced tumor immune infiltration and decreased tumor burden (Cortez et al. 2020). These results solidify USP22 as a highly attractive potential tumor therapeutic.

3.16 USP21 Prevents Generation of Th1-Like Treg Cells

Another USP family member has been established as a FoxP3 regulator, ubiquitin-specific peptidase 21 (USP21). USP21 was first discovered through a human placenta cDNA library as a novel regulator of cell growth through both as a DUB and as a NEDD8-specific isopeptidase (Gong et al. 2000). More recently, USP21 was implicated in Treg phenotypic stability and FoxP3 expression through the deubiquitination of GATA3 (Zhang et al. 2013). Although GATA3 is the master regulator for Th2 cell differentiation and function, it also plays a major role in nTreg development and function (Wang et al. 2011; Wohlfert et al. 2011). Treg-specific loss of GATA3 results in spontaneous autoimmunity through defective Treg suppressive capabilities (Wang et al. 2011). Therefore, by stabilizing GATA3 levels through its DUB function, USP21 indirectly supports Foxp3 expression and Treg function (Zhang et al. 2013).

Directly, USP21 prevents the generation of T-helper-1-like Treg cells through the deubiquitination of FoxP3 itself (Li et al. 2016). Aged mice with USP21 ablation in Tregs displayed autoimmune symptoms such as lymphocytic infiltration into peripheral organs as well as spontaneous T-cell activation. Particularly, these mice demonstrated excessive Th1 skewing of Treg cells through both heightened Treg expression of IFNγ and lower expression of FoxP3. RNAseq of the skewed Tregs displayed impaired transcriptional ability of Treg signature genes, with a trend toward expression of genes controlling Teff cell fate. Furthermore, about 35% were direct FoxP3 targets, suggesting that USP22 regulates the function of Tregs mainly through FoxP3 (Li et al. 2016). This finding, however, could likely be due to the fact that the aged mice are already highly inflamed, creating an optimal environment for FoxP3 loss. Functionally, USP21-null Tregs had a significantly impaired suppressive capacity in both in vitro on T_{eff} cell proliferation and in vivo EAE model of multiple sclerosis. Furthermore, when transferred into EAE-bearing mice, the USP21-null Tregs quickly began to lose FoxP3, suggesting that USP21 is important for Treg stability as well as function. Although the authors demonstrate USP21 DUB function on FoxP3, all experiments were done in an overexpression system. To further cement USP21 as an important FoxP3 regulator, investigation of the acute loss of USP21 on FoxP3 protein stability on endogenous Tregs should be explored. Furthermore, to state that USP21 loss functions primarily through FoxP3, the consequence of the re-introduction of FoxP3 through retroviral transduction in the USP21-null Tregs would have to be tested.

3.17 USP7 Inhibition Results in Severe Treg Instability

Ubiquitin-specific peptidase 7 (USP7) was the first DUB implicated in Treg-suppressive function. Although not highly expressed in Treg cells, USP7 was found to regulate Treg function through direct interaction and deubiquitination of FoxP3 (Li et al. 2016). In co-ordinance with the other DUBs of FoxP3, the loss of USP7 in Tregs resulted in a loss of suppressive function in vitro and in vivo. Strikingly, both a pan-DUB inhibitor and shRNA against USP7 significantly reduced Treg function in both colitis and tumor models in vivo, suggesting that DUB targeting is an effective strategy to break Treg-enforced tolerance in the cancer setting (Chauhan et al. 2012; Wang et al. 2016). Singularly, USP7 resulted in both a drop in FoxP3 MFI and a decrease in Treg percentage, leading to severe autoimmunity and premature death in the USP7f/fFoxP3YFP-Cre inducible mice. Therefore, many of the experiments were conducted in a tamoxifen inducible model, speaking to the severity of USP7 deletion and its importance in Treg function. It is important to note that therapeutic inhibition of USP7 may result in extremely high levels of autoimmunity in cancer patients upon administration. So, although USP7 is an important regulator of FoxP3 and therefore Treg function, its inhibition may have too strong of adverse effects for it to be an attractive potential therapeutic.

3.18 Dysregulation of Treg Ubiquitin Pathway in Human Diseases

As part of important pathways regulating Treg function, many of these proteins are found to be dysregulated in immune diseases and cancer (Table 3.1). For example, the elevated expressions of the E3 ubiquitin ligases,

Table 3.1 Association of the dysregulated ubiquitin pathways in Treg with inflammatory diseases

Name	Function	Disease
Stub1	E3 ligase	Stub1 expression is increased in CD4+ T cells from systemic lupus erythematosus patients (SLE) (Guo et al. 2016).
		Stub1 expression is upregulated both in the airway epithelium of patients with asthma and in lung tissues of individuals with early stage mild chronic obstructive pulmonary disease (COPD) (Wei et al. 2013)
TRAF6	E3 ligase	Genetic variants at TRAF6 loci are associated with rheumatoid arthritis (RA) risk (BIRAC Consortium et al. 2009).
		Association of multiple TRAF6 SNPs in association with SLE (Namjou et al. 2012)
Cbl-b	E3 ligase	Small yet significant reduction of Cbl-b expression in T lymphocytes from patients with SLE, with an association to the 2126(A/G) gene polymorphism (Doníz-Padilla et al. 2011).
		Cbl-b SNP variant rs9657904 is associated with multiple sclerosis (MS) (Sanna et al. 2010)
		Whole-exome sequencing of a pedigree identified Cbl-b as a possible asthma susceptibility variant (DeWan et al. 2012)
RNF20	E3 ligase	Patient colon samples from both ulcerative colitis and colorectal tumors revealed downregulated levels of RNF20 and H2Bub (Tarcic et al. 2016)
		RNF20 displays tumor suppressor features (Shema et al. 2008) and is suppressed, along with H2Bub, in many cancers (Chernikova et al. 2012; Hahn et al. 2012; Prenzel et al. 2011; Wang et al. 2013)
Hrd1	E3 ligase	Elevated Hrd1 mRNA expression in $CD4^+$ T cells from the MS patients (Xu et al. 2016).
		Hrd1 is highly expressed in the rheumatoid synovium of RA patients (Amano et al. 2003)
HIF	VHL	Hypoxia-driven HIF stabilization within the tumor microenvironment impedes anti-tumor immunity (Facciabene et al. 2011; Horikawa et al. 2017; Ivashkiv 2020; Rivera et al. 2015; Westendorf et al. 2017)
ITCH	E3 ligase	ITCH deficiency causes syndromic multisystem autoimmune disease (Lohr et al. 2010)
GRAIL	E3 ligase	Elevated GRAIL levels seen in remising ulcerative colitis patients (Egawa et al. 2008)
UBC13	E2 ligase	UBC13 levels elevated in $CD4^+$ T cells from SLE patients (Guo et al. 2016).
		Interactive network of genes including UBC13, IL-4, and IL-4R important in childhood asthma with UBC13 expression increased in airway epithelium of asthmatic patients and in the lungs of individuals with COPD (Holtzman et al. 2014)
USP22	DUB	Acts as an oncogene through P53 antagonization and is correlated with poor prognosis in many cancers (Glinsky 2006; Jiang et al. 2018; Li et al. 2014; Lv et al. 2011; Piao et al. 2012; Zhang et al. 2008)
USP21	DUB	Elevated USP21 levels in Tregs from asthmatic patients, resulting in a FoxP3/GATA3 imbalance (Chen et al. 2018)
USP7	DUB	Significant overexpression correlated with disease severity in patients with SLE, particularly through the stabilization of IFNγRI (Yu et al. 2017).
		Overexpressed in various cancers (Nicholson and Kumar 2011)

which consequently facilitates FoxP3 protein degradation, have been detected in T cells from patients with autoimmune inflammatory diseases including systemic lupus erythematosus, rheumatoid arthritis, multiple sclerosis, chronic obstructive pulmonary disease and asthma. Continued characterization of these pathways within the context of disease will shed light on ways to a better understanding of inflammatory disease pathogenesis and more stringently control over Treg function, possibly shedding light on new therapies.

3.19 Conclusion and Perspectives

It is clear that the dynamic nature of FoxP3 expression is pivotal to proper immune regulation. These studies reveal the importance of transcriptional to posttranslational modifications to preserving tolerance and Treg function. The breadth of direct DUBs and E3 ligases of FoxP3 suggest a strong need for conditional regulation of Treg suppression. Enlightening both the direct and the indirect ubiquitin-mediated pathways that downregulate FoxP3 reveal means to alter Treg functional capabilities. Specifically, the targeting of these regulatory molecules may be an effective way to either elicit or break tolerance in autoimmune or anti-tumor immune therapies.

Acknowledgments This study was supported by the NIH grant DK120330 and CA232347 to DF, as well as F31 CA220801-03 to EM.

References

Akimova T, Zhang T, Negorev D, Singhal S, Stadanlick J, Rao A, Annunziata M et al (2017) Human lung tumor FOXP+ Tregs upregulate four 'Treg-locking' transcription factors. JCI Insight 2(16):e94075

Akiyama T, Maeda S, Yamane S, Ogino K, Kasai M, Kajiura F, Matsumoto M et al (2005) Dependence of self-tolerance on TRAF6-directed development of thymic stroma. Science 308(5719):248–251

Amano T, Yamasaki S, Yagishita N, Tsuchimochi K, Shin H, Kawahara K, Aratani S et al (2003) Synoviolin/Hrd1, an E3 ubiquitin ligase, as a novel pathogenic factor for arthropathy. Genes Dev 17 (19):2436–2449

Anandasabapathy N, Ford GS, Bloom D, Holness C, Paragas V, Seroogy C, Skrenta H et al (2003) GRAIL an E3 ubiquitin ligase that inhibits cytokine gene transcription is expressed in anergic CD4+ T cells. Immunity 18(4):535–547

Angelin A, Gil-de-Gómez L, Dahiya S, Jiao J, Guo L, Levine MH, Wang Z et al (2017) Foxp3 reprograms T cell metabolism to function in low-glucose, high-lactate environments. Cell Metab 25(6):1282–1293.e7

Bachmaier K, Krawczyk C, Kozieradzki I, Kong Y-Y, Sasaki T, Oliveira-dos-Santos A, Mariathasan S et al (2000) Negative regulation of lymphocyte activation and autoimmunity by the molecular adaptor Cbl-b. Nature 403(6766):211–216

Bai Y, Yang C, Hu K, Elly C, Liu Y-C (2004) Itch E3 ligase-mediated regulation of TGF-β signaling by modulating Smad2 phosphorylation. Mol Cell 15 (5):825–831

Ballinger CA, Connell P, Wu Y, Hu Z, Thompson LJ, Yin L-Y, Patterson C (1999) Identification of CHIP, a novel tetratricopeptide repeat-containing protein that interacts with heat shock proteins and negatively regulates chaperone functions. Mol Cell Biol 19 (6):4535–4545

Bettelli E, Carrier Y, Gao W, Korn T, Strom TB, Oukka M, Weiner HL et al (2006) Reciprocal developmental pathways for the generation of pathogenic effector TH17 and regulatory T cells. Nature 441 (7090):235–238

BIRAC Consortium, Year Consortium, Raychaudhuri S, Thomson BP, Remmers EF, Eyre S, Hinks A et al (2009) Genetic variants at CD28, PRDM1 and CD2/CD58 are associated with rheumatoid arthritis risk. Nat Genet 41(12):1313–1318

Cejas PJ, Walsh MC, Pearce EL, Han D, Harms GM, Artis D, Turka LA et al (2010) TRAF6 inhibits Th17 differentiation and TGF-beta-mediated suppression of IL-2. Blood 115(23):4750–4757

Chang J-H, Xiao Y, Hu H, Jin J, Yu J, Zhou X, Wu X et al (2012) Ubc13 maintains the suppressive function of regulatory T cells and prevents their conversion into effector-like T cells. Nat Immunol 13(5):481–490

Chauhan D, Tian Z, Nicholson B, Kumar KGS, Zhou B, Carrasco R, McDermott JL et al (2012) A small molecule inhibitor of ubiquitin-specific protease-7 induces apoptosis in multiple myeloma cells and overcomes bortezomib resistance. Cancer Cell 22(3):345–358

Chen Z, Barbi J, Bu S, Yang H-Y, Li Z, Gao Y, Jinasena D et al (2013) The ubiquitin ligase Stub1 negatively modulates regulatory T cell suppressive activity by promoting degradation of the transcription factor Foxp3. Immunity 39(2):272–285

Chen T, Hou X, Ni Y, Du W, Han H, Yu Y, Shi G (2018) The imbalance of FOXP3/GATA3 in regulatory T cells from the peripheral blood of asthmatic patients. J Immunol Res 2018:1–10

Chernikova SB, Razorenova OV, Higgins JP, Sishc BJ, Nicolau M, Dorth JA, Chernikova DA et al (2012) Deficiency in mammalian histone H2B ubiquitin ligase

Bre1 (Rnf20/Rnf40) leads to replication stress and chromosomal instability. Cancer Res 72(8):2111–2119

Chiang YJ, Kole HK, Brown K, Naramura M, Fukuhara S, Hu R-J, Jang IK et al (2000) Cbl-b regulates the CD28 dependence of T-cell activation. Nature 403 (6766):216–220

Chiffoleau E, Kobayashi T, Walsh MC, King CG, Walsh PT, Hancock WW, Choi Y et al (2003) TNF receptor-associated factor 6 deficiency during hemopoiesis induces Th2-polarized inflammatory disease. J Immunol 171(11):5751–5759

Ciechanover A, Elias S, Heller H, Hershko A (1982) 'Covalent affinity' purification of ubiquitin-activating enzyme. J Biol Chem 257(5):2537–2542

Ciehanover A, Hod Y, Hershko A (1978) A heat-stable polypeptide component of an ATP-dependent proteolytic system from reticulocytes. Biochem Biophys Res Commun 81(4):1100–1105

Clark CE, Hingorani SR, Mick R, Combs C, Tuveson DA, Vonderheide RH (2007) Dynamics of the immune reaction to pancreatic cancer from inception to invasion. Cancer Res 67(19):9518–9527

Cortez JT, Montauti E, Shifrut E, Gatchalian J, Zhang Y, Shaked O, Xu Y et al (2020) CRISPR screen in regulatory T cells reveals modulators of Foxp3. Nature 582 (7812):416–420

Cyr DM, Höhfeld J, Patterson C (2002) Protein quality control: U-box-containing E3 ubiquitin ligases join the fold. Trends Biochem Sci 27(7):368–375

Dang EV, Barbi J, Yang H-Y, Jinasena D, Yu H, Zheng Y, Bordman Z et al (2011) Control of TH17/Treg balance by hypoxia-inducible factor 1. Cell 146(5):772–784

De Simone M, Arrigoni A, Rossetti G, Gruarin P, Ranzani V, Politano C, Bonnal RJP et al (2016) Transcriptional landscape of human tissue lymphocytes unveils uniqueness of tumor-infiltrating T regulatory cells. Immunity 45(5):1135–1147

DeWan AT, Egan KB, Hellenbrand K, Sorrentino K, Pizzoferrato N, Walsh KM, Bracken MB (2012) Whole-exome sequencing of a pedigree segregating asthma. BMC Med Genet 13(1):95

Doníz-Padilla L, Martínez-Jiménez V, Niño-Moreno P, Abud-Mendoza C, Hernández-Castro B, González-Amaro R, Layseca-Espinosa E et al (2011) Expression and function of Cbl-b in T cells from patients with systemic lupus erythematosus, and detection of the 2126 A/G Cblb gene polymorphism in the Mexican mestizo population. Lupus 20(6):628–635

Egawa S, Iijima H, Shinzaki S, Nakajima S, Wang J, Kondo J, Ishii S et al (2008) Upregulation of GRAIL is associated with remission of ulcerative colitis. Am J Physiol Gastrointest Liver Physiol 295(1):G163–G169

Facciabene A, Peng X, Hagemann IS, Balint K, Barchetti A, Wang L-P, Gimotty PA et al (2011) Tumour hypoxia promotes tolerance and angiogenesis via CCL28 and Treg cells. Nature 475(7355):226–230

Fontenot JD, Gavin MA, Rudensky AY (2003) Foxp3 programs the development and function of CD4+CD25+ regulatory T cells. Nat Immunol 4 (4):330–336

Fraile JM, Quesada V, Rodríguez D, Freije JMP, López-Otín C (2011) Deubiquitinases in cancer: new functions and therapeutic options. Oncogene 31 (19):2373–2388

Fukushima T, Matsuzawa S-i, Kress CL, Bruey JM, Krajewska M, Lefebvre S, Zapata JM et al (2007) Ubiquitin-conjugating enzyme Ubc13 is a critical component of TNF receptor-associated factor (TRAF)-mediated inflammatory responses. Proc Natl Acad Sci 104(15):6371–6376

Gao Y, Tang J, Chen W, Li Q, Nie J, Lin F, Wu Q et al (2015) Inflammation negatively regulates FOXP3 and regulatory T-cell function via DBC1. Proc Natl Acad Sci 112(25):E3246–E3254

Gerondakis S, Banerjee A, Grigoriadis G, Vasanthakumar A, Gugasyan R, Sidwell T, Grumont RJ (2012) NF-κB subunit specificity in hemopoiesis. Immunol Rev 246(1):272–285

Glinsky GV (2006) Genomic models of metastatic cancer: functional analysis of death-from-cancer signature genes reveals aneuploid, anoikis-resistant, metastasis-enabling phenotype with altered cell cycle control and activated PcG protein chromatin silencing pathway. Cell Cycle 5(11):1208–1216

Glinsky GV, Berezovska O, Glinskii AB (2005) Microarray analysis identifies a death-from-cancer signature predicting therapy failure in patients with multiple types of cancer. J Clin Investig 115(6):1503–1521

Gong L, Kamitani T, Millas S, Yeh ETH (2000) Identification of a novel isopeptidase with dual specificity for ubiquitin- and NEDD8-conjugated proteins. J Biol Chem 275(19):14212–14216

Guan S-Y, Leng R-X, Tao J-H, Li X-P, Ye D-Q, Olsen N, Zheng SG, Pan H-F (2017) Hypoxia-inducible factor-1α: a promising therapeutic target for autoimmune diseases. Expert Opin Ther Targets 21(7):715–723

Guo Y, Zhao M, Lu Q (2016) Transcription factor RFX1 is ubiquitinated by E3 ligase STUB1 in systemic lupus erythematosus. Clin Immunol 169:1–7

Hahn MA, Dickson K-A, Jackson S, Clarkson A, Gill AJ, Marsh DJ (2012) The tumor suppressor CDC73 interacts with the ring finger proteins RNF20 and RNF40 and is required for the maintenance of histone 2B monoubiquitination. Hum Mol Genet 21 (3):559–568

Hanna J, Leggett DS, Finley D (2003) Ubiquitin depletion as a key mediator of toxicity by translational inhibitors. Mol Cell Biol 23(24):9251–9261

Harada Y, Harada Y, Elly C, Ying G, Paik J-H, DePinho RA, Liu Y-C (2010) Transcription factors Foxo3a and Foxo1 couple the E3 ligase Cbl-b to the induction of Foxp3 expression in induced regulatory T cells. J Exp Med 207(7):1381–1391

Holtzman MJ, Byers DE, Alexander-Brett J, Wang X (2014) The role of airway epithelial cells and innate immune cells in chronic respiratory disease. Nat Rev Immunol 14(10):686–698

Horikawa N, Abiko K, Matsumura N, Hamanishi J, Baba T, Yamaguchi K, Yoshioka Y et al (2017) Expression of vascular endothelial growth factor in ovarian cancer inhibits tumor immunity through the accumulation of myeloid-derived suppressor cells. Clin Cancer Res 23(2):587–599

Horwitz DA, Zheng SG, Gray JD (2008) Natural and TGF-β–induced Foxp3+CD4+ CD25+ regulatory T cells are not mirror images of each other. Trends Immunol 29(9):429–435

Hsiao H-W, Hsu T-S, Liu W-H, Hsieh W-C, Chou T-F, Wu Y-J, Jiang S-T et al (2015) Deltex1 antagonizes HIF-1α and sustains the stability of regulatory T cells in vivo. Nat Commun 6(1):6353

Isomura I, Palmer S, Grumont RJ, Bunting K, Hoyne G, Wilkinson N, Banerjee A et al (2009) c-Rel is required for the development of thymic Foxp3+ CD4 regulatory T cells. J Exp Med 206(13):3001–3014

Ivashkiv LB (2020) The hypoxia–lactate axis tempers inflammation. Nat Rev Immunol 20:85–86

Jiang S, Song C, Gu X, Wang M, Miao D, Lv J, Liu Y (2018) Ubiquitin-specific peptidase 22 contributes to colorectal cancer stemness and chemoresistance via Wnt/β-catenin pathway. Cell Physiol Biochem 46 (4):1412–1422

Jin H, Park Y, Elly C, Liu Y-C (2013) Itch expression by Treg cells controls Th2 inflammatory responses. J Clin Investig 123(11):4923–4934

Johnson ES, Ma PCM, Ota IM, Varshavsky A (1995) A proteolytic pathway that recognizes ubiquitin as a degradation signal. J Biol Chem 270(29):17442–17456

Kerdiles YM, Stone EL, Beisner DL, McGargill MA, Ch'en IL, Stockmann C, Katayama CD et al (2010) Foxo transcription factors control regulatory T cell development and function. Immunity 33(6):890–904

Kim HT, Kim KP, Lledias F, Kisselev AF, Scaglione KM, Skowyra D, Gygi SP et al (2007) Certain pairs of ubiquitin-conjugating enzymes (E2s) and ubiquitin-protein ligases (E3s) synthesize nondegradable forked ubiquitin chains containing all possible isopeptide linkages. J Biol Chem 282(24):17375–17386

Kirisako T, Kamei K, Murata S, Kato M, Fukumoto H, Kanie M, Sano S et al (2006) A ubiquitin ligase complex assembles linear polyubiquitin chains. EMBO J 25(20):4877–4887

Kobayashi T, Walsh PT, Walsh MC, Speirs KM, Chiffoleau E, King CG, Hancock WW et al (2003) TRAF6 is a critical factor for dendritic cell maturation and development. Immunity 19(3):353–363

Kong N, Lan Q, Chen M, Zheng T, Su W, Wang J, Yang Z et al (2012) Induced T regulatory cells suppress osteoclastogenesis and bone erosion in collagen-induced arthritis better than natural T regulatory cells. Ann Rheum Dis 71(9):1567–1572

Li L, Osdal T, Ho Y, Chun S, McDonald T, Agarwal P, Lin A et al (2014) SIRT1 activation by a c-MYC oncogenic network promotes the maintenance and drug resistance of human FLT3-ITD acute myeloid leukemia stem cells. Cell Stem Cell 15(4):431–446

Li Y, Lu Y, Wang S, Han Z, Zhu F, Ni Y, Liang R et al (2016) USP21 prevents the generation of T-helper-1-like Treg cells. Nat Commun 7(1):13559

Lin Z, Yang H, Kong Q, Li J, Lee S-M, Gao B, Dong H et al (2012) USP22 antagonizes p53 transcriptional activation by deubiquitinating Sirt1 to suppress cell apoptosis and is required for mouse embryonic development. Mol Cell 46(4):484–494

Lin Z, Tan C, Qiu Q, Kong S, Yang H, Zhao F, Liu Z et al (2015) Ubiquitin-specific protease 22 is a deubiquitinase of CCNB1. Cell Discov 1(1):15028

Liston A, Lu L-F, O'Carroll D, Tarakhovsky A, Rudensky AY (2008) Dicer-dependent microRNA pathway safeguards regulatory T cell function. J Exp Med 205 (9):1993–2004

Lohr NJ, Molleston JP, Strauss KA, Torres-Martinez W, Sherman EA, Squires RH, Rider NL et al (2010) Human ITCH E3 ubiquitin ligase deficiency causes syndromic multisystem autoimmune disease. Am J Hum Genet 86(3):447–453

Lu L, Lan Q, Li Z, Zhou X, Gu J, Li Q, Wang J et al (2014) Critical role of all-trans retinoic acid in stabilizing human natural regulatory T cells under inflammatory conditions. Proc Natl Acad Sci 111(33):E3432–E3440

Lv L, Xiao X, Gu Z, Zeng F, Huang L, Jiang G (2011) Silencing USP22 by asymmetric structure of interfering RNA inhibits proliferation and induces cell cycle arrest in bladder cancer cells. Mol Cell Biochem 346(1–2):11–21

MacKenzie DA, Schartner J, Lin J, Timmel A, Jennens-Clough M, Fathman CG, Seroogy CM (2007) GRAIL is up-regulated in CD4 + CD25 + T regulatory cells and is sufficient for conversion of T cells to a regulatory phenotype. J Biol Chem 282(13):9696–9702

Mahajan D, Wang Y, Qin X, Wang Y, Zheng G, Wang YM, Alexander SI et al (2006) CD4+CD25+ regulatory T cells protect against injury in an innate murine model of chronic kidney disease. J Am Soc Nephrol 17 (10):2731–2741

Maloy KJ, Salaun L, Cahill R, Dougan G, Saunders NJ, Powrie F (2003) CD4+CD25+ TR cells suppress innate immune pathology through cytokine-dependent mechanisms. J Exp Med 197(1):111–119

Mao F, Kong H, Zhao Y, Peng L, Chen W, Zhang J, Cheng P et al (2017) Increased tumor-infiltrating CD45RA−CCR7− regulatory T-cell subset with immunosuppressive properties foster gastric cancer progress. Cell Death Dis 8(8):e3002–e3002

Martinon F, Chen X, Lee A-H, Glimcher LH (2010) TLR activation of the transcription factor XBP1 regulates innate immune responses in macrophages. Nat Immunol 11(5):411–418

Muto G, Kotani H, Kondo T, Morita R, Tsuruta S, Kobayashi T, Luche H et al (2013) TRAF6 is essential for maintenance of regulatory T cells that suppress Th2 type autoimmunity. PLoS One 8(9):e74639

Namjou B, Choi C-B, Harley ITW, Alarcón-Riquelme ME, Network B, Kelly JA, Glenn SB et al (2012) Evaluation of TRAF6 in a large multiancestral lupus cohort. Arthritis Rheum 64(6):1960–1969

Ni X, Kou W, Gu J, Wei P, Wu X, Peng H, Tao J et al (2019) TRAF 6 directs FOXP 3 localization and

facilitates regulatory T-cell function through K63-linked ubiquitination. EMBO J 38(9):e99766, available at: https://doi.org/10.15252/embj.201899766
Nicholson B, Kumar KGS (2011) The multifaceted roles of USP7: new therapeutic opportunities. Cell Biochem Biophys 60(1–2):61
Nurieva RI, Zheng S, Jin W, Chung Y, Zhang Y, Martinez GJ, Reynolds JM et al (2010) The E3 ubiquitin ligase GRAIL regulates T cell tolerance and regulatory T cell function by mediating T cell receptor-CD3 degradation. Immunity 32(5):670–680
Peng J, Schwartz D, Elias JE, Thoreen CC, Cheng D, Marsischky G, Roelofs J et al (2003) A proteomics approach to understanding protein ubiquitination. Nat Biotechnol 21(8):921–926
Piao S, Liu Y, Hu J, Guo F, Ma J, Sun Y, Zhang B (2012) USP22 is useful as a novel molecular marker for predicting disease progression and patient prognosis of oral squamous cell carcinoma. PLoS One 7(8): e42540
Plitas G, Konopacki C, Wu K, Bos PD, Morrow M, Putintseva EV, Chudakov DM et al (2016) Regulatory T cells exhibit distinct features in human breast cancer. Immunity 45(5):1122–1134
Prenzel T, Begus-Nahrmann Y, Kramer F, Hennion M, Hsu C, Gorsler T, Hintermair C et al (2011) Estrogen-dependent gene transcription in human breast cancer cells relies upon proteasome-dependent monoubiquitination of histone H2B. Cancer Res 71 (17):5739–5753
Rech AJ, Mick R, Martin S, Recio A, Aqui NA, Powell DJ, Colligon TA et al (2012) CD25 blockade depletes and selectively reprograms regulatory T cells in concert with immunotherapy in cancer patients. Sci Transl Med 4(134):134ra62
Rivera LB, Meyronet D, Hervieu V, Frederick MJ, Bergsland E, Bergers G (2015) Intratumoral myeloid cells regulate responsiveness and resistance to antiangiogenic therapy. Cell Rep 11(4):577–591
Robb R, Munck A, Smith K (1981) T cell growth factor receptors. Quantitation, specificity, and biological relevance. J Exp Med 154(5):1455–1474
Sakaguchi S, Sakaguchi N, Asano M, Itoh M, Toda M (1995) Immunologic self-tolerance maintained by activated T cells expressing IL-2 receptor alpha-chains (CD25). Breakdown of a single mechanism of self-tolerance causes various autoimmune diseases. J Immunol 155(3):1151–1164
Sanna S, Pitzalis M, Zoledziewska M, Zara I, Sidore C, Murru R, Whalen MB et al (2010) Variants within the immunoregulatory CBLB gene are associated with multiple sclerosis. Nat Genet 42(6):495–497
Sawant DV, Yano H, Chikina M, Zhang Q, Liao M, Liu C, Callahan DJ et al (2019) Adaptive plasticity of IL-10+ and IL-35+ Treg cells cooperatively promotes tumor T cell exhaustion. Nat Immunol 20(6):724–735
Schaer DA, Budhu S, Liu C, Bryson C, Malandro N, Cohen A, Zhong H et al (2013) GITR pathway activation abrogates tumor immune suppression through loss of regulatory T-cell lineage stability. Cancer Immunol Res 1(5):320–331
Schmidt-Supprian M, Tian J, Grant EP, Pasparakis M, Maehr R, Ovaa H, Ploegh HL et al (2004) Differential dependence of CD4+CD25+ regulatory and natural killer-like T cells on signals leading to NF-κB activation. Proc Natl Acad Sci U S A 101(13):4566–4571
Shema E, Tirosh I, Aylon Y, Huang J, Ye C, Moskovits N, Raver-Shapira N et al (2008) The histone H2B-specific ubiquitin ligase RNF20/hBRE1 acts as a putative tumor suppressor through selective regulation of gene expression. Genes Dev 22(19):2664–2676
Shi LZ, Wang R, Huang G, Vogel P, Neale G, Green DR, Chi H (2011) HIF1α–dependent glycolytic pathway orchestrates a metabolic checkpoint for the differentiation of TH17 and Treg cells. J Exp Med 208 (7):1367–1376
Shimo Y, Yanai H, Ohshima D, Qin J, Motegi H, Maruyama Y, Hori S et al (2011) TRAF6 directs commitment to regulatory T cells in thymocytes. Genes Cells 16(4):437–447
Singh B, Read S, Asseman C, Malmström V, Mottet C, Stephens LA, Stepankova R et al (2001) Control of intestinal inflammation by regulatory T cells. Immunol Rev 182(1):190–200
Suresh B, Lee J, Kim H, Ramakrishna S (2016) Regulation of pluripotency and differentiation by deubiquitinating enzymes. Cell Death Differ 23(8):1257–1264
Tai X, Cowan M, Feigenbaum L, Singer A (2005) CD28 costimulation of developing thymocytes induces Foxp3 expression and regulatory T cell differentiation independently of interleukin 2. Nat Immunol 6 (2):152–162
Takahashi R, Nishimoto S, Muto G, Sekiya T, Tamiya T, Kimura A, Morita R et al (2011) SOCS1 is essential for regulatory T cell functions by preventing loss of Foxp3 expression as well as IFN-γ and IL-17A production. J Exp Med 208(10):2055–2067
Tang Q, Henriksen KJ, Boden EK, Tooley AJ, Ye J, Subudhi SK, Zheng XX et al (2003) Cutting edge: CD28 controls peripheral homeostasis of CD4 + CD25 + regulatory T cells. J Immunol 171 (7):3348–3352
Tarcic O, Pateras IS, Cooks T, Shema E, Kanterman J, Ashkenazi H, Boocholez H et al (2016) RNF20 links histone H2B ubiquitylation with inflammation and inflammation-associated cancer. Cell Rep 14 (6):1462–1476
Tenno T, Fujiwara K, Tochio H, Iwai K, Morita EH, Hayashi H, Murata S et al (2004) Structural basis for distinct roles of Lys63- and Lys48-linked polyubiquitin chains. Genes Cells 9(10):865–875
van Loosdregt J, Vercoulen Y, Guichelaar T, Gent YYJ, Beekman JM, van Beekum O, Brenkman AB et al (2009) Regulation of Treg functionality by acetylation-mediated Foxp3 protein stabilization. Blood 115(5):965–974
van Loosdregt J, Brunen D, Fleskens V, Pals CEGM, Lam EWF, Coffer PJ (2011) Rapid temporal control of

Foxp3 protein degradation by sirtuin-1. PLoS One 6 (4):e19047
van Loosdregt J, Fleskens V, Fu J, Brenkman AB, Bekker CPJ, Pals CEGM, Meerding J et al (2013) Stabilization of the transcription factor Foxp3 by the deubiquitinase USP7 increases Treg-cell-suppressive capacity. Immunity 39(2):259–271
Venuprasad K, Huang H, Harada Y, Elly C, Subramaniam M, Spelsberg T, Su J et al (2008) The E3 ubiquitin ligase Itch regulates expression of transcription factor Foxp3 and airway inflammation by enhancing the function of transcription factor TIEG1. Nat Immunol 9(3):245–253
Vignali DAA, Collison LW, Workman CJ (2008) How regulatory T cells work. Nat Rev Immunol 8(7):523–532
Wang Y, Su MA, Wan YY (2011) An essential role of the transcription factor GATA-3 for the function of regulatory T cells. Immunity 35(3):337–348
Wang G, Gao Y, Li L, Jin G, Cai Z, Chao J-I, Lin H-K (2012) K63-linked ubiquitination in kinase activation and cancer. Front Oncol 2:5
Wang Z-J, Yang J-L, Wang Y-P, Lou J-Y, Chen J, Liu C, Guo L-D (2013) Decreased histone H2B monoubiquitination in malignant gastric carcinoma. World J Gastroenterol 19(44):8099–8107
Wang L, Kumar S, Dahiya S, Wang F, Wu J, Newick K, Han R et al (2016) Ubiquitin-specific protease-7 inhibition impairs Tip60-dependent Foxp3+ T-regulatory cell function and promotes antitumor immunity. EBioMedicine 13:99–112
Wei Q, Sha Y, Bhattacharya A, Fattah EA, Bonilla D, Jyothula SSSK, Pandit L et al (2013) Regulation of IL-4 receptor signaling by STUB1 in lung inflammation. Am J Respir Crit Care Med 189(1):16–29
Westendorf AM, Skibbe K, Adamczyk A, Buer J, Geffers R, Hansen W, Pastille E et al (2017) Hypoxia enhances immunosuppression by inhibiting CD4+ effector T cell function and promoting Treg activity. Cell Physiol Biochem 41(4):1271–1284
Wilkinson KD (1997) Regulation of ubiquitin-dependent processes by deubiquitinating enzymes. FASEB J 11 (14):1245–1256
Wohlfert EA, Grainger JR, Bouladoux N, Konkel JE, Oldenhove G, Ribeiro CH, Hall JA et al (2011) GATA3 controls Foxp3+ regulatory T cell fate during inflammation in mice. J Clin Investig 121 (11):4503–4515
Woo EY, Chu CS, Goletz TJ, Schlienger K, Yeh H, Coukos G, Rubin SC et al (2001) Regulatory CD4(+) CD25(+) T cells in tumors from patients with early-stage non-small cell lung cancer and late-stage ovarian cancer. Cancer Res 61(12):4766–4772
Xu P, Duong DM, Seyfried NT, Cheng D, Xie Y, Robert J, Rush J et al (2009) Quantitative proteomics reveals the function of unconventional ubiquitin chains in proteasomal degradation. Cell 137(1):133–145
Xu Y, Zhao F, Qiu Q, Chen K, Wei J, Kong Q, Gao B et al (2016) The ER membrane-anchored ubiquitin ligase Hrd1 is a positive regulator of T-cell immunity. Nat Commun 7(1):12073
Xu Y, Melo-Cardenas J, Zhang Y, Gau I, Wei J, Montauti E, Zhang Y et al (2019) The E3 ligase Hrd1 stabilizes Tregs by antagonizing inflammatory cytokine–induced ER stress response. JCI Insight 4 (5):e121887
Yu Y, Su Z, Wang Z, Xu H (2017) USP7 is associated with greater disease activity in systemic lupus erythematosus via stabilization of the IFNα receptor. Mol Med Rep 16(2):2274–2280
Zhang X-Y, Varthi M, Sykes SM, Phillips C, Warzecha C, Zhu W, Wyce A et al (2008) The putative cancer stem cell marker USP22 is a subunit of the human SAGA complex required for activated transcription and cell-cycle progression. Mol Cell 29(1):102–111
Zhang J, Chen C, Hou X, Gao Y, Lin F, Yang J, Gao Z et al (2013) Identification of the E3 deubiquitinase ubiquitin-specific peptidase 21 (USP21) as a positive regulator of the transcription factor GATA3. J Biol Chem 288(13):9373–9382
Zhang Y, Chen Z, Luo X, Wu B, Li B, Wang B (2016) Cimetidine down-regulates stability of Foxp3 protein via Stub1 in Treg cells. Hum Vaccin Immunother 12 (10):2512–2518
Zhao Y, Guo H, Qiao G, Zucker M, Langdon WY, Zhang J (2015) E3 ubiquitin ligase Cbl-b regulates thymic-derived CD4+CD25+ regulatory T cell development by targeting Foxp3 for ubiquitination. J Immunol 194 (4):1639–1645
Zheng SG, Gray JD, Ohtsuka K, Yamagiwa S, Horwitz DA (2002) Generation ex vivo of TGF-β-producing regulatory T cells from CD4+CD25− precursors. J Immunol 169(8):4183–4189
Zheng SG, Wang JH, Gray JD, Soucier H, Horwitz DA (2004) Natural and induced CD4+CD25+ cells educate CD4+CD25− cells to develop suppressive activity: the role of IL-2, TGF-β, and IL-10. J Immunol 172 (9):5213–5221
Zheng SG, Wang JH, Stohl W, Kim KS, Gray JD, Horwitz DA (2006) TGF-β requires CTLA-4 early after T cell activation to induce FoxP3 and generate adaptive CD4+CD25+ regulatory cells. J Immunol 176 (6):3321–3329
Zheng SG, Wang J, Horwitz DA (2008) Cutting edge: Foxp3+CD4+CD25+ regulatory T cells induced by IL-2 and TGF-β are resistant to Th17 conversion by IL-6. J Immunol 180(11):7112–7116
Zhou X, Jeker LT, Fife BT, Zhu S, Anderson MS, McManus MT, Bluestone JA (2008) Selective miRNA disruption in T reg cells leads to uncontrolled autoimmunity. J Exp Med 205(9):1983–1991
Zhou X, Kong N, Wang J, Fan H, Zou H, Horwitz D, Brand D et al (2010) Cutting edge: all-trans retinoic acid sustains the stability and function of natural regulatory T cells in an inflammatory milieu. J Immunol 185(5):2675–2679

Ubiquitin-Dependent Regulation of Treg Function and Plasticity

4

Yi Dong and Fan Pan

Abstracts

As an indispensable part of peripheral tolerance, regulatory T (Treg) cells play an important role in immune homeostasis by suppressing other immune cells. Behind this function is a complex network of transcription factors and signaling cascades that regulates the function and plasticity of regulatory T cells. Among these, Forkhead box P3 (Foxp3) is considered as the master transcription factor, and its stability will influence the function and viability of Treg cells. Because of this, understanding the mechanisms that regulate Foxp3 and its co-regulators will provide more understanding to Treg cells and uncover more targets to manipulate Treg cells in treating autoimmune diseases, organ transplantation, and tumor. Interestingly, several recent studies show that ubiquitin-dependent pathways are important regulators of Foxp3, which suggest both great scientific and therapeutic values. In this chapter, we cover emerging evidence of ubiquitin-dependent, posttranslational regulation of Treg function and plasticity.

Keywords

Treg · Ubiquitin · Posttranslational regulation · Foxp3

4.1 Introduction

Through evolution, the immune system has been equipped with some highly efficient killing machines to defend against both external and internal threats. However, this raises an issue that the immune system may be overpowered and misdirected to target the self. Fortunately, the immune homeostasis is secured by immune tolerance. Among different mechanisms of immune tolerance, regulatory T (Treg) cells play an important role in mediating the intensity of the immune response and suppressing self-reactive leukocytes (Sakaguchi et al. 2008).

While several subsets of T cells display suppressive functions, the most characterized are those $CD4^+$ cells that constitutively express CD25 and the transcription factor Forkhead box protein 3 (Foxp3). These cells utilize multiple mechanisms to suppress immune activation. For example, they produce anti-inflammatory cytokines like IL-10, TGF-β, and IL-35 to inhibit effector T cells; they express coinhibitory molecules like CTLA-4 and LAG3 to inhibit DC

Y. Dong
Department of Cell Biology, Johns Hopkins University School of Medicine, Baltimore, MD, USA

F. Pan (✉)
Center for Cancer Immunology Research, Shenzhen Institutes of Advanced Technology, Chinese Academy of Sciences, Shenzhen, Guangdong, China
e-mail: fan.pan@siat.ac.cn

S.-G. Zheng (ed.), *T Regulatory Cells in Human Health and Diseases*, Advances in Experimental Medicine and Biology 1278, https://doi.org/10.1007/978-981-15-6407-9_4

maturation and function; they disrupt the survival of other leukocytes by scavenging growth factors like IL-2 (Vignali et al. 2008). Losing or disrupting the function and the number of these Treg cells will induce severe autoimmune diseases (Sakaguchi et al. 2008).

After decades of studies, we now understand that Foxp3+ Treg is composed of a heterogeneric pool of cells from different origins (Hori 2011). Not surprisingly, they appear with different proliferation, stability, and suppression potentials (Miyara et al. 2009). These properties are regulated not only by the stability of the master transcription factor Foxp3 (Williams and Rudensky 2007; Lu et al. 2014) but also by a network of coregulators with certain redundant functions (Pan et al. 2009; Fu et al. 2012; Zheng et al. 2009). As a result, altering the stability of Foxp3 and the interactions with these coregulators can modulate the immune response at different contexts (Darce et al. 2012; Bettini et al. 2012).

So far, multiple layers of regulation have been identified, including both transcriptional (Luo and Li 2013) and translational controls (Bjur et al. 2013). Within these regulations, ubiquitin-dependent modification is of high interests because of its unique properties. First, proteins can be modified for different functions. They can be either poly-ubiquitinated for proteasome-dependent degradation or monoubiquitinated for signaling cascades (Gao and Karin 2005). Second, these modifications involve highly specific E3 ligases, which provide new insights into protein functions and potentials targets for new drugs (Bielskiene et al. 2015).

Since both Treg cells and ubiquitination are very broad topics, in this chapter, we focus on a number of ubiquitin-dependent pathways that regulate Treg function and plasticity. In terms of Treg function, we summarize recent discoveries on factors that mediate Foxp3 stability directly or indirectly. In terms of Treg plasticity, we emphasize the factors that control Th17–iTreg balance.

4.2 A Brief History About Regulatory T Cells and Foxp3

Like other great discoveries in science, the discovery of regulatory T cells is fairly tortuous. Back in 1969, Nishizuka and Sakakura showed that neonatal thymectomy of normal mice at about 3 days (3d NTx) would induce ovarian dysgenesis, which suggests that some T cells may be responsible for this autoimmunity (Nishizuka and Sakakura 1969). In addition, in 1970, Gershon and Kondo also suggested that a subtype of T cells named suppressor T cells may be able to suppress immune response (Gershon and Kondo 1970). However, it suffered from a lot of skepticisms during the 1980s, mostly because of three reasons: failed to find suppression related I-J region in the *MHC* gene (Kronenberg et al. 1983), unbale to identify reliably markers for cell sorting and cloning (Bloom et al. 1992), and unable to explain the molecular basis of cell suppression (Kappler et al. 1987).

A dramatic turn happened in 1995 when Sakaguchi identified CD25 (IL-2 receptor α chain) as a reliable marker for CD4+ suppressive T cells in normal naïve mice (Sakaguchi et al. 1995). Not only did they found that removal of these CD4+CD25+ T cells induce severe autoimmune disease in mice, but also they confirmed that these T cells are developed in thymus and constitutes 5–10% of peripheral CD4+ T cells in adults (Asano et al. 1996). Further research from Sakaguchi and other teams established the in vitro functional assays for natural Treg (nTreg) cells (Takahashi et al. 1998; Thornton and Shevach 1998) and characterized the importance of IL-2 as the key cytokine for nTreg maintenance (Willerford et al. 1995; Suzuki et al. 1999; Almeida et al. 2002; Malek et al. 2002; Setoguchi et al. 2005).

Another milestone in regulatory T cells happened in 2001, when the transcription factor Forkhead box P3 (Foxp3) was identified as the disease-causative gene in *Scurfy* mice (Brunkow et al. 2001; Wildin et al. 2001). Mutation in

Foxp3 gene leads to severe autoimmune diseases not only in mice but also in human, which is called IPEX syndrome (immune dysregulation, polyendocrinopathy, enteropathy, and X-linked syndrome) (Wildin et al. 2001; Bennett et al. 2001). Soon, in 2003, Sakaguchi, Rudensky, and Ramsdell confirmed that Foxp3 is the unique transcription factor that programs the development and function of regulatory T cells (Hori et al. 2003; Fontenot et al. 2003; Khattri et al. 2003). Later, genetically engineered Foxp3 reporter mice were made by Rudensky and his colleges, which makes it convenient to study the development and the signaling in Treg cells (Fontenot et al. 2005; Kim et al. 2007).

In the next decade, our understanding of Treg cells expanded a lot. For example, naïve CD4+ T cells can be differentiated into Foxp3+ Treg cells under TGF-β environment (Chen et al. 2003; Zheng et al. 2002, 2004) but would be skewed into IL-17 secreting Th17 lineage under the influence of IL-6 (Veldhoen et al. 2006; Bettelli et al. 2006). In terms of development, the Foxp3 landscape is locked in and maintained by multiple transcription factors such as NFAT and Eos (Pan et al. 2009; Fu et al. 2012; Zheng et al. 2007a; Wu et al. 2006). In terms of expansion, Foxp3+ Treg cells can undergo expansion with the help of certain mature DC in a CD80/CD86-dependent manner (Yamazaki et al. 2003). Besides these, our knowledge on Treg cells is still expanding, especially in the field of induced Treg (iTreg) cells (Horwitz et al. 2008).

4.3 Natural Regulatory T (nTreg) and Induced Regulatory T (iTreg) Cells

After Foxp3 is ultimately identified as the master transcription factor of Treg cells (Zheng and Rudensky 2007), more insights come into Treg classification. As the name suggests, natural Treg (nTreg) or thymic Treg (tTreg) cells are those developed in the thymus and induced Treg (iTreg) cells are those differentiated from peripheral naïve conventional T (Tconv) cells (Curotto de Lafaille and Lafaille 2009). Since there are multiple subsets of iTreg cells, we cover only those Foxp3 expressing cells induced from CD4+CD25−Foxp3− cells. In this section, we review and compare the difference between nTreg and iTreg cells in terms of cell development, signaling, TCR specificity, and stability.

Unlike nTreg which is developed from CD4+CD8− single positive T cells in the thymus, iTreg comes from peripheral or lymphoid CD4+ Tconv cells in the presence of TGF-β and IL-2 (Chen et al. 2003; Zheng et al. 2002, 2007b; Davidson et al. 2007). Besides, other requirements should be met for a functionally suppressive iTreg to develop in vivo, including strong antigen presentation without optimal co-stimulation, IL-35, and retinoic acid (Thorstenson and Khoruts 2001; Apostolou and von Boehmer 2004; Cobbold et al. 2004; Mucida et al. 2005; Zheng et al. 2006; Lu et al. 2011). In addition, other reports also show that commensal microbiota also contributes to iTreg development (Round and Mazmanian 2010; Atarashi et al. 2011; Geuking et al. 2011). Because of these characteristics, iTreg typically appears or resides in either tolerogenic or chronic inflammatory environments.

As of signaling, there are some difference between nTreg and iTreg cells. Although both nTreg and iTreg cells need IL-2 to maintain cell survival and suppressive function (Hori 2010; Roncador et al. 2005), TGF-β is critical for iTreg generation (Curotto de Lafaille and Lafaille 2009; Ouyang et al. 2010). Studies have shown that TGF-β activates Smad3, which binds to CNS1 (Conserved Non-coding Sequence 1) enhancer of the *Foxp3* gene (Schlenner et al. 2012; Josefowicz et al. 2012). Additionally, activated PI3K-AKT-mTOR signaling dampens iTreg generation and destabilize Foxp3 expression (Haxhinasto et al. 2008; Sauer et al. 2008; Zhang et al. 2013a).

In terms of TCR specificity, nTreg and iTreg cells tends to have different tendencies. Since nTreg cells are developed during negative selection in the thymus (Jordan et al. 2001), their T-cell receptor tends to recognize more self-antigens (Relland et al. 2009). On the contrary, since iTreg cells are induced from conventional T cells, they tend to have higher affinity toward foreign antigens (Relland et al. 2012). Although the magnitude of their intersection is still under debate, based on CDR3 (complementarity determining region 3) studies, they tend to have different repertories (Relland et al. 2012; Wong et al. 2007; Hsieh et al. 2006).

There are some debates in cell stability. In general, Foxp3 is believed to be critical in order to maintain Treg identity (Li and Zheng 2015). However, although nTreg cells have relatively stable Foxp3 expression, they are not stable under inflammatory condition and can be converted into Th17 lineage in the presence of IL-6 (Zheng et al. 2008; Kong et al. 2012a). On the contrary, iTreg cells have relatively less stable Foxp3 expression partially due to lack of demethylation at CNS2 region, aka TSDR region (Treg-specific demethylated region) (Zheng et al. 2010). Yet, they are relatively stable under inflammatory condition and is able to maintain their suppression capability (Kong et al. 2012b). Recent studies have also demonstrated that both nTreg and iTreg have a different biological feature under high salt diet (Hernandez et al. 2015; Luo et al. 2019).

4.4 Mechanisms of Ubiquitin-Dependent Regulation

Ubiquitin is an 8.6-kDa small molecule protein that is discovered in 1975 (Goldstein et al. 1975). It involves in an ATP-dependent posttranslational modification process in all eukaryotic cells called ubiquitination (Ciehanover et al. 1978). In this process, one ubiquitin protein is attached to a substrate protein at a time with the help of three types of adapter enzymes, namely E1 (ubiquitin-activating enzyme), E2 (ubiquitin-conjugating enzyme), and E3 (ubiquitin ligase) (Pickart and Eddins 2004). In human, there are two E1s, 35 E2s, and more than 600 of E3s (Li et al. 2008). Despite the large abundance in the genome, E3 ligase can be categorized into four subgroups: HECT, RING, U-Box, and PHD finger (Nakayama and Nakayama 2006).

The detailed mechanism is described below. To start with, ubiquitin is activated in a two-step reaction by an E1. First, E1 binds to both ubiquitin and ATP and modifies the C-terminus of the ubiquitin. Second, the acyl-adenylated ubiquitin is transferred to the active site that includes a cysteine residue and release the AMP. This step will form a thioester bond between ubiquitin and E1 (Pickart 2001). Then, E2 binds to both activated ubiquitin and the E1 and conjugates the ubiquitin via a trans-thio-esterification reaction (van Wijk and Timmers 2010). Finally, E3 recognizes the target protein and catalyzes the ligation process by creating an iso-peptide bond between the C-terminal glycine residue of ubiquitin and the lysine residue of the target protein (Komander and Rape 2012).

The substrate protein can be either monoubiquitinated or polyubiquitinated (Behrends and Harper 2011). As the name suggests, monoubiquitination involves one ubiquitin molecule to one substrate protein residue. This typically involves cell signaling and membrane trafficking (Ikeda and Dikic 2008; Hicke 2001). Similarly, polyubiquitination involves multiple ubiquitin attached subsequently to the substrate protein (Komander 2009). All seven residues and the N-terminus have been reported as sights for polyubiquitination, including K6, K11, K27, K29, K33, K48, K63, and M1 (Komander and Rape 2012). Among these, K48 and K63 polyubiquitination are well characterized. K48 polyubiquitination involves at least four ubiquitin molecules at a time to the target protein. This leads to target protein recognition and degradation by the 26S proteasome (Lecker et al. 2006). K63 polyubiquitination is not associated with proteasomal degradation, but it involves different cell signaling events (Miranda and Sorkin 2007).

The opposite side of ubiquitination is deubiquitination. This relies on deubiquitinating enzymes (DUBs) to cleave both the isopeptide bonds between ubiquitin and the target and the peptide bonds between ubiquitin molecules (Reyes-Turcu et al. 2009). These cysteine proteases are important for both ubiquitin recycling and signaling (Nijman et al. 2005).

4.5 Ubiquitin-Dependent Regulation of Treg Function

4.5.1 Itch

Itch is a HECT domain-type E3 ligase that plays an important role in generation and maintenance of Treg cells. It is named after the "Itchy" phenotype observed from Itch-deficient mice. These mice suffer from severe systemic and progressive autoimmune disease especially in skin. In addition, these mice are prone to T-cell hyperproliferation and Th2-biased inflammation (Venuprasad et al. 2006).

The mechanism of this Itch-mediated T-cell response has been studied in two parts. First, Itch has been linked to the phosphorylation and activation of Smad2/3, which is downstream of TGF-β/TGF-β receptor (Bai et al. 2004). Second, Itch is involved in monoubiquitination of a transcription factor known as TIEG1 (TGF-β Induced Early Gene 1) (Venuprasad et al. 2008). This monoubiquitination leads to TIEG1 translocation from cytoplasm to the nucleus. Subsequently, TIEG1 binds to the promoter region of Foxp3, which stabilizes the expression of Foxp3 (Fig. 4.1). Reflecting this, ITCH−/− and TIEG1−/− mice have poor iTreg function and are unable to suppress airway inflammation (Jin et al. 2013).

4.5.2 Ubc13

Ubc13 is a K63-type E2 ubiquitin-conjugating enzyme that regulates NFκB and TNFα signaling. In Treg cells, Ubc13 plays an important role in suppressing T-effector genes from other T-cell lineages (Fukushima et al. 2007). For example, mice with Ubc13 deficiency suffer from systemic inflammation with functionally defective Treg cells that express pro-inflammatory cytokines like IFNγ and IL-17 instead of IL-10 (Chang et al. 2012). One of the known mechanisms is that Ubc13 targets IKK-β for deactivation in Treg cells. IKK-β is a subunit of IKK (IκB kinase) that phosphorylates IκB (inhibitor of NFκB) for degradation (Fig. 4.1). In this way, NFκB is activated to upregulate pro-inflammatory cytokine expression and downregulates SOCS1 (Chang et al. 2012; Takahashi et al. 2011).

4.5.3 Stub1

Stub1 (STIP1 homology and U-box containing protein 1) aka CHIP (C terminus of HSC70-Interacting Protein) is a U-box domain-type E3 ubiquitin ligase that interacts with a chaperone molecule named Hsp70 (Chen et al. 2013). Early research shows that Stub1 enhances Hsp70 induction and promotes ubiquitination upon acute stress (Ballinger et al. 1999). In the context of Treg cells, it has been reported that Stub1 mediates Foxp3 turnover, especially when exposed to stress signals like LPS and proinflammatory cytokines (Chen et al. 2013).

Recently, our group and collaborators have identified further mechanisms behind Stub1-mediated Treg function. At homeostasis, Foxp3 protein is subject to constant turnover (Chen et al. 2013; Fleskens et al. 2019). One of the mechanisms is through Stub1-mediated K48-type polyubiquitination (Chen et al. 2013). Upon stress, Stub1 is upregulated in Treg cells which reduces both the quantity of total Foxp3 and the suppression function of Treg cells (Chen et al. 2013). In addition, other groups indicate that Stub1 ubiquitinates the CARMA1 in a non-degradative manner (Wang et al. 2013) (Fig. 4.1). As CARMA1 is important in NFκB signaling that initiates Foxp3 transcription, it suggests another role of Stub1 in the development of nTreg in thymus (Wang et al. 2013).

4.6 Ubiquitin-Dependent Regulation of Treg Plasticity

4.6.1 Cbl-b

Cbl-b (Casitas-B-lineage lymphoma protein-b) is a RING-type E3 ligase that plays an important role in setting up the threshold for T-cell activation. It is critical for iTreg formation and

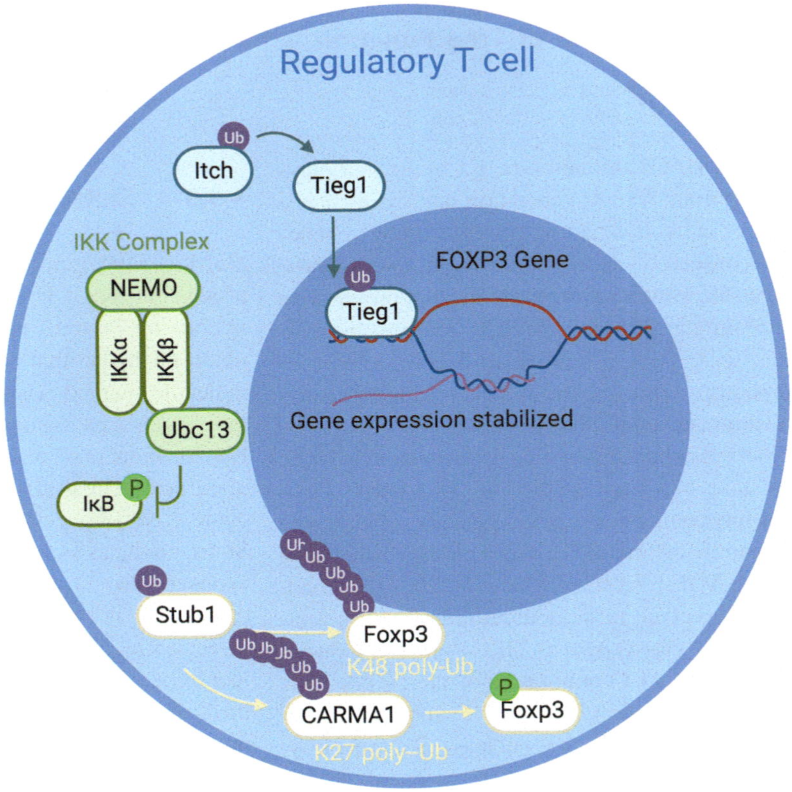

Fig. 4.1 Ubiquitin-dependent regulation of Treg function. The function of regulatory T cells can be regulated by multiple mechanisms. Here, we summarize some of the mechanisms that we listed in the review. In brief, ubiquitination leads to both transcriptional and translational regulations. This can be done by regulating *FOXP3* gene expression, Foxp3 protein turnover, and pathways that affect Treg function

plasticity, which has been reported in autoimmune diseases like type 1 diabetes (T1D) (Yokoi et al. 2008) and systemic lupus erythematosus (SLE) (Romo-Tena et al. 2018).

The mechanism of Cbl-b regulation involves restricting activation signaling. Upon normal T-cell activation, Cbl-b itself is K48-type polyubiquitinated for degradation (Bachmaier et al. 2000). When T cell is activated with insufficient amount of co-stimulation or with co-inhibitory signals like CTLA-4, Cbl-b is stabilized (Li et al. 2004; Chiang et al. 2000). In this case, the p85 regulatory subunit of PI3K is polyubiquitinated at K63 by Cbl-b, blocking it from associating with the ITAM (Immunoreceptor Tyrosine-based Activation Motif) and CD28. Thus, PI3K-related activation signaling will be blocked, including PI3K-ATK-mTOR pathway (Chiang et al. 2000) (Fig. 4.2). As explained earlier, this will favor the formation of iTreg cells (Haxhinasto et al. 2008; Sauer et al. 2008; Zhang et al. 2013a).

4.6.2 TRAF6

TRAF6 (TNF receptor-associated factor 6) is a RING-type E3 ligase that mediates signaling from IL-1 receptor, TLRs, and TNF receptor superfamily (Kobayashi et al. 2003). It plays an important role in activation or "licensing" of antigen-presenting cells to produce inflammatory cytokines and to activate T cells (Kobayashi et al. 2004). Mice with TRAF6 deficiency will result in systematical autoimmunity and robust T-cell activation (Chiffoleau et al. 2003; Akiyama et al. 2005).

Further studies show that TRAF6 mice harbor enhanced Th17 commitment and reduced Treg population (Cejas et al. 2010). TRAF6-specific

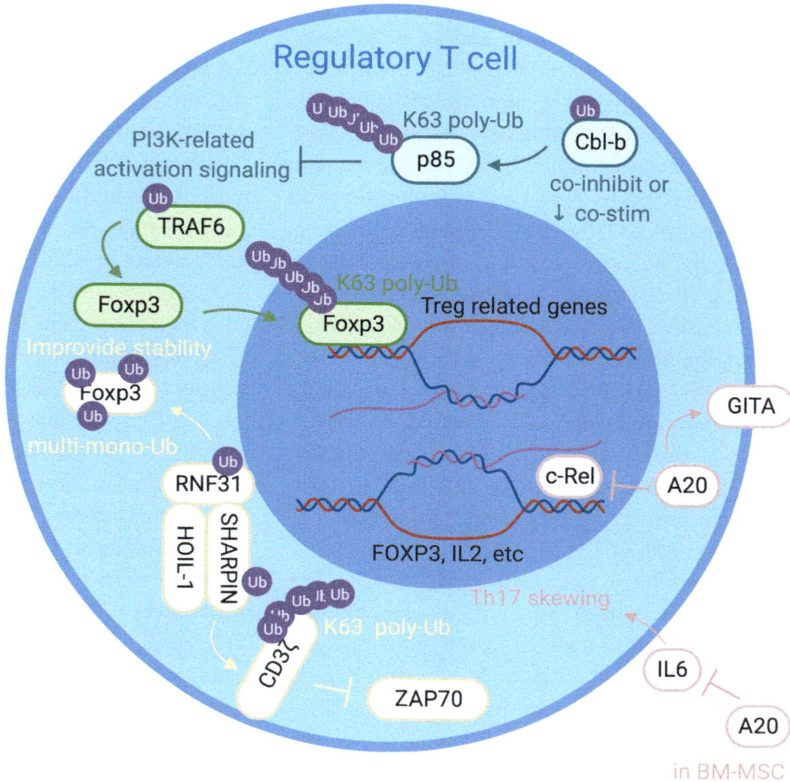

Fig. 4.2 Ubiquitin-dependent regulation of Treg plasticity. The plasticity of regulatory T cells can be regulated by multiple mechanisms. Here, we summarize some of the mechanisms that we listed in the review. In brief, ubiquitination can affect T-cell activation pathways, favoring the development of regulatory T cells in thymus. Additionally, it can also render critical cytokines and transcription factors, regulating the balance between Th17 and Treg cells

knockout (Foxp3^{cre}/TRAF6^{flox}) Treg cells have reduced stability and upregulated effector T-cell associated transcription factors and cytokines, including GATA3, Tbet, and IL-4 (Shimo et al. 2011; Muto et al. 2013). These findings suggest that TRAF6 stabilizes Treg cells by improving its lineage commitment. However, the detailed mechanism remains uncertain.

Recently, our group found an important link between TRAF6 and Foxp3 localization. We identified that TRAF6 mediates a K63-type polyubiquitination on Foxp3 protein at Lysine 262 that does not interfere with other K48-type polyubiquitination (Ni et al. 2019) (Fig. 4.2). In addition, TRAF6-deficient Treg cells have aberrantly more Foxp3 localized in the cytoplasm than in the nucleus (our unpublished data), which partially explains its instability and provides a potential target for immunotherapies.

4.6.3 RNF31

RNF31 (ring finger protein 31, aka HOIP) is an E3 ligase of LUBAC (linear ubiquitin chain assembly complex) that is responsible for linear polyubiquitination (Stieglitz et al. 2012). It is auto-inhibited before binding with the other two proteins of LUBAC: HOIL-1 and SHARPIN (Kirisako et al. 2006; Tokunaga et al. 2011). While HOIL-1 is another E3 ligase that is important for polyubiquitination and degradation of SOCS6 (Bayle et al. 2006), SHARPIN is a ubiquitin-binding protein that links ubiquitin and E3 ligase RNF31 (Gerlach et al. 2011). As an important regulator of immune responses, LUBAC involves in multiple pathways including TCR, BCR, NOD, TLR, and TNFR (Gerlach et al. 2011; Zinngrebe et al. 2016; Damgaard et al. 2012; Rodgers et al. 2014; Sasaki et al. 2013; Haas et al. 2009; Park et al. 2016).

In Treg cells, components of LUBAC are critical for both its development and its suppressive function. For example, SHARPIN deficiency mice have diminished a number of Treg cells and reduced Treg suppressive function (Park et al. 2016; Redecke et al. 2016). The mechanism is that SHARPIN is K63-type polyubiquitinated and binds to CD3ζ upon TCR activation. It regulates the negative selection of nTreg cells by inhibiting the interaction between CD3ζ and Zap70 (Park et al. 2016) (Fig. 4.2). In addition, it skews the homeostasis between Treg and Th17 cells as SHARPIN-deficient mice have more IL-17-producing Treg cells (Park et al. 2016). Despite this reduced Treg population, SHARPIN-deficient mice are still able to form SP (single positive) cells during thymic development (Teh et al. 2016).

On the contrary, mice with HOIL-1 or RNF31 knockout suffers from impaired late-stage thymocyte differentiation and could not form matured SP cells (Teh et al. 2016). Mice with RNF31 knockout in Foxp3+ cells have very few Treg cells and would develop Foxp3-deficient scurfy phenotype (Teh et al. 2016). A more recent study indicates that not only does RNF31 regulate TCR signaling but also it regulates multi-monoubiquitination to Foxp3 protein directly through atypical ubiquitin chains (Zhu et al. 2018). Using shRNA to knockdown human Treg cells, researchers observed reduced Foxp3 protein levels and increased Th1-like phenotypes (Zhu et al. 2018). Besides this, they found that intratumoral Treg cells tend to have upregulated RNF31, which suggests that RNF31 is also involved in Treg function (Zhu et al. 2018).

4.6.4 A20

A20, aka TNFAIP3 (TNF-α-induced protein 3), is an anti-inflammatory zinc finger protein that regulates TNF, TLR, NOD, and NFκB singling pathways (Lin et al. 2008; Wertz et al. 2004). It is considered as a ubiquitin-modification protein because it has both E3 ligase and DUB (deubiquitinase) functions (Evans et al. 2004). The N-terminal domain of A20 is an OTU (ovarian tumor) family DUB that cleaves K63-type polyubiquitinated substrate, and the C-terminal domain is a seven-C2/C2-zinc finger structure that serves as an E3 ligase that polyubiquitinates the same substrate at its K48 site (Wertz et al. 2004; Lu et al. 2013). Interestingly, it is a highly specific enzyme, and the two known targets are TRAF6 and RIP1 (Lin et al. 2008; Evans et al. 2004).

A20 influences Treg plasticity from two different perspectives. On the one hand, A20 inhibits nTreg development by rendering NFκB signaling and upregulating GITR (Fischer et al. 2017). It has been shown that A20 knockout mice have upregulated NFκB activation and c-Rel nuclear translocation in nTreg cells (Fischer et al. 2017). Despite this, these nTreg cells hold normal suppressive functions (Fischer et al. 2017) (Fig. 4.2).

On the other hand, A20 protects Treg plasticity from external mechanisms (Catrysse et al. 2014; Luo et al. 2020). Studies from autoimmune diseases like RA (rheumatoid arthritis) and colitis have shown that A20 from nonimmune cells can influence the plasticity of Treg cells (Feng et al. 2018; Hu et al. 2019). In the RA study, researchers have shown that A20 inhibits IL-6 secretion from BM-MSCs (bone marrow mesenchymal stem cells) (Fig. 4.2). By restoring A20 expression in these cells, their IL-6 secretion will reduce, which will promote Treg skewing from Th17 lineage (Feng et al. 2018). In the colitis study, researchers have shown similar conclusion by feeding nanoparticle-coated A20 recombinant vector to DSS-colitis mice, which reduces colonic tissue damage and promotes Treg cells (Hu et al. 2019).

4.7 Deubiquitinases: The Janus of Ubiquitin-Dependent Regulation

There is no doubt that E3 ligases regulate both the function and the plasticity of Treg cells by regulating Foxp3 directly or its related cofactors. Given the mechanism of E3 ligase, it is not surprising that ubiquitination process can be reversible. By removing the ubiquitin tags from ubiquitinated proteins, their function and fates can be modified as well (Nijman et al. 2005). In reality, such enzymes are named deubiquitinases (DUBs), and several of them play important roles regulating Treg cells. Here, we will review some of the most recent discoveries and briefly recap some of the known ones that we reviewed before (Pan and Barbi 2014; Barbi et al. 2015).

4.7.1 USP7

As described earlier, Foxp3 level can be regulated through polyubiquitination (van Loosdregt et al. 2011). Because of this, it stands to reason that deubiquitination would be a mechanism to preserve the Foxp3 level at a homeostatic level. One of DUB under this principle is known as USP7 (ubiquitin-specific processing protease 7). Recently, it is reported that USP7 is important for Treg function by preserving Foxp3 expression level (van Loosdregt et al. 2013) (Fig. 4.3). Using a pan-DUB inhibitor, these authors used mouse adoptive transfer colitis model to confirm that DUBs modulate Treg suppression function in vivo. In addition, in vitro studies show that USP7 deubiquitinates Foxp3, which increases its protein expression. Furthermore, using shRNA to knockdown USP7, they confirmed that USP7 improves Treg suppression (van Loosdregt et al. 2013). These results confirmed the importance of DUBs in Treg regulation and suggests the potential of other DUBs in Treg regulation.

4.7.2 USP21

Another deubiquitinase that was recently characterized is USP21. It is first identified as a dispensable positive regulator of GATA3 that prevents its ubiquitination and degradation in Treg cells (Zhang et al. 2013b; Pannu et al. 2015). Later studies have shown a feedback regulation of USP21 that functions on Foxp3 (Li et al. 2016). On the one hand, Foxp3 can bind to the promoter region of *USP21* gene to upregulate its expression (Li et al. 2016). On the other hand, *USP21* promotes Treg stability and function by deubiquitinating K48-type polyubiquitination (Li et al. 2016) (Fig. 4.3). Treg cells with USP21 deficiency are prone to lose Foxp3 protein and acquire Th1-like phenotype upon stress signals (Li et al. 2016).

4.7.3 USP44

Recently, we and our collaborators found an additional DUB that plays important roles in Treg cells—USP44. USP44 is originally known to play important roles in cell cycle regulation (Stegmeier et al. 2007; Visconti et al. 2012), stem cell differentiation (Fuchs et al. 2012), and centrosome positioning (Zhang et al. 2012). Here, we identified that USP44 is upregulated in Treg cell in response to TGF-β signaling (our unpublished data). Interestingly, USP44 acts cooperatively with USP7 but targets specifically to K48-type polyubiquitinated Foxp3 (Fig. 4.3). Treg cells with USP44 knockdown have decreased Foxp3 expression and disrupted suppression capability (our unpublished data). Above all, these findings provide additional approaches to regulate ubiquitin-regulated Treg function and plasticity, which improved our knowledge understating the signaling network of Treg cells. Additionally, they provide more clinical insights as USP7 and USP44 can be potential drug targets to treat autoimmune disease and cancer.

4.8 Concluding Remarks

With persistent endeavor, our knowledge on regulations on Treg is growing every day. Among these, ubiquitin-mediated posttranslational modification on Treg cells is rising. In this chapter, we only covered some of the most recent discoveries on ubiquitin-mediated regulations with an emphasis on Foxp3 and its cofactors. Clearly, the homeostasis of Foxp3 is regulated by factors that stabilizes Foxp3 (like TRAF6, USP7, and USP44) and factors that degrades Foxp3 (like Stub1). This reflects a complex network of ubiquitination on different residues and in different formats

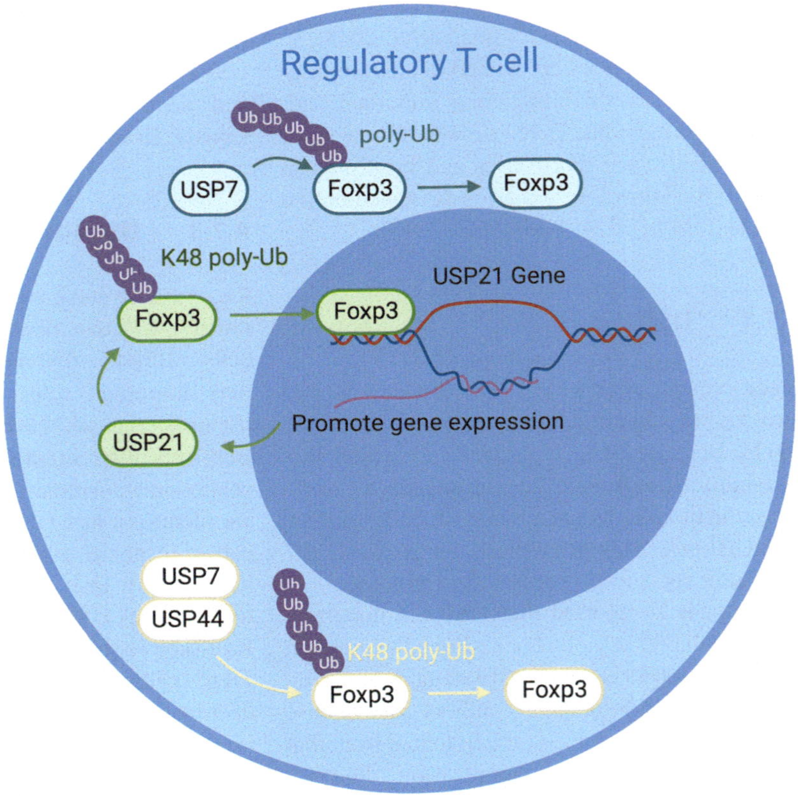

Fig. 4.3 Deubiquitinases: the Janus of ubiquitin-dependent regulation. Deubiquitinase plays important roles in both the function and plasticity of Treg cells, mostly by preventing Foxp3 from ubiquitin-dependent degradation. Specifically, USP21 and USP44 favor K48 polyubiquitination while USP7 favors multiple polyubiquitination

(ubiquitination vs. deubiquitination; monoubiquitination vs. polyubiquitination; K48-type vs. K63-type).

Indeed, besides TGFβ and NFκB signaling, there are many other pathways that contribute to Treg function and plasticity, including cytokines, metabolites, and environmental stress. Due to the limitation of our knowledge, we did not cover much in this chapter. However, they are equally interesting as they provide additional information on mechanisms that regulate Treg cells. Here, we would like to include two additional molecules/ pathways that may involve ubiquitin-mediated regulation with some knowns and unknowns.

4.8.1 HIF-1

HIF-1 (hypoxia-inducible factor-1) is a transcription factor that is known as the oxygen sensor and the mediator of hypoxic responses (Semenza 2007). It plays an important role in deciding the crossroad of Th17-iTreg differentiation. During cell development, TGFβ is required by both proinflammatory Th17 cells and immunosuppressive iTreg cells (Zhou et al. 2008). When naïve CD4+ T cells are treated with ample amount of TGFβ, cells will reach a crossroad where they express both Foxp3 and RORγt (Zhou et al. 2008; Dang et al. 2011). At this time, they require additional factors to be skewed into either one of the lineages. Upon these factors, HIF-1 promotes Th17 commitment (Dang et al. 2011).

Studies have shown that this is mediated in a posttranslational mechanism where Foxp3 protein is degraded in a HIF-1-dependent manner (Shi et al. 2011). In addition, HIF-1 has been examined to display physical interaction with Foxp3 (Dang et al. 2011). Furthermore, experiments have been shown to confirm that both knocking out critical factors of HIF-1 turnover (DTX1) and chemically inhibiting the proteasome can stabilize Foxp3 levels (Dang et al. 2011; Hsiao et al. 2015), confirming that HIF-1 mediates Foxp3 degradation using a ubiquitin-mediated degradation mechanism.

So far, the exact mechanism behind this ubiquitin-mediated Foxp3 degradation is unclear, and the E3 ligase that involves is unknown. One of the hypotheses is that it is regulated by HIF-1α, which includes a PHD domain that serves as a binding site for VHL-box protein to recruit Elongin-B/Elongin-C/Cullin-2/RBX1 E3 ubiquitin complex for K48-type polyubiquitination (Lecker et al. 2006; Semenza 2007). When HIF-1 binds to Foxp3, this may lead to co-degradation by proteasome (Dang et al. 2011). Further investigations are needed to elucidate the exact mechanism.

4.8.2 YAP/Mst1/Mst2 from Hippo Signaling Pathway

Yap is a transcriptional regulator in Hippo signaling pathway that mediates cell proliferation and organ development (Dong et al. 2007; Yu et al. 2015). It is often upregulated in cancers and plays an important role in tumor progression (Pan 2010). Surprisingly, among all CD4+ T cell subsets, YAP is only upregulated in Treg cells (Ni et al. 2018). Recent studies have shown that it plays a critical role in Treg function. Ni et al. show that YAP enhances the TGFβ/Smad signaling through activins (Ni et al. 2018). In addition, knocking out YAP in Treg cells will dampen their suppression function but not impair their survival (Ni et al. 2018). This work clearly demonstrates the importance of Yap as an amplifier in Treg function.

Besides YAP, Mst1, and Mst2, two Hippo kinases are reported to be important for Treg plasticity (Creasy et al. 1996; Creasy and Chernoff 1995). Shi et al. show that Mst1 and Mst2 sense and amplify IL-2-STAT5 signaling in Treg cells through the small GTPase Rac, which is critical for their plasticity and survival (Shi et al. 2018). Mice with Mst1/Mst2 deficiency will lose their Treg population and develop systemic Th1-dominant autoimmunity (Shi et al. 2018). Thus, these two studies demonstrate the importance of non-canonical Hippo pathway as a mediator of Treg function and plasticity.

Despite that no E3 ligase has been reported to regulate Yap or Mst1/2 in Treg cells, several E3 ligases have been reported to regulate the Hippo pathway in general. For example, ITCH (an E3 ligase) has been reported to regulate Lats1 (Salah and Aqeilan 2011), which phosphorylates Yap for $SCF^{\beta\text{-}TRCP}$ (an E3 ligase) mediated, K48-type polyubiquitinated degradation (Zhao et al. 2010). Additionally, in neurons, Stub1 (an E3 ligase) has been reported to mediate the degradation of Mst1, which prevents Mst1 from been phosphorylated by c-Abl that leads to oxidative stress-induced apoptosis (Paul and Ghosh 2014; Xiao et al. 2011). Above all, future studies are needed to decipher the role of Hippo pathway-related E3 ligases in Treg cells.

4.8.3 Clinical Implication

Treg cells play important roles in autoimmunity and cancer. In autoimmune diseases, Treg cells typically failed to suppress self-reactive effector cells, leading to the death of healthy somatic cells and even tissue damage. On the other hand, in cancer settings, Tregs tend to be enriched within the tumor microenvironment and suppress effector cells from removing tumor cells. In these two cases, Treg cells play opposing roles: patients need more suppressive Treg cells in autoimmune diseases and less in cancer settings. However, a careful balance must be made to restore immune homeostasis because enhancing excessive amount of suppressive Treg cells to autoimmune patients can induce cancer and removing too many Treg cells in cancer patients can induce autoimmunity (Kim et al. 2006). Due to this reason, we need a safer and controllable approach to manipulate the number and the function of Treg cells.

As we discussed previously, ubiquitination is a reversible process that mediates cell signaling. Because of this, it becomes a great target for drug discovery. In Treg settings, it offers a unique approach to manipulate the function and quantity of Treg cells temporarily and reversibly. When treating cancer, for example, a temporal disruption of Treg function may be sufficient for immune system to eliminate the tumor cells without inducing autoimmunity. After the treatment is done, immune homeostasis can be quickly restored by removing the drugs. In this way, doctors can not only save patients but also restore their immune balance in the long run.

So far, our knowledge of ubiquitin-dependent pathways and the molecules within these pathways is still limited. As a rising field, more efforts and supports are needed to reveal more therapeutic potentials in the future.

References

Akiyama T, Maeda S, Yamane S, Ogino K, Kasai M, Kajiura F et al (2005) Dependence of self-tolerance on TRAF6-directed development of thymic stroma. Science 308(5719):248–251. https://doi.org/10.1126/science.1105677

Almeida AR, Legrand N, Papiernik M, Freitas AA (2002) Homeostasis of peripheral CD4+ T cells: IL-2R alpha and IL-2 shape a population of regulatory cells that controls CD4+ T cell numbers. J Immunol 169 (9):4850–4860

Apostolou I, von Boehmer H (2004) In vivo instruction of suppressor commitment in naive T cells. J Exp Med 199(10):1401–1408. https://doi.org/10.1084/jem.20040249

Asano M, Toda M, Sakaguchi N, Sakaguchi S (1996) Autoimmune disease as a consequence of developmental abnormality of a T cell subpopulation. J Exp Med 184(2):387–396

Atarashi K, Tanoue T, Shima T, Imaoka A, Kuwahara T, Momose Y et al (2011) Induction of colonic regulatory T cells by indigenous Clostridium species. Science 331 (6015):337–341. https://doi.org/10.1126/science.1198469

Bachmaier K, Krawczyk C, Kozieradzki I, Kong YY, Sasaki T, Oliveira-dos-Santos A et al (2000) Negative regulation of lymphocyte activation and autoimmunity by the molecular adaptor Cbl-b. Nature 403 (6766):211–216. https://doi.org/10.1038/35003228

Bai Y, Yang C, Hu K, Elly C, Liu YC (2004) Itch E3 ligase-mediated regulation of TGF-beta signaling by modulating smad2 phosphorylation. Mol Cell 15 (5):825–831. https://doi.org/10.1016/j.molcel.2004.07.021

Ballinger CA, Connell P, Wu Y, Hu Z, Thompson LJ, Yin LY et al (1999) Identification of CHIP, a novel tetratricopeptide repeat-containing protein that interacts with heat shock proteins and negatively regulates chaperone functions. Mol Cell Biol 19 (6):4535–4545

Barbi J, Pardoll DM, Pan F (2015) Ubiquitin-dependent regulation of Foxp3 and Treg function. Immunol Rev 266(1):27–45. https://doi.org/10.1111/imr.12312

Bayle J, Lopez S, Iwai K, Dubreuil P, De Sepulveda P (2006) The E3 ubiquitin ligase HOIL-1 induces the polyubiquitination and degradation of SOCS6 associated proteins. FEBS Lett 580(11):2609–2614. https://doi.org/10.1016/j.febslet.2006.03.093

Behrends C, Harper JW (2011) Constructing and decoding unconventional ubiquitin chains. Nat Struct Mol Biol 18(5):520–528. https://doi.org/10.1038/nsmb.2066

Bennett CL, Christie J, Ramsdell F, Brunkow ME, Ferguson PJ, Whitesell L et al (2001) The immune dysregulation, polyendocrinopathy, enteropathy, X-linked syndrome (IPEX) is caused by mutations of FOXP3. Nat Genet 27(1):20–21. https://doi.org/10.1038/83713

Bettelli E, Carrier Y, Gao W, Korn T, Strom TB, Oukka M et al (2006) Reciprocal developmental pathways for the generation of pathogenic effector TH17 and regulatory T cells. Nature 441(7090):235–238. https://doi.org/10.1038/nature04753

Bettini ML, Pan F, Bettini M, Finkelstein D, Rehg JE, Floess S et al (2012) Loss of epigenetic modification driven by the Foxp3 transcription factor leads to regulatory T cell insufficiency. Immunity 36(5):717–730. https://doi.org/10.1016/j.immuni.2012.03.020

Bielskiene K, Bagdoniene L, Mozuraitiene J, Kazbariene B, Janulionis E (2015) E3 ubiquitin ligases as drug targets and prognostic biomarkers in melanoma. Medicina (Kaunas) 51(1):1–9. https://doi.org/10.1016/j.medici.2015.01.007

Bjur E, Larsson O, Yurchenko E, Zheng L, Gandin V, Topisirovic I et al (2013) Distinct translational control in CD4+ T cell subsets. PLoS Genet 9(5):e1003494. https://doi.org/10.1371/journal.pgen.1003494

Bloom BR, Salgame P, Diamond B (1992) Revisiting and revising suppressor T cells. Immunol Today 13 (4):131–136. https://doi.org/10.1016/0167-5699(92)90110-S

Brunkow ME, Jeffery EW, Hjerrild KA, Paeper B, Clark LB, Yasayko SA et al (2001) Disruption of a new forkhead/winged-helix protein, scurfin, results in the fatal lymphoproliferative disorder of the scurfy mouse. Nat Genet 27(1):68–73. https://doi.org/10.1038/83784

Catrysse L, Vereecke L, Beyaert R, van Loo G (2014) A20 in inflammation and autoimmunity. Trends Immunol 35(1):22–31. https://doi.org/10.1016/j.it.2013.10.005

Cejas PJ, Walsh MC, Pearce EL, Han D, Harms GM, Artis D et al (2010) TRAF6 inhibits Th17 differentiation and TGF-beta-mediated suppression of IL-2. Blood 115 (23):4750–4757. https://doi.org/10.1182/blood-2009-09-242,768

Chang JH, Xiao Y, Hu H, Jin J, Yu J, Zhou X et al (2012) Ubc13 maintains the suppressive function of regulatory T cells and prevents their conversion into effector-like T cells. Nat Immunol 13(5):481–490. https://doi.org/10.1038/ni.2267

Chen W, Jin W, Hardegen N, Lei KJ, Li L, Marinos N et al (2003) Conversion of peripheral CD4+ CD25- naive T cells to CD4+ CD25+ regulatory T cells by TGF-beta induction of transcription factor Foxp3. J Exp Med 198(12):1875–1886. https://doi.org/10.1084/jem.20030152

Chen Z, Barbi J, Bu S, Yang HY, Li Z, Gao Y et al (2013) The ubiquitin ligase Stub1 negatively modulates regulatory T cell suppressive activity by promoting degradation of the transcription factor Foxp3. Immunity 39(2):272–285. https://doi.org/10.1016/j.immuni.2013.08.006

Chiang YJ, Kole HK, Brown K, Naramura M, Fukuhara S, Hu RJ et al (2000) Cbl-b regulates the CD28 dependence of T-cell activation. Nature 403(6766):216–220. https://doi.org/10.1038/35003235

Chiffoleau E, Kobayashi T, Walsh MC, King CG, Walsh PT, Hancock WW et al (2003) TNF receptor-associated factor 6 deficiency during hemopoiesis induces Th2-polarized inflammatory disease. J Immunol 171(11):5751–5759

Ciehanover A, Hod Y, Hershko A (1978) A heat-stable polypeptide component of an ATP-dependent proteolytic system from reticulocytes. Biochem Biophys Res Commun 81(4):1100–1105

Cobbold SP, Castejon R, Adams E, Zelenika D, Graca L, Humm S et al (2004) Induction of foxP3+ regulatory T cells in the periphery of T cell receptor transgenic mice tolerized to transplants. J Immunol 172(10):6003–6010

Creasy CL, Chernoff J (1995) Cloning and characterization of a member of the MST subfamily of Ste20-like kinases. Gene 167(1–2):303–306

Creasy CL, Ambrose DM, Chernoff J (1996) The Ste20-like protein kinase, Mst1, dimerizes and contains an inhibitory domain. J Biol Chem 271(35):21049–21053

Curotto de Lafaille MA, Lafaille JJ (2009) Natural and adaptive foxp3+ regulatory T cells: more of the same or a division of labor? Immunity 30(5):626–635. https://doi.org/10.1016/j.immuni.2009.05.002

Damgaard RB, Nachbur U, Yabal M, Wong WW, Fiil BK, Kastirr M et al (2012) The ubiquitin ligase XIAP recruits LUBAC for NOD2 signaling in inflammation and innate immunity. Mol Cell 46(6):746–758. https://doi.org/10.1016/j.molcel.2012.04.014

Dang EV, Barbi J, Yang HY, Jinasena D, Yu H, Zheng Y et al (2011) Control of T(H)17/T(reg) balance by hypoxia-inducible factor 1. Cell 146(5):772–784. https://doi.org/10.1016/j.cell.2011.07.033

Darce J, Rudra D, Li L, Nishio J, Cipolletta D, Rudensky AY et al (2012) An N-terminal mutation of the Foxp3 transcription factor alleviates arthritis but exacerbates diabetes. Immunity 36(5):731–741. https://doi.org/10.1016/j.immuni.2012.04.007

Davidson TS, DiPaolo RJ, Andersson J, Shevach EM (2007) Cutting edge: IL-2 is essential for TGF-beta-mediated induction of Foxp3+ T regulatory cells. J Immunol 178(7):4022–4026

Dong J, Feldmann G, Huang J, Wu S, Zhang N, Comerford SA et al (2007) Elucidation of a universal size-control mechanism in Drosophila and mammals. Cell 130(6):1120–1133. https://doi.org/10.1016/j.cell.2007.07.019

Evans PC, Ovaa H, Hamon M, Kilshaw PJ, Hamm S, Bauer S et al (2004) Zinc-finger protein A20, a regulator of inflammation and cell survival, has de-ubiquitinating activity. Biochem J 378(Pt 3):727–734. https://doi.org/10.1042/BJ20031377

Feng Z, Zhai Y, Zheng Z, Yang L, Luo X, Dong X et al (2018) Loss of A20 in BM-MSCs regulates the Th17/Treg balance in rheumatoid arthritis. Sci Rep 8(1):427. https://doi.org/10.1038/s41598-017-18,693-0

Fischer JC, Otten V, Kober M, Drees C, Rosenbaum M, Schmickl M et al (2017) A20 restrains thymic regulatory T cell development. J Immunol 199(7):2356–2365. https://doi.org/10.4049/jimmunol.1602102

Fleskens V, Minutti CM, Wu X, Wei P, Pals C, McCrae J et al (2019) Nemo-like kinase drives Foxp3 stability and is critical for maintenance of immune tolerance by regulatory T cells. Cell Rep 26(13):3600–12 e6. https://doi.org/10.1016/j.celrep.2019.02.087

Fontenot JD, Gavin MA, Rudensky AY (2003) Foxp3 programs the development and function of CD4+ CD25+ regulatory T cells. Nat Immunol 4(4):330–336. https://doi.org/10.1038/ni904

Fontenot JD, Rasmussen JP, Williams LM, Dooley JL, Farr AG, Rudensky AY (2005) Regulatory T cell lineage specification by the forkhead transcription factor foxp3. Immunity 22(3):329–341. https://doi.org/10.1016/j.immuni.2005.01.016

Fu W, Ergun A, Lu T, Hill JA, Haxhinasto S, Fassett MS et al (2012) A multiply redundant genetic switch 'locks in' the transcriptional signature of regulatory T cells. Nat Immunol 13(10):972–980. https://doi.org/10.1038/ni.2420

Fuchs G, Shema E, Vesterman R, Kotler E, Wolchinsky Z, Wilder S et al (2012) RNF20 and USP44 regulate stem cell differentiation by modulating H2B monoubiquitylation. Mol Cell 46(5):662–673. https://doi.org/10.1016/j.molcel.2012.05.023

Fukushima T, Matsuzawa S, Kress CL, Bruey JM, Krajewska M, Lefebvre S et al (2007) Ubiquitin-conjugating enzyme Ubc13 is a critical component of TNF receptor-associated factor (TRAF)-mediated inflammatory responses. Proc Natl Acad Sci U S A 104(15):6371–6376. https://doi.org/10.1073/pnas.0700548104

Gao M, Karin M (2005) Regulating the regulators: control of protein ubiquitination and ubiquitin-like modifications by extracellular stimuli. Mol Cell 19(5):581–593. https://doi.org/10.1016/j.molcel.2005.08.017

Gerlach B, Cordier SM, Schmukle AC, Emmerich CH, Rieser E, Haas TL et al (2011) Linear ubiquitination prevents inflammation and regulates immune signalling. Nature 471(7340):591–596. https://doi.org/10.1038/nature09816

Gershon RK, Kondo K (1970) Cell interactions in the induction of tolerance: the role of thymic lymphocytes. Immunology 18(5):723–737

Geuking MB, Cahenzli J, Lawson MA, Ng DC, Slack E, Hapfelmeier S et al (2011) Intestinal bacterial colonization induces mutualistic regulatory T cell responses. Immunity 34(5):794–806. https://doi.org/10.1016/j.immuni.2011.03.021

Goldstein G, Scheid M, Hammerling U, Schlesinger DH, Niall HD, Boyse EA (1975) Isolation of a polypeptide that has lymphocyte-differentiating properties and is probably represented universally in living cells. Proc Natl Acad Sci U S A 72(1):11–15

Haas TL, Emmerich CH, Gerlach B, Schmukle AC, Cordier SM, Rieser E et al (2009) Recruitment of the linear ubiquitin chain assembly complex stabilizes the TNF-R1 signaling complex and is required for TNF-mediated gene induction. Mol Cell 36(5):831–844. https://doi.org/10.1016/j.molcel.2009.10.013

Haxhinasto S, Mathis D, Benoist C (2008) The AKT-mTOR axis regulates de novo differentiation of CD4+ Foxp3+ cells. J Exp Med 205(3):565–574. https://doi.org/10.1084/jem.20071477

Hernandez AL, Kitz A, Wu C, Lowther DE, Rodriguez DM, Vudattu N et al (2015) Sodium chloride inhibits the suppressive function of FOXP3+ regulatory T cells. J Clin Invest 125(11):4212–4222. https://doi.org/10.1172/JCI81151

Hicke L (2001) Protein regulation by monoubiquitin. Nat Rev Mol Cell Biol 2(3):195–201. https://doi.org/10.1038/35056583

Hori S (2010) Developmental plasticity of Foxp3+ regulatory T cells. Curr Opin Immunol 22(5):575–582. https://doi.org/10.1016/j.coi.2010.08.004

Hori S (2011) Stability of regulatory T-cell lineage. Adv Immunol 112:1–24. https://doi.org/10.1016/B978-0-12-387,827-4.00001-2

Hori S, Nomura T, Sakaguchi S (2003) Control of regulatory T cell development by the transcription factor Foxp3. Science 299(5609):1057–1061. https://doi.org/10.1126/science.1079490

Horwitz DA, Zheng SG, Gray JD (2008) Natural and TGF-beta-induced Foxp3(+)CD4(+) CD25(+) regulatory T cells are not mirror images of each other. Trends Immunol 29(9):429–435. https://doi.org/10.1016/j.it.2008.06.005

Hsiao HW, Hsu TS, Liu WH, Hsieh WC, Chou TF, Wu YJ et al (2015) Deltex1 antagonizes HIF-1alpha and sustains the stability of regulatory T cells in vivo. Nat Commun 6:6353. https://doi.org/10.1038/ncomms7353

Hsieh CS, Zheng Y, Liang Y, Fontenot JD, Rudensky AY (2006) An intersection between the self-reactive regulatory and nonregulatory T cell receptor repertoires. Nat Immunol 7(4):401–410. https://doi.org/10.1038/ni1318

Hu T, Hu W, Ma L, Zeng X, Liu J, Cheng B et al (2019) pVAX1-A20 alleviates colitis in mice by promoting regulatory T cells. Dig Liver Dis 51(6):790–797. https://doi.org/10.1016/j.dld.2018.11.005

Ikeda F, Dikic I (2008) Atypical ubiquitin chains: new molecular signals. 'Protein modifications: beyond the usual suspects' review series. EMBO Rep 9(6):536–542. https://doi.org/10.1038/embor.2008.93

Jin HS, Park Y, Elly C, Liu YC (2013) Itch expression by Treg cells controls Th2 inflammatory responses. J Clin Invest 123(11):4923–4934. https://doi.org/10.1172/JCI69355

Jordan MS, Boesteanu A, Reed AJ, Petrone AL, Holenbeck AE, Lerman MA et al (2001) Thymic selection of CD4+ CD25+ regulatory T cells induced by an agonist self-peptide. Nat Immunol 2(4):301–306. https://doi.org/10.1038/86302

Josefowicz SZ, Niec RE, Kim HY, Treuting P, Chinen T, Zheng Y et al (2012) Extrathymically generated regulatory T cells control mucosal TH2 inflammation. Nature 482(7385):395–399. https://doi.org/10.1038/nature10772

Kappler JW, Roehm N, Marrack P (1987) T cell tolerance by clonal elimination in the thymus. Cell 49(2):273–280

Khattri R, Cox T, Yasayko SA, Ramsdell F (2003) An essential role for scurfin in CD4+ CD25+ T regulatory cells. Nat Immunol 4(4):337–342. https://doi.org/10.1038/ni909

Kim R, Emi M, Tanabe K (2006) Cancer immunosuppression and autoimmune disease: beyond immunosuppressive networks for tumour immunity. Immunology 119(2):254–264. https://doi.org/10.1111/j.1365-2567.2006.02430.x

Kim JM, Rasmussen JP, Rudensky AY (2007) Regulatory T cells prevent catastrophic autoimmunity throughout the lifespan of mice. Nat Immunol 8(2):191–197. https://doi.org/10.1038/ni1428

Kirisako T, Kamei K, Murata S, Kato M, Fukumoto H, Kanie M et al (2006) A ubiquitin ligase complex assembles linear polyubiquitin chains. EMBO J 25(20):4877–4887. https://doi.org/10.1038/sj.emboj.7601360

Kobayashi T, Walsh PT, Walsh MC, Speirs KM, Chiffoleau E, King CG et al (2003) TRAF6 is a critical factor for dendritic cell maturation and development. Immunity 19(3):353–363

Kobayashi T, Walsh MC, Choi Y (2004) The role of TRAF6 in signal transduction and the immune response. Microbes Infect 6(14):1333–1338. https://doi.org/10.1016/j.micinf.2004.09.001

Komander D (2009) The emerging complexity of protein ubiquitination. Biochem Soc Trans 37(Pt 5):937–953. https://doi.org/10.1042/BST0370937

Komander D, Rape M (2012) The ubiquitin code. Annu Rev Biochem 81:203–229. https://doi.org/10.1146/annurev-biochem-060310-170,328

Kong N, Lan Q, Chen M, Wang J, Shi W, Horwitz DA et al (2012a) Antigen-specific transforming growth factor beta-induced Treg cells, but not natural Treg cells, ameliorate autoimmune arthritis in mice by shifting the Th17/Treg cell balance from Th17 predominance to Treg cell predominance. Arthritis Rheum 64(8):2548–2558. https://doi.org/10.1002/art.34513

Kong N, Lan Q, Su W, Chen M, Wang J, Yang Z et al (2012b) Induced T regulatory cells suppress osteoclastogenesis and bone erosion in collagen-induced arthritis better than natural T regulatory cells. Ann Rheum Dis 71(9):1567–1572. https://doi.org/10.1136/annrheumdis-2011-201,052

Kronenberg M, Steinmetz M, Kobori J, Kraig E, Kapp JA, Pierce CW et al (1983) RNA transcripts for I-J polypeptides are apparently not encoded between the I-A and I-E subregions of the murine major histocompatibility complex. Proc Natl Acad Sci U S A 80 (18):5704–5708

Lecker SH, Goldberg AL, Mitch WE (2006) Protein degradation by the ubiquitin-proteasome pathway in normal and disease states. J Am Soc Nephrol 17 (7):1807–1819. https://doi.org/10.1681/ASN.2006010083

Li X, Zheng Y (2015) Regulatory T cell identity: formation and maintenance. Trends Immunol 36 (6):344–353. https://doi.org/10.1016/j.it.2015.04.006

Li D, Gal I, Vermes C, Alegre ML, Chong AS, Chen L et al (2004) Cutting edge: Cbl-b: one of the key molecules tuning CD28- and CTLA-4-mediated T cell costimulation. J Immunol 173(12):7135–7139

Li W, Bengtson MH, Ulbrich A, Matsuda A, Reddy VA, Orth A et al (2008) Genome-wide and functional annotation of human E3 ubiquitin ligases identifies MULAN, a mitochondrial E3 that regulates the organelle's dynamics and signaling. PLoS One 3(1): e1487. https://doi.org/10.1371/journal.pone.0001487

Li Y, Lu Y, Wang S, Han Z, Zhu F, Ni Y et al (2016) USP21 prevents the generation of T-helper-1-like Treg cells. Nat Commun 7:13559. https://doi.org/10.1038/ncomms13559

Lin SC, Chung JY, Lamothe B, Rajashankar K, Lu M, Lo YC et al (2008) Molecular basis for the unique deubiquitinating activity of the NF-kappaB inhibitor A20. J Mol Biol 376(2):526–540. https://doi.org/10.1016/j.jmb.2007.11.092

Lu L, Ma J, Li Z, Lan Q, Chen M, Liu Y et al (2011) All-trans retinoic acid promotes TGF-beta-induced Tregs via histone modification but not DNA demethylation on Foxp3 gene locus. PLoS One 6(9):e24590. https://doi.org/10.1371/journal.pone.0024590

Lu TT, Onizawa M, Hammer GE, Turer EE, Yin Q, Damko E et al (2013) Dimerization and ubiquitin mediated recruitment of A20, a complex deubiquitinating enzyme. Immunity 38(5):896–905. https://doi.org/10.1016/j.immuni.2013.03.008

Lu L, Lan Q, Li Z, Zhou X, Gu J, Li Q et al (2014) Critical role of all-trans retinoic acid in stabilizing human natural regulatory T cells under inflammatory conditions. Proc Natl Acad Sci U S A 111(33):E3432–E3440. https://doi.org/10.1073/pnas.1408780111

Luo CT, Li MO (2013) Transcriptional control of regulatory T cell development and function. Trends Immunol 34(11):531–539. https://doi.org/10.1016/j.it.2013.08.003

Luo Y, Xue Y, Wang J, Dang J, Fang Q, Huang G et al (2019) Negligible effect of sodium chloride on the development and function of TGF-beta-induced CD4 (+) Foxp3(+) regulatory T cells. Cell Rep 26 (7):1869–79 e3. https://doi.org/10.1016/j.celrep.2019.01.066

Luo XQ, Zhong JW, Qiu SY, Zhi M, Yang LQ, Zhou YL et al (2020) A20-OVA nanoparticles inhibit allergic asthma in a murine model. Inflammation 43 (3):953–961. https://doi.org/10.1007/s10753-020-01181-5

Malek TR, Yu A, Vincek V, Scibelli P, Kong L (2002) CD4 regulatory T cells prevent lethal autoimmunity in IL-2Rbeta-deficient mice. Implications for the nonredundant function of IL-2. Immunity 17(2):167–178

Miranda M, Sorkin A (2007) Regulation of receptors and transporters by ubiquitination: new insights into surprisingly similar mechanisms. Mol Interv 7 (3):157–167. https://doi.org/10.1124/mi.7.3.7

Miyara M, Yoshioka Y, Kitoh A, Shima T, Wing K, Niwa A et al (2009) Functional delineation and differentiation dynamics of human CD4+ T cells expressing the FoxP3 transcription factor. Immunity 30(6):899–911. https://doi.org/10.1016/j.immuni.2009.03.019

Mucida D, Kutchukhidze N, Erazo A, Russo M, Lafaille JJ, Curotto de Lafaille MA (2005) Oral tolerance in the absence of naturally occurring Tregs. J Clin Invest 115 (7):1923–1933. https://doi.org/10.1172/JCI24487

Muto G, Kotani H, Kondo T, Morita R, Tsuruta S, Kobayashi T et al (2013) TRAF6 is essential for maintenance of regulatory T cells that suppress Th2 type autoimmunity. PLoS One 8(9):e74639. https://doi.org/10.1371/journal.pone.0074639

Nakayama KI, Nakayama K (2006) Ubiquitin ligases: cell-cycle control and cancer. Nat Rev Cancer 6 (5):369–381. https://doi.org/10.1038/nrc1881

Ni X, Tao J, Barbi J, Chen Q, Park BV, Li Z et al (2018) YAP is essential for Treg-mediated suppression of antitumor immunity. Cancer Discov 8(8):1026–1043. https://doi.org/10.1158/2159-8290.CD-17-1124

Ni X, Kou W, Gu J, Wei P, Wu X, Peng H et al (2019) TRAF6 directs FOXP3 localization and facilitates regulatory T-cell function through K63-linked ubiquitination. EMBO J 38(9):e99766. https://doi.org/10.15252/embj.201899766

Nijman SM, Luna-Vargas MP, Velds A, Brummelkamp TR, Dirac AM, Sixma TK et al (2005) A genomic and functional inventory of deubiquitinating enzymes. Cell 123(5):773–786. https://doi.org/10.1016/j.cell.2005.11.007

Nishizuka Y, Sakakura T (1969) Thymus and reproduction: sex-linked dysgenesia of the gonad after neonatal thymectomy in mice. Science 166(3906):753–755

Ouyang W, Beckett O, Ma Q, Li MO (2010) Transforming growth factor-beta signaling curbs thymic negative

selection promoting regulatory T cell development. Immunity 32(5):642–653. https://doi.org/10.1016/j.immuni.2010.04.012
Pan D (2010) The hippo signaling pathway in development and cancer. Dev Cell 19(4):491–505. https://doi.org/10.1016/j.devcel.2010.09.011
Pan F, Barbi J (2014) Ubiquitous points of control over regulatory T cells. J Mol Med (Berl) 92(6):555–569. https://doi.org/10.1007/s00109-014-1156-z
Pan F, Yu H, Dang EV, Barbi J, Pan X, Grosso JF et al (2009) Eos mediates Foxp3-dependent gene silencing in CD4+ regulatory T cells. Science 325 (5944):1142–1146. https://doi.org/10.1126/science.1176077
Pannu J, Belle JI, Forster M, Duerr CU, Shen S, Kane L et al (2015) Ubiquitin specific protease 21 is dispensable for normal development, hematopoiesis and lymphocyte differentiation. PLoS One 10(2):e0117304. https://doi.org/10.1371/journal.pone.0117304
Park Y, Jin HS, Lopez J, Lee J, Liao L, Elly C et al (2016) SHARPIN controls regulatory T cells by negatively modulating the T cell antigen receptor complex. Nat Immunol 17(3):286–296. https://doi.org/10.1038/ni.3352
Paul I, Ghosh MK (2014) The E3 ligase CHIP: insights into its structure and regulation. Biomed Res Int 2014:918183. https://doi.org/10.1155/2014/918183
Pickart CM (2001) Mechanisms underlying ubiquitination. Annu Rev Biochem 70:503–533. https://doi.org/10.1146/annurev.biochem.70.1.503
Pickart CM, Eddins MJ (2004) Ubiquitin: structures, functions, mechanisms. Biochim Biophys Acta 1695 (1–3):55–72. https://doi.org/10.1016/j.bbamcr.2004.09.019
Redecke V, Chaturvedi V, Kuriakose J, Hacker H (2016) SHARPIN controls the development of regulatory T cells. Immunology 148(2):216–226. https://doi.org/10.1111/imm.12604
Relland LM, Mishra MK, Haribhai D, Edwards B, Ziegelbauer J, Williams CB (2009) Affinity-based selection of regulatory T cells occurs independent of agonist-mediated induction of Foxp3 expression. J Immunol 182(3):1341–1350
Relland LM, Williams JB, Relland GN, Haribhai D, Ziegelbauer J, Yassai M et al (2012) The TCR repertoires of regulatory and conventional T cells specific for the same foreign antigen are distinct. J Immunol 189(7):3566–3574. https://doi.org/10.4049/jimmunol.1102646
Reyes-Turcu FE, Ventii KH, Wilkinson KD (2009) Regulation and cellular roles of ubiquitin-specific deubiquitinating enzymes. Annu Rev Biochem 78:363–397. https://doi.org/10.1146/annurev.biochem.78.082307.091526
Rodgers MA, Bowman JW, Fujita H, Orazio N, Shi M, Liang Q et al (2014) The linear ubiquitin assembly complex (LUBAC) is essential for NLRP3 inflammasome activation. J Exp Med 211 (7):1333–1347. https://doi.org/10.1084/jem.20132486
Romo-Tena J, Rajme-Lopez S, Aparicio-Vera L, Alcocer-Varela J, Gomez-Martin D (2018) Lys63-polyubiquitination by the E3 ligase casitas B-lineage lymphoma-b (Cbl-b) modulates peripheral regulatory T cell tolerance in patients with systemic lupus erythematosus. Clin Exp Immunol 191(1):42–49. https://doi.org/10.1111/cei.13054
Roncador G, Brown PJ, Maestre L, Hue S, Martinez-Torrecuadrada JL, Ling KL et al (2005) Analysis of FOXP3 protein expression in human CD4+ CD25+ regulatory T cells at the single-cell level. Eur J Immunol 35(6):1681–1691. https://doi.org/10.1002/eji.200526189
Round JL, Mazmanian SK (2010) Inducible Foxp3+ regulatory T-cell development by a commensal bacterium of the intestinal microbiota. Proc Natl Acad Sci U S A 107(27):12204–12209. https://doi.org/10.1073/pnas.0909122107
Sakaguchi S, Sakaguchi N, Asano M, Itoh M, Toda M (1995) Immunologic self-tolerance maintained by activated T cells expressing IL-2 receptor alpha-chains (CD25). Breakdown of a single mechanism of self-tolerance causes various autoimmune diseases. J Immunol 155(3):1151–1164
Sakaguchi S, Yamaguchi T, Nomura T, Ono M (2008) Regulatory T cells and immune tolerance. Cell 133 (5):775–787. https://doi.org/10.1016/j.cell.2008.05.009
Salah Z, Aqeilan RI (2011) WW domain interactions regulate the Hippo tumor suppressor pathway. Cell Death Dis 2:e172. https://doi.org/10.1038/cddis.2011.53
Sasaki Y, Sano S, Nakahara M, Murata S, Kometani K, Aiba Y et al (2013) Defective immune responses in mice lacking LUBAC-mediated linear ubiquitination in B cells. EMBO J 32(18):2463–2476. https://doi.org/10.1038/emboj.2013.184
Sauer S, Bruno L, Hertweck A, Finlay D, Leleu M, Spivakov M et al (2008) T cell receptor signaling controls Foxp3 expression via PI3K, Akt, and mTOR. Proc Natl Acad Sci U S A 105(22):7797–7802. https://doi.org/10.1073/pnas.0800928105
Schlenner SM, Weigmann B, Ruan Q, Chen Y, von Boehmer H (2012) Smad3 binding to the foxp3 enhancer is dispensable for the development of regulatory T cells with the exception of the gut. J Exp Med 209(9):1529–1535. https://doi.org/10.1084/jem.20112646
Semenza GL (2007) Hypoxia-inducible factor 1 (HIF-1) pathway. Sci STKE 2007(407):cm8. https://doi.org/10.1126/stke.4072007cm8
Setoguchi R, Hori S, Takahashi T, Sakaguchi S (2005) Homeostatic maintenance of natural Foxp3(+) CD25 (+) CD4(+) regulatory T cells by interleukin (IL)-2 and induction of autoimmune disease by IL-2 neutralization. J Exp Med 201(5):723–735. https://doi.org/10.1084/jem.20041982
Shi LZ, Wang R, Huang G, Vogel P, Neale G, Green DR et al (2011) HIF1alpha-dependent glycolytic pathway

orchestrates a metabolic checkpoint for the differentiation of TH17 and Treg cells. J Exp Med 208 (7):1367–1376. https://doi.org/10.1084/jem.20110278
Shi H, Liu C, Tan H, Li Y, Nguyen TM, Dhungana Y et al (2018) Hippo kinases Mst1 and Mst2 sense and amplify IL-2R-STAT5 signaling in regulatory T cells to establish stable regulatory activity. Immunity 49 (5):899–914 e6. https://doi.org/10.1016/j.immuni.2018.10.010
Shimo Y, Yanai H, Ohshima D, Qin J, Motegi H, Maruyama Y et al (2011) TRAF6 directs commitment to regulatory T cells in thymocytes. Genes Cells 16 (4):437–447. https://doi.org/10.1111/j.1365-2443.2011.01500.x
Stegmeier F, Rape M, Draviam VM, Nalepa G, Sowa ME, Ang XL et al (2007) Anaphase initiation is regulated by antagonistic ubiquitination and deubiquitination activities. Nature 446(7138):876–881. https://doi.org/10.1038/nature05694
Stieglitz B, Morris-Davies AC, Koliopoulos MG, Christodoulou E, Rittinger K (2012) LUBAC synthesizes linear ubiquitin chains via a thioester intermediate. EMBO Rep 13(9):840–846. https://doi.org/10.1038/embor.2012.105
Suzuki H, Zhou YW, Kato M, Mak TW, Nakashima I (1999) Normal regulatory alpha/beta T cells effectively eliminate abnormally activated T cells lacking the interleukin 2 receptor beta in vivo. J Exp Med 190 (11):1561–1572
Takahashi T, Kuniyasu Y, Toda M, Sakaguchi N, Itoh M, Iwata M et al (1998) Immunologic self-tolerance maintained by CD25+ CD4+ naturally anergic and suppressive T cells: induction of autoimmune disease by breaking their anergic/suppressive state. Int Immunol 10(12):1969–1980
Takahashi R, Nishimoto S, Muto G, Sekiya T, Tamiya T, Kimura A et al (2011) SOCS1 is essential for regulatory T cell functions by preventing loss of Foxp3 expression as well as IFN-{gamma} and IL-17A production. J Exp Med 208(10):2055–2067. https://doi.org/10.1084/jem.20110428
Teh CE, Lalaoui N, Jain R, Policheni AN, Heinlein M, Alvarez-Diaz S et al (2016) Linear ubiquitin chain assembly complex coordinates late thymic T-cell differentiation and regulatory T-cell homeostasis. Nat Commun 7:13353. https://doi.org/10.1038/ncomms13353
Thornton AM, Shevach EM (1998) CD4+ CD25+ immunoregulatory T cells suppress polyclonal T cell activation in vitro by inhibiting interleukin 2 production. J Exp Med 188(2):287–296
Thorstenson KM, Khoruts A (2001) Generation of anergic and potentially immunoregulatory CD25+ CD4 T cells in vivo after induction of peripheral tolerance with intravenous or oral antigen. J Immunol 167 (1):188–195
Tokunaga F, Nakagawa T, Nakahara M, Saeki Y, Taniguchi M, Sakata S et al (2011) SHARPIN is a component of the NF-kappaB-activating linear ubiquitin chain assembly complex. Nature 471 (7340):633–636. https://doi.org/10.1038/nature09815
van Loosdregt J, Brunen D, Fleskens V, Pals CE, Lam EW, Coffer PJ (2011) Rapid temporal control of Foxp3 protein degradation by sirtuin-1. PLoS One 6(4): e19047. https://doi.org/10.1371/journal.pone.0019047
van Loosdregt J, Fleskens V, Fu J, Brenkman AB, Bekker CP, Pals CE et al (2013) Stabilization of the transcription factor Foxp3 by the deubiquitinase USP7 increases Treg-cell-suppressive capacity. Immunity 39(2):259–271. https://doi.org/10.1016/j.immuni.2013.05.018
van Wijk SJ, Timmers HT (2010) The family of ubiquitin-conjugating enzymes (E2s): deciding between life and death of proteins. FASEB J 24(4):981–993. https://doi.org/10.1096/fj.09-136,259
Veldhoen M, Hocking RJ, Atkins CJ, Locksley RM, Stockinger B (2006) TGFbeta in the context of an inflammatory cytokine milieu supports de novo differentiation of IL-17-producing T cells. Immunity 24 (2):179–189. https://doi.org/10.1016/j.immuni.2006.01.001
Venuprasad K, Elly C, Gao M, Salek-Ardakani S, Harada Y, Luo JL et al (2006) Convergence of Itch-induced ubiquitination with MEKK1-JNK signaling in Th2 tolerance and airway inflammation. J Clin Invest 116(4):1117–1126. https://doi.org/10.1172/JCI26858
Venuprasad K, Huang H, Harada Y, Elly C, Subramaniam M, Spelsberg T et al (2008) The E3 ubiquitin ligase Itch regulates expression of transcription factor Foxp3 and airway inflammation by enhancing the function of transcription factor TIEG1. Nat Immunol 9(3):245–253. https://doi.org/10.1038/ni1564
Vignali DA, Collison LW, Workman CJ (2008) How regulatory T cells work. Nat Rev Immunol 8 (7):523–532. https://doi.org/10.1038/nri2343
Visconti R, Palazzo L, Della Monica R, Grieco D (2012) Fcp1-dependent dephosphorylation is required for M-phase-promoting factor inactivation at mitosis exit. Nat Commun 3:894. https://doi.org/10.1038/ncomms1886
Wang S, Li Y, Hu YH, Song R, Gao Y, Liu HY et al (2013) STUB1 is essential for T-cell activation by ubiquitinating CARMA1. Eur J Immunol 43 (4):1034–1041. https://doi.org/10.1002/eji.201242554
Wertz IE, O'Rourke KM, Zhou H, Eby M, Aravind L, Seshagiri S et al (2004) De-ubiquitination and ubiquitin ligase domains of A20 downregulate NF-kappaB signalling. Nature 430(7000):694–699. https://doi.org/10.1038/nature02794
Wildin RS, Ramsdell F, Peake J, Faravelli F, Casanova JL, Buist N et al (2001) X-linked neonatal diabetes mellitus, enteropathy and endocrinopathy syndrome is the human equivalent of mouse scurfy. Nat Genet 27(1):18–20. https://doi.org/10.1038/83707
Willerford DM, Chen J, Ferry JA, Davidson L, Ma A, Alt FW (1995) Interleukin-2 receptor alpha chain regulates the size and content of the peripheral lymphoid compartment. Immunity 3(4):521–530

Williams LM, Rudensky AY (2007) Maintenance of the Foxp3-dependent developmental program in mature regulatory T cells requires continued expression of Foxp3. Nat Immunol 8(3):277–284. https://doi.org/10.1038/ni1437

Wong J, Obst R, Correia-Neves M, Losyev G, Mathis D, Benoist C (2007) Adaptation of TCR repertoires to self-peptides in regulatory and nonregulatory CD4+ T cells. J Immunol 178(11):7032–7041

Wu Y, Borde M, Heissmeyer V, Feuerer M, Lapan AD, Stroud JC et al (2006) FOXP3 controls regulatory T cell function through cooperation with NFAT. Cell 126 (2):375–387. https://doi.org/10.1016/j.cell.2006.05.042

Xiao L, Chen D, Hu P, Wu J, Liu W, Zhao Y et al (2011) The c-Abl-MST1 signaling pathway mediates oxidative stress-induced neuronal cell death. J Neurosci 31 (26):9611–9619. https://doi.org/10.1523/JNEUROSCI.0035-11.2011

Yamazaki S, Iyoda T, Tarbell K, Olson K, Velinzon K, Inaba K et al (2003) Direct expansion of functional CD25+ CD4+ regulatory T cells by antigen-processing dendritic cells. J Exp Med 198(2):235–247. https://doi.org/10.1084/jem.20030422

Yokoi N, Fujiwara Y, Wang HY, Kitao M, Hayashi C, Someya T et al (2008) Identification and functional analysis of CBLB mutations in type 1 diabetes. Biochem Biophys Res Commun 368(1):37–42. https://doi.org/10.1016/j.bbrc.2008.01.032

Yu FX, Zhao B, Guan KL (2015) Hippo pathway in organ size control, tissue homeostasis, and cancer. Cell 163 (4):811–828. https://doi.org/10.1016/j.cell.2015.10.044

Zhang Y, Foreman O, Wigle DA, Kosari F, Vasmatzis G, Salisbury JL et al (2012) USP44 regulates centrosome positioning to prevent aneuploidy and suppress tumorigenesis. J Clin Invest 122(12):4362–4374. https://doi.org/10.1172/JCI63084

Zhang P, Tey SK, Koyama M, Kuns RD, Olver SD, Lineburg KE et al (2013a) Induced regulatory T cells promote tolerance when stabilized by rapamycin and IL-2 in vivo. J Immunol 191(10):5291–5303. https://doi.org/10.4049/jimmunol.1301181

Zhang J, Chen C, Hou X, Gao Y, Lin F, Yang J et al (2013b) Identification of the E3 deubiquitinase ubiquitin-specific peptidase 21 (USP21) as a positive regulator of the transcription factor GATA3. J Biol Chem 288(13):9373–9382. https://doi.org/10.1074/jbc.M112.374744

Zhao B, Li L, Tumaneng K, Wang CY, Guan KL (2010) A coordinated phosphorylation by Lats and CK1 regulates YAP stability through SCF(beta-TRCP). Genes Dev 24(1):72–85. https://doi.org/10.1101/gad.1843810

Zheng Y, Rudensky AY (2007) Foxp3 in control of the regulatory T cell lineage. Nat Immunol 8(5):457–462. https://doi.org/10.1038/ni1455

Zheng SG, Gray JD, Ohtsuka K, Yamagiwa S, Horwitz DA (2002) Generation ex vivo of TGF-beta-producing regulatory T cells from CD4+ CD25- precursors. J Immunol 169(8):4183–4189. https://doi.org/10.4049/jimmunol.169.8.4183

Zheng SG, Wang JH, Gray JD, Soucier H, Horwitz DA (2004) Natural and induced CD4+ CD25+ cells educate CD4+ CD25- cells to develop suppressive activity: the role of IL-2, TGF-beta, and IL-10. J Immunol 172(9):5213–5221. https://doi.org/10.4049/jimmunol.172.9.5213

Zheng SG, Wang JH, Stohl W, Kim KS, Gray JD, Horwitz DA (2006) TGF-beta requires CTLA-4 early after T cell activation to induce FoxP3 and generate adaptive CD4+ CD25+ regulatory cells. J Immunol 176 (6):3321–3329. https://doi.org/10.4049/jimmunol.176.6.3321

Zheng Y, Josefowicz SZ, Kas A, Chu TT, Gavin MA, Rudensky AY (2007a) Genome-wide analysis of Foxp3 target genes in developing and mature regulatory T cells. Nature 445(7130):936–940. https://doi.org/10.1038/nature05563

Zheng SG, Wang J, Wang P, Gray JD, Horwitz DA (2007b) IL-2 is essential for TGF-beta to convert naive CD4+ CD25- cells to CD25+ Foxp3+ regulatory T cells and for expansion of these cells. J Immunol 178 (4):2018–2027. https://doi.org/10.4049/jimmunol.178.4.2018

Zheng SG, Wang J, Horwitz DA (2008) Cutting edge: Foxp3+ CD4+ CD25+ regulatory T cells induced by IL-2 and TGF-beta are resistant to Th17 conversion by IL-6. J Immunol 180(11):7112–7116

Zheng Y, Chaudhry A, Kas A, deRoos P, Kim JM, Chu TT et al (2009) Regulatory T-cell suppressor program co-opts transcription factor IRF4 to control T(H) 2 responses. Nature 458(7236):351–356. https://doi.org/10.1038/nature07674

Zheng Y, Josefowicz S, Chaudhry A, Peng XP, Forbush K, Rudensky AY (2010) Role of conserved non-coding DNA elements in the Foxp3 gene in regulatory T-cell fate. Nature 463(7282):808–812. https://doi.org/10.1038/nature08750

Zhou L, Lopes JE, Chong MM, Ivanov II, Min R, Victora GD et al (2008) TGF-beta-induced Foxp3 inhibits T (H)17 cell differentiation by antagonizing RORgammat function. Nature 453(7192):236–240. https://doi.org/10.1038/nature06878

Zhu F, Yi G, Liu X, Zhu F, Zhao A, Wang A et al (2018) Ring finger protein 31-mediated atypical ubiquitination stabilizes forkhead box P3 and thereby stimulates regulatory T-cell function. J Biol Chem 293 (52):20099–20111. https://doi.org/10.1074/jbc.RA118.005802

Zinngrebe J, Rieser E, Taraborrelli L, Peltzer N, Hartwig T, Ren H, et al (2016) --LUBAC deficiency perturbs TLR3 signaling to cause immunodeficiency and autoinflammation. J Exp Med 213(12):2671–2689. https://doi.org/10.1084/jem.20160041.

Metabolic Choice Tunes Foxp3+ Regulatory T Cell Function

5

Xiaoxia Wang, Hao Cheng, Yige Shen, and Bin Li

Abstract

Metabolic programs and dynamic nutrient signaling can direct cell biological function. Cellular metabolism and biological function are coordinated to cell activity. Regulatory T cells (Foxp3+ Tregs) expressing the key transcription factor FOXP3 play critical roles in the maintenance of immune tolerance and in the control of immune homeostasis. A bundle of data demonstrated that Foxp3+ Tregs were influenced and regulated by Toll-cell receptor (TCR) and costimulatory signals, cytokine conditions and metabolic changes, including metabolites, etc. In this context, Foxp3+ Tregs are impacted by different environmental conditions and metabolic differences associated with diverse transcriptional patterns, which, in turn, display a high degree of plasticity and tissue specificity. During the past decades, significant progresses have been made in understanding the correlation between metabolic changes and manipulation of Foxp3+ Treg function. Taken together, this chapter aims to summarize the important advances in the fields, decipher what metabolic ways are involved in Foxp3+ Tregs, and how metabolism modulates Foxp3 expression, stability, and suppressive functions, which may provide a potential pace on lightening up Foxp3+ Treg-mediated immune functions.

Keywords

Metabolism · Foxp3+ Tregs · Immunomodulatory functions

Abbreviations

3-HAA	3-hydroxyanthranilic acid
AhR	Aryl hydrocarbon receptor
BCAAs	Branched-chain amino acids
CBM	CARMA1-BCL10-MALT1
FAO	Fatty acid oxidation
GlcNAc	*N*-acetylglucosamine
GLS	Glutaminase
HDAC	Histone deacetylase
IDO	Indoleamine 2,3-dioxygenase
LNAA	Large neutral amino acids
MS	Multiple sclerosis
NO	Nitric oxide

X. Wang · B. Li (✉)
Shanghai Institute of Immunology, Department of Immunology and Microbiology, Shanghai Jiao Tong University School of Medicine, Shanghai Jiao Tong University, Shanghai, China
e-mail: wangxiaoxia777@shsmu.edu.cn; binli@shsmu.edu.cn

H. Cheng
Bio-X Institutes, Shanghai Jiao Tong University, Shanghai, China
e-mail: chenghaowj@sjtu.edu.cn

Y. Shen
Shanghai Jiao Tong University School of Medicine, Shanghai Jiao Tong University, Shanghai, China
e-mail: sun-girl@sjtu.edu.cn

S.-G. Zheng (ed.), *T Regulatory Cells in Human Health and Diseases*, Advances in Experimental Medicine and Biology 1278, https://doi.org/10.1007/978-981-15-6407-9_5

OXPHOS	Oxidative phosphorylation
PARP1	Poly (ADP-ribose) polymerase 1
PIM1	Proto-oncogene serine/threonine-protein kinase
TCR	Toll-cell receptor

5.1 Introduction

Regulatory T cells (Foxp3+ Tregs), widely known as Foxp3-expressing T cells, are a special subset of CD4+ T cells, which are critically involved in the maintenance of immune tolerance to self-antigen and in the control of autoimmunity (Wing and Sakaguchi 2010; Josefowicz et al. 2012). Foxp3+ Tregs are classified into at least two subsets: one derived from thymus is called thymus or natural Tregs and the other subsets arose from peripheral conversion of CD4+CD25− T cells (Horwitz et al. 2008; Zheng et al. 2002). Both of them respond to inflammatory signals to exert the immunomodulatory role in sustaining immune homeostasis. Metabolic fitness is intimately linked to Foxp3+ Treg fate and function. The metabolic state and the functional capability are dynamically adapted and balanced upon different environment cues in Foxp3+ Tregs.

5.2 Treg Cell Metabolism State and Function

Experimental evidence has indicated that Treg activation and differentiation rely on multiple signaling pathways. The integration of multiple cell signals can directly affect Treg cell transcriptional programs, which were involved in Foxp3+ Tregs proliferation, cytokines production, and energy metabolism. However, how Treg cells finely integrate functional capability with cellular metabolism under certain environment milieu remains elusive (Chen et al. 2019) (Fig. 5.1). In this context, it must be noted that metabolic states in Tregs are totally different compared with effector T cells. Previous studies have also reported that Tregs and effector T cells have different tendencies in using glycolysis and fatty acid oxidation (FAO) (Procaccini et al. 2016).

However, which metabolic energy do Treg cells prefer to utilize is still controversial due to different conditions and species difference in recent years. Most studies supported that glycolysis has detrimental effects on Treg functions; however, under certain conditions, aerobic glycolysis has been reported to favor Foxp3 expression. In vitro, human regulatory T cells (iTreg) induced from naïve CD4+ T cells are highly proliferative, and the FOXP3 splicing variants containing exon 2 controlled suppressive capability of human iTreg is tightly relied on glycolysis through the glycolytic enzyme enolase-1 (De Rosa et al. 2015). Human activated Tregs uniquely accelerate glucose consumption for their suppression compared with effector T cells. TLR8 signaling inhibits glycolytic enzyme expression and suppresses glycolysis metabolism through the downregulation of mTORC1-HIF1α pathway, resulting in the reversal of human Treg cell suppressive function. Furthermore, findings also suggested that human Treg cells could trigger senescence in responder T cells through DNA damage induction under shortage of glucose in the tumor suppressive environment (Li et al. 2019). Ex vivo, highly proliferative human iTreg showed active mTOR activity and expressed more Glut1 (Michalek et al. 2011). The metabolic profiles of human Treg cells are distinct from murine Treg cells shown to use lipid oxidation as a primary metabolic pathway (Michalek et al. 2011; Buck Michael et al. 2016; Pearce et al. 2009; Shi et al. 2011). Mouse-induced iTreg showed less Glut1 expression and slower glycolytic rate compared with Th2 and Th17 induced in vitro, and the results revealed that glycolysis inhibition by 2-DG treatment could promote mouse iTreg differentiation but prohibit Teff proliferation. In contrast, selective inhibition mitochondrial oxidation by rotenone sharply reduced Treg differentiation. Metabolite level analysis also indicates that Tregs are inclined to pyruvate oxidation and Th17 cells prefer glutamine oxidation though TCA cycle intermediates in Tregs are equivalent to those in Th17 cells (Gerriets et al. 2015). Further, TLR

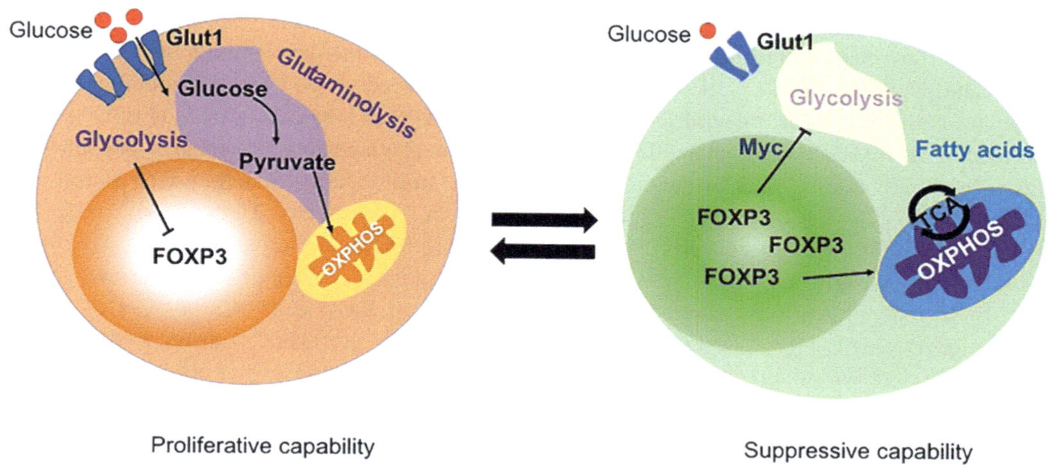

Fig. 5.1 In proliferative state, Treg cells are prone to use glycolysis metabolism, although glycolysis ways have been reported to inhibit FOXP3 expression. By contrast, FOXP3 expression itself could suppress glycolysis ability and enhance Treg cell function by triggering OXPHOS

signaling that increased glycolysis through PI3K-AKT-mTORC1 axis would promote Treg cell proliferation but impair Treg cell suppressive capability. Reciprocally, forced Treg cells key transcription factor Foxp3 diminished glycolysis and anabolic metabolism while promoted fatty acid oxidation (FAO). Glut1 overexpression would rescue Treg numbers while restrain their immunosuppressive function. These demonstrate that there exists a dynamic regulation that TLR signal and Foxp3 expression couple glucose metabolism to modulate the rapid switch of Treg cell proliferative capability and suppressive function in order to adapt to the local environment perturbs. Thus, we could understand that why in chronic autoimmune diseases, inflammatory signaling triggers Treg cell accumulation but fails to suppress inflammation effectively (Kong et al. 2012a, b; Zhou et al. 2010; Lu et al. 2014). While in tumor microenvironment, though low glucose condition, high lactate level might enhance the suppressive function of Treg cells (Gerriets et al. 2016). Another reason is that Foxp3 expression suppressed Myc expression and made the Treg cells adapt from glycolysis to oxidative phosphorylation state, which enables Treg cells survive well than effector T cells, meanwhile high lactate impairs effector T cell function through LDH-mediated NAD depletion in low glucose, lactate-rich cancer milieu (Angelin et al. 2017)

Glycolysis is primarily activated in Treg cells through mTOR and tends to suppress Foxp3 expression and Treg lineage stability. mTORC1 activity is high in human and murine Tregs (Procaccini et al. 2010; Zeng et al. 2013), which restrains TCR or IL-2 stimulation-induced proliferation in vitro (Procaccini et al. 2010). The activation of mTOR is tightly linked to diverse cellular processes including glycolysis metabolism. Treg cells are also activated by upstream signals that drive mTOR activation (Sauer et al. 2008), and activation of the PI3K/AKT/mTOR signaling axis inhibits Foxo transcription levels and promotes Hif1α, which destabilizes the master Treg cell transcription factor Foxp3 (Shi et al. 2011; Ouyang et al. 2012; Dang et al. 2011). One major downstream effect of PI3K signaling is induction of aerobic glycolysis, which is increasingly emerging as a key control mechanism of Treg function. Excessive PI3K activity is detrimental to Tregs since loss of PTEN diminishes Treg functions in mice (Huynh et al. 2015; Shrestha et al. 2015). HIF1α is responsible for the glycolytic response downstream of mTOR (Shi et al. 2011). As rapamycin treatment was similar to HIF1α deficiency at altering Th17 and Treg cell differentiation (Delgoffe et al. 2009; Powell and Delgoffe 2010). HIF1α-dependent glycolytic pathway is an integral component downstream of mTOR to mediate Th17 and Treg cell differentiation. HIF1α-deficient and 2-DG-treated T cells have clearly established a negative role for HIF1α-induced metabolic reprogramming in Treg cell development (Shi et al. 2011),which influence the nutrient availability inducing effector T cell responses. In addition, the mTOR activation in effector T cells also affects Treg cell differentiation and functions. Several groups have found that rapamycin treatment for short time could induce de novo expression of Foxp3 from naïve T cells and acute treatment could drive functional Treg cell expansion in vitro. While chronic rapamycin stimulation does not induce Treg proliferation in the absence of exogenous IL-2. Indeed, IL-2 is crucial for the differentiation of Tregs (Zheng et al. 2007; Davidson et al. 2007). However, rapamycin treatment does not fully block mTORC1 functions, and there are some limitations to assess the true role of mTORC1 in functions. Recent papers have utilized genetic modes to ascertain the function of mTORC1 in Tregs. The first finding demonstrated that mTOR−/− T cells spontaneously develop into iTregs in the absence of IL-2 and TGF-β (Delgoffe et al. 2009). Genetic deletion mTORC1, but not mTORC2, ablated Treg cell suppressive functions and mice that bearing Raptor-deficient Tregs drove fatal autoimmune disease in vivo (Park et al. 2013). Consistent with this observation, Tsc1 deficiency in T cells leads to a loss of Foxp3+ T cells in the periphery, and Tsc1-deficient Tregs lose suppressive function in vivo. These studies supported the notion that mTORC1 is a positive regulator of Treg functions. A bunch of studies indicate that mTORC1 inhibition in Tregs may disrupt immune homeostasis and induce autoimmunity or hyperinflammation. However, clinical targeting mTORC1 by inhibitor in some solid tumor environment may enhance antitumor immune responses by suppressing Treg function. A study demonstrated by Kishore et al. (2017) provides new evidence that Treg cells could use glycolytic machineries to migrate into the inflamed sites mediated by the enzyme glucokinase induced via PI3K-mTORC2 pathway, although these Treg cells arrived at inflammatory location continue to use oxidative phosphorylation to proliferate and maintain their suppressive function. Moreover, Tregs from relapsing-remitting multiple sclerosis patients have altered IL-2 signaling, which drives excessive mTOR signaling in these cells (Carbone et al. 2013). This defect is correlated with reduced Treg proliferation and Foxp3 expression (Carbone et al. 2013), further supporting the contrary notion that excessive mTOR signaling dampens Treg responses. Nonetheless, as Tregs express IL-2 receptor, it is likely that low dose of IL-2 also directly targets Treg cells to improve their frequency and functional capacity in the disease setting (Koreth et al. 2011; Saadoun et al. 2011; Ye et al. 2018).

Rictor deletion in Tregs alone results in no gross abnormalities, unlike those observed in mice bearing Raptor-deficient Tregs (Zeng et al. 2013). Thus, mTORC2 does not appear to have a dominant role in maintaining Treg functions in vivo or in promoting iTreg generation. Moreover, mTORC1 was found to be essential for mouse Treg cell homeostasis and suppressive function via promotion of lipid and cholesterol biosynthesis (Zeng et al. 2013). Furthermore, mitochondrial respiratory and oxidative phosphor ylation are also critical for mouse Treg cell proliferation and suppressive function. ALL link hyperactivation of mTOR to dysregulated T cell responses in autoimmunity (Carbone et al. 2013; Daley et al. 2013), underscoring the importance of addressing the role mTOR signaling serves in Treg biology. Metabolic programs used by various types of Treg cells for survival and function are far more flexible than what was originally anticipated (He et al. 2017). It would be interesting to further characterize how metabolism regulates Foxp3+ Treg development or functions in different disease conditions.

Emerging studies have demonstrated that Treg cells exhibit a unique metabolic profile characterized by an increase in mitochondrial metabolism relative to other CD4+ effector subsets (Michalek et al. 2011; Gerriets et al. 2015). Tregs, like memory CD8+ T cells, rely on FAO for their basal metabolism but utilize some degree of aerobic glycolysis to properly execute their suppressive functions. High OXPHOS (oxidative phosphorylation) activity facilitates Treg function in low-glucose and high-lactate environments in peripheral organs, such as the large intestine (Grzes et al. 2017). Deletion of the metabolic sensor Stk11 in Tregs disrupts mitochondrial fitness and metabolites (Fu et al. 2019). Recent studies have demonstrated that the better survival and proliferation ability than effector T cells in the glucose restricted tumor environment are ascribed to the utilization of both glycolysis and fatty acid oxidation. Targeting metabolism for cancer therapy has been investigated for several decades (Luengo et al. 2017). It is noted that awareness of the metabolic dynamics of Tregs in tumor may provide a strategy for cancer immunotherapy. The gut microbiota-derived metabolites and short chain fatty acids could maintain the colonic Treg pools and protect against colitis through downregulation of histone deacetylase 6 (HDAC6) and histone deacetylase (HDAC9) (Smith et al. 2013). A study reported that Treg cell-specific deletion of mitochondrial respiratory chain complex III impaired Treg suppressive functions and resulted in the development of fatty inflammatory disease early on the stage. Furthermore, loss of complex III in Treg cells increased DNA methylation, which suggests that mitochondrial complex III is essential for Treg cell-mediated immunomodulatory function (Weinberg et al. 2019). And genetic ablation of Tfam in Tregs, which is vital for mitochondrial respiration and controls mitochondrial DNA replication, transcription and packaging, reduced Treg cell proliferation under glucose deprivation condition in vitro. Tfam deletion enhanced methylation in the TSDR of Foxp3 locus, which could promote Foxp3 expression and stability. Furthermore, they demonstrated that loss of Tfam in Tregs affects Treg cell homing and stability, resulting in tissue inflammation in colitis, but enhances tumor rejection (Chai et al. 2019).

T cells upon activation would quickly modulate their metabolic programs to match the demands for mediating immune response. Previous data demonstrated that activated T cells increase cellular amino acid uptake to meet the requirements of protein synthesis for proliferation, the demands of nucleotide synthesis, energy metabolism, and redox control for immune function. Importantly, more and more researchers have revealed that amino acids not only serve as building blocks for protein biosynthesis but also participate in coupling environment signals and metabolic homeostasis (see summary in Box 5.1).

Glutamine is the most abundant amino acids in serum, making it more easily available and potential resource for cells, also lymphocyte T cells. Glutamine donates an amine group, and its first product glutamate also serve as glutathione synthesis. Glutamate can be converted into a-ketoglutarate, which is an important intermediate in the classical TCA cycle. Previous studies have referred that mitogen-induced T cell proliferation and cytokine production require high levels of glutamine, and T cells are highly sensitive to glutamine (Carr et al. 2010). T-cell activation needs selective glutamine import, accompanied with increased expression of glutamine transporter genes. Neutral amino acid transporter ASCT2 is Na+-dependent transporter, whose expression regulated downstream of TCR and CD28 signaling by the CARMA1-BCL10-MALT1 (CBM) complex. Glutamine uptake relies on ASCT2 and ASCT2 deficiency in T cells led to decreased mTORC1 activity and decreased Th1 and Th17 responses (Nakaya et al. 2014). Another study demonstrated that glutaminolysis integrates with glycolysis to facilitate T cell function. Targeting glutaminase (GLS) would restrain Th17 response yet enhance Th1 response (Johnson et al. 2018). In T cells, reduced branching induced by aerobic glycolysis and glutaminolysis promotes Th17 differentiation over Treg differentiation and the salvage of *N*-acetylglucosamine (GlcNAc) would promote Treg differentiation while blocking Th17 skewing, which was further validated in experimental autoimmune encephalomyelitis, a mouse model of multiple sclerosis (MS), as well as autoimmune diabetes in the non-obese diabetic mouse model (Araujo et al. 2017). Moreover, lack of glutamine affects the differentiation of naïve T cells into distinct subsets. In the skewing condition, glutamine deprivation results in more Th1 cell generation instead of Foxp3+ Treg cells (Klysz et al. 2015).

Except glutamine, L-arginine is a common amino acid for protein synthesis and also as a precursor for multiple metabolites, including polyamines and nitric oxide (NO) that have strong immunomodulatory properties (Grohmann and Bronte 2010). T cell intracellular L-arginine is rapidly metabolized upon antigen activation. Elevating L-arginine levels in activated T cells skewed the metabolism shifting from glycolysis to oxidative phosphorylation, mediating the reprogramming of T cells toward high survival (Geiger et al. 2016).

A single system L transporter, Slc7a5, coupled large neutral amino acids (LNAA) uptake in activated T cells. Evidence showed that pharmacological blockade of system L transport function hindered T cell proliferation (Schulte et al. 2018). Pathogen infection and TCR stimulation induced more LNAAs into effector T cells via system transporters, including leucine (Sinclair et al. 2013). Loss of Slc7a5 in T cells prevented naïve CD4+ T cells skewing into Th1 and Th17 cells but showed unimpaired Treg cell differentiation and showed obvious defect in glucose transport and glutamine uptake in response to immunological activation (Sinclair et al. 2013)

Kynurenine, the tryptophan metabolite, has critical immunomodulatory properties, which is also a ligand for aryl hydrocarbon receptor (AHR). Slc7a5 also facilitates kynurenine transporting across T cells triggered by TCR or proinflammatory cytokines (Sinclair et al. 2018)

Tryptophan, catabolized by an enzyme indoleamine 2,3-dioxygenase (IDO), was reported to play roles in modulating T cells function in vitro (Ramalingam et al. 2012; Huang et al. 2017). In the absence of tryptophan in the culture medium, activated T cells would arrest at mid-G1 phase and become more susceptible to apoptosis partially via Fas signaling (Lee et al. 2002). Thus, mature antigen-presenting cells expressing IDO could induce activated T cells to cell death prior to cell cycle progression in specific condition. And IDO-deficient mice developed severe EAE symptom with reduced regulatory T-cell response (Yan et al. 2010). Tryptophan catabolite 3-hydroxyanthranilic acid (3-HAA) promoted Treg cell differentiation while inhibited Th1 and Th17 percentage (Krause et al. 2011)

Branched-chain amino acids (BCAAs) including leucine, isoleucine, and valine are among the most hydrophobic of amino acids. BCAAs play an important role in protein turnover and glucose metabolism. Reduced BCAAs induced defective Foxp3+ Treg proliferation and decreased Foxp3+ Treg number in vivo. Meanwhile, Treg cell-specific Slc3a2-deficient mice impaired Treg percentage due to reduced isoleucine content. Also, significantly decreased isoleucine in Slc3a2-deficient Treg cells displayed impaired suppressive function. All data demonstrated BCAAs could maintain Treg cell property via mTORC1-mediated immune signals (Ikeda et al. 2017). mTOR signals are propagated and tuned to impart Treg cell programs and functions in vivo (Gerriets et al. 2016; Zeng et al. 2013). Amino acids, especially arginine and leucine, are reported to be crucial to license and maintain TCR-dependent mTORC1 activity. Hao Shi et al. (2019) demonstrated that amino acids combined with TCR signals act as a costimulatory-like signal to activate mTORC1 activity via RagA/B or Rheb1/2, which can affect effector Treg accumulation and function. Thus, amino acid abundance and concentration may be sufficient nutrients available for mTORC1 activation underlined Treg cell-mediated immune tolerance.

Box 5.1: Amino acids not only supply for protein synthesis but also can meet the other requirements for cellular immune function

Glutamine

- Glutamine is the most abundant amino acid in serum and can be uptake via ASCT2 (Nakaya et al. 2014).
- Glutamine deprivation inhibited Treg cell differentiation (Klysz et al. 2015).

(continued)

Box 5.1 (continued)

L-Arginine

- L-Arginine can be precursors for multiple metabolites, including polyamines and nitric oxide (NO) that have strong immunomodulatory properties (Grohmann and Bronte 2010).
- Loss of Slc7a5, a single system L transporter for L-arginine, prevented Th1 and Th17 skewing, but not Treg cell generation (Schulte et al. 2018).

Tryptophan

- Tryptophan, catabolized by an enzyme indoleamine 2,3-dioxygenase (IDO); IDO-deficient mice showed reduced regulatory T cell response (Yan et al. 2010).
- Tryptophan catabolite 3-hydroxyanthranilic acid (3-HAA) promoted Treg cell differentiation (Krause et al. 2011).

Branched chain amino acids (BCAAs)

- Branched chain amino acids including valine, leucine, and isoleucine could activate mTOR pathway (Shi et al. 2019).
- Treg cell-specific Slc3a2-deficient mice impaired Treg percentage and suppressed Treg cell immunomodulatory functions (Ikeda et al. 2017).

5.3 Cellular Metabolism, Epigenetic Regulation, and Treg Cells Function

Metabolites have been previously investigated to regulate Treg cell development and function. Lactate promotes NF-κB activity that, in turn, enhances Foxp3 expression and induces Treg polarization from naive T cells (Comito et al. 2019), although NF-κB signal is complicated for Tregs development and function (Yang et al. 2018, 2019). Genetic knockdown of citrate synthase, elevated nuclear acetyl-CoA levels, increased histone acetylation at the FOXP3 promotor and induction of FOXP3 transcription (Hawse et al. 2019). Vitamin A metabolite *all trans*-retinoic acid induces histone acetylation at the *Foxp3* gene promoter and expression of the Foxp3 protein in CD4 T cells (Kang et al. 2007; Lu et al. 2011). In addition, all *trans*-retinoic acids also displayed potent activities on sustaining nTreg stability and function under inflammatory condition (Lu et al. 2011; Lu et al. 2014). SCFAs, namely acetate and propionate, were recently shown to promote accumulation of Treg cells in the colon through activating GPR43 protein (Smith et al. 2013). Microbial fermentation product, butyrate, could also induce the differentiation of colonic Treg cells in naive mice and increase systemic Tregs in mice suffering from autoimmuneuveitis (Furusawa et al. 2013; Chen et al. 2017). The epigenetic regulation of *Foxp3* gene transcriptions, interactions, and modifications between components of the FOXP3 complex together regulate Treg cell activity. The transcription factor FOXP3 is essential for Treg cell development and differentiation. The function of FOXP3 is regulated at multiply levels accurately and complicatedly, including transcriptional (Li and Greene 2008), translational, and posttranslational modification (Li et al. 2007). All these regulations determine the stability, plasticity, and functional activity of Treg cells, which manipulate the suppressive functions of Foxp3+ Treg cells. Plenty of signaling and pathways have been reported to participate into the regulation of transcriptional and translational levels, such as TCR signaling, IL2-STAT5 pathway, PI3K-AKT signaling, and even microRNAs. Most importantly, FOXP3 could crosstalk with many proteins or enzymes to be modified through posttranslational (Li and Greene 2008; Samanta et al. 2008). Modification ways of FOXP3 have been well studied, including phosphorylation, acetylation, ubiquitination, etc. PIM1 (proto-oncogene serine/threonine-protein kinase Pim-1) phosphorylates FOXP3 at Ser422, directly reducing Treg cell suppressive capability (Li et al. 2014), otherwise inhibition of PIM1 activity by Kaempferol could promote

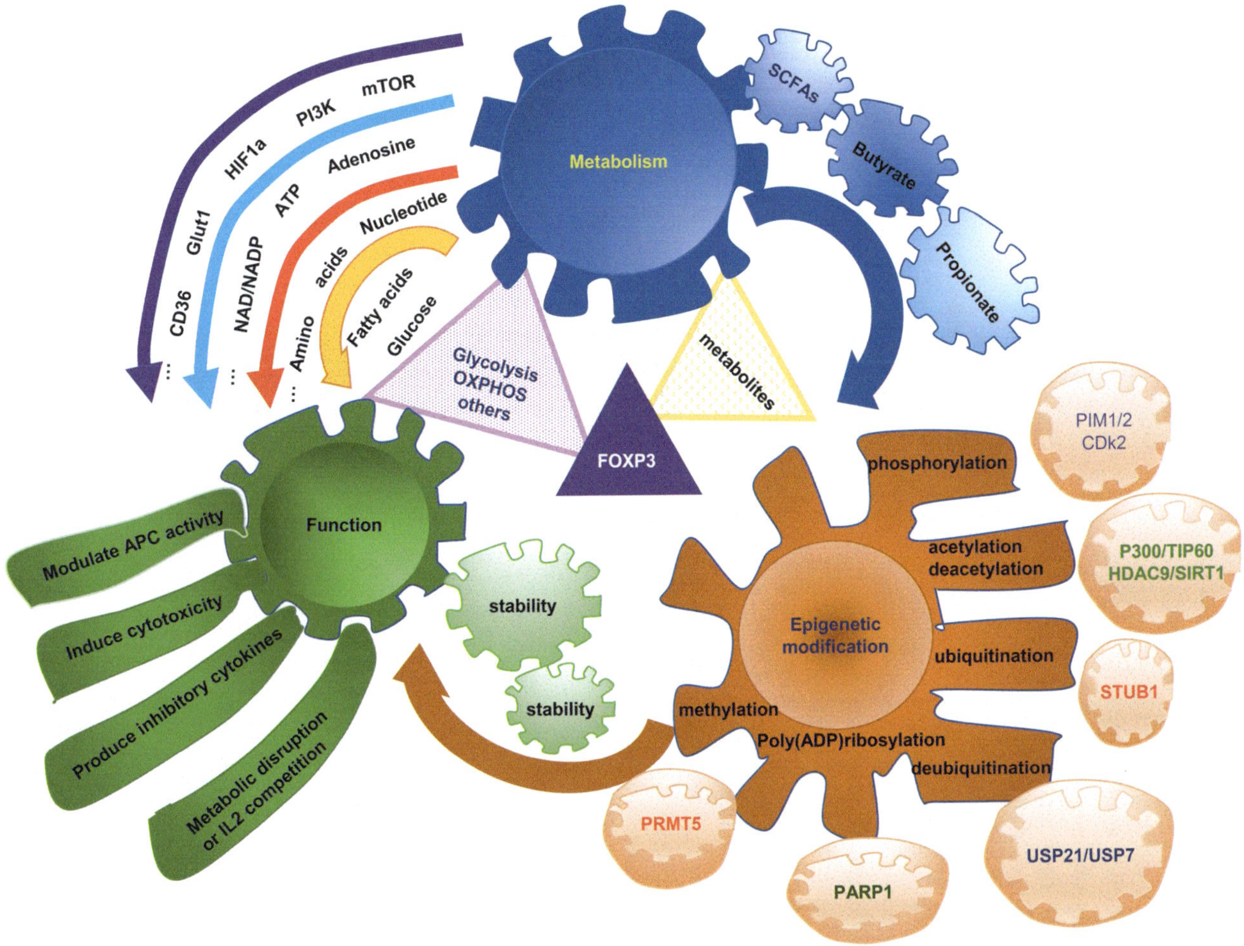

Fig. 5.2 In the process of immunity training, epigenetic modifications on FOXP3 lead to stronger gene changes at different levels, including transcription, translation, and posttranslational, further regulating Treg cell stability and manipulating FOXP3+ Treg cell function. Moreover, metabolites could also affect some proteins and then regulate epigenome through diverse modifications, such as histone acetylation and ubiquitination. Meanwhile, Treg cells that has exerted different functions under certain conditions would adopt proper metabolic ways and produce matching metabolites

Foxp3 transcriptional activity (Lin et al. 2015). Another paper has also reported cyclin-dependent kinase 2 could interact and phosphorylate FOXP3 at the repressor domain and negatively regulate the stability and activity of FOXP3 (Chunder et al. 2012). Two acetyltransferases, Tip60 and p300, have been reported as enzymes that are responsible for FOXP3 acetylation (Jorg et al. 2010). Tip60 interacted with FOXP3 and regulated its inhibition on IL2 expression; p300 could be recruited to acetylate FOXP3 (Liu et al. 2013). On the contrary, two lysine deacetylases, HDAC9 and SIRT1, can associate with FOXP3, downregulate FOXP3 acetylation level, and prohibit Treg cell activity (Beier et al. 2011; de Zoeten et al. 2010). On the other aspect, lipopolysaccharide-induced increases in E3 ubiquitin ligase STUB1 expression could promote FOXP3 polyubiquitination and degradation, decreasing Foxp3+ Treg cell suppressive ability to Teff proliferation (Chen et al. 2013). E3 deubiquitinase USP21 stabilizes FOXP3 protein by mediating its deubiquitination and maintains the expression of Treg signature genes (Li et al. 2016). Meanwhile, PARP-1 (poly(ADP-ribose) polymerase 1) could interact with FOXP3 and induce poly(ADP-ribose) polymerase modification of FOXP3, further promoting FOXP3 polyubiquitination mediated by STUB1 (Luo et al. 2015). Loosdregt and colleagues demonstrated that ectopic expression of the deubiquitinase USP7 oppositely decreased FOXP3 polyubiquitination and increased its stability (van Loosdregt et al. 2013). Now more and more researchers have paid attention to the inner metabolic switch and metabolite changes on Treg cell epigenetic regulations and vice visa. It was reported that butyrate, a short-chain fatty acid, boosted generation of Treg cells dependent on intronic enhancer CNS1 of FOXP3, and another SCFA, propionate could induce Treg cell generation from the periphery through histone deacetylase (HDAC) inhibition (Arpaia et al. 2013). Also a report has shown that acetate increased acetylation at the Foxp3 promoter through HDAC9 inhibition, finally promoting Treg cell numbers and suppressing allergic airways disease after high-fiber or acetate feeding (Thorburn et al. 2015). Collectively, metabolic changes, epigenetic regulation, and Foxp3+ Treg cell function, the three are coordinated, interdependent, and mutually conditional, that is metabolic changes modulate epigenetic modification of FOXP3 and affect Treg cell function; in turn, Treg cell functional switch would reach to metabolic adaption, further impacting epigenetic perturbs, like DNA methylation and chromatin confirmation (Fig. 5.2).

5.4 Conclusion

The metabolic state of Treg cells is determined by different conditions and external signals. Metabolic fitness is intimately linked to T cell fate and function. Although elevated glycolysis is commonly associated with Treg activation and thought to be detrimental to Treg lineage stability, fatty acid oxidation preferred by activated Tregs is supposed to be helpful for their suppressive capability. And tTreg tendency for cholesterol synthesis may be an interesting area of discovery. Research on amino acid metabolism and related metabolic enzymes has also been revealing the moonlighting roles in Treg biology. Metabolic interventions have been investigated for cancer therapy for several years and targeting metabolism may offer a promising new approach by modulating Treg function to break the tolerance in certain tumor therapy.

References

Angelin A, Gil-de-Gomez L, Dahiya S, Jiao J, Guo L, Levine MH, Wang Z, Quinn WJ 3rd, Kopinski PK, Wang L, Akimova T, Liu Y, Bhatti TR, Han R, Laskin BL, Baur JA, Blair IA, Wallace DC, Hancock WW, Beier UH (2017) Foxp3 reprograms T cell metabolism to function in low-glucose, high-lactate environments. Cell Metab 25(6):1282–1293 e1287. https://doi.org/10.1016/j.cmet.2016.12.018

Araujo L, Khim P, Mkhikian H, Mortales C-L, Demetriou M (2017) Glycolysis and glutaminolysis cooperatively control T cell function by limiting metabolite supply to N-glycosylation. eLife 6:e21330. https://doi.org/10.7554/eLife.21330

Arpaia N, Campbell C, Fan X, Dikiy S, van der Veeken J, deRoos P, Liu H, Cross JR, Pfeffer K, Coffer PJ,

Rudensky AY (2013) Metabolites produced by commensal bacteria promote peripheral regulatory T-cell generation. Nature 504(7480):451–455. https://doi.org/10.1038/nature12726

Beier UH, Wang L, Bhatti TR, Liu Y, Han R, Ge G, Hancock WW (2011) Sirtuin-1 targeting promotes Foxp3+ T-regulatory cell function and prolongs allograft survival. Mol Cell Biol 31(5):1022–1029. https://doi.org/10.1128/MCB.01206-10

Buck Michael D, O'Sullivan D, Klein Geltink Ramon I, Curtis Jonathan D, Chang C-H, Sanin David E, Qiu J, Kretz O, Braas D, van der Windt Gerritje JW, Chen Q, Huang Stanley C-C, O'Neill Christina M, Edelson Brian T, Pearce Edward J, Sesaki H, Huber Tobias B, Rambold Angelika S, Pearce Erika L (2016) Mitochondrial dynamics controls T cell fate through metabolic programming. Cell 166(1):63–76. https://doi.org/10.1016/j.cell.2016.05.035

Carbone F, De Rosa V, Carrieri PB, Montella S, Bruzzese D, Porcellini A, Procaccini C, La Cava A, Matarese G (2013) Regulatory T cell proliferative potential is impaired in human autoimmune disease. Nat Med 20(1):69–74. https://doi.org/10.1038/nm.3411

Carr EL, Kelman A, Wu GS, Gopaul R, Senkevitch E, Aghvanyan A, Turay AM, Frauwirth KA (2010) Glutamine uptake and metabolism are coordinately regulated by ERK/MAPK during T lymphocyte activation. J Immunol 185(2):1037–1044. https://doi.org/10.4049/jimmunol.0903586

Chai M, Jiang M, Vergnes L, Fu X, de Barros SC, Doan NB, Huang W, Chu J, Jiao J, Herschman H, Crooks GM, Reue K, Huang J (2019) Stimulation of hair growth by small molecules that activate autophagy. Cell Rep 27(12):3413–3421.e3413. https://doi.org/10.1016/j.celrep.2019.05.070

Chen Z, Barbi J, Bu S, Yang H-Y, Li Z, Gao Y, Jinasena D, Fu J, Lin F, Chen C, Zhang J, Yu N, Li X, Shan Z, Nie J, Gao Z, Tian H, Li Y, Yao Z, Zheng Y, Park Benjamin V, Pan Z, Zhang J, Dang E, Li Z, Wang H, Luo W, Li L, Semenza Gregg L, Zheng S-G, Loser K, Tsun A, Greene Mark I, Pardoll Drew M, Pan F, Li B (2013) The ubiquitin ligase Stub1 negatively modulates regulatory T cell suppressive activity by promoting degradation of the transcription factor Foxp3. Immunity 39(2):272–285. https://doi.org/10.1016/j.immuni.2013.08.006

Chen X, Su W, Wan T, Yu J, Zhu W, Tang F, Liu G, Olsen N, Liang D, Zheng SG (2017) Sodium butyrate regulates Th17/Treg cell balance to ameliorate uveitis via the Nrf2/HO-1 pathway. Biochem Pharmacol 142:111–119. https://doi.org/10.1016/j.bcp.2017.06.136

Chen Y, Colello J, Jarjour W, Zheng SG (2019) Cellular metabolic regulation in the differentiation and function of regulatory T cells. Cells 8(2):188. https://doi.org/10.3390/cells8020188

Chunder N, Wang L, Chen C, Hancock WW, Wells AD (2012) Cyclin-dependent kinase 2 controls peripheral immune tolerance. J Immunol 189(12):5659–5666. https://doi.org/10.4049/jimmunol.1202313

Comito G, Iscaro A, Bacci M, Morandi A, Ippolito L, Parri M, Montagnani I, Raspollini MR, Serni S, Simeoni L, Giannoni E, Chiarugi P (2019) Lactate modulates CD4(+) T-cell polarization and induces an immunosuppressive environment, which sustains prostate carcinoma progression via TLR8/miR21 axis. Oncogene 38(19):3681–3695. https://doi.org/10.1038/s41388-019-0688-7

Daley SR, Coakley KM, Hu DY, Randall KL, Jenne CN, Limnander A, Myers DR, Polakos NK, Enders A, Roots C, Balakishnan B, Miosge LA, Sjollema G, Bertram EM, Field MA, Shao Y, Andrews TD, Whittle B, Barnes SW, Walker JR, Cyster JG, Goodnow CC, Roose JP (2013) Rasgrp1 mutation increases naïve T-cell CD44 expression and drives mTOR-dependent accumulation of Helios+ T cells and autoantibodies. eLife 2:e01020. https://doi.org/10.7554/eLife.01020

Dang EV, Barbi J, Yang HY, Jinasena D, Yu H, Zheng Y, Bordman Z, Fu J, Kim Y, Yen HR, Luo W, Zeller K, Shimoda L, Topalian SL, Semenza GL, Dang CV, Pardoll DM, Pan F (2011) Control of T(H)17/T(reg) balance by hypoxia-inducible factor 1. Cell 146(5):772–784. https://doi.org/10.1016/j.cell.2011.07.033

Davidson TS, DiPaolo RJ, Andersson J, Shevach EM (2007) Cutting edge: IL-2 is essential for TGF-beta-mediated induction of Foxp3+ T regulatory cells. J Immunol 178(7):4022–4026. https://doi.org/10.4049/jimmunol.178.7.4022

De Rosa V, Galgani M, Porcellini A, Colamatteo A, Santopaolo M, Zuchegna C, Romano A, De Simone S, Procaccini C, La Rocca C, Carrieri PB, Maniscalco GT, Salvetti M, Buscarinu MC, Franzese A, Mozzillo E, La Cava A, Matarese G (2015) Glycolysis controls the induction of human regulatory T cells by modulating the expression of FOXP3 exon 2 splicing variants. Nat Immunol 16(11):1174–1184. https://doi.org/10.1038/ni.3269

de Zoeten EF, Wang L, Sai H, Dillmann WH, Hancock WW (2010) Inhibition of HDAC9 increases T regulatory cell function and prevents colitis in mice. Gastroenterology 138(2):583–594. https://doi.org/10.1053/j.gastro.2009.10.037

Delgoffe GM, Kole TP, Zheng Y, Zarek PE, Matthews KL, Xiao B, Worley PF, Kozma SC, Powell JD (2009) The mTOR kinase differentially regulates effector and regulatory T cell lineage commitment. Immunity 30(6):832–844. https://doi.org/10.1016/j.immuni.2009.04.014

Fu Z, Ye J, Dean JW, Bostick JW, Weinberg SE, Xiong L, Oliff KN, Chen ZE, Avram D, Chandel NS, Zhou L (2019) Requirement of mitochondrial transcription factor A in tissue-resident regulatory T cell maintenance and function. Cell Rep 28(1):159–171.e154. https://doi.org/10.1016/j.celrep.2019.06.024

Furusawa Y, Obata Y, Fukuda S, Endo TA, Nakato G, Takahashi D, Nakanishi Y, Uetake C, Kato K, Kato T, Takahashi M, Fukuda NN, Murakami S, Miyauchi E, Hino S, Atarashi K, Onawa S, Fujimura Y, Lockett T, Clarke JM, Topping DL, Tomita M, Hori S, Ohara O, Morita T, Koseki H, Kikuchi J, Honda K, Hase K, Ohno H (2013) Commensal microbe-derived butyrate induces the differentiation of colonic regulatory T cells. Nature 504(7480):446–450. https://doi.org/10.1038/nature12721

Geiger R, Rieckmann JC, Wolf T, Basso C, Feng Y, Fuhrer T, Kogadeeva M, Picotti P, Meissner F, Mann M, Zamboni N, Sallusto F, Lanzavecchia A (2016) L-arginine modulates T cell metabolism and enhances survival and anti-tumor activity. Cell 167 (3):829–842 e813. https://doi.org/10.1016/j.cell.2016.09.031

Gerriets VA, Kishton RJ, Nichols AG, Macintyre AN, Inoue M, Ilkayeva O, Winter PS, Liu X, Priyadharshini B, Slawinska ME, Haeberli L, Huck C, Turka LA, Wood KC, Hale LP, Smith PA, Schneider MA, MacIver NJ, Locasale JW, Newgard CB, Shinohara ML, Rathmell JC (2015) Metabolic programming and PDHK1 control CD4+ T cell subsets and inflammation. J Clin Invest 125(1):194–207. https://doi.org/10.1172/JCI76012

Gerriets VA, Kishton RJ, Johnson MO, Cohen S, Siska PJ, Nichols AG, Warmoes MO, de Cubas AA, MacIver NJ, Locasale JW, Turka LA, Wells AD, Rathmell JC (2016) Foxp3 and Toll-like receptor signaling balance Treg cell anabolic metabolism for suppression. Nat Immunol 17(12):1459–1466. https://doi.org/10.1038/ni.3577

Grohmann U, Bronte V (2010) Control of immune response by amino acid metabolism. Immunol Rev 236:243–264. https://doi.org/10.1111/j.1600-065X.2010.00915.x

Grzes KM, Field CS, Pearce EJ (2017) Treg cells survive and thrive in inhospitable environments. Cell Metab 25 (6):1213–1215. https://doi.org/10.1016/j.cmet.2017.05.012

Hawse WF, Cattley RT, Wendell SG (2019) Cutting edge: TCR signal strength regulates acetyl-CoA metabolism via AKT. J Immunol 203(11):2771–2775. https://doi.org/10.4049/jimmunol.1900749

He N, Fan W, Henriquez B, Yu RT, Atkins AR, Liddle C, Zheng Y, Downes M, Evans RM (2017) Metabolic control of regulatory T cell (Treg) survival and function by Lkb1. Proc Natl Acad Sci U S A 114 (47):12542–12547. https://doi.org/10.1073/pnas.1715363114

Horwitz DA, Zheng SG, Gray JD (2008) Natural and TGF-beta-induced Foxp3(+)CD4(+) CD25(+) regulatory T cells are not mirror images of each other. Trends Immunol 29(9):429–435. https://doi.org/10.1016/j.it.2008.06.005

Huang F, Chen M, Chen W, Gu J, Yuan J, Xue Y, Dang J, Su W, Wang J, Zadeh HH, He X, Rong L, Olsen N, Zheng SG (2017) Human gingiva-derived mesenchymal stem cells inhibit xeno-graft-versus-host disease via CD39-CD73-adenosine and IDO signals. Front Immunol 8:68. https://doi.org/10.3389/fimmu.2017.00068

Huynh A, DuPage M, Priyadharshini B, Sage PT, Quiros J, Borges CM, Townamchai N, Gerriets VA, Rathmell JC, Sharpe AH, Bluestone JA, Turka LA (2015) Control of PI(3) kinase in Treg cells maintains homeostasis and lineage stability. Nat Immunol 16 (2):188–196. https://doi.org/10.1038/ni.3077

Ikeda K, Kinoshita M, Kayama H, Nagamori S, Kongpracha P, Umemoto E, Okumura R, Kurakawa T, Murakami M, Mikami N, Shintani Y, Ueno S, Andou A, Ito M, Tsumura H, Yasutomo K, Ozono K, Takashima S, Sakaguchi S, Kanai Y, Takeda K (2017) Slc3a2 mediates branched-chain amino-acid-dependent maintenance of regulatory T cells. Cell Rep 21(7):1824–1838. https://doi.org/10.1016/j.celrep.2017.10.082

Johnson MO, Wolf MM, Madden MZ, Andrejeva G, Sugiura A, Contreras DC, Maseda D, Liberti MV, Paz K, Kishton RJ, Johnson ME, de Cubas AA, Wu P, Li G, Zhang Y, Newcomb DC, Wells AD, Restifo NP, Rathmell WK, Locasale JW, Davila ML, Blazar BR, Rathmell JC (2018) Distinct regulation of Th17 and Th1 cell differentiation by glutaminase-dependent metabolism. Cell 175(7):1780–1795 e1719. https://doi.org/10.1016/j.cell.2018.10.001

Jorg VL et al (2010) Regulation of Treg functionality by acetylation-mediated Foxp3 protein stabilization. Immunology 115(5):965–974. https://doi.org/10.1182/blood-2009-02-207118

Josefowicz SZ, Lu LF, Rudensky AY (2012) Regulatory T cells: mechanisms of differentiation and function. Annu Rev Immunol 30:531–564. https://doi.org/10.1146/annurev.immunol.25.022106.141623

Kang SG, Lim HW, Andrisani OM, Broxmeyer HE, Kim CH (2007) Vitamin A metabolites induce gut-homing FoxP3+ regulatory T cells. J Immunol 179 (6):3724–3733. https://doi.org/10.4049/jimmunol.179.6.3724

Kishore M, Cheung KCP, Fu H, Bonacina F, Wang G, Coe D, Ward EJ, Colamatteo A, Jangani M, Baragetti A, Matarese G, Smith DM, Haas R, Mauro C, Wraith DC, Okkenhaug K, Catapano AL, De Rosa V, Norata GD, Marelli-Berg FM (2017) Regulatory T cell migration is dependent on glucokinase-mediated glycolysis. Immunity 47(5):875–889.e810. https://doi.org/10.1016/j.immuni.2017.10.017

Klysz D, Tai X, Robert PA, Craveiro M, Cretenet G, Oburoglu L, Mongellaz C, Floess S, Fritz V, Matias MI, Yong C, Surh N, Marie JC, Huehn J, Zimmermann V, Kinet S, Dardalhon V, Taylor N (2015) Glutamine-dependent alpha-ketoglutarate production regulates the balance between T helper 1 cell and regulatory T cell generation. Sci Signal 8(396): ra97. https://doi.org/10.1126/scisignal.aab2610

Kong N, Lan Q, Chen M, Wang J, Shi W, Horwitz DA, Quesniaux V, Ryffel B, Liu Z, Brand D, Zou H, Zheng

SG (2012a) Antigen-specific transforming growth factor beta-induced Treg cells, but not natural Treg cells, ameliorate autoimmune arthritis in mice by shifting the Th17/Treg cell balance from Th17 predominance to Treg cell predominance. Arthritis Rheum 64 (8):2548–2558. https://doi.org/10.1002/art.34513

Kong N, Lan Q, Su W, Chen M, Wang J, Yang Z, Park R, Dagliyan G, Conti PS, Brand D, Liu Z, Stohl W, Zou H, Zheng SG (2012b) Induced T regulatory cells suppress osteoclastogenesis and bone erosion in collagen-induced arthritis better than natural T regulatory cells. Ann Rheum Dis 71(9):1567–1572. https://doi.org/10.1136/annrheumdis-2011-201052

Koreth J, Matsuoka K, Kim HT, McDonough SM, Bindra B, Alyea EP 3rd, Armand P, Cutler C, Ho VT, Treister NS, Bienfang DC, Prasad S, Tzachanis D, Joyce RM, Avigan DE, Antin JH, Ritz J, Soiffer RJ (2011) Interleukin-2 and regulatory T cells in graft-versus-host disease. N Engl J Med 365 (22):2055–2066. https://doi.org/10.1056/NEJMoa1108188

Krause D, Suh HS, Tarassishin L, Cui QL, Durafourt BA, Choi N, Bauman A, Cosenza-Nashat M, Antel JP, Zhao ML, Lee SC (2011) The tryptophan metabolite 3-hydroxyanthranilic acid plays anti-inflammatory and neuroprotective roles during inflammation: role of hemeoxygenase-1. Am J Pathol 179(3):1360–1372. https://doi.org/10.1016/j.ajpath.2011.05.048

Lee GK, Park HJ, Macleod M, Chandler P, Munn DH, Mellor AL (2002) Tryptophan deprivation sensitizes activated T cells to apoptosis prior to cell division. Immunology 107(4):452–460. https://doi.org/10.1046/j.1365-2567.2002.01526.x

Li B, Greene MI (2008) Special regulatory T-cell review: FOXP3 biochemistry in regulatory T cells--how diverse signals regulate suppression. Immunology 123(1):17–19. https://doi.org/10.1111/j.1365-2567.2007.02774.x

Li B, Samanta A, Song X, Iacono KT, Bembas K, Tao R, Basu S, Riley JL, Hancock WW, Shen Y, Saouaf SJ, Greene MI (2007) FOXP3 interactions with histone acetyltransferase and class II histone deacetylases are required for repression. Proc Natl Acad Sci U S A 104 (11):4571–4576. https://doi.org/10.1073/pnas.0700298104

Li Z, Lin F, Zhuo C, Deng G, Chen Z, Yin S, Gao Z, Piccioni M, Tsun A, Cai S, Zheng SG, Zhang Y, Li B (2014) PIM1 kinase phosphorylates the human transcription factor FOXP3 at serine 422 to negatively regulate its activity under inflammation. J Biol Chem 289(39):26872–26881. https://doi.org/10.1074/jbc.M114.586651

Li Y, Lu Y, Wang S, Han Z, Zhu F, Ni Y, Liang R, Zhang Y, Leng Q, Wei G, Shi G, Zhu R, Li D, Wang H, Zheng SG, Xu H, Tsun A, Li B (2016) USP21 prevents the generation of T-helper-1-like Treg cells. Nat Commun 7:13559. https://doi.org/10.1038/ncomms13559

Li L, Liu X, Sanders KL, Edwards JL, Ye J, Si F, Gao A, Huang L, Hsueh EC, Ford DA, Hoft DF, Peng G (2019) TLR8-mediated metabolic control of human Treg function: a mechanistic target for cancer immunotherapy. Cell Metab 29(1):103–123.e105. https://doi.org/10.1016/j.cmet.2018.09.020

Lin F, Luo X, Tsun A, Li Z, Li D, Li B (2015) Kaempferol enhances the suppressive function of Treg cells by inhibiting FOXP3 phosphorylation. Int Immunopharmacol 28(2):859–865. https://doi.org/10.1016/j.intimp.2015.03.044

Liu Y, Wang L, Predina J, Han R, Beier UH, Wang L-CS, Kapoor V, Bhatti TR, Akimova T, Singhal S, Brindle PK, Cole PA, Albelda SM, Hancock WW (2013) Inhibition of p300 impairs Foxp3+ T regulatory cell function and promotes antitumor immunity. Nat Med 19 (9):1173–1177. https://doi.org/10.1038/nm.3286

Lu L, Ma J, Li Z, Lan Q, Chen M, Liu Y, Xia Z, Wang J, Han Y, Shi W, Quesniaux V, Ryffel B, Brand D, Li B, Liu Z, Zheng SG (2011) All-trans retinoic acid promotes TGF-beta-induced Tregs via histone modification but not DNA demethylation on Foxp3 gene locus. PLoS One 6(9):e24590. https://doi.org/10.1371/journal.pone.0024590

Lu L, Lan Q, Li Z, Zhou X, Gu J, Li Q, Wang J, Chen M, Liu Y, Shen Y, Brand DD, Ryffel B, Horwitz DA, Quismorio FP, Liu Z, Li B, Olsen NJ, Zheng SG (2014) Critical role of all-trans retinoic acid in stabilizing human natural regulatory T cells under inflammatory conditions. Proc Natl Acad Sci U S A 111(33):E3432–E3440. https://doi.org/10.1073/pnas.1408780111

Luengo A, Gui DY, Vander Heiden MG (2017) Targeting metabolism for cancer therapy. Cell Chem Biol 24 (9):1161–1180. https://doi.org/10.1016/j.chembiol.2017.08.028

Luo X, Nie J, Wang S, Chen Z, Chen W, Li D, Hu H, Li B (2015) Poly(ADP-ribosyl)ation of FOXP3 protein mediated by PARP-1 protein regulates the function of regulatory T cells. J Biol Chem 290(48):28675–28682. https://doi.org/10.1074/jbc.M115.661611

Michalek RD, Gerriets VA, Jacobs SR, Macintyre AN, MacIver NJ, Mason EF, Sullivan SA, Nichols AG, Rathmell JC (2011) Cutting edge: distinct glycolytic and lipid oxidative metabolic programs are essential for effector and regulatory CD4+ T cell subsets. J Immunol 186(6):3299–3303. https://doi.org/10.4049/jimmunol.1003613

Nakaya M, Xiao Y, Zhou X, Chang JH, Chang M, Cheng X, Blonska M, Lin X, Sun SC (2014) Inflammatory T cell responses rely on amino acid transporter ASCT2 facilitation of glutamine uptake and mTORC1 kinase activation. Immunity 40(5):692–705. https://doi.org/10.1016/j.immuni.2014.04.007

Ouyang W, Liao W, Luo CT, Yin N, Huse M, Kim MV, Peng M, Chan P, Ma Q, Mo Y, Meijer D, Zhao K, Rudensky AY, Atwal G, Zhang MQ, Li MO (2012) Novel Foxo1-dependent transcriptional programs

control T(reg) cell function. Nature 491 (7425):554–559. https://doi.org/10.1038/nature11581
Park Y, Jin H-S, Lopez J, Elly C, Kim G, Murai M, Kronenberg M, Liu Y-C (2013) TSC1 regulates the balance between effector and regulatory T cells. J Clin Investig 123(12):5165–5178. https://doi.org/10.1172/jci69751
Pearce EL, Walsh MC, Cejas PJ, Harms GM, Shen H, Wang LS, Jones RG, Choi Y (2009) Enhancing CD8 T-cell memory by modulating fatty acid metabolism. Nature 460(7251):103–107. https://doi.org/10.1038/nature08097
Powell JD, Delgoffe GM (2010) The mammalian target of rapamycin: linking T cell differentiation, function, and metabolism. Immunity 33(3):301–311. https://doi.org/10.1016/j.immuni.2010.09.002
Procaccini C, De Rosa V, Galgani M, Abanni L, Cali G, Porcellini A, Carbone F, Fontana S, Horvath TL, La Cava A, Matarese G (2010) An oscillatory switch in mTOR kinase activity sets regulatory T cell responsiveness. Immunity 33(6):929–941. https://doi.org/10.1016/j.immuni.2010.11.024
Procaccini C, Carbone F, Di Silvestre D, Brambilla F, De Rosa V, Galgani M, Faicchia D, Marone G, Tramontano D, Corona M, Alviggi C, Porcellini A, La Cava A, Mauri P, Matarese G (2016) The proteomic landscape of human ex vivo regulatory and conventional T cells reveals specific metabolic requirements. Immunity 44(3):712. https://doi.org/10.1016/j.immuni.2016.02.022
Ramalingam R, Larmonier CB, Thurston RD, Midura-Kiela MT, Zheng SG, Ghishan FK, Kiela PR (2012) Dendritic cell-specific disruption of TGF-beta receptor II leads to altered regulatory T cell phenotype and spontaneous multiorgan autoimmunity. J Immunol 189(8):3878–3893. https://doi.org/10.4049/jimmunol.1201029
Saadoun D, Rosenzwajg M, Joly F, Six A, Carrat F, Thibault V, Sene D, Cacoub P, Klatzmann D (2011) Regulatory T-cell responses to low-dose interleukin-2 in HCV-induced vasculitis. N Engl J Med 365 (22):2067–2077. https://doi.org/10.1056/NEJMoa1105143
Samanta A, Li B, Song X, Bembas K, Zhang G, Katsumata M, Saouaf SJ, Wang Q, Hancock WW, Shen Y, Greene MI (2008) TGF-beta and IL-6 signals modulate chromatin binding and promoter occupancy by acetylated FOXP3. Proc Natl Acad Sci U S A 105 (37):14023–14027. https://doi.org/10.1073/pnas.0806726105
Sauer S, Bruno L, Hertweck A, Finlay D, Leleu M, Spivakov M, Knight ZA, Cobb BS, Cantrell D, O'Connor E, Shokat KM, Fisher AG, Merkenschlager M (2008) T cell receptor signaling controls Foxp3 expression via PI3K, Akt, and mTOR. Proc Natl Acad Sci U S A 105(22):7797–7802. https://doi.org/10.1073/pnas.0800928105
Schulte ML, Fu A, Zhao P, Li J, Geng L, Smith ST, Kondo J, Coffey RJ, Johnson MO, Rathmell JC, Sharick JT, Skala MC, Smith JA, Berlin J, Washington MK, Nickels ML, Manning HC (2018) Pharmacological blockade of ASCT2-dependent glutamine transport leads to antitumor efficacy in preclinical models. Nat Med 24(2):194–202. https://doi.org/10.1038/nm.4464
Shi LZ, Wang R, Huang G, Vogel P, Neale G, Green DR, Chi H (2011) HIF1alpha-dependent glycolytic pathway orchestrates a metabolic checkpoint for the differentiation of TH17 and Treg cells. J Exp Med 208 (7):1367–1376. https://doi.org/10.1084/jem.20110278
Shi H, Chapman NM, Wen J, Guy C, Long L, Dhungana Y, Rankin S, Pelletier S, Vogel P, Wang H, Peng J, Guan KL, Chi H (2019) Amino acids license kinase mTORC1 activity and Treg cell function via small G proteins Rag and Rheb. Immunity 51(6):1012–1027 e1017. https://doi.org/10.1016/j.immuni.2019.10.001
Shrestha S, Yang K, Guy C, Vogel P, Neale G, Chi H (2015) Treg cells require the phosphatase PTEN to restrain TH1 and TFH cell responses. Nat Immunol 16(2):178–187. https://doi.org/10.1038/ni.3076
Sinclair LV, Rolf J, Emslie E, Shi YB, Taylor PM, Cantrell DA (2013) Control of amino-acid transport by antigen receptors coordinates the metabolic reprogramming essential for T cell differentiation. Nat Immunol 14 (5):500–508. https://doi.org/10.1038/ni.2556
Sinclair LV, Neyens D, Ramsay G, Taylor PM, Cantrell DA (2018) Single cell analysis of kynurenine and System L amino acid transport in T cells. Nat Commun 9(1):1981. https://doi.org/10.1038/s41467-018-04366-7
Smith PM, Howitt MR, Panikov N, Michaud M, Gallini CA, Bohlooly YM, Glickman JN, Garrett WS (2013) The microbial metabolites, short-chain fatty acids, regulate colonic Treg cell homeostasis. Science 341 (6145):569–573. https://doi.org/10.1126/science.1241165
Thorburn AN, McKenzie CI, Shen S, Stanley D, Macia L, Mason LJ, Roberts LK, Wong CHY, Shim R, Robert R, Chevalier N, Tan JK, Mariño E, Moore RJ, Wong L, McConville MJ, Tull DL, Wood LG, Murphy VE, Mattes J, Gibson PG, Mackay CR (2015) Evidence that asthma is a developmental origin disease influenced by maternal diet and bacterial metabolites. Nat Commun 6(1):7320. https://doi.org/10.1038/ncomms8320
van Loosdregt J, Fleskens V, Fu J, Brenkman Arjan B, Bekker Cornelis PJ, Pals Cornelieke EGM, Meerding J, Berkers Celia R, Barbi J, Gröne A, Sijts Alice JAM, Maurice Madelon M, Kalkhoven E, Prakken Berent J, Ovaa H, Pan F, Zaiss Dietmar MW, Coffer Paul J (2013) Stabilization of the transcription factor Foxp3 by the deubiquitinase USP7 increases Treg-cell-suppressive capacity. Immunity 39(2):259–271. https://doi.org/10.1016/j.immuni.2013.05.018
Weinberg SE, Singer BD, Steinert EM, Martinez CA, Mehta MM, Martinez-Reyes I, Gao P, Helmin KA,

Abdala-Valencia H, Sena LA, Schumacker PT, Turka LA, Chandel NS (2019) Mitochondrial complex III is essential for suppressive function of regulatory T cells. Nature 565(7740):495–499. https://doi.org/10.1038/s41586-018-0846-z

Wing K, Sakaguchi S (2010) Regulatory T cells exert checks and balances on self tolerance and autoimmunity. Nat Immunol 11(1):7–13. https://doi.org/10.1038/ni.1818

Yan Y, Zhang G-X, Gran B, Fallarino F, Yu S, Li H, Cullimore ML, Rostami A, Xu H (2010) IDO upregulates regulatory T cells via tryptophan catabolite and suppresses encephalitogenic T cell responses in experimental autoimmune encephalomyelitis. J Immunol 185(10):5953–5961. https://doi.org/10.4049/jimmunol.1001628

Yang S, Wang J, Brand DD, Zheng SG (2018) Role of TNF-TNF receptor 2 signal in regulatory T cells and its therapeutic implications. Front Immunol 9:784. https://doi.org/10.3389/fimmu.2018.00784

Yang S, Xie C, Chen Y, Wang J, Chen X, Lu Z, June RR, Zheng SG (2019) Differential roles of TNFalpha-TNFR1 and TNFalpha-TNFR2 in the differentiation and function of CD4(+)Foxp3(+) induced Treg cells in vitro and in vivo periphery in autoimmune diseases. Cell Death Dis 10(1):27. https://doi.org/10.1038/s41419-018-1266-6

Ye C, Brand D, Zheng SG (2018) Targeting IL-2: an unexpected effect in treating immunological diseases. Signal Transduct Target Ther 3:2. https://doi.org/10.1038/s41392-017-0002-5

Zeng H, Yang K, Cloer C, Neale G, Vogel P, Chi H (2013) mTORC1 couples immune signals and metabolic programming to establish T(reg)-cell function. Nature 499(7459):485–490. https://doi.org/10.1038/nature12297

Zheng SG, Gray JD, Ohtsuka K, Yamagiwa S, Horwitz DA (2002) Generation ex vivo of TGF-beta-producing regulatory T cells from CD4+CD25- precursors. J Immunol 169(8):4183–4189. https://doi.org/10.4049/jimmunol.169.8.4183

Zheng SG, Wang J, Wang P, Gray JD, Horwitz DA (2007) IL-2 is essential for TGF-beta to convert naive CD4+CD25- cells to CD25+Foxp3+ regulatory T cells and for expansion of these cells. J Immunol 178(4):2018–2027

Zhou X, Kong N, Wang J, Fan H, Zou H, Horwitz D, Brand D, Liu Z, Zheng SG (2010) Cutting edge: all-trans retinoic acid sustains the stability and function of natural regulatory T cells in an inflammatory milieu. J Immunol 185(5):2675–2679. https://doi.org/10.4049/jimmunol.1000598

Treg Cells and Epigenetic Regulation

6

Joseph A. Bellanti and Dongmei Li

Abstract

The discovery of the epigenetic regulation of Treg cells, a cell population with fundamental immunoregulatory properties, has shed considerable insights into an understanding of the role of these cells in health and disease. Research over the past several years has shown that the interaction of Treg cells with the gut microbiota are critical not only for the development of Treg function in health but also for abnormalities of Treg function that play a critical role in the pathogenesis of human diseases such as the allergic diseases, the autoimmune disorders, and cancer. The equilibrium between phenotypic plasticity and stability of Treg cells is defined by the fine-tuned transcriptional and epigenetic events required to ensure stable expression of Foxp3 in Treg cells. In this chapter, we discuss the molecular events that control *Foxp3* gene expression and address the importance of DNA methylation as an important molecular switch that regulates the genetic expression of Treg induction and the possible implications of these findings for the treatment of human diseases characterized by abnormalities of Treg cell function.

Keywords

T reg cells = T regulatory cells · Epigenetics · DNA methylation · Gut = gastrointestinal tract · Microbiome · Microbiota

6.1 Introduction

The book title *T Regulatory Cells in Human Health and Diseases*, chosen by the editor-in-chief, strikes at the heart of major global efforts that are being directed to the translation of principles and mechanisms of basic immunology to clinical application using techniques and tools that address current critical needs of medicine. This chapter reviews fundamental concepts of the epigenetic regulation of the immune response by Treg cells that control and shape the immune response by interaction with the microbiota together with a discussion of abnormalities of Treg cell function which play a major role in human disease. In keeping with this mandate,

J. A. Bellanti (✉)
Department of Pediatrics, Georgetown University Medical Center, Washington, DC, USA

Department of Microbiology-Immunology, Georgetown University Medical Center, Washington, DC, USA

International Center for Interdisciplinary Studies of Immunology (ICISI), Georgetown University Medical Center, Washington, DC, USA
e-mail: bellantj@georgetown.edu

D. Li
Department of Microbiology-Immunology, Georgetown University Medical Center, Washington, DC, USA

International Center for Interdisciplinary Studies of Immunology (ICISI), Georgetown University Medical Center, Washington, DC, USA

S.-G. Zheng (ed.), *T Regulatory Cells in Human Health and Diseases*, Advances in Experimental Medicine and Biology 1278, https://doi.org/10.1007/978-981-15-6407-9_6

the topic of epigenetic regulation of Treg cells has seen a supernova of research contributions that offer a unique potential for disease prevention by modalities that affect gene expression through epigenetic intervention.

6.2 Treg Cell Differentiation and Epigenetics

6.2.1 Definition of Epigenetics

Epigenetics plays an important role in the control of gene expression (Rodenhiser and Mann 2006). The term "epigenetics" was introduced by Conrad Hal Waddington, a leading embryologist and geneticist (Slack 2002) in the late 1950s. According to this new epigenetic concept, gene function is controlled by mechanisms which do not change the basic nucleotide sequence of the DNA molecule (Waddington 1957; Rodenhiser and Mann 2006). With the advent of gene sequencing technologies and other tools for the study of genetics, investigative efforts have shifted the focus of investigation from the structure of the DNA molecule to mechanisms "beyond" the DNA molecule in our understanding of complex diseases. The goal of epigenetic studies is to provide the long sought-after link between genetics and environmental factors that can trigger clinical disorders arising from inappropriate immune responses.

Applications of this discovery have already been made, and these extend throughout the entire lifespan of the human from embryonic and fetal development issues, as well as to the management and treatment of a vast number of immunologically mediated diseases occurring in later life that include allergic diseases, autoimmune disorders, and cancer (Waddington 1957; Rodenhiser and Mann 2006; Bellanti 2019). As a framework for this presentation and for ease of discussion, epigenetics exerts its effects through the interaction of environment, various susceptibility genes, and immunologic development, e.g., age (Bellanti 2012) (Fig. 6.1).

6.2.2 Three Mechanisms of Epigenetics

In eukaryotes, DNA molecules are packed in nucleosomes that are assembled in a linear arrangement to form chromatin. Structurally, each nucleosome contains two loops of DNA that encircle a histone protein (Fig. 6.2) (Bellanti 2019) and exists in two forms: (1) euchromatin (active form) in which the nucleosomes are spaced apart and (2) heterochromatin (inactive form) in which the nucleosomes are closely packed, that affect metabolic activity. For gene transcription in euchromatin to be active, there must be sufficient space between nucleosomes to allow insertion of transcription factors into the DNA molecule. Conversely, closely packed nucleosomes in heterochromatin limit access to the transcription factors for gene transcription and do not permit gene expression to proceed.

Epigenetic modifications not only alter chromatin into a transcriptionally repressive or permissive configuration but also serve as signals to other chromatin-modifying complexes. At least three epigenetic mechanisms have been described: DNA methylation, histone modification, and non-coding RNA-mediated gene silencing. In DNA methylation, a methyl group can be added directly to a cytosine residue by a DNA methyltransferase (DNMTs). This DNA modification occurs at regions rich in CpG residues—known as CpG islands—which are often found in regulatory regions of the DNA molecule, as described below for the Treg-specific demethylated region (TSDR) in the CpG-rich promoter region of Foxp3. Gene expression is largely influenced by the methylation status of these islands. Posttranslational modifications of histone proteins through acetylation and methylation often take place in the tail of histone proteins, which regulate gene transcription, chromosome packaging, and DNA repair. Finally, the short non-coding RNAs, especially microRNAs (miRNAs), are shown to silence the gene expression by their involvement in heterochromatin formation, histone modification, and DNA methylation.

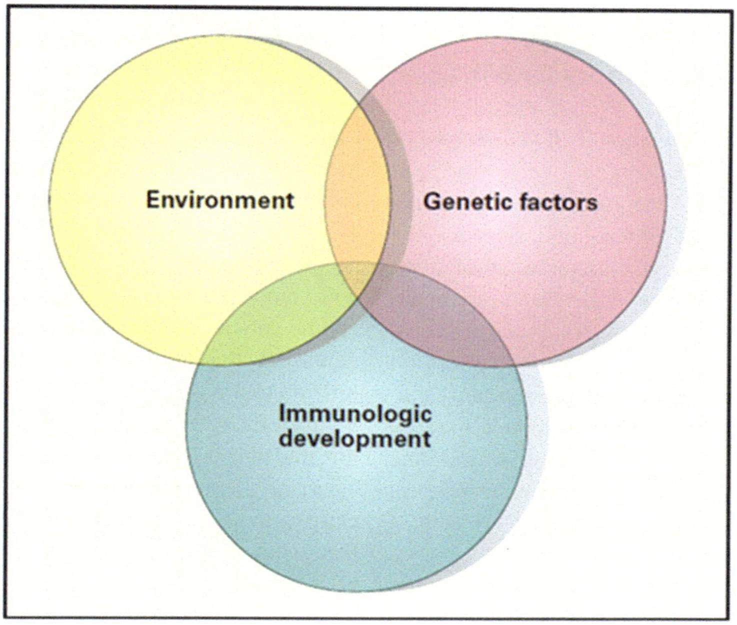

Fig. 6.1 Schematic representation of how epigenetics exerts its effects through the interaction of environment, susceptibility genes, and immunologic development, e.g., age. (*Reproduced with permission from Bellanti, JA (Ed). Immunology IV: Clinical Applications in Health and Disease. I Care Press, Washington, DC, 2012* (Bellanti 2012))

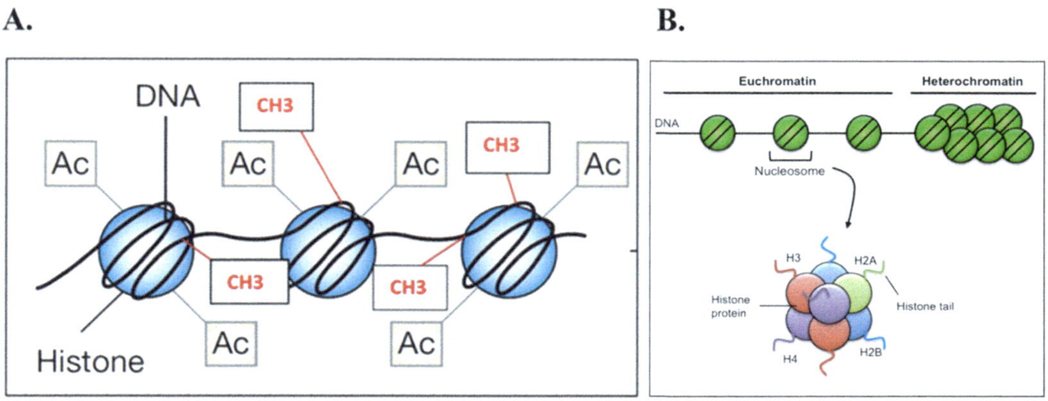

Fig. 6.2 Schematic representation of multiple levels of DNA packaging. (**a**) Nucleosomes consist of histone proteins (shown in blue) surrounded by two loops of DNA and, when assembled as beads on a string, form chromatin. (**b**) Chromatin exists in two forms: (1) a metabolically active form called euchromatin with the nucleosomes spatially separated (required for insertion of transcription factors for access to DNA to permit transcription) and (2) a metabolically inert form of chromatin called heterochromatin with the nucleosomes closely packed. CH_3 = methyl groups attached to a DNA strand after DNA methylation; *Ac* acetyl groups attached to histone proteins after histone acetylation. (*Reproduced with permission from Bellanti JA. Genetics/epigenetics/allergy: The gun is loaded … but what pulls the trigger? Allergy Asthma Proc. 2019; 40:76–83* (Bellanti 2019))

6.2.3 Immunologic Balance: The Balance of Pro-versus Anti-inflammatory Cytokines

The immune system is anatomically composed of collection of lymphoid organs, cells, and cell products designed to recognize and remove foreignness. As with all physiologic systems, the immune system may be considered as an adaptive response in which the body attempts to maintain homeostasis between the internal and external environments. For ease of discussion, the total composite functionality of the total immune capability of the host to all foreign matter may be viewed in three phases: (1) the innate immune responses, (2) the adaptive immune responses, and (3) the tissue-damaging responses (Fig. 6.3) (Bellanti 2012).

This model is based on the following assumptions concerning immune responsiveness: (1) the foreign substance (e.g., immunogen) drives the immune response, which remains active only as long as the foreign substance persists and will cease after its removal; and (2) the extent of the progression through these three phases is determined by the efficiency of elimination of the foreign substance, which is controlled by genetic and epigenetic factors. The immune system in health is achieved by the efficient removal of the foreign agent by the innate system alone or together with the adaptive immune system; inefficient removal of the foreign agent, with persistence into the third phase represents the immune system in disease (Bellanti 2012). It is now clear that cytokines are the important immune system signaling molecules par excellence that maintain the immunologic balance between the external and internal environments and accomplish this function by either promoting or inhibiting inflammatory responses (Fig. 6.4) (Bellanti 2012). A balanced production of pro- and anti-inflammatory cytokines in response to a foreign configuration results in immunologic equilibrium and defines the immune system in health in contrast to the immunologic imbalance seen when an overproduction of pro-inflammatory cytokines and/or an inadequate production of anti-inflammatory cytokines create the disequilibrium, which represents the immune system in disease, e.g., allergic diseases and autoimmune disorders (Fig. 6.4) (Bellanti 2012). Thus, cytokines are reactive molecules that can both promote and counteract inflammation; this dual set of seemingly opposing functions is accomplished because the same molecules are able to react with receptors that share homologies and that can be activated or turned off, depending on the nature of the foreign insult and the resulting inflammatory response. The development of highly effective biologic agents in recent years directed at cytokines are revolutionizing the management of the allergic and other immunologically mediated diseases refractory to other therapeutic regimens (Bellanti 2020).

6.2.4 The Two Major Pathways of T-Cell Differentiation: The CD4$^+$ T Helper (Th) and the CD8$^+$ T Cytotoxic (Tc) Populations

There are two major pathways of T-cell differentiation: the CD4$^+$ T helper (Th) and the CD8$^+$ T cytotoxic (Tc) populations. Figure 6.5 shows a schematic representation of the two major arms of the T-cell system that include the CD4$^+$ T helper (Th) and the CD8$^+$ T cytotoxic (Tc) populations. The CD4$^+$ Th cell population is made up of several effector subsets that include the Th1, Th2, Th17, and Treg populations. The CD4$^+$ Th cells and their subsets are the key choreographers of adaptive immune responses in mammalian hosts, and each of their effector subsets facilitate a distinct form of antimicrobial immunity. While Th1 cells eliminate intracellular microbial pathogens such as viruses, Th2 and Th17 subsets promote resistance to intestinal helminths and extracellular bacteria and fungi. The Treg cells, on the other hand, suppress an overreactive immune and inflammatory reaction arising from Th1, Th2, and Th17 hyperreactions. Treg cells can also be differentiated further into distinct subsets according to integrin $\alpha_E\beta_7$ expression. The

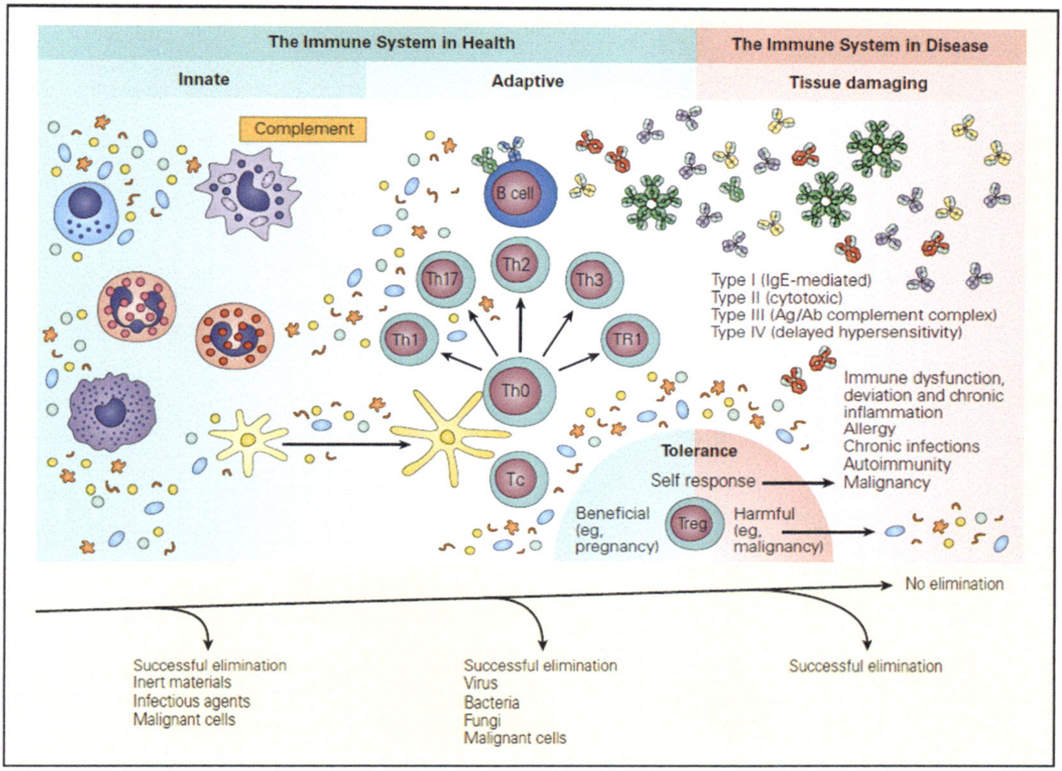

Fig. 6.3 Schematic representation of the total immune capability of the host based on efficiency of elimination of foreign matter. (*Reproduced with permission from Bellanti, JA (Ed). Immunology IV: Clinical Applications in Health and Disease. I Care Press, Washington, DC, 2012* (Bellanti 2012))

α_E-positive subsets (CD25$^+$ and CD25$^-$) have been found to be the most potent suppressors of inflammatory processes in disease models such as antigen-induced arthritis, where they act by expressing high levels of E/P-selectin-binding ligands and multiple adhesion molecules as well as inducing the formation of receptors for inflammatory chemokines (Huehn et al. 2004).

The CD8$^+$ T-cell population is similarly comprised of subpopulations that include Tc1 and Tc2 subsets. Tc1 cells are active in the destruction of virally infected or malignant cells. During their ontogeny in the thymus and their subsequent differentiation in secondary lymphoid tissues, CD4$^+$ T cells and CD8$^+$ T cells undergo a sequence of complex and mechanistically distinct processes that result in the acquisition of helper or regulatory repertoire as well as cytotoxic activity, respectively.

6.2.5 Cytokines Produced Following CD4$^+$ T-Cell Differentiation

One of the major hallmark properties of Th differentiation is the amazing capacity of subset flexibility and plasticity rendered by the capacity of these cells to produce cytokines. At present, lineage commitment is defined by the signature cytokines that differentiated cells secrete (i.e., IFN-γ, IL-4, and IL-17, for Th1, Th2, and Th17 cells, respectively). Figure 6.6 shows the repertoire of cytokines produced following CD4$^+$ T-cell differentiation using autoimmunity as a model (Schrezenmeier and Dörner 2020). During

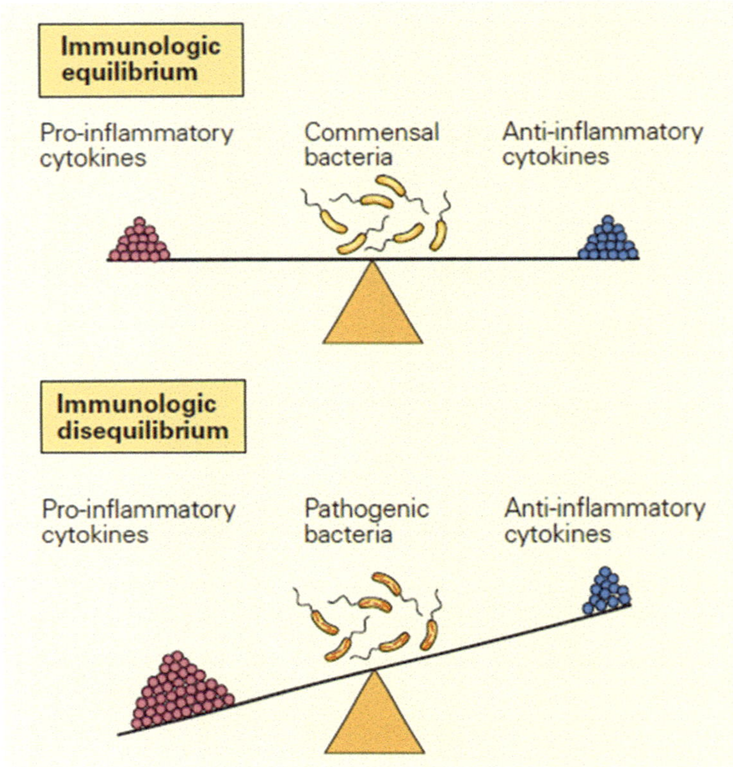

Fig. 6.4 Schematic representation of the opposing pro- and anti-inflammatory functions of cytokines in maintaining immunologic equilibrium. A balanced production of pro- and anti-inflammatory cytokines in response to a foreign configuration results in immunologic equilibrium and the immune system in health (upper panel) in contrast to the immunologic imbalance seen when an overproduction of pro-inflammatory cytokines and/or an inadequate production of anti-inflammatory cytokines creates the disequilibrium, representing the immune system in disease (lower panel). (*Reproduced with permission from Bellanti, JA (Ed). Immunology IV: Clinical Applications in Health and Disease. I Care Press, Washington, DC, 2012* (Bellanti 2012))

autoimmunity cellular debris is released from dying cells which can activate Toll-like receptor 7 (TLR7) and TLR7 signaling pathways in plasmacytoid dendritic cells (pDCs) and other immune antigen-presenting cells (APCs), including monocytes, macrophages, and B cells, resulting in the activation of multiple cell types and secretion of various pro-inflammatory cytokines. Following binding of ligand with TLR7 and TLR9, activation of TLR signaling in plasmacytoid dendritic cells (pDCs) leads to the production of cytokines IL-1, IFN-α, and IL-6. Following subsequent antigen processing and MHC class II presentation of processed antigenic peptides (by pDCs and B cells) to T cells, the activation and differentiation of $CD4^+$ Th1 cells produce TNF-α, IL-1, IFN-γ, and IL-6, and B cells release BAFF (B-cell activating factor).

6.2.6 Molecular Mechanisms of Treg Cells in Immune Regulation

Of the multitude of cell–cell interactions that control T-cell reactivity of particular importance are those that comprise a subset of T cells, termed regulatory T, or Treg, cells, that co-express CD4 and CD25, i.e., the IL-2 receptor α chain. These

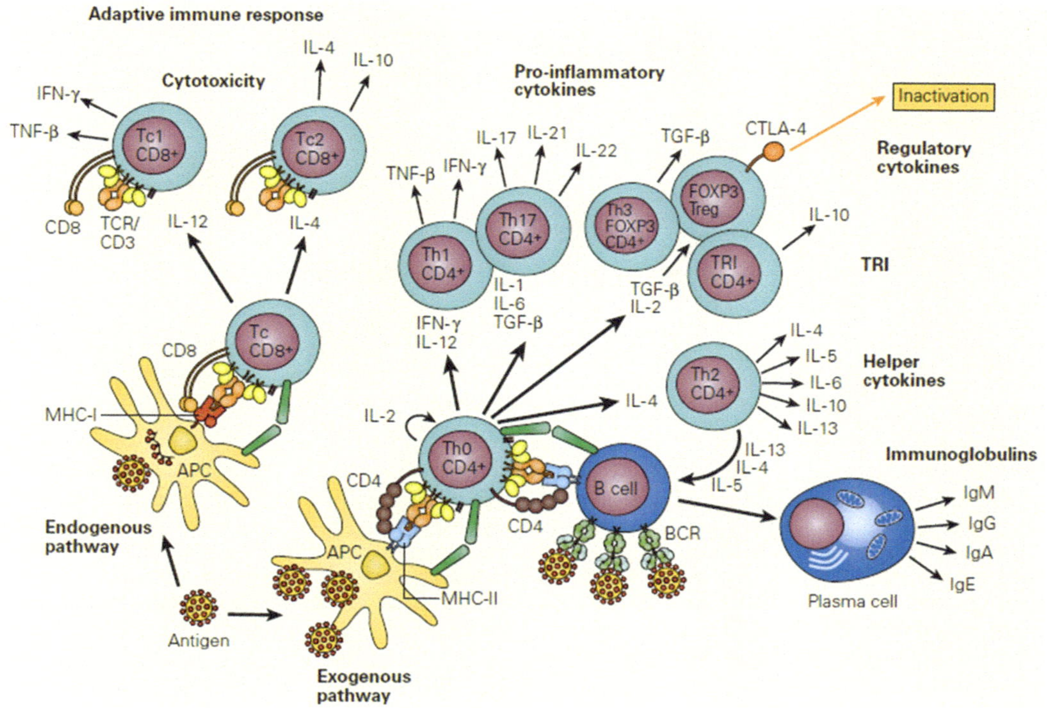

Fig. 6.5 Schematic representation of the two major pathways of T-cell differentiation: the CD4$^+$ T helper (Th) and the CD8$^+$ T cytotoxic (Tc) populations and their subsets. Following uptake and processing of an antigen by an APC shown in the figure as a dendritic cell, peptide is presented either to the CD8$^+$ population in the context of MHC-I or to the CD4$^+$ subpopulation in the context of MHC-II following which a cascading set of cellular lymphoproliferative and differentiative steps are initiated under the inductive influence of cytokines that ultimately determine their effector functions. (*Reproduced with permission from Bellanti, JA (Ed). Immunology IV: Clinical Applications in Health and Disease. I Care Press, Washington, DC, 2012* (Bellanti 2012))

naturally occurring Treg cells inhibit responses of other T cells and are produced in the thymus, probably during a stage of negative selection and not only display anergic properties in vitro but also can suppress CD4$^+$CD25$^+$ T cells in vivo, probably through a cytokine-independent direct cell–cell interaction. Table 6.1 shows several subsets of Treg cells capable of controlling effector T-cell responses. These include both naturally occurring Tregs (labeled nTreg) and induced Treg (iTreg) cells (Table 6.1). The nTreg cells originate directly from thymic precursors and iTreg cells—which include Tr1 cells and Th3 regulatory T-cell subsets, differentiated from peripheral T helper (Th) cell precursors through the actions of different cytokines.

The discovery of the transcription factor Forkhead box-p3 (Foxp3) has shed some light into an understanding of the molecular mechanisms of Treg cells, which form a key cell population with immunoregulatory functions (Fig. 6.5). The maintenance of stable expression of Foxp3 in Treg cells requires the involvement of finely tuned transcriptional and epigenetic events, the alterations of which are associated with human diseases (Sekiya et al. 2016). Epigenetic mechanisms are largely externally induced and are particularly critical for the maintenance of Treg cell function. The mechanism(s) by which Treg cells carry out their immunoregulatory function are twofold: (1) either through direct cell–cell interaction or (2) through the secretion of cytokines, e.g., transforming growth factor

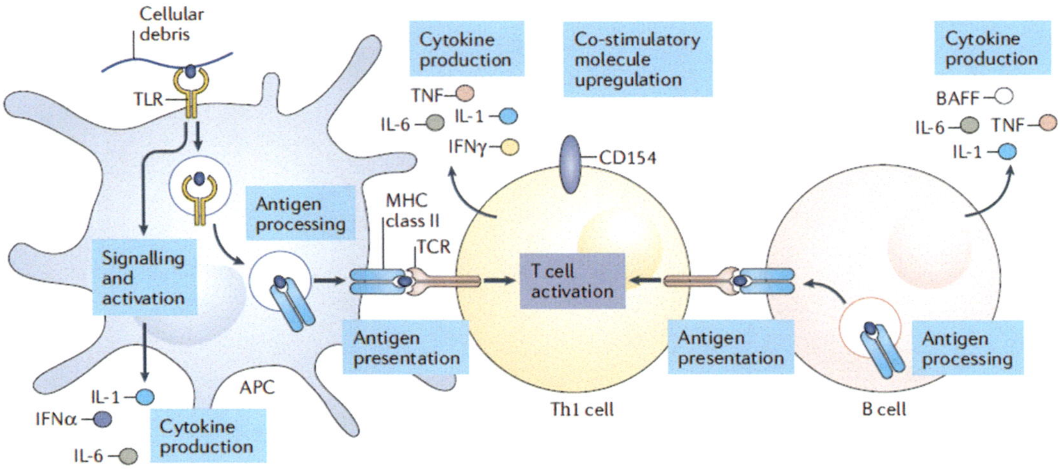

Fig. 6.6 Schematic representation of the repertoire of cytokines produced following CD4⁺ T-cell differentiation using autoimmunity as a model. During autoimmunity cellular debris released from dying cells can activate Toll-like receptor 7 (TLR7) and TLR7 signaling pathways in plasmacytoid dendritic cells (pDCs) and other immune antigen-presenting cells (APCs), including monocytes, macrophages and B cells, resulting in the activation of multiple cell types and secretion of various pro-inflammatory cytokines by pDCs, Th1 cells, and B cells. (*Modified and reproduced with permission from Schrezenmeier E, Dörner T* Schrezenmeier and Dörner 2020))

(TGF)-β and IL-10 as shown in Fig. 6.7. The critical role of TGF-β in Treg function was described in several seminal studies (Zheng et al. 2002, 2004, 2006, 2007; Gu et al. 2014; Xu et al. 2016).

The molecular mechanisms involved in the development and maintenance of Tregs require the participation of essential regulators such as transcription factors—NR4a and Smad2/3 (Sekiya et al. 2016), and the transcription factor Ets-1 (Polansky et al. 2010). In addition to these,

Table 6.1 CD4+ T cells with regulatory activity

Regulatory T cell	Characteristic
nTreg	$CD25^+$ $FOXP3^+$ thymus-derived
	Not dependent on IL-10 for biologic activity
	Mediates self-tolerance/prevents autoimmune disease
iTreg	Peripheral-derived Treg cells
	Dependent on CTLA-4 for suppressive activity
	$FOXP3^+$ $CD25^+$
	TGF-β responsible for their induction
Th3	Peripheral-derived regulatory T cells
	$FOXP3^-$
	Characterized by TGF-β production
	Mediates mucosal tolerance/antigen-specific IgA production
Tr1	Peripheral-derived Treg cells
	$FOXP3^-$ or $FOXP3^+$ at times
	Characterized by IL-10 production
	Possibly derived from Th1-like/Th2-like lymphocytes or naïve T cells ± CD25 expression (reflecting their effector function-activation)

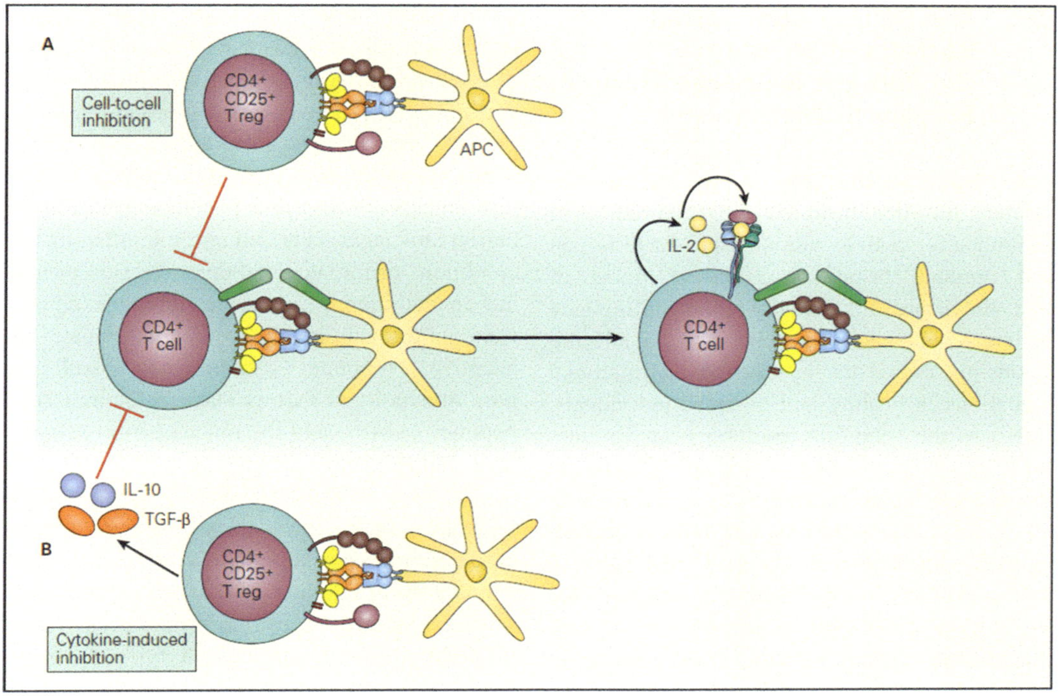

Fig. 6.7 Schematic representation of two mechanisms of immune regulation by Treg cells. Panel **a**: Self-antigen specific $CD4^+CD25^+$ T cells can directly suppress the activation of self-reactive T cells through cell–cell interaction. Panel **b**: Anti-inflammatory cytokines such as IL-10 and TGF-β that are secreted by other immunoregulatory T cells can also suppress the immune response

epigenetic modification in the CpG-rich Treg-specific demethylated region (TSDR) in the Foxp3 locus offers more critical molecular mechanism for activation of Treg cells via Foxp3 expression. The unique FOXP3 promoter demethylation profile in nTregs is partially methylated in iTreg, suggesting that a demethylated pattern is a prerequisite for stable FOXP3 expression and a suppressive phenotype (Janson et al. 2008; Floess et al. 2007). In this case, epigenetic regulation operates by the methylation status of CpG motifs within the TSDR region, which provide binding sites for essential transcription factors, such as CREB/ATF and NF-κB (Polansky et al. 2010). In $CD4^+$ T cells, the DNA methyltransferases DNMT1 and DNMT3b reside within the Foxp3 locus and function to methylate CpG residues, thereby repressing Foxp3 expression in $CD4^+$ cells, even though complete demethylation of this site is required for stable Foxp3 expression (Lal and Bromberg 2009). The binding of Ets-1 to the demethylated form of TSDR in ex vivo isolated Tregs contributes to the stabilization of Foxp3 expression in the Treg lineage (Sekiya et al. 2016), but the methylated TSDR in conventional $CD4^+$ T cells remains undifferentiated. Finally, the expression of IL-10 from Treg requires epigenetic regulation via demethylation of the intronic enhancer of STAT5 (I-SRE-4) in the IL-2-STAT5 signaling axis (Tsuji-Takayama et al. 2008). An example of another pathway for exploration of Treg function deals with the maintenance of the suppressive function of regulatory T cells by a cAMP-adenosine feedback loop (Su et al. 2019) and its application to mast cell activity and allergic disease (Su et al. 2012).

6.3 The Role of Commensal Bacteria in Shaping the Immune Response Through Epigenetic Mechanisms

Over the past several years, a growing body of evidence has demonstrated the importance of metabolically active commensal bacteria on local immune responses. However, most of these studies are still virtually at the starting gate and are concerned primarily with identifying responsible organisms by comparing microbial populations in healthy and diseased individuals. Dysbiosis, the condition resulting from a reduction in the diversity and variations in the quantity of some particular commensals, contributes to pathogenesis of a wide range of diseases including many immune and metabolic disorders in humans. These abnormal immune conditions are frequently accompanied by overreactive chronic inflammatory responses that have been shown to be etiologic triggers for colon cancer and autoimmune disease. These conditions are now known to be associated with a deficiency of Treg cells located at local anatomic sites, e.g., the GI tract.

A major investigative effort of several laboratories is being directed to the study of the Treg-suppressive effects of probiotic candidates and the specific mechanisms by which these microbes exert their immune suppressive effects. The continued study of how these microbes affect innate or adaptive immune responses mediating these effects is expected to yield a harvest of improved therapeutic regimens directed to either promoting or inhibiting inflammation through the immune suppressive effects of Treg cells.

6.3.1 Local Gut Immune Responses

The local immune responses in the gut represent approximately 70% of those of the entire immune system (Vighi et al. 2008). These include hundreds of species of symbiotic microbes inhabiting the mammalian gastrointestinal tract, which comprise five major phyla (the two major phyla Firmicutes and Bacteroidetes as well as the significantly less abundant Proteobacteria, Actinobacteria, and Fusobacteria). The diversity of microbes in the gut serves as a highly important epigenetic network that profoundly shapes the host immune system (Kamada et al. 2013). While some microbes coexist peacefully with the host, others invade and penetrate the inherent barriers to cause acute and chronic inflammatory conditions or initiate an allergic reaction or initiate an autoimmune response. Immune equilibrium in the gut involves both the innate and adaptive immune systems as well as non-immunologic factors such as antimicrobial peptides (AMPs) and the mucus barrier (Hooper et al. 2012).

The local immune response in the gut varies with the commensal species, and their specific type of immune stimulatory responses—i.e., stimulatory or inhibitory—that can paradoxically often be manifested by the same species under different disease conditions. A typically useful model for study of this diversity is the study of immune responses of each organism in a germfree mice model system with monocolonization (Geva-Zatorsky et al. 2017). However, a major limitation of this model is its inability to express possible interactive effects between commensal species in vivo. Nevertheless, given two facts—that germ-free mice show a vastly undeveloped immune system such that effector T cells are skewed toward a Th2 phenotype (Surana and Kasper 2014) and that the number of lymphocytes and the aggressiveness of cytokine expression are repressed in antibiotic-treated animals (Hill et al. 2010)—the molecular effects of commensal bacteria on immunomodulation have been extremely useful for the study of disease pathogenesis and have been a source of inspiration for researchers hoping to unearth the root causes of many diseases. For example, segmented filamentous bacteria (SFB) elicit a robust Th17 response (Ivanov et al. 2009); next, a glycosphingolipid from *Bacteroides fragilis* inhibits invariant natural killer T-cell differentiation (An et al. 2014); and finally, specific subsets of $CD4^{+}Foxp3^{+}$ regulatory T cells are induced by the presence of specific individual microbes or groups of microbes (Sefik et al. 2015; Atarashi et al.

2013). The importance of the immunomodulating roles of probiotics in health and disease has been addressed in many well-done reviews (Lazar et al. 2018; Gareau et al. 2010; Sanders et al. 2019).

6.3.2 Classification and Function of Microbe-Induced CD4$^+$ Treg in the Gut

CD4$^+$Foxp3$^+$ regulatory T cells play pivotal roles in balancing immune response. CD4$^+$ Treg cells identified by producing IL-10 in the gut are a heterogeneous group that consists of thymus-derived natural tTreg cells as well as peripheral pTreg cells. The tTreg cell population traces its origin to the thymus and a development that utilizes a thymic selection process. The remainder of the Treg subsets are developed from circulated naive CD4$^+$ T cells that encounter harmless antigens in the gut and include the Foxp3$^-$CD4$^+$ Tr1 cell subset and two subsets of specialized Foxp3$^+$CD4$^+$ pTreg—the colonic Foxp3$^+$RORγt$^+$ Treg cells specific to microbiota populations and small intestinal Foxp3$^+$RORγt$^-$ Treg cells for food antigens (Luu et al. 2017).

The function of Treg cells has wide application ranging from the prevention of dysregulated mucosal immune responses in healthy hosts, to the treatment of chronic immunological disorders such as celiac disease, food allergies, and inflammatory bowel seen in patients with autoimmune disorders. Functional studies have also suggested that Treg cells are also involved in several other processes such as the control of microbial diversity in the gut by immunoglobulin A and the support of tissue repair in response to intestinal tissue damage (Luu et al. 2017). Remarkably, Treg induction is one of the few adaptive immune responses that can be directly induced by symbiotic microbes in the gut.

6.3.3 Mechanism(s) of Induction of Gut CD4$^+$ Treg Cells by Commensal Microbiota

The microbial involvement with Treg induction in the gastrointestinal tract are best demonstrated in patients with abnormal states of immune overreaction (Wong et al. 2014). Both anaerobic and facultative aerobic gut organisms are well known to serve as probiotics for Treg induction. Colonic regulatory T cells (pTreg) induced by colonic residential bacteria (especially anaerobic organisms) are key effectors in colon disease. However, the potential for Treg induction varies with species, and even within strains of the same species. The mechanism of Treg induction is also very different for each given beneficial organism.

Currently, only a limited number of microbes such as facultative anaerobic *Clostridium perfringens* (Sefik et al. 2015) and lactic acid-producing *Lactobacillus* spp. and a few anaerobic microbes have been confirmed to have Treg-inducing effects in the colon as well as in association with breast feeding (Melnik et al. 2016). The maturation Treg cells in the colon have best been demonstrated by a studies with *Bacteroides fragilis* and several commensal *Clostridium* strains (Atarashi et al. 2011). In these studies, polysaccharide A (PSA), an immunomodulatory molecule produced by *B. fragilis*, was shown to mediate the conversion of CD4$^+$ T cells into Foxp3$^+$ Treg cells that produce IL-10 during commensal colonization (Round and Mazmanian 2010).

While the general influence of microbiota on intestinal homeostasis mediated by Treg cells is well appreciated, it remains unclear whether these effects are mediated by microorganisms themselves or by metabolic products produced by them (Arpaia et al. 2013; Bollrath and Powrie 2013). Recently, soluble microbial metabolites such as short chain fatty acids (SCFAs) and PSAs have been generally acknowledged to play crucial roles in controlling differentiation and functional activity of colonic pTreg cells (Furusawa et al. 2013; Smith et al. 2013).

In a *Citrobacter rodentium*-induced mouse diarrhea model, *Lactobacillus acidophilus* was shown to counteract diarrheal symptoms by completely blocking the induction of IL-1β, IFN-γ, and CXCL1 and inhibiting STAT3 phosphorylation (Kumar et al. 2016). A recent report by Hrdý et al. found that *Lactobacillus reuteri* 5454 and *Bifidobacterium animalis ssp. lactis* 5764 improved colitis while differentially impacting dendritic cell maturation and antimicrobial responses (Hrdý et al. 2020). They found that induction of Treg cells by *L. reuteri* 5454 in an in vitro DC/CD4 T-cell co-culture decreased the levels of inflammatory markers such as lipocalin-2, IL-1β, IL-6, and TNF-α (Hrdý et al. 2020) and thereby protected against colitis. Another example of microbial-induced Treg effects is provided by a study of coeliac disease (CD), a gluten-derived autoimmune enteropathy, in which a FoxP3 isoform mechanism was found to be initiated by microbial metabolites produced by butyrate-producing bacterial flora (Serena et al. 2017). In patients with active CD disease, intestinal biopsies showed increased expression of a FoxP3Δ2 isoform over a full-length FoxP3 isomeric segment. Such alternatively spliced FoxP3Δ2 isoforms, in contrast, were not capable of downregulating Th17-driven immune responses in the small intestine.

However, Treg effects from other studies are not always in agreement. While some studies demonstrate a favorable immune response with lactobacilli, others fail to show significant improvement in disease progression using the American College of Rheumatology (ACR) response criteria for RA (Vaghef-Mehrabany et al. 2014; Bedaiwi and Inman 2014; Nenonen et al. 1998). Therefore, the therapeutic potential of *Lactobacillus* spp., as a probiotic treatment for rheumatoid arthritis (RA) is still inconclusive.

6.3.4 Role of Anaerobic Microbes in Shaping Gut Immune Responses

A number of symbiotic organisms found in the colon, such as the rod-shaped nonmotile obligate anaerobes, are important microbes for bile metabolism, lipid metabolism, and vitamin synthesis in the host. Therefore, any alterations in the quantity of these microbes can lead to increased susceptibility to the occurrence of metabolic disorders (e.g., type 2 diabetes) and have also been linked to the development of chronic inflammatory disorders such as irritable bowel syndrome (IBS). Patients with rheumatoid arthritis (RA) also exhibit decreased gut microbial diversity, a finding which correlates with disease manifestations and autoantibody levels (Chen et al. 2016). In the colon, nearly one-quarter of the bacteria, encompassing a diversity of species, has been shown to induce RORγ^{+}Helios^{-} Tregs in a monocolonized model (Geva-Zatorsky et al. 2017). The species of the phylum Actinobacteria with the most pronounced capacity for Treg induction is *Bifidobacterium longum*, a well-known probiotic found in both the small intestine and colon that displays strong Treg induction via the PSA-DC mechanism. *Bifidobacterium longum* also has differential effects on innate and adaptive immune responses by promoting a colonic Th1 (IFNγ^{+}) reaction together with strong simultaneous stimulation of IL-22 production by innate lymphoid cells (Geva-Zatorsky et al. 2017). Enhanced IL-22 and IL-17A production in colon tissue was also shown by *B. animalis ssp. lactis* 5764 in a mouse model study where the protective roles of probiotics were assessed in chemically and biologically induced acute colitis where this microbe had a minor impact on Treg induction (Hrdý et al. 2020).

A decreased quantity of *Faecalibacterium prausnitzii* was documented in patients with RA (Chen et al. 2016), Crohn's disease (Sokol et al. 2008), obesity (Newton et al. 2015), asthma, and major depressive disorders (Jiang et al. 2015). However, the frequency of this microbe was found to be increased in patients with psoriasis (Odoñer et al. 2018). *Faecalibacterium* is part of the *Clostridium* cluster (phylum Firmicutes), gram-positive obligate anaerobe genus. *F. prausnitzii* is the sole known member in the genus, accounting for greater than 5% of the bacteria in the intestine. The production of butyrate and other SCFAs through the fermentation of dietary fiber is the signature feature of this microbe which has been frequently recommended for probiotic therapy for human IBD and IBS (Miquel et al. 2013). *F. prausnitzii* exhibits anti-inflammatory effects on cellular and TNBS colitis models. A report evaluating the molecular anti-inflammatory mechanism(s) of *F. prausnitzii* showed that the supernatant of cell cultures co-cultured with PMBC abolished NF-κB activity and was associated with significantly lower levels of IL-12 and IFN-γ and elevated IL-10 secretion (Sokol et al. 2008). These results indicate that the secreted metabolites are the effectors to block NF-κB activation. The human colonic mucosa also contains regulatory type 1-like (Tr1-like, Foxp3^{-} IL-10 secreting) T cells, which specifically respond to *Faecalibacterium prausnitzii* but not to the related *Clostridia*. Following TLR4 stimulation, human dendritic cells (DC) produce a unique array of potent Tr1/Treg polarizing molecules including an elevated expression of IL-10, IL-27, CD39, IDO-1, and PDL-1 (Alameddine et al. 2019). The same bacterium, in contrast, was shown to inhibit TLR2/6 triggering IL-12 (p35 and p40) and TNFα.

Treg effects by other bacterial species have not been well-characterized. For example, a taxon-level analysis carried out on patients with arthritis demonstrated an expansion of Actinobacteria, including *Collinsella* and *Eggerthella* (Chen et al. 2016). An important finding of this study was a strong correlation of *Collinsella* with high levels of alpha-aminoadipic acid, the proinflammatory cytokine IL-17A, and the chemokines CXCL1 and CXCL5, which may result in the recruitment of neutrophils and activation of NFkB in patients with RA. *Collinsella aerofaciens,* belonging to the phylum Actinobacteria, is the major utilizer of lactose in the human colon. The fermentation of carbohydrates leads to the production of hydrogen, ethanol, short-chain fatty acids, and lactate in the human colon (Kageyama et al. 1999). Similar to *Fusobacteria* spp. and *Bifidobacterium*, *Collinsella* can change host plasma cholesterol levels and modify the host bile acids to modulate the virulence and pathogenicity of enteric pathogens (Rajilić-Stojanović and de Vos 2014). *Eggerthella lenta* is another organism that was detected with more abundance in RA patients than in healthy people (Chen et al. 2016). Although the impact of *Eggerthella* on the immune response in the gut has not been fully determined, metabolically, it participates in the ornithine–urea cycle in the gut by using ornithine as a substrate to generate energy. It has been proposed that the increased levels of citrullination in the gut may interfere with the process under which antibodies might be produced (Chen et al. 2016).

The phylum Fusobacteria, a gram-negative obligate anaerobe, has been shown to have significant immunologic effects in clinical observation and animal model studies. *Fusobacterium nucleatum*, for example, is prevalent in patients with colorectal carcinoma and among some patients with inflammatory bowel disease (Kostic et al. 2013; Strauss et al. 2011), periodontal disease (Bolstad et al. 1996; Signat et al. 2011), and skin ulcers (Dryden 2010). The immune effects of *F. varium* include reducing both T4 (CD4^{+}) and T8 (CD8^{+}) populations, increasing colonic DN (CD4^{-}CD8^{-}TCRβ^{+}) cells more than any other microbe, suppressing C-type lectins of the Reg3 antimicrobial family (a natural peptide of epithelial cells), and inhibiting *P450* gene family and *trans*-retinoid acid metabolism in the hosts (Geva-Zatorsky et al. 2017).

6.4 Studies from Our Own Laboratory

6.4.1 In Vitro Induction of T Regulatory Cells by a Methylated CpG DNA Sequence in Humans: Potential Therapeutic Applications in Allergic and Autoimmune Diseases

Allergic and autoimmune diseases comprise a group of immunologically mediated disorders caused by aberrant immune responses in which CD25+ Forkhead box P3-positive (FOXP3+) T regulatory (Treg) cells that normally suppress inflammatory events are often poorly functioning. This has stimulated an intense investigative effort to find ways of increasing Tregs as a method of therapy for these conditions.

In a pilot study carried out in our laboratory, we investigated the epigenetic effects of methylated DNA on lymphoproliferative and Treg responses (Lawless et al. 2018). In this

Table 6.2 Base methylated motifs in three bacteria (*unpublished data, Reproduced with permission from 'Dongmei Li,* et al.)

	Type of motif	Motif string	Type	Cent Pos	# m5C/CpG	nDetction	nGenome	Fraction	Mean score	Mean Ipd ratio	Objective score
B. longum	Bl-TYPE 1	GAGGAC	m6A	5		1486	1490	1.0	241.5	5.32	357,963
	Bl-TYPE 2	RGCGGCGCC	m5C	3	**62**	62	160	0.38	144.6	2.38	1740
	Bl-TYPE 3	SNCNNTGGCGCC	m4C	3		40	203	0.20	133.9	3.04	1241
L. rhamnosus	Lr-TYPE 1	YACAGC	m6A	4		1734	1741	1.0	633.0	6.3	1,093,712
	Lr-TYPE 2	GTRAAT	m6A	5		1797	3227	0.56	141.2	2.2	149,827
		Total m6A				3531					
	Lr-TYPE 3	GNNCNNGNGANCNCGT	m5C	14		113	276	0.41	166.9	1.9	8445
	Lr-TYPE 4	GCGTNNNNSNGCGNG	m5C	12		18	120	0.15	162.6	1.6	531
	m5C/CpG motifs				**131**						
	Lr-TYPE 5	CCNNGGNCTNNTANNC	m5C	16		5	5	1.0	191.2	1.6	956
	Lr-TYPE 6	HNCTNCNCNNGCNNNNGC	m5C	12		19	119	0.16	169.1	2.0	616
	Lr-TYPE 7	CGNGNNNGNNCANGCT	m5C	15		19	123	0.15	158.0	1.5	559
		Total m5C				**174**					
E. coli strain B	Ec-TYPE 1	GATC	m6A	2		37,806	37,896	1.0	342.9	4.9	12,934,375
	Ec-TYPE 2	AGCANNNNNNNNTCA	m6A	4		686	692	1.0	303.2	4.7	206,380
	Ec-TYPE 3	TGANNNNNNNNTGCT									
	m6A	4		688	692	1.0	308.9	4.5	183,844		
	m5C/CpG motifs				**0**						

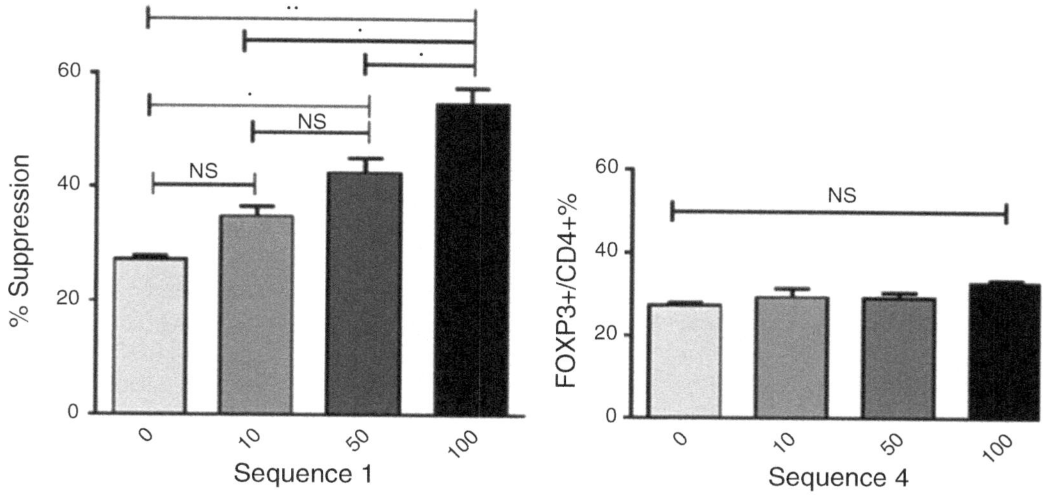

Fig. 6.8 Comparative studies of FoxP3 expression by methylated **Sequence 1** (DNA-1) and unmethylated **Sequence 4** (DNA-4). Sequence 1 shows greater Forkhead box P3 (FOXP3) expression in CD4+ T cells than sequence 4 in 5-day cultures, and it also increases FoxP3 in a dose-dependent responsive manner. *NS* nonsignificance. $*p < 0.05$; $**p < 0.01$. (*Reproduced with permission from Lawless OJ, Bellanti JA, Brown ML*, et al, (Lawless et al. 2018))

study, we evaluated the capacity of synthetic methylated DNA CpG ODN and unmethylated DNA CpG ODN sequences to induce FoxP3$^+$ Treg cells in human PMBC CD4$^+$ cells.

We showed that CT-DNA and a synthetic methylated DNA 8-mer sequence could suppress antigen-, mitogen-, and alloantigen-induced lymphoproliferation in vitro when measured by [3*H*]-thymidine. The synthetic methylated DNA CpG ODN but not the unmethylated CpG ODN sequence is seen to promote FoxP3 expression in human CD4$^+$ T cells in the presence of TGF-β and IL-2 (Fig. 6.8). The induction of FoxP3 suppressor cells is dose dependent and offers a potential clinical therapeutic application in allergic and autoimmune and inflammatory diseases.

These preliminary studies demonstrate that the use of this methylated CpG ODN offers a broad clinical application as a novel therapeutic method for Treg induction. Further, because of its low cost and small size, the CpG ODN should easy to deliver via nasal, respiratory, gastrointestinal routes, and/or by injection, all of which are routes of administration that are important for vaccine delivery to target sites responsible for respiratory, gastrointestinal, and systemic forms of allergic and autoimmune diseases.

6.4.2 Studies of the Treg-Inducing Capacity of DNA from *Bifidobacterium longum* subsp. *infantis* and *Lactobacillus rhamnosus:* Implications for Microbial DNA Immunotherapy and Treg Induction

Based upon our preliminary findings that (1) DNA methylation may be an important determinant of immune reactivity and (2) immunostimulatory CpGs of microbial DNA have been found to be largely unmethylated and proinflammatory in contrast to normal mammalian DNA that has methylated CpG moieties that are both noninflammatory and lack immunogenicity, we hypothesized that microbiota with a greater degree of DNA methylation would have a greater potential for Treg induction in the gut than those with a lower degree of methylation. As described previously, commensal gut microbiota within the gastrointestinal microbiome can favorably affect host health by directly regulating the induction of regulatory T cells (Tregs). Among the commensal microbiota known to have beneficial health benefits related to induction of Treg cells are the lactic acid-producing bacteria *Bifidobacterium longum* subsp. *infantis* and *Lactobacillus rhamnosus*. In a separate set of experiments, we tested this hypothesis by studying the Treg-inducing capacity of purified genomic DNA (gDNA) preparations from *B. longum* subsp. *infantis* and *L. rhamnosus,* compared to that of a gDNA preparation from a pathogenic *E. coli* strain B (Li et al. 2020).

We showed that gDNA from *B. longum* is a potent Treg-inducer that operates in a dose-dependent response with a dose-threshold of 20 μg gDNA. No similar correlation was found in gDNA from *L. rhamnosus* or *E. coli.* Moreover, the gDNA from *Bifidobacterium longum* was associated with a unique bacterial motif not found in *L. rhamnosus* or *E. coli* strain B. We found 62 m5C hits in *B. longum* genome; all of which were m5C motifs sharing the same m5C motif (RGCpGGCGCC) in this species at an average 144 modQV level as shown in Table 6.2. By contrast, a total of 174 m5C contexts were detected in *L. rhamnosus*, consisting of five types of motifs with CpG, CpT, and CpN modification, corresponding to a weak Treg-inducing phenotype.

Our findings add an additional layer in our understanding of the complex host–microbiota interactions responsible for Treg induction and offer an alternative approach to Treg induction. The anti-inflammatory capacity of methylated bacterial DNA may be harnessed by its endogenous content of suppressive methylated motifs that may provide a novel therapeutic approach to the treatment of immunologic diseases in which Treg populations are diminished—such as the allergic and autoimmune diseases.

6.5 Prospective Clinical Applications of Treg Cells and Epigenetic Regulation

It is now becoming clear that microbiota and their metabolites in the host gut can regulate immune cells and cytokines and the expression level of Treg cells via epigenetic modifications. The Th17/Treg (T helper 17/regulatory T cell) ratio as a biomarker, for example, has been linked to the development and progression of inflammatory diseases, insulin resistance, and systemic lupus erythematosus (SLE) (Rahbar Saadat et al. 2019), which is potentially mediated by epigenetic mechanisms. For example, SCFAs produced by gut microbiota promote the differentiation of naïve T cell into Treg by suppressing histone deacetylases (HDACs) (Chen et al. 2017). Targeting epigenetic changes via HDAC inhibitors (Cantley et al. 2015) and miRNA-based therapies has been applied in both human and animal models of inflammatory or autoimmune diseases. However, the effectiveness of HDAC inhibitors as anti-inflammatory agents depends on cell type and dosage, and the data are at times contradictory (Hull et al. 2016).

The fact that methylated CpG-rich islands of the Foxp3 in conventional $CD4^+$ is bounded by DNMT1, DNMT3b, MeCP2, and MBD2 provides a therapeutic idea to enhance Treg subset through the methyltransferase inhibitor. The fact that DNA 5-aza-2′-deoxycytidine (Aza), a DNMT inhibitor, together with TGF-β, resulted in a strong and stable induction of Foxp3 in human and in mouse $CD4^+$ cells, indicates that the epigenetic regulation of Foxp3 can be predictably controlled (Lal et al. 2009; Polansky et al. 2008). In addition, tolerogenic vaccination by DEC-205-mediated targeting of peptide-agonist ligands to DC in vivo converts naive T cells into Treg cells that display both stable Foxp3 expression and demethylated TSDR (Polansky et al. 2008). DEC-205, a multilectin receptor expressed by a variety of cells including DC, has been identified as a receptor for CpG ODN (Lahoud et al. 2012). The impact of this receptor on a

Table 6.3 Microbial metabolites or components that are implicated in disease (*Modified and reproduced with permission from Rooks MG, Garrett WS. Gut microbiota, metabolites and host immunity. Nat Rev Immunol. 2016; 16:341–52*)

Human disease and preclinical models	Microbial metabolites or components
Allergic and immune disorders	
• Asthma	SCFAs
• Inflammatory bowel disease	SCFAs
	B vitamins
Cancer	
• Colorectal cancer	SCFAs
	B vitamins
	*N*1,*N*12-diacetylspermine
Gynecological and reproductive disorders	
• Bacterial vaginosis and other sexually transmitted infections	Polyamines
	HBP
• Preterm labor	SCFAs
Metabolic disorders	
• Cardiovascular disease	TMAO
• Kidney disease	SCFAs
	p-Cresol
• Obesity and metabolic syndrome	TMAO
• Type 2 diabetes	TMAO
Neurological disorders	
• Autism spectrum disorder	4-EPS
• Central nervous system dysfunction	SCFAs
Other gastrointestinal disorders	
• Infectious colitis (*Clostridium difficile*)	Bile acids

EPS 4-ethyl phenol sulfate, *HBP* D-glycero-β-D-manno-heptose-1,7-biphosphate, *SCFAs* short-chain fatty acids, *TMAO* trimethylamine *N*-oxide

synthetic CpG ODN vaccine could be significant when used as an adjuvant.

Recently identified participation of gut microbiota in the remodeling of the epigenome of immune cells has opened new paths toward operative therapeutic strategies. These data are mostly collected from animal models, in which the most promising dietary interventions include the use of prebiotics and probiotics (Luo et al. 2017). Prebiotics are defined as fermented ingredients such as non-digestible saccharides. Supplementation with PSA can increase the population of butyrogenic strains in the gut that produce SCFAs such as butyrate. Increased levels of SCFAs eventually remodel the T-cell epigenome to favor Treg differentiation (Smith et al. 2013). Probiotics, a common ingredient in yogurt, are live organisms which, when consumed by the host, provide beneficial health effects as described in Sect. 6.3.

In summary, studies of the epigenetic roles of microbial communities, and the effects of metabolites and components on T-cell induction and subsequently on immune homeostasis and immune-mediated diseases are a promising area for future research. Table 6.3 shows the results of studies that continue to uncover the role of an altered microbiome in human disorders and diseases, as well as in preclinical models, and to uncover microbial metabolites that have been implicated in their pathogenesis. In addition to the role of extracellular metabolites produced by these microbiota, our recent findings suggest that molecular composition and degree of methylation of the genetic DNA of these microbes also play a role in the induction of Treg cells that may provide a novel therapeutic modality to increase these important immunomodulatory cells through the use of vaccines.

References

Alameddine J, Godefroy E, Papargyris L, Sarrabayrouse G, Tabiasco J, Bridonneau C et al (2019) *Faecalibacterium prausnitzii* skews human DC to prime IL10-producing T cells through TLR2/6/JNK signaling and IL-10, IL-27, CD39, and IDO-1 induction. Front Immunol 10:143

An D, Oh SF, Olszak T, Neves JF, Avci FY, Erturk-Hasdemir D et al (2014) Sphingolipids from a symbiotic microbe regulate homeostasis of host intestinal natural killer T cells. Cell 156:123–133

Arpaia N, Campbell C, Fan X, Dikiy S, van der Veeken J, deRoos P et al (2013) Metabolites produced by commensal bacteria promote peripheral regulatory T-cell generation. Nature 504:451–455

Atarashi K, Tanoue T, Shima T, Imaoka A, Kuwahara T, Momose Y et al (2011) Induction of colonic regulatory T cells by indigenous *Clostridium* species. Science 331:337–341

Atarashi K, Tanoue T, Oshima K, Suda W, Nagano Y, Nishikawa H et al (2013) Treg induction by a rationally selected mixture of *Clostridia* strains from the human microbiota. Nature 500:232–236

Bedaiwi MK, Inman RD (2014) Microbiome and probiotics: link to arthritis. Curr Opin Rheumatol 2 (6):410–415

Bellanti JA (ed) (2012) Immunology IV: clinical applications in health and disease. I Care Press, Bethesda

Bellanti JA (2019) Genetics/epigenetics/allergy: the gun is loaded . . . but what pulls the trigger? Allergy Asthma Proc 40:76–83

Bellanti JA (2020) Phenotypic classification of asthma based upon a new T2-high and T2-low endotypic classification: it all began with Rackemann. J Precision Resp Med 3:9–20

Bollrath J, Powrie F (2013) Immunology. Feed your Tregs more fiber. Science 341:463–464

Bolstad AI, Jensen HB, Bakken V (1996) Taxonomy, biology, and periodontal aspects of *Fusobacterium nucleatum*. Clin Microbiol Rev 9:55–71

Cantley MD, Fairlie DP, Bartold PM, Marino V, Gupta PK, Haynes DR (2015) Inhibiting histone deacetylase 1 suppresses both inflammation and bone loss in arthritis. Rheumatology 54:1713–1723

Chen J, Wright K, Davis JM, Jeraldo P, Marietta EV, Murray J et al (2016) An expansion of rare lineage intestinal microbes characterizes rheumatoid arthritis. Genome Med 8:43

Chen B, Sun L, Zhang X (2017) Integration of microbiome and epigenome to decipher the pathogenesis of autoimmune diseases. J Autoimmun 83:31–42

Dryden MS (2010) Complicated skin and soft tissue infection. J Antimicrob Chemother 65(Suppl 3):iii35–iii44

Floess S, Freyer J, Siewert C, Baron U, Olek S, Polansky J et al (2007) Epigenetic control of the foxp3 locus in regulatory T cells. PLoS Biol 5:e38

Furusawa Y, Obata Y, Fukuda S, Endo TA, Nakato G, Takahashi D et al (2013) Commensal microbe-derived butyrate induces the differentiation of colonic regulatory T cells. Nature 504:446–450

Gareau MG, Sherman PM, Walker WA (2010) Probiotics and the gut microbiota in intestinal health and disease. Nat Rev Gastroenterol Hepatol 7:503–514

Geva-Zatorsky N, Sefik E, Kua L, Pasman L, Tan TG, Ortiz-Lopez A et al (2017) Mining the human gut microbiota for immunomodulatory organisms. Cell 168:928–943

Gu J, Lu L, Chen M, Xu L, Lan Q, Li Q et al (2014) TGF-β-induced CD4+Foxp3+ T cells attenuate acute graft-versus-host disease by suppressing expansion and killing of effector CD8+ cells. J Immunol 193:3388–3397

Hill DA, Hoffmann C, Abt MC, Du Y, Kobuley D, Kirn TJ et al (2010) Metagenomic analyses reveal antibiotic-induced temporal and spatial changes in intestinal microbiota with associated alterations in immune cell homeostasis. Mucosal Immunol 3:148–158

Hooper LV, Littman DR, Macpherson AJ (2012) Interactions between the microbiota and the immune system. Science 336:1268–1273

Hrdý J, Alard J, Couturier-Maillard A, Boulard O, Boutillier D, Delacre M et al (2020) *Lactobacillus reuteri* 5454 and *Bifidobacterium animalis ssp. lactis 5764* improve colitis while differentially impacting dendritic cells maturation and antimicrobial responses. Sci Rep 10:5345

Huehn J, Siegmund K, Lehmann JC, Siewert C, Haubold U, Feuerer M et al (2004) Developmental stage, phenotype, and migration distinguish naive- and effector memory-like CD4+ regulatory T cells. J Exp Med 199:303–313

Hull EE, Montgomery MR, Leyva KJ (2016) HDAC inhibitors as epigenetic regulators of the immune system: impacts on cancer therapy and inflammatory diseases. Biomed Res Int 2016:8797206

Ivanov II, Atarashi K, Manel N, Brodie EL, Shima T, Karaoz U et al (2009) Induction of intestinal Th17 cells by segmented filamentous bacteria. Cell 139:485–498

Janson PC, Winerdal ME, Marits P, Thörn M, Ohlsson R, Winqvist O (2008) FOXP3 promoter demethylation reveals the committed Treg population in humans. PLoS One 3:e1612

Jiang H, Ling Z, Zhang Y, Mao H, Ma Z, Yin Y et al (2015) Altered fecal microbiota composition in patients with major depressive disorder. Brain Behav Immun 48:186–194

Kageyama A, Benno Y, Nakase T (1999) Phylogenetic and phenotypic evidence for the transfer of *Eubacterium aerofaciens* to the genus *Collinsella* as *Collinsella aerofaciens* gen. nov., comb. nov. Int J Syst Bacteriol 49(Pt 2):557–565

Kamada N, Seo SU, Chen GY, Núñez G (2013) Role of the gut microbiota in immunity and inflammatory disease. Nat Rev Immunol 13:321–335

Kostic AD, Chun E, Robertson L, Glickman JN, Gallini CA, Michaud M et al (2013) *Fusobacterium nucleatum* potentiates intestinal tumorigenesis and modulates the tumor-immune microenvironment. Cell Host Microbe 14:207–215

Kumar A, Anbazhagan AN, Coffing H, Chatterjee I, Priyamvada S, Gujral T et al (2016) *Lactobacillus acidophilus* counteracts inhibition of NHE3 and DRA expression and alleviates diarrheal phenotype in mice infected with *Citrobacter rodentium*. Am J Physiol Gastrointest Liver Physiol 311:G817–G826

Lahoud MH, Ahmet F, Zhang JG, Meuter S, Policheni AN, Kitsoulis S et al (2012) DEC-205 is a cell surface receptor for CpG oligonucleotides. Proc Natl Acad Sci U S A 109:16270–16275

Lal G, Bromberg JS (2009) Epigenetic mechanisms of regulation of Foxp3 expression. Blood 114:3727–3735

Lal G, Zhang N, van der Touw W, Ding Y, Ju W, Bottinger EP et al (2009) Epigenetic regulation of Foxp3 expression in regulatory T cells by DNA methylation. J Immunol 182:259–273

Lawless OJ, Bellanti JA, Brown ML, Sandberg K, Umans JG, Zhou L, Chen W, Wang J, Wang K, Zheng SG (2018) In vitro induction of T regulatory cells by a methylated CpG DNA sequence in humans: potential therapeutic applications in allergic and autoimmune diseases. Allergy Asthma Proc 39:143–152

Lazar V, Ditu LM, Pircalabioru GG, Gheorghe I, Curutiu C, Holban AM et al (2018) Aspects of gut microbiota and immune system interactions in infectious diseases, immunopathology, and cancer. Front Immunol 9:1830

Li D, Cheng J, Zhu Z, Catalfamo M, Goerlitz D, Lawless OJ et al (2020) Treg-inducing capacity of genomic DNA of Bifidobacterium longum subsp. infantis. Allergy Asthma Proc 41:372–385

Luo A, Leach ST, Barres R, Hesson LB, Grimm MC, Simar D (2017) The microbiota and epigenetic regulation of T helper 17/regulatoryT cells: in search of a balanced immune system. Front Immunol 8:417

Luu M, Steinhoff U, Visekruna A (2017) Functional heterogeneity of gut-resident regulatory T cells. Clin Transl Immunol 6:e156

Melnik BC, John SM, Carrera-Bastos P, Schmitz G (2016) Milk: a postnatal imprinting system stabilizing FoxP3 expression and regulatory T cell differentiation. Clin Transl Allergy 6:18

Miquel S, Martín R, Rossi O, Bermúdez-Humarán LG, Chatel JM, Sokol H et al (2013) *Faecalibacterium prausnitzii* and human intestinal health. Curr Opin Microbiol 16:255–261

Nenonen MT, Helve TA, Rauma AL, Hanninen OO (1998) Uncooked, lactobacilli-rich, vegan food and rheumatoid arthritis. Br J Rheumatol 37:274–281

Newton RJ, McLellan SL, Dila DK, Vineis JH, Morrison HG, Eren AM et al (2015) Sewage reflects the microbiomes of human populations. mBio 6:e02574

Odoñer FM, Ramírez-Bosca A, Climent E, Carrión-Gutierrez M, Guerrero M, Pérez-Orquín JM et al (2018) Gut microbial composition in patients with psoriasis. Sci Rep 8:3812

Polansky JK, Kretschmer K, Freyer J, Floess S, Garbe A, Baron U et al (2008) DNA methylation controls Foxp3 gene expression. Eur J Immunol 38:1654–1663

Polansky JK, Schreiber L, Thelemann C, Ludwig L, Krüger M, Baumgrass R et al (2010) Methylation matters: binding of Ets-1 to the demethylated Foxp3 gene contributes to the stabilization of Foxp3 expression in regulatory T cells. J Mol Med (Berl) 88:1029–1040

Rahbar Saadat Y, Hejazian M, Bastami M, Hosseinian Khatibi SM, Ardalan M, Zununi VS (2019) The role of microbiota in the pathogenesis of lupus: dose it impact lupus nephritis? Pharmacol Res 139:191–198

Rajilić-Stojanović M, de Vos WM (2014) The first 1000 cultured species of the human gastrointestinal microbiota. FEMS Microbiol Rev 38:996–1047

Rodenhiser D, Mann M (2006) Epigenetics and human disease: translating basic biology into clinical applications. CMAJ 174:341–348

Round JL, Mazmanian SK (2010) Inducible Foxp3+ regulatory T-cell development by a commensal bacterium of the intestinal microbiota. Proc Natl Acad Sci U S A 107:12204–12209

Sanders ME, Merenstein DJ, Reid G, Gibson GR, Rastall RA (2019) Probiotics and prebiotics in intestinal health and disease: from biology to the clinic. Nat Rev Gastroenterol Hepatol 16:605–616

Schrezenmeier E, Dörner T (2020) Mechanisms of action of hydroxychloroquine and chloroquine. Implications for rheumatology. Nat Rev Rheumatol 16:155–166

Sefik E, Geva-Zatorsky N, Oh S, Konnikova L, Zemmour D, McGuire AM et al (2015) MUCOSAL IMMUNOLOGY. Individual intestinal symbionts induce a distinct population of $ROR\gamma^+$ regulatory T cells. Science 349:993–997

Sekiya T, Nakatsukasa H, Lu Q, Yoshimura A (2016) Roles of transcription factors and epigenetic modifications in differentiation and maintenance of regulatory T cells. Microbes Infect 18(6):378–386

Serena G, Yan S, Camhi S, Patel S, Lima RS, Sapone A et al (2017) Proinflammatory cytokine interferon-γ and microbiome-derived metabolites dictate epigenetic switch between forkhead box protein 3 isoforms in coeliac disease. Clin Exp Immunol 187:490–506

Signat B, Roques C, Poulet P, Duffaut D (2011) *Fusobacterium nucleatum* in periodontal health and disease. Mol Biol 13:25–36

Slack JM (2002) Conrad Hal Waddington: the last renaissance biologist? Nat Rev Genet 3:889–895

Smith PM, Howitt MR, Panikov N, Michaud M, Gallini CA, Bohlooly YM et al (2013) The microbial metabolites, short-chain fatty acids, regulate colonic Treg cell homeostasis. Science 341:569–573

Sokol H, Pigneur B, Watterlot L, Lakhdari O, Bermúdez-Humarán LG, Gratadoux JJ et al (2008) *Faecalibacterium prausnitzii* is an anti-inflammatory commensal bacterium identified by gut microbiota analysis of Crohn disease patients. Proc Natl Acad Sci U S A 105:16731–16736

Strauss J, Kaplan GG, Beck PL, Rioux K, Panaccione R, Devinney R et al (2011) Invasive potential of gut mucosa-derived *Fusobacterium nucleatum* positively correlates with IBD status of the host. Inflamm Bowel Dis 17:1971–1978

Su W, Fan H, Chen M, Wang J, Brand D, He X et al (2012) Induced CD4+ forkhead box protein-positive T cells inhibit mast cell function and established contact hypersensitivity through TGF-β1. J Allergy Clin Immunol 130:444–52.e7

Su W, Chen X, Zhu W, Yu J, Li W, Li Y et al (2019) The cAMP-adenosine feedback loop maintains the suppressive function of regulatory T cells. J Immunol 203:1436–1446

Surana NK, Kasper DL (2014) Deciphering the tête-à-tête between the microbiota and the immune system. J Clin Invest 124(10):4197–4203

Tsuji-Takayama K, Suzuki M, Yamamoto M, Harashima A, Okochi A, Otani T et al (2008) The production of IL-10 by human regulatory T cells is enhanced by IL-2 through a STAT5-responsive intronic enhancer in the IL-10 locus. J Immunol 181:3897–3905

Vaghef-Mehrabany E, Alipour B, Homayouni-Rad A, Sharif SK, Asghari-Jafarabadi M, Zavvari S (2014) Probiotic supplementation improves inflammatory status in patients with rheumatoid arthritis. Nutrition 30:430–435

Vighi G, Marcucci F, Sensi L, Di Cara G, Frati F (2008) Allergy and the gastrointestinal system. Clin Exp Immunol 153(Suppl 1):3–6

Waddington C (1957) The study of the genes. Allen and Unwin, London

Wong D, Aoki F, Rubinstein E (2014) Bacteremia caused by *Eggerthella lenta* in an elderly man with a gastrointestinal malignancy: a case report. Can J Infect Dis Med Microbiol 25:e85–e86

Xu A, Liu Y, Chen W, Wang J, Xue Y, Huang F et al (2016) TGF-β-induced regulatory T cells directly suppress b cell responses through a noncytotoxic mechanism. J Immunol 196:3631–3641

Zheng SG, Gray JD, Ohtsuka K, Yamagiwa S, Horwitz DA (2002) Generation ex vivo of TGF-beta-producing regulatory T cells from $CD4^{+}CD25^{-}$ precursors. J Immunol 169:4183–4189

Zheng SG, Wang JH, Gray JD, Soucier H, Horwitz DA (2004) Natural and induced $CD4^{+}CD25^{+}$ cells educate $CD4^{+}CD25^{-}$ cells to develop suppressive activity: the role of IL-2, TGF-beta, and IL-10. J Immunol 172:5213–5221

Zheng SG, Wang JH, Stohl W, Kim KS, Gray JD, Horwitz DA (2006) TGF-beta requires CTLA-4 early after T cell activation to induce FoxP3 and generate adaptive $CD4^{+}CD25^{+}$ regulatory cells. J Immunol 176:3321–3329

Zheng SG, Wang J, Wang P, Gray JD, Horwitz DA (2007) IL-2 is essential for TGF-beta to convert naive CD4 +CD25- cells to CD25+Foxp3+ regulatory T cells and for expansion of these cells. J Immunol 178:2018–2027

TCR/ITK Signaling in Type 1 Regulatory T cells

7

Michael C. McGee, Avery August, and Weishan Huang

Abstract

Type 1 regulatory T (Tr1) cells can modulate inflammation through multiple direct and indirect molecular and cellular mechanisms and have demonstrated potential for anti-inflammatory therapies. Tr1 cells do not express the master transcription factor of conventional regulatory T cells, Foxp3, but express high levels of the immunomodulatory cytokine, IL-10. IL-2-inducible T-cell kinase (ITK) is conserved between mouse and human and is highly expressed in T cells. ITK signaling downstream of the T-cell receptor (TCR) is critical for T-cell subset differentiation and function. Upon activation by TCR, ITK is critical for Ras activation, leading to downstream activation of MAPKs and upregulation of IRF4, which further enable Tr1 cell differentiation and suppressive function. We summarize here the structure, signaling pathway, and function of ITK in T-cell lineage designation, with an emphasis on Tr1 cell development and function.

Keywords

ITK · TCR · Tr1 cells · Ras · IRF4

7.1 Introduction

Regulatory T cells are critical in promoting self-tolerance and preventing immunopathology from excessive inflammation. Transcription factor Foxp3 is a well-recognized lineage specification factor of conventional T regulatory (Treg) cells. $Foxp3^+$ conventional Treg cells express Foxp3 and CD25 as identifying markers and are important immune regulators that promote self-tolerance and immune homeostasis in human and mouse (Bennett et al. 2001; Brunkow et al. 2001; Chatila et al. 2000; Fontenot et al. 2003; Gambineri et al. 2003; Hori et al. 2003; Khattri et al. 2003; Sakaguchi et al. 1995; Shevach et al. 2001; Wildin et al. 2001). Type 1 regulatory T (Tr1) cells, on the other hand, lack the expression of Foxp3 and CD25, but express high levels of the immunomodulatory cytokine IL-10, which can suppress inflammatory responses associated with type 2 cytokines (IL-4/5/13) and IL-17 responses

M. C. McGee
Department of Pathobiological Sciences, School of Veterinary Medicine, Louisiana State University, Baton Rouge, LA, USA

A. August
Department of Microbiology and Immunology, College of Veterinary Medicine, Cornell University, Ithaca, NY, USA

W. Huang (✉)
Department of Pathobiological Sciences, School of Veterinary Medicine, Louisiana State University, Baton Rouge, LA, USA

Department of Microbiology and Immunology, College of Veterinary Medicine, Cornell University, Ithaca, NY, USA
e-mail: huang1@lsu.edu

S.-G. Zheng (ed.), *T Regulatory Cells in Human Health and Diseases*, Advances in Experimental Medicine and Biology 1278, https://doi.org/10.1007/978-981-15-6407-9_7

(Gagliani et al. 2013; Gol-Ara et al. 2012; Huber et al. 2011; Okamura et al. 2012). Tr1 cells directly suppress the effector cell response through the secretion of IL-10, granzyme-dependent killing of antigen-presenting cells, and expression of coinhibitory receptors such as CTLA-4 and PD-1 (Akdis 2008; Haringer et al. 2009; Huber et al. 2011; Magnani et al. 2011). Indirect immune suppression by Tr1 cells has been shown as well, such as depletion of proinflammatory extracellular ATP by CD39 expressed on Tr1 cells (Mascanfroni et al. 2015).

Tr1 cells have been shown to suppress the development of diseases such as allergic asthma, colitis, and atopic dermatitis in animal models (Ahangarani et al. 2009; Groux et al. 1997; Volz et al. 2014). Antigen-specific tolerance correlates positively with Tr1 numbers during hematopoietic stem cell transplant and immunotherapy (Bohm et al. 2015; Serafini et al. 2009). Their ability to regulate inflammation and induce tolerance makes Tr1 cells a promising candidate for immunotherapies (Bohm et al. 1998; Gol-Ara et al. 2012; Mobs et al. 2010; Roncarolo et al. 2014; Volz et al. 2014; Zeng et al. 2015). However, this potent regulatory activity has the potential to "cut both ways" by hindering the protective response toward pathogens or tumors. Patients with chronic hepatitis C possess higher numbers of virus-specific Tr1 cells than patients who spontaneously clear the infection (Brady et al. 2003; MacDonald et al. 2002). In a mouse model of *Bordetella pertussis*, pathogen-specific Tr1 cells have been shown to suppress the protective Th1 response (McGuirk et al. 2002). Tr1 cells isolated from tumors display immunosuppressive functions ex vivo (Bergmann et al. 2008; Pedroza-Gonzalez et al. 2015). Thus, understanding the signaling pathways that drive Tr1 differentiation and function is of great interest. While much is known about signaling pathways involved in Treg cell development and function (Sakaguchi et al. 2008; Vignali et al. 2008), significantly less is known about the molecular mechanisms regulating Tr1 cell development and how they function against inflammation.

Members of the Tec family of nonreceptor tyrosine kinases are critical in signaling pathways of the immune system (August and Ragin 2012; Berg et al. 2005; Gilfillan and Rivera 2009). IL-2-inducible T-cell kinase (ITK) is the predominant Tec family kinase expressed in T cells and is a signaling mediator downstream of the T-cell receptor (TCR) (August et al. 2002). ITK plays major roles in modulating T-cell development, activation, differentiation, and function (Gomez-Rodriguez et al. 2014, 2016; Huang et al. 2014; Kannan et al. 2015; Schaeffer et al. 2000, 2001). The role of ITK in the differentiation and function of several T helper (Th) lineages is well studied (August and Ragin 2012; Gomez-Rodriguez et al. 2014; Huang et al. 2014; Kannan et al. 2013; Miller et al. 2004). However, it has only been recently explored in Tr1 cells (Huang et al. 2017). This chapter summarizes the role of ITK in TCR signaling and Th cell differentiation, with an emphasis on the role of ITK in Tr1 cell differentiation and function.

7.2 Structure of ITK

ITK is a member of the Tec family kinases and is expressed by mast cells and T cells, with a predominant preference in T cells (Andreotti et al. 2010; August et al. 2002). Tec family includes ITK, BTK, TEC, BMX, and TXK and is the second largest non-receptor protein-tyrosine kinase family (second to Src family) (Takesono et al. 2002). These kinases are classified based on the unique Tec-Homology (TH) domain, which is composed of a zinc-binding Btk-homology (BH) motif and/or proline-rich regions (PRR). ITK and its family members share a high homology in structure consisting of, from protein N terminal to C terminal, pleckstrin homology (PH), TH (BH + PRR), Src-homology (SH) 3, SH2, and kinase domains (Fig. 7.1a) (Felices et al. 2007). At the steady state in resting T cells, ITK normally resides in the cytosolic compartment, and the PH domain is critical in recruiting ITK to the plasma membrane upon TCR activation (August et al. 1997). The SH3 and SH2 domains are important for mediating protein–protein interactions between ITK and other components of the TCR signaling complex,

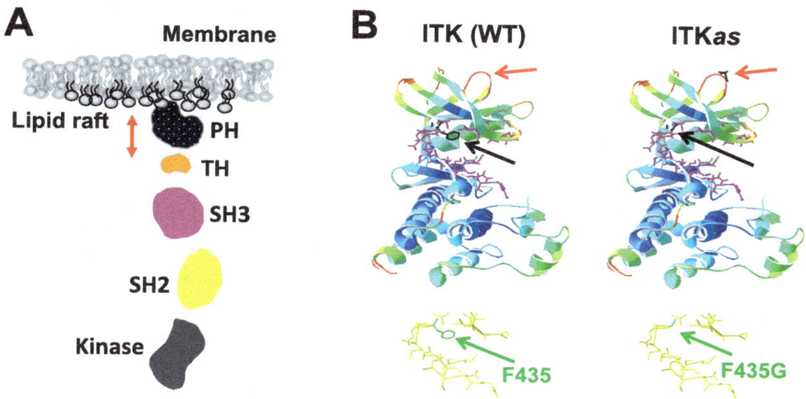

Fig. 7.1 Structure of ITK. (**a**) Schematic of structure of ITK: from N to C terminus, there are PH, TH, SH3, SH3, and kinase domains. (**b**) Structures of kinase domain of WT and allele sensitive (*as*) mutant of ITK (upper panel). ITK*as* harbors an F435G mutant (black and green arrows) in the ATP binding pocket (lower panel) of ITK kinase domain along with a ΔA429 (red arrows) to preserve the ITK kinase activity of ITK*as*. (Modified from the authors' previous publication (Kannan et al. 2015))

most notably SLP-76 (Berg et al. 2005; Bunnell et al. 2000; Su et al. 1999). In addition, these SH2 and SH3 protein–protein interactions are critical for kinase-independent functions in regulating actin cytoskeleton rearrangement (Dombroski et al. 2005; Grasis et al. 2003).

The C-terminal kinase domain has high levels of structural similarity between ITK and other Tec family members, making it challenging to identify small molecule inhibitors that would be specific for ITK over the other Tec kinases. The presence of unique gatekeeper residues in the kinase domain may enable the development of an allele-sensitive kinase domain that allows temporal inhibition of the kinase activity using small molecule inhibitors (Bishop et al. 1998). A bulky gatekeeper residue, F435, has been identified in ITK, and substitution of this residue with the short chain glycine (F435G) allows for bulkier ATP analogs to efficiently compete for the ATP binding and prevent activation of this allele sensitive mutant of ITK (ITK*as*) (Fig. 7.1b) (Shokat and Velleca 2002). These bulkier ATP analogs include modified derivatives of the Src-kinase inhibitor PP1 (1-(1,1-dimethylethyl)-3-(4-methylphenyl)-1*H*-pyrazolo[3,4-d]pyrimidin-4-amine) such as 3MB-PP1 (1-(1,1-dimethylethyl)-3-[(3-methylphenyl)methyl]-1*H*-pyrazolo[3,4-d]pyrimidin-4-amine), 1NM-PP1 (1-(1,1-dimethylethyl)-3-(1-naphthalenylmethyl)-1*H*-pyrazolo[3,4-d]pyrimidin-4-amine), and 1-NA-PP1 (1-(1,1-dimethylethyl)-3-(1-naphthalenyl)-1*H*-pyrazolo[3,4-d]pyrimidin-4-amine), that have unique selectivity for kinases that are modified by substitution of similarly located gatekeeper residues in other kinases (Bishop et al. 2000).

7.3 TCR/ITK Signaling Pathway

The TCR is stimulated upon interaction with the peptide/MHC complex on antigen-presenting cells. This interaction results in the phosphorylation of Lck. Lck phosphorylates the ITAMs on the cytoplasmic CD3ζ chain which allows for recruitment and phosphorylation of ZAP-70. Activated ZAP-70 phosphorylates the adaptor proteins LAT and SLP-76 which forms a crucial scaffold for the TCR signaling complex or signalosome (Werlen and Palmer 2002). Lck also phosphorylates phosphatidylinositol 3-kinase (PI3K) which generates phosphatidylinositol (3,4,5)-trisphosphate (PIP3) lipids in the plasma membrane. ITK is recruited to the plasma membrane by PIP3 via the PH domain (August et al. 1997) where it is phosphorylated by Lck. This event allows ITK to interact with the TCR signalosome via phosphorylated tyrosine 145 on SLP-76 via its SH2 domain (Su et al. 1999). Once associated with the signalosome, ITK activates PLCγ-1 via phosphorylation. Activated PLCγ-1 converts membrane phosphatidylinositol 4,5-bisphosphate (PIP2) into the second messengers inositol trisphosphate (IP3) and diacylglycerol (DAG) (Schaeffer et al. 1999). IP3 stimulates IP3R on the endoplasmic reticulum (ER), resulting in release of STIM1-mediated ER calcium (Ca^{2+}) stores into the cytoplasm. Depletion of ER Ca^{2+} stores results in an influx of extracellular Ca^{2+} through calcium release-activated channels such as Orai1 on the plasma membrane (Smith-Garvin et al. 2009). Increased intracellular Ca^{2+} levels activate calcineurin which dephosphorylates and promotes nuclear translocation of NFAT (Smith-Garvin et al. 2009). Production of DAG leads to the activation of Ras/Raf1/MAPK pathway (Smith-Garvin et al. 2009), which can further lead to activation and upregulation of transcription factor AP-1 and IRF family members (Huang et al. 2017; Schaeffer et al. 2001). DAG also recruits PKCθ to the plasma membrane where it regulates NF-κB activation and nuclear translocation (Smith-Garvin et al. 2009). This TCR/ITK signaling pathway is detailed in Fig. 7.2.

TCR signaling still occurs in the absence of ITK, but with attenuated strength (August and Ragin 2012; Schaeffer et al. 2001), which makes ITK more a signal amplifier rather than a signal switch during T-cell activation (August and Ragin 2012). In the absence of ITK, activation of the MAPK pathway and downstream AP-1 and IRF4 transcription factors is reduced, as well as the calcium flux and NFAT activity (Huang et al. 2017; Schaeffer et al. 2001); NF-κB activation is also impaired, but to a lesser extent than MAPK and NFAT pathways (Fowell et al. 1999; Schaeffer et al. 2001). The attenuation of these signaling pathways has consequences for T-cell subset differentiation and function.

7.4 Kinase-Independent ITK Function

ITK also has kinase-independent functions during TCR activation. Upon T-cell activation, ITK directly interacts with Vav, a regulator of actin polymerization, via the SH2 domain (Dombroski et al. 2005). In mouse models of *Itk* deficiency, T-cell-specific expression of ITK with mutations in the SH2 and SH3 domain display impaired actin polymerization along with reduced recruitment of Vav to the signaling complex during T-cell activation (Dombroski et al. 2005; Grasis et al. 2003). However, ITK kinase dead mutants do not show this deficiency, suggesting that ITK acts as part of a scaffold important for cytoskeleton reorganization independently of its kinase activity (Fig. 7.2) (Dombroski et al. 2005; Grasis et al. 2003; Hao et al. 2006; Qi et al. 2011; Sahu et al. 2008).

7.5 Function of ITK in T Helper Cell Designation

The role of ITK in T-cell subset differentiation and function has been well studied over the past two decades (Fig. 7.3). Naïve CD4 cells preferentially differentiate into IFN-γ-producing Th1 cells in the absence of ITK, partly due to its negative regulation of Tbet expression (Kannan et al. 2013; Miller et al. 2004). *Itk*-deficient mice have defects in Th2 differentiation and function and were reported to be resistant to developing allergic airway inflammation (Au-Yeung et al. 2006; Fowell et al. 1999; Kannan et al. 2013). ITK plays a crucial role in regulating the balance between Th17 and Treg cell differentiation in part via regulating sensitivity to IL-2 (Gomez-Rodriguez et al. 2014). Under Th17-polarizing conditions, ITK signals suppress Treg cell differentiation while promoting Th17 differentiation (Gomez-Rodriguez et al. 2014; Huang et al. 2014). Importantly, impaired calcium/NFAT signaling in $Itk^{-/-}$ $CD4^{+}$ cells results in decreased cytokine production by both Th17 and Th2 cells (Fowell et al. 1999; Gomez-Rodriguez et al. 2009). ITK has also been shown to be critical in Th9

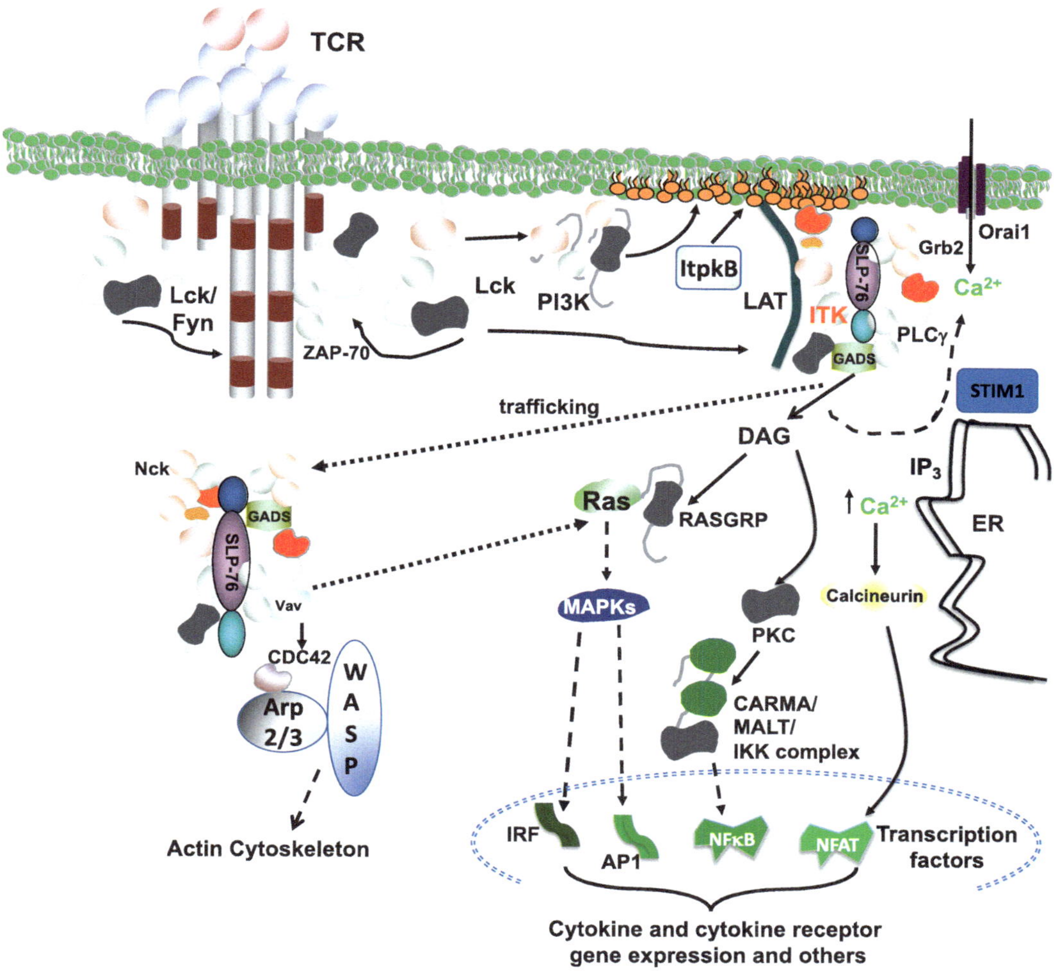

Fig. 7.2 Scheme of TCR signaling through ITK. (Modified from the authors' previous publication (Kannan et al. 2012))

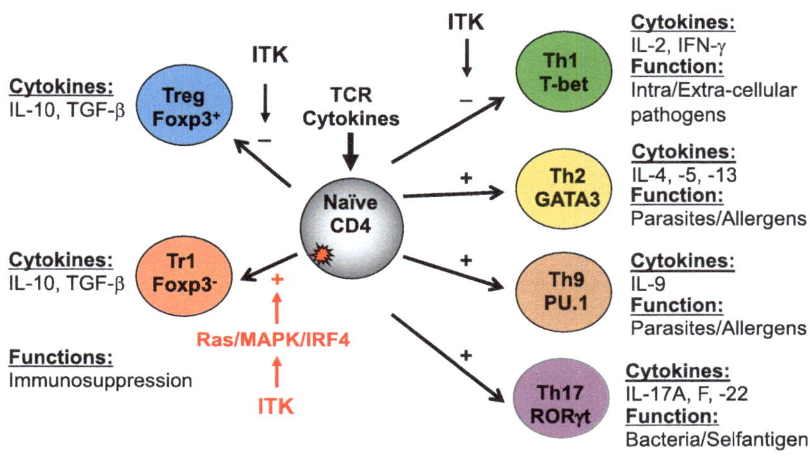

Fig. 7.3 ITK function in Th and regulatory T-cell differentiation and function. Note that ITK signals through Ras/MAPK/IRF4 are critical for Tr1 cell development and suppressive function

differentiation by regulating induction of IL-2/STAT5 signaling leading to upregulation of IRF4 (Gomez-Rodriguez et al. 2016). The roles of ITK in regulating cytokine production in Th1, Th2, and Th17 cells are dependent on the kinase activity of ITK (Kannan et al. 2015).

7.6 TCR/ITK → Ras/MAPK/IRF4 Pathway in Tr1 Cells

Naïve CD4$^+$ cells can differentiate into Tr1 cells upon TCR stimulation in the presence of IL-27 (Fitzgerald et al. 2007). Th17 cells are also capable of transdifferentiating into Tr1 cells upon resolution of inflammation (Gagliani et al. 2015). IL-27 signals through STAT1 and STAT3 to induce a network of transcription factors, including IRF4, Blimp-1, Ahr, and c-Maf, that controls Tr1 differentiation (Apetoh et al. 2010; Cretney et al. 2011; Pot et al. 2011; Stumhofer et al. 2007). The absence of ITK or ITK kinase activity results in severely impaired upregulation of IL-10, Tr1 cell associated surface markers, and transcription factors under Tr1 cell-inducing conditions both in vivo and in vitro. Indeed, anti-CD3 antibody treatment induces significantly less Tr1 cells in *Itk*$^{-/-}$ compared to WT mice. The absence of ITK also reduced the development of Tr1 cells in models of parasitic and viral infections. Inhibition of ITK kinase activity significantly reduced Th17 to Tr1 cell trans differentiation in vitro. Furthermore, ITK kinase activity is critical for upregulating the expression of transcription factors IRF4, Ahr, and Blimp-1, all of which are involved in regulating Tr1 cell development, in both mouse and human, although there are some nuances between mouse and human (e.g., Maf appears to be downstream of ITK in mice, but not human Tr1 cell differentiation). Downstream of TCR/ITK, Ras/MAPK pathway is upregulated, further leading to upregulation of IRF4 expression. Using ITK*as*-expressing Tr1 cells co-cultured with TCR-activated WT responder T cells, ITK*as* kinase activity was shown to be critical in the suppressive function of Tr1 cells against responder T-cell expansion. This reduction in suppression coincided with a decrease in IL-10 production. Importantly, *Itk*$^{-/-}$ Tr1 cell differentiation could be rescued when IRF4 was reintroduced via retroviral transduction, and the resultant IRF4-expressing *Itk*$^{-/-}$ Tr1 cells were functionally suppressive. Taken together, these data show that the TCR/ITK → Ras/MAPK/IRF4 signaling pathway is critical in both Tr1 differentiation and suppressive function (Fig. 7.3) (Huang et al. 2017).

7.7 Conclusions

Regulatory T cells serve to protect against autoimmunity and restrict immunopathology but may also prevent eradication of pathogens or tumors. While the TCR/ITK → Ras/MAPK/IRF4 pathway is crucial for Tr1 cell differentiation and function, the exact mechanism of Ras activation and signaling leading to upregulation of IRF4 has not yet been determined. Proof-of-concept clinical trials with Tr1 cells have demonstrated the safety and feasibility of this approach and indicated some preclinical benefits in graft-versus-host disease during hematopoietic stem cell transplantation (Bacchetta et al. 2014). In mouse models, IL-10-producing T cells exhibited anti-inflammatory effects in host immune rejection to organ transplants (Mfarrej et al. 2017), inflammatory bowel diseases (Clemente-Casares et al. 2016; Desreumaux et al. 2012), allergic asthma (Tousa et al. 2017), food allergy (Pellerin et al. 2018), dermatitis (Volz et al. 2014), type 1 diabetes (Clemente-Casares et al. 2016), autoimmune encephalitis (Clemente-Casares et al. 2016), arthritis (Clemente-Casares et al. 2016), acute immunopathology due to infection (Huang et al. 2017), and other related inflammatory conditions, as well as in shaping vaccine-induced immune responses (Ndure and Flanagan 2014). A more complete understanding of the signaling pathways that promote Tr1 cell differentiation and function may allow for the development of improved strategies to modulate the immune responses under different disease conditions.

Acknowledgments A.A. received research support from 3M Corporation. W.H. received research support from MegaRobo Technologies Corporation. M.C.M. declares no competing financial interests. Research related to this review topic in the authors' laboratories was supported in part by grants from the National Institutes of Health (R01AI120701, R01AI138570, R56AI146226, R21AI137822, P20GM130555-6610, R21AI129422, R21AI138497, and R35ES028244 (to Gary Perdew)). M.C.M. is a recipient of the Pathobiological Sciences Graduate Program Fellowship in the School of Veterinary Medicine at the Louisiana State University.

References

Ahangarani RR, Janssens W, VanderElst L, Carlier V, VandenDriessche T, Chuah M, Weynand B, Vanoirbeek JA, Jacquemin M, Saint-Remy JM (2009) In vivo induction of type 1-like regulatory T cells using genetically modified B cells confers long-term IL-10-dependent antigen-specific unresponsiveness. J Immunol 183:8232–8243

Akdis M (2008) T-cell tolerance to inhaled allergens: mechanisms and therapeutic approaches. Expert Opin Biol Ther 8:769–777

Andreotti AH, Schwartzberg PL, Joseph RE, Berg LJ (2010) T-cell signaling regulated by the Tec family kinase, Itk. Cold Spring Harb Perspect Biol 2:a002287

Apetoh L, Quintana FJ, Pot C, Joller N, Xiao S, Kumar D, Burns EJ, Sherr DH, Weiner HL, Kuchroo VK (2010) The aryl hydrocarbon receptor interacts with c-Maf to promote the differentiation of type 1 regulatory T cells induced by IL-27. Nat Immunol 11:854–861

August A, Ragin MJ (2012) Regulation of T-cell responses and disease by tec kinase Itk. Int Rev Immunol 31:155–165

August A, Sadra A, Dupont B, Hanafusa H (1997) Src-induced activation of inducible T cell kinase (ITK) requires phosphatidylinositol 3-kinase activity and the Pleckstrin homology domain of inducible T cell kinase. Proc Natl Acad Sci U S A 94:11227–11232

August A, Fischer A, Hao S, Mueller C, Ragin M (2002) The Tec family of tyrosine kinases in T cells, amplifiers of T cell receptor signals. Int J Biochem Cell Biol 34:1184–1189

Au-Yeung BB, Katzman SD, Fowell DJ (2006) Cutting edge: Itk-dependent signals required for CD4+ T cells to exert, but not gain, Th2 effector function. J Immunol 176:3895–3899

Bacchetta R, Lucarelli B, Sartirana C, Gregori S, Lupo Stanghellini MT, Miqueu P, Tomiuk S, Hernandez-Fuentes M, Gianolini ME, Greco R et al (2014) Immunological outcome in haploidentical-HSC transplanted patients treated with IL-10-anergized donor T cells. Front Immunol 5:16

Bennett CL, Brunkow ME, Ramsdell F, O'Briant KC, Zhu Q, Fuleihan RL, Shigeoka AO, Ochs HD, Chance PF (2001) A rare polyadenylation signal mutation of the FOXP3 gene (AAUAAA-->AAUGAA) leads to the IPEX syndrome. Immunogenetics 53:435–439

Berg LJ, Finkelstein LD, Lucas JA, Schwartzberg PL (2005) Tec family kinases in T lymphocyte development and function. Annu Rev Immunol 23:549–600

Bergmann C, Strauss L, Wang Y, Szczepanski MJ, Lang S, Johnson JT, Whiteside TL (2008) T regulatory type 1 cells in squamous cell carcinoma of the head and neck: mechanisms of suppression and expansion in advanced disease. Clin Cancer Res 14:3706–3715

Bishop AC, Shah K, Liu Y, Witucki L, Kung C, Shokat KM (1998) Design of allele-specific inhibitors to probe protein kinase signaling. Curr Biol 8:257–266

Bishop AC, Ubersax JA, Petsch DT, Matheos DP, Gray NS, Blethrow J, Shimizu E, Tsien JZ, Schultz PG, Rose MD et al (2000) A chemical switch for inhibitor-sensitive alleles of any protein kinase. Nature 407:395–401

Bohm W, Thoma S, Leithauser F, Moller P, Schirmbeck R, Reimann J (1998) T cell-mediated, IFN-gamma-facilitated rejection of murine B16 melanomas. J Immunol 161:897–908

Bohm L, Maxeiner J, Meyer-Martin H, Reuter S, Finotto S, Klein M, Schild H, Schmitt E, Bopp T, Taube C (2015) IL-10 and regulatory T cells cooperate in allergen-specific immunotherapy to ameliorate allergic asthma. J Immunol 194:887–897

Brady MT, MacDonald AJ, Rowan AG, Mills KH (2003) Hepatitis C virus non-structural protein 4 suppresses Th1 responses by stimulating IL-10 production from monocytes. Eur J Immunol 33:3448–3457

Brunkow ME, Jeffery EW, Hjerrild KA, Paeper B, Clark LB, Yasayko SA, Wilkinson JE, Galas D, Ziegler SF, Ramsdell F (2001) Disruption of a new forkhead/winged-helix protein, scurfin, results in the fatal lymphoproliferative disorder of the scurfy mouse. Nat Genet 27:68–73

Bunnell SC, Diehn M, Yaffe MB, Findell PR, Cantley LC, Berg LJ (2000) Biochemical interactions integrating Itk with the T cell receptor-initiated signaling cascade. J Biol Chem 275:2219–2230

Chatila TA, Blaeser F, Ho N, Lederman HM, Voulgaropoulos C, Helms C, Bowcock AM (2000) JM2, encoding a fork head-related protein, is mutated in X-linked autoimmunity-allergic disregulation syndrome. J Clin Invest 106:R75–R81

Clemente-Casares X, Blanco J, Ambalavanan P, Yamanouchi J, Singha S, Fandos C, Tsai S, Wang J, Garabatos N, Izquierdo C et al (2016) Expanding antigen-specific regulatory networks to treat autoimmunity. Nature 530:434–440

Cretney E, Xin A, Shi W, Minnich M, Masson F, Miasari M, Belz GT, Smyth GK, Busslinger M, Nutt SL, Kallies A (2011) The transcription factors Blimp-1 and IRF4 jointly control the differentiation and function of effector regulatory T cells. Nat Immunol 12:304–311

Desreumaux P, Foussat A, Allez M, Beaugerie L, Hebuterne X, Bouhnik Y, Nachury M, Brun V, Bastian H, Belmonte N et al (2012) Safety and efficacy of antigen-specific regulatory T-cell therapy for patients with refractory Crohn's disease. Gastroenterology 143(1207–1217):e1202

Dombroski D, Houghtling RA, Labno CM, Precht P, Takesono A, Caplen NJ, Billadeau DD, Wange RL, Burkhardt JK, Schwartzberg PL (2005) Kinase-independent functions for Itk in TCR-induced regulation of Vav and the actin cytoskeleton. J Immunol 174:1385–1392

Felices M, Falk M, Kosaka Y, Berg LJ (2007) Tec kinases in T cell and mast cell signaling. Adv Immunol 93:145–184

Fitzgerald DC, Zhang GX, El-Behi M, Fonseca-Kelly Z, Li H, Yu S, Saris CJ, Gran B, Ciric B, Rostami A (2007) Suppression of autoimmune inflammation of the central nervous system by interleukin 10 secreted by interleukin 27-stimulated T cells. Nat Immunol 8:1372–1379

Fontenot JD, Gavin MA, Rudensky AY (2003) Foxp3 programs the development and function of CD4+CD25+ regulatory T cells. Nat Immunol 4:330–336

Fowell DJ, Shinkai K, Liao XC, Beebe AM, Coffman RL, Littman DR, Locksley RM (1999) Impaired NFATc translocation and failure of Th2 development in Itk-deficient CD4+ T cells. Immunity 11:399–409

Gagliani N, Magnani CF, Huber S, Gianolini ME, Pala M, Licona-Limon P, Guo B, Herbert DR, Bulfone A, Trentini F et al (2013) Coexpression of CD49b and LAG-3 identifies human and mouse T regulatory type 1 cells. Nat Med 19:739–746

Gagliani N, Amezcua Vesely MC, Iseppon A, Brockmann L, Xu H, Palm NW, de Zoete MR, Licona-Limon P, Paiva RS, Ching T et al (2015) Th17 cells transdifferentiate into regulatory T cells during resolution of inflammation. Nature 523:221–225

Gambineri E, Torgerson TR, Ochs HD (2003) Immune dysregulation, polyendocrinopathy, enteropathy, and X-linked inheritance (IPEX), a syndrome of systemic autoimmunity caused by mutations of FOXP3, a critical regulator of T-cell homeostasis. Curr Opin Rheumatol 15:430–435

Gilfillan AM, Rivera J (2009) The tyrosine kinase network regulating mast cell activation. Immunol Rev 228:149–169

Gol-Ara M, Jadidi-Niaragh F, Sadria R, Azizi G, Mirshafiey A (2012) The role of different subsets of regulatory T cells in immunopathogenesis of rheumatoid arthritis. Arthritis 2012:805875

Gomez-Rodriguez J, Sahu N, Handon R, Davidson TS, Anderson SM, Kirby MR, August A, Schwartzberg PL (2009) Differential expression of interleukin-17A and -17F is coupled to T cell receptor signaling via inducible T cell kinase. Immunity 31:587–597

Gomez-Rodriguez J, Wohlfert EA, Handon R, Meylan F, Wu JZ, Anderson SM, Kirby MR, Belkaid Y, Schwartzberg PL (2014) Itk-mediated integration of T cell receptor and cytokine signaling regulates the balance between Th17 and regulatory T cells. J Exp Med 211:529–543

Gomez-Rodriguez J, Meylan F, Handon R, Hayes ET, Anderson SM, Kirby MR, Siegel RM, Schwartzberg PL (2016) Itk is required for Th9 differentiation via TCR-mediated induction of IL-2 and IRF4. Nat Commun 7:10857

Grasis JA, Browne CD, Tsoukas CD (2003) Inducible T cell tyrosine kinase regulates actin-dependent cytoskeletal events induced by the T cell antigen receptor. J Immunol 170:3971–3976

Groux H, O'Garra A, Bigler M, Rouleau M, Antonenko S, de Vries JE, Roncarolo MG (1997) A CD4+ T-cell subset inhibits antigen-specific T-cell responses and prevents colitis. Nature 389:737–742

Hao S, Qi Q, Hu J, August A (2006) A kinase independent function for Tec kinase ITK in regulating antigen receptor induced serum response factor activation. FEBS Lett 580:2691–2697

Haringer B, Lozza L, Steckel B, Geginat J (2009) Identification and characterization of IL-10/IFN-gamma-producing effector-like T cells with regulatory function in human blood. J Exp Med 206:1009–1017

Hori S, Nomura T, Sakaguchi S (2003) Control of regulatory T cell development by the transcription factor Foxp3. Science 299:1057–1061

Huang W, Jeong AR, Kannan AK, Huang L, August A (2014) IL-2-inducible T cell kinase tunes T regulatory cell development and is required for suppressive function. J Immunol 193:2267–2272

Huang W, Solouki S, Koylass N, Zheng SG, August A (2017) ITK signalling via the Ras/IRF4 pathway regulates the development and function of Tr1 cells. Nat Commun 8:15871

Huber S, Gagliani N, Esplugues E, O'Connor W Jr, Huber FJ, Chaudhry A, Kamanaka M, Kobayashi Y, Booth CJ, Rudensky AY et al (2011) Th17 cells express interleukin-10 receptor and are controlled by Foxp3(-) and Foxp3+ regulatory CD4+ T cells in an interleukin-10-dependent manner. Immunity 34:554–565

Kannan A, Huang W, Huang F, August A (2012) Signal transduction via the T cell antigen receptor in naive and effector/memory T cells. Int J Biochem Cell Biol 44:2129–2134

Kannan AK, Sahu N, Mohanan S, Mohinta S, August A (2013) IL-2-inducible T-cell kinase modulates TH2-mediated allergic airway inflammation by suppressing IFN-gamma in naive CD4+ T cells. J Allergy Clin Immunol 132(811–820):e811–e815

Kannan A, Lee Y, Qi Q, Huang W, Jeong AR, Ohnigian S, August A (2015) Allele-sensitive mutant, Itkas, reveals that Itk kinase activity is required for Th1, Th2, Th17, and iNKT-cell cytokine production. Eur J Immunol 45:2276–2285

Khattri R, Cox T, Yasayko SA, Ramsdell F (2003) An essential role for Scurfin in CD4+CD25+ T regulatory cells. Nat Immunol 4:337–342

MacDonald AJ, Duffy M, Brady MT, McKiernan S, Hall W, Hegarty J, Curry M, Mills KH (2002) CD4 T helper type 1 and regulatory T cells induced against the

same epitopes on the core protein in hepatitis C virus-infected persons. J Infect Dis 185:720–727
Magnani CF, Alberigo G, Bacchetta R, Serafini G, Andreani M, Roncarolo MG, Gregori S (2011) Killing of myeloid APCs via HLA class I, CD2 and CD226 defines a novel mechanism of suppression by human Tr1 cells. Eur J Immunol 41:1652–1662
Mascanfroni ID, Takenaka MC, Yeste A, Patel B, Wu Y, Kenison JE, Siddiqui S, Basso AS, Otterbein LE, Pardoll DM et al (2015) Metabolic control of type 1 regulatory T cell differentiation by AHR and HIF1-alpha. Nat Med 21:638–646
McGuirk P, McCann C, Mills KH (2002) Pathogen-specific T regulatory 1 cells induced in the respiratory tract by a bacterial molecule that stimulates interleukin 10 production by dendritic cells: a novel strategy for evasion of protective T helper type 1 responses by Bordetella pertussis. J Exp Med 195:221–231
Mfarrej B, Tresoldi E, Stabilini A, Paganelli A, Caldara R, Secchi A, Battaglia M (2017) Generation of donor-specific Tr1 cells to be used after kidney transplantation and definition of the timing of their in vivo infusion in the presence of immunosuppression. J Transl Med 15:40
Miller AT, Wilcox HM, Lai Z, Berg LJ (2004) Signaling through Itk promotes T helper 2 differentiation via negative regulation of T-bet. Immunity 21:67–80
Mobs C, Slotosch C, Loffler H, Jakob T, Hertl M, Pfutzner W (2010) Birch pollen immunotherapy leads to differential induction of regulatory T cells and delayed helper T cell immune deviation. J Immunol 184:2194–2203
Ndure J, Flanagan KL (2014) Targeting regulatory T cells to improve vaccine immunogenicity in early life. Front Microbiol 5:477
Okamura T, Fujio K, Sumitomo S, Yamamoto K (2012) Roles of LAG3 and EGR2 in regulatory T cells. Ann Rheum Dis 71(Suppl 2):i96–i100
Pedroza-Gonzalez A, Zhou G, Vargas-Mendez E, Boor PP, Mancham S, Verhoef C, Polak WG, Grunhagen D, Pan Q, Janssen H et al (2015) Tumor-infiltrating plasmacytoid dendritic cells promote immunosuppression by Tr1 cells in human liver tumors. Onco Targets Ther 4:e1008355
Pellerin L, Jenks JA, Chinthrajah S, Dominguez T, Block W, Zhou X, Noshirvan A, Gregori S, Roncarolo MG, Nadeau KC, Bacchetta R (2018) Peanut-specific type 1 regulatory T cells induced in vitro from allergic subjects are functionally impaired. J Allergy Clin Immunol 141(202–213):e208
Pot C, Apetoh L, Awasthi A, Kuchroo VK (2011) Induction of regulatory Tr1 cells and inhibition of T(H)17 cells by IL-27. Semin Immunol 23:438–445
Qi Q, Xia M, Bai Y, Yu S, Cantorna M, August A (2011) Interleukin-2-inducible T cell kinase (Itk) network edge dependence for the maturation of iNKT cell. J Biol Chem 286:138–146
Roncarolo MG, Gregori S, Bacchetta R, Battaglia M (2014) Tr1 cells and the counter-regulation of immunity: natural mechanisms and therapeutic applications. Curr Top Microbiol Immunol 380:39–68
Sahu N, Mueller C, Fischer A, August A (2008) Differential sensitivity to Itk kinase signals for T helper 2 cytokine production and chemokine-mediated migration. J Immunol 180:3833–3838
Sakaguchi S, Sakaguchi N, Asano M, Itoh M, Toda M (1995) Immunologic self-tolerance maintained by activated T cells expressing IL-2 receptor alpha-chains (CD25). Breakdown of a single mechanism of self-tolerance causes various autoimmune diseases. J Immunol 155:1151–1164
Sakaguchi S, Yamaguchi T, Nomura T, Ono M (2008) Regulatory T cells and immune tolerance. Cell 133:775–787
Schaeffer EM, Debnath J, Yap G, McVicar D, Liao XC, Littman DR, Sher A, Varmus HE, Lenardo MJ, Schwartzberg PL (1999) Requirement for Tec kinases Rlk and Itk in T cell receptor signaling and immunity. Science 284:638–641
Schaeffer EM, Broussard C, Debnath J, Anderson S, McVicar DW, Schwartzberg PL (2000) Tec family kinases modulate thresholds for thymocyte development and selection. J Exp Med 192:987–1000
Schaeffer EM, Yap GS, Lewis CM, Czar MJ, McVicar DW, Cheever AW, Sher A, Schwartzberg PL (2001) Mutation of Tec family kinases alters T helper cell differentiation. Nat Immunol 2:1183–1188
Serafini G, Andreani M, Testi M, Battarra M, Bontadini A, Biral E, Fleischhauer K, Marktel S, Lucarelli G, Roncarolo MG, Bacchetta R (2009) Type 1 regulatory T cells are associated with persistent split erythroid/lymphoid chimerism after allogeneic hematopoietic stem cell transplantation for thalassemia. Haematologica 94:1415–1426
Shevach EM, McHugh RS, Piccirillo CA, Thornton AM (2001) Control of T-cell activation by CD4+ CD25+ suppressor T cells. Immunol Rev 182:58–67
Shokat K, Velleca M (2002) Novel chemical genetic approaches to the discovery of signal transduction inhibitors. Drug Discov Today 7:872–879
Smith-Garvin JE, Koretzky GA, Jordan MS (2009) T cell activation. Annu Rev Immunol 27:591–619
Stumhofer JS, Silver JS, Laurence A, Porrett PM, Harris TH, Turka LA, Ernst M, Saris CJ, O'Shea JJ, Hunter CA (2007) Interleukins 27 and 6 induce STAT3-mediated T cell production of interleukin 10. Nat Immunol 8:1363–1371
Su YW, Zhang Y, Schweikert J, Koretzky GA, Reth M, Wienands J (1999) Interaction of SLP adaptors with the SH2 domain of Tec family kinases. Eur J Immunol 29:3702–3711
Takesono A, Finkelstein LD, Schwartzberg PL (2002) Beyond calcium: new signaling pathways for Tec family kinases. J Cell Sci 115:3039–3048

Tousa S, Semitekolou M, Morianos I, Banos A, Trochoutsou AI, Brodie TM, Poulos N, Samitas K, Kapasa M, Konstantopoulos D et al (2017) Activin-A co-opts IRF4 and AhR signaling to induce human regulatory T cells that restrain asthmatic responses. Proc Natl Acad Sci U S A 114:E2891–E2900

Vignali DA, Collison LW, Workman CJ (2008) How regulatory T cells work. Nat Rev Immunol 8:523–532

Volz T, Skabytska Y, Guenova E, Chen KM, Frick JS, Kirschning CJ, Kaesler S, Rocken M, Biedermann T (2014) Nonpathogenic bacteria alleviating atopic dermatitis inflammation induce IL-10-producing dendritic cells and regulatory Tr1 cells. J Invest Dermatol 134:96–104

Werlen G, Palmer E (2002) The T-cell receptor signalosome: a dynamic structure with expanding complexity. Curr Opin Immunol 14:299–305

Wildin RS, Ramsdell F, Peake J, Faravelli F, Casanova JL, Buist N, Levy-Lahad E, Mazzella M, Goulet O, Perroni L et al (2001) X-linked neonatal diabetes mellitus, enteropathy and endocrinopathy syndrome is the human equivalent of mouse scurfy. Nat Genet 27:18–20

Zeng H, Zhang R, Jin B, Chen L (2015) Type 1 regulatory T cells: a new mechanism of peripheral immune tolerance. Cell Mol Immunol 12:566–571

Adipose Tissue T Regulatory Cells: Implications for Health and Disease

8

David Bradley, Dharti Shantaram, Alan Smith, and Willa A. Hsueh

Abstract

Obesity dramatically increases the risk of numerous conditions, including type 2 diabetes mellitus and other components of the metabolic syndrome. Pro-inflammatory changes that occur in adipose tissue are critical to the pathogenesis of these obesity-induced complications. Adipose tissue is one of the body's largest endocrine organs, and the cells that comprise the adipose tissue immunoenvironment secrete multiple factors (including adipokines and cytokines) that impact systemic metabolism. In particular, immunosuppressive regulatory T cells (Tregs) decline in obesity, partly in response to its complex interaction with adipocytes, and this decline contributes to disruption of the typical homeostasis observed in lean adipose tissue. Although the regulation of Treg differentiation, function, and enrichment is incompletely understood, factors including various cell-surface co-stimulatory molecules, certain lipid species, and cytokines such as PPARγ, adiponectin, and leptin are important mediators. It is also clear that there may be depot-specific differences in Tregs, rendering adipose tissue Tregs distinct from lymphoid or circulating Tregs, with implications on maintenance and functionality. While most of these findings are derived from studies in murine models, comparatively little is known about the human adipose tissue Treg signature, which requires further investigation.

Keywords

Obesity · Adipose tissue · Adaptive immunity · Regulatory T cells · Insulin resistance

8.1 Introduction

More than two thirds of the US population is overweight or obese (Ogden et al. 2014), which increases the risk for type 2 diabetes (T2D) and other components of the metabolic syndrome including elevated plasma triglycerides, low levels of high-density lipoprotein cholesterol, and hypertension, all predisposing to cardiovascular disease (Osborn and Olefsky 2012). Obesity is a principal cause of both insulin resistance (IR) and diminished beta (β)-cell function, the two major factors involved in the pathogenesis of T2D (DeFronzo 2004). Accordingly, the rate of T2D has substantially increased (Khan et al. 2009) in parallel with the rates of obesity. In addition to T2D, obesity predisposes to multiple other abnormalities including atherosclerosis and myocardial infarction, heart failure, stroke and

D. Bradley (✉) · D. Shantaram · A. Smith · W. A. Hsueh
Diabetes and Metabolism Research Center, Division of Endocrinology, Diabetes and Metabolism, Department of Internal Medicine, The Ohio State University, Columbus, OH, USA
e-mail: David.Bradley@osumc.edu

S.-G. Zheng (ed.), *T Regulatory Cells in Human Health and Diseases*, Advances in Experimental Medicine and Biology 1278, https://doi.org/10.1007/978-981-15-6407-9_8

dementia, obstructive sleep apnea, nonalcoholic steatohepatitis and liver cirrhosis, certain cancers (including breast, colon, prostate, hepatocellular carcinoma, and others), peripheral arterial disease, and more; all of which increase morbidity and mortality, and substantially escalate healthcare costs. Inflammation stemming from adipose tissue (AT) changes that occur in obesity is an important underlying driver of these obesity-induced complications; yet, there are no anti-inflammatory-based therapies for obesity. Therefore, this chapter is focused on mechanisms of AT inflammation in obesity, particularly early changes following high fat feeding, which involve adipocyte/T-cell interactions, leading to loss of immunosuppressive regulatory T cells (Tregs), and disruption of systemic metabolism.

8.2 Adipocytes at the Crossroads of Immunity and Metabolism

In mammals, there are three distinct types of AT: white AT (WAT), whose classical function is energy storage of fat (i.e., triglyceride); brown AT (BAT), which plays a major role in energy dissipation through thermogenesis; and beige adipose tissue, which are largely depots of subcutaneous fat that can burn energy similar to BAT, but has different developmental origins than BAT (Gesta et al. 2007). WAT is distributed in different parts of the body, with subcutaneous AT (SAT) located predominantly in the abdomen, legs, and buttocks, and visceral AT (VAT) located in intra-abdominal depots including within the omentum, mesenterium, and perirenal areas (Gesta et al. 2007; Tchkonia et al. 2013).

Adipose tissue, one of the body's largest endocrine organs, secretes a multitude of metabolically active adipokines (from WAT adipocytes), batokines (from BAT adipocytes), cytokines, and chemokines (MacDougald and Burant 2007). Over 1000 unique proteins/peptides have been identified to be expressed in mouse adipocytes with 100 differentially expressed in WAT vs. BAT adipocytes (Hotamisligil et al. 1995, 1993). Cytokines derived from WAT, such as tumor necrosis factor alpha (TNF-α) and interleukin (IL)-6, influence obesity-induced insulin resistance in both humans and mice. A critical link between WAT and immune function was first demonstrated by Hotamisligil et al., who showed that pro-inflammatory factors (cytokines) were higher in the VAT of obese mice and that these factors had a direct role in decreasing insulin sensitivity (Rotter et al. 2003; Sabio et al. 2008). These cytokines substantially enhance AT lipolysis, with resultant free fatty acid (FFA) release into the circulation (Boden 2006; Holland et al. 2011; Shulman 2000). These cytokines substantially enhance AT lipolysis, with resultant free fatty acid (FFA) release into the circulation (Boden 2006; Holland et al. 2011; Shulman 2000) that is subsequently taken up by skeletal muscle and liver impairing the ability of insulin to stimulate muscle glucose uptake (Kelley et al. 1993) and suppress hepatic endogenous glucose production (EGP) (Ferrannini et al. 1983). Skeletal muscle accounts for ~70–80% of postprandial glucose uptake (DeFronzo et al. 1981), and the liver is the major site of endogenous glucose production (EGP). Therefore, the AT microenvironment is now recognized as a major determinant of systemic insulin action, leading to tissue uptake of lipids which can impair the ability of insulin to stimulate muscle glucose uptake (Kelley et al. 1993) and suppress hepatic endogenous glucose production (EGP) (Ferrannini et al. 1983). Skeletal muscle accounts for ~70–80% of postprandial glucose uptake (DeFronzo et al. 1981), and the liver is the major site of endogenous glucose production (EGP). Therefore, the AT microenvironment is now recognized as a major determinant of systemic insulin action (Boden 2006; Heilbronn and Campbell 2008). Early in obesity, inflammatory gene expression is selectively induced in WAT, not liver or muscle (Isomaa et al. 2001). Early in obesity, inflammatory gene expression is selectively induced in WAT, not liver or muscle (Isomaa et al. 2001).

Adipokines and batokines can act in an endocrine, paracrine, and autocrine manner and affect diverse processes including glucose homeostasis and insulin sensitivity, inflammation, energy expenditure, appetite control, tissue repair, and other processes (Scheja and Heeren 2019). More recently, AT has been shown to employ another means of communication with other tissues by emitting exosomes, secretory vesicles, containing adipokines, microRNAs, lipids, and other substances. Exosomes can act locally or be transported via blood to enter cells via non-receptor-mediated mechanisms and regulate their function (Crewe et al. 2018; Flaherty et al. 2019; Pan et al. 2019; Thomou et al. 2017). Adipocytes are considered an important source of exosomes (Kita et al. 2019).

Adipocytes not only regulate energy balance, but are increasingly recognized as critical immune cells, standing at the crossroads of metabolism and immunity (Fig. 8.1). This crossroad is important in evolution, since energy derived from metabolic processes is required for survival and the critical need to combat infection. Adiponectin and leptin are hormones specifically produced by adipocytes that mediate their immunometabolic effects, linking body weight and immune cell regulation. Adiponectin circulates at higher plasma concentrations than any of the other adipokines and is produced by adipocytes of any color. Children and young adults, compared to older adults, have relatively higher levels of adiponectin, which has anti-inflammatory effects, enhances insulin sensitivity, decreases liver fat and inflammation, increases beta-oxidation, and lowers gluconeogenesis (Stern et al. 2016). The anti-inflammatory effect has primarily been described in macrophages in which adiponectin suppresses the proinflammatory M1 phenotype, as well as proinflammatory cytokine production, and enhances the phagocytic M2 phenotype and IL-10 production (Luo and Liu 2016; Wolf et al. 2004; Xuan et al. 2016). However, there is also evidence that adiponectin has proinflammatory effects on macrophages, which may depend on the circulating isoform (Choi et al. 2020). Little is known about adiponectin's effects on T cells. In obesity, adiponectin secretion and blood levels are attenuated, while leptin secretion is enhanced.

Leptin suppresses appetite, but there is resistance to this action in obesity leading to elevated circulating levels in obese individuals (Moon et al. 2013). Leptin is also pro-inflammatory impacting nearly all immune cells, but these actions are not suppressed in obesity (La Cava 2017). The structure of leptin and its receptor suggests that it belongs to the class I cytokine superfamily, which includes cytokines such as IL-1, IL-6, IL-12, and G-CSF (Perez-Perez et al. 2017). Adipocytes are a significant source of leptin within AT, yet other cells including lymphocytes have been shown to produce and secrete leptin (Alwarawrah et al. 2018; De Rosa et al. 2007; El Karim et al. 2009). The leptin receptor is ubiquitously expressed within tissue with individual cell types expressing differing isoforms of the receptor (Myers 2004). Receptor stimulation has been shown to activate the Janus Kinase/Signal Transducer and Activator of Transcription (JAK-STAT), phosphoinositide 3-kinase (PI3K), and mitogen-activated protein kinase (MAPK) signaling pathways, generally producing a proinflammatory response across all immune cell types (Perez-Perez et al. 2017). The monocyte response to leptin includes increased phagocytosis, proliferation, and increased expression of activation markers and pro-inflammatory cytokines such as TNF-α, IL-6, and IL-12 (Gainsford et al. 1996; Loffreda et al. 1998; Mancuso et al. 2001; Zarkesh-Esfahani et al. 2001). Dendritic cells increase expression of major histocompatibility complex (MHC) class II and co-stimulatory receptors and increased expression of inflammatory cytokines (Lam et al. 2006). Leptin promotes migration of granulocytes as well as increased phagocytosis in neutrophils (Ottonello et al. 2004). It also promotes differentiation, maturation, and cytotoxicity of natural killer (NK) cells (Loffreda et al. 1998). The effects of leptin on T-cell populations can vary depending on which specific T-cell subsets are involved. Leptin increases differentiation toward inflammatory type 1 helper T cells (Th1) and has been shown to be required for Th17 differentiation (Naylor and Petri 2016). It can act as a negative regulator of type 2 helper T cells (Th2) and Treg differentiation as mice with genetic deficiency in leptin have reduced Th1 cells and increased percentages of Tregs and Th2 cells (De Rosa et al. 2007; Naylor and Petri 2016). In VAT, leptin-deficient mice have

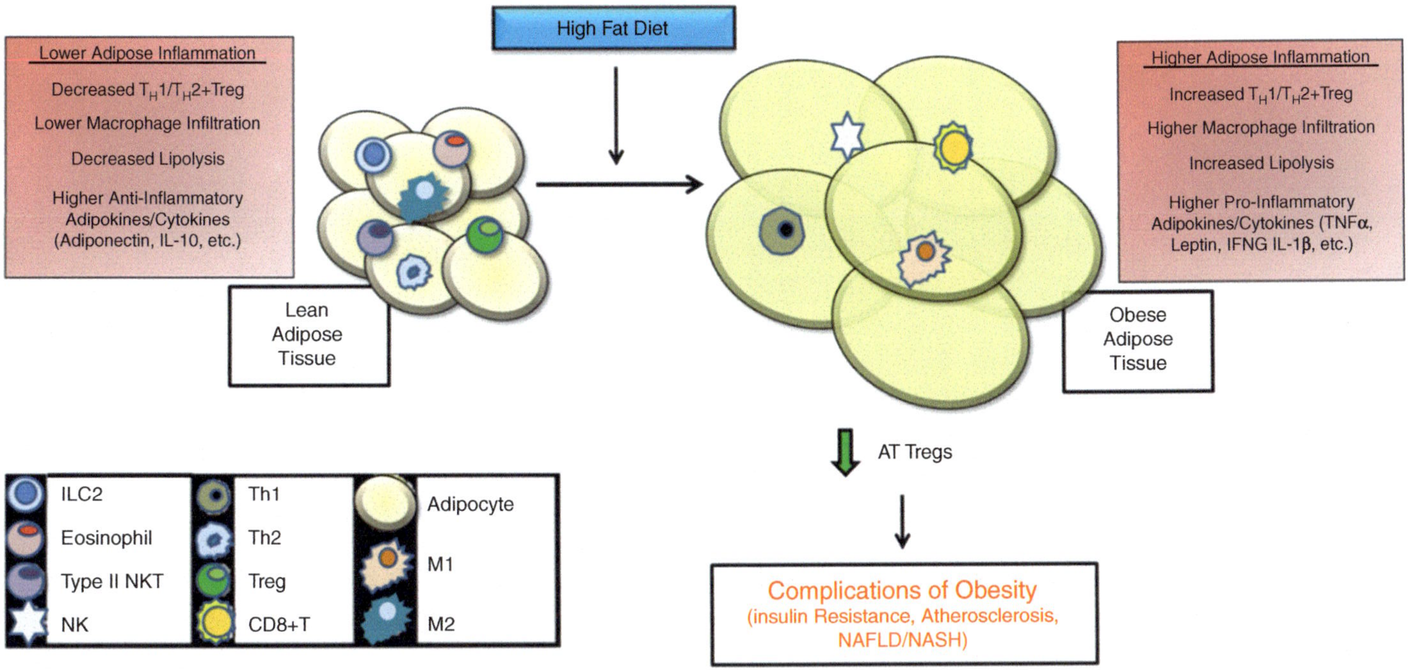

Fig. 8.1 Differences between obese and lean adipose tissue and obesity-induced changes in the adipose tissue microenvironment. *ILC2* innate lymphoid type 2 cell, *NKT* natural killer T cell, *Th1* type 1 helper T cell, *Th2* type 2 helper T cell, *Treg* regulatory T cell, *M1* M1 type macrophage, *M2* M2 type macrophage

reduced loss of Tregs, despite obesity, and less inflammation (Deng et al. 2013). These inflammatory effects of leptin are opposed by IL-4 and IL-10 from Th2 cells and Tregs (Biswas and Mantovani 2010). The inflammatory profile generated by leptin is further evidenced by studies finding increased leptin in an array of autoimmune and inflammatory conditions such as metabolic disorders, rheumatoid arthritis, systemic lupus erythematosus, multiple sclerosis, allergic disease, and bacterial infections (Perez-Perez et al. 2017).

Brown AT (BAT) is recruited in response to multiple stimuli, including cold acclimation and sympathetic stimulation, among others (Scheele and Wolfrum 2020). Up to 30% of BAT is comprised of brown adipocytes (Muller et al. 2016), which secrete factors termed batokines that modulate local adipogenesis and angiogenesis and stimulate afferent signaling to the central nervous system (Scheele and Wolfrum 2020). Batokines are also increasingly recognized to impact systemic metabolism, partly by increasing energy uptake of glucose and lipids. In adults, metabolically active BAT content negatively associates with higher body mass index (Saito et al. 2009). Dysfunctional WAT in states of obesity (Large et al. 1999) and obesity-associated insulin resistance (Reynisdottir et al. 1994) is accompanied by a reduction in lipolysis and higher lipid accumulation. A similar process may also occur in BAT (Saito et al. 2009; van Marken Lichtenbelt et al. 2009). However, the secretome appears to be distinct from WAT, with over 100 proteins exclusively quantified in brown adipocytes, including ependymin-related protein 1 (EPDR1), which plays a key role in thermogenic determination during adipogenesis (Deshmukh et al. 2019). The batokine 12,13-diHOME has also been shown to reduce the metabolic dysfunction that occurs with both aging and exercise, via increasing skeletal muscle fatty acid uptake and oxidation, without an appreciable effect on glucose uptake (Stanford et al. 2018).

The complex immune cell interactions within BAT may also be affected by batokines with potential ramifications on metabolism. In obese mice fed HFD, several inflammatory markers are upregulated (TNF-α, IL-1β, IL-6) in the absence of significant macrophage infiltration. This occurs despite an increase in monocyte chemoattractant protein-1 (MCP-1) expression (Alcala et al. 2017; Fitzgibbons et al. 2011). Although data is lacking, these findings suggest that other immune cells within BAT, including T cells, could potentially play a role in the pro-inflammatory state of BAT in obesity. Batokines are involved in the crosstalk between BAT and other critical organs involved in systemic metabolism. Neuregulin 4 (NRG4) has been shown to limit diet-induced insulin resistance not only by increasing fatty acid oxidation in BAT but by reducing hepatic de novo lipogenesis (Wang et al. 2019). Although data in human subjects is lacking, in murine models, loss of the transcription factor interferon regulatory factor 4 (IRF4) from BAT results in decreased transcription of skeletal muscle myostatin, reduced mitochondrial function, diminished exercise capacity (Kong et al. 2018), and a reduction in insulin-stimulated glucose uptake (Steculorum et al. 2016). In mice, fibroblast growth factor 21 (FGF21) derived from BAT (Quesada-Lopez et al. 2016) has effects on weight loss and glycemic control (Samms et al. 2015; Veniant et al. 2015), indicating that it may be integral to a potential liver-BAT crosstalk.

8.3 Tregs Help Maintain Adipose Tissue Homeostasis

Lean AT in mice is composed of a highly balanced system of diverse immune cells and anti-inflammatory mediators (IL-4 and IL-13) that maintain normal adipose storage, endocrine function, and ultimately, systemic insulin action (Fig. 8.1). Studies in mice suggest that eosinophils, innate lymphoid type 2 cells (ILC2s), and type II natural killer T cells (NKTs) work in concert with Th2 cells, Tregs, and alternatively activated macrophages (M2) to maintain a distinct anti-inflammatory milieu in AT (Baker et al. 2017; Cheng et al. 2009; Hams et al. 2013; Huang et al. 2015; Ricardo-Gonzalez et al. 2010; Sano et al. 2003). Several studies have shown that Tregs are enriched in VAT from lean mice (Feuerer et al. 2009). Notably, Tregs comprise 50% of $CD4^+$ adipose resident T cells

(ARTs) in VAT of lean mice, compared to 5–15% of the CD4$^+$ cells in spleen and lymphoid tissue (Feuerer et al. 2009). This VAT accumulation results from enhanced proliferation and survival and not recruitment from the circulation (Kolodin et al. 2015). Recent studies suggest that Foxp3+ Tregs migrate from the thymus early in life, then pass through the spleen where they begin to assume some of the transcriptomic features of VAT Tregs. Finally, they migrate to the VAT where they assume a VAT-specific signature (Li et al. 2018). Depletion of VAT Tregs in lean mice induces pro-inflammatory gene expression of inflammatory mediators (TNF-a, IL-6, and CCL5, among others) and impairs insulin signaling in VAT and liver, indicating a role for VAT Tregs in maintenance of AT and metabolic homeostasis (Feuerer et al. 2009).

ILC2 cells, which possess lymphocyte morphology, but lack expression of surface markers ("lineage negative") (Yu et al. 2018), are found in all AT depots and also importantly contribute to AT homeostasis (Benezech and Jackson-Jones 2019). They promote Treg accumulation in AT by expression of the Inducible T-Cell Costimulator (ICOS) ligand which binds to ICOS, a co-regulatory receptor expressed in Tregs (Molofsky et al. 2015). ILC2 secretion of IL-5 is also critical for eosinophil recruitment into AT, and secretion of IL-13 is important for the development and maintenance of alternatively activated macrophages (Molofsky et al. 2013; Nussbaum et al. 2013). ILC2 additionally promotes adipocyte beiging by the secretion of a methionine-enkephalin peptide that enhances UCP-1 in adipocytes (Brestoff and Artis 2015) and by the stimulation of adipocyte proliferation and differentiation into beige adipocytes (Lee et al. 2015).

8.4 Adipocytes and Macrophages Contribute to the Loss of at Tregs in Obesity: Implications for Systemic Metabolism

The homeostatic milieu of AT dramatically changes in the setting of obesity. These changes have major implications for systemic metabolism and other obesity-related comorbidities and are largely determined by the AT immune cell microenvironment. Obesity dramatically decreases the VAT Treg percentage to ~20% of CD4+ ARTs (Deng et al. 2013). This loss is one mechanism to explain how obesity induces insulin resistance, since adoptive transfer of Tregs to obese, insulin-resistant mice improves insulin action, underscoring the role of AT Tregs in insulin sensitivity (Freigang et al. 2013). As such, AT Treg cell abundance and function have emerged as a key determinant of both AT inflammation and systemic metabolism, and there is a strong interest in understanding mechanisms that regulate Tregs in AT. The drop in AT Tregs is accompanied by a greater abundance of pro-inflammatory interferon gamma (IFNγ)-secreting CD4+ Th1 and CD8+ T cells (Deng et al. 2017; Maeda et al. 2014).

IFNγ impairs Treg differentiation from naïve T cells, but not Treg immunosuppressive function (Deng et al. 2017). Mice with genetic loss of the IFNγ receptor do not display a drop in AT Tregs and do not become insulin resistant despite obesity. In addition to its role in innate immunity, the adipocyte also plays a critical role in adaptive immunity. Adipocyte expression and activity of the class II major histocompatibility complex (MHCII) pathway increase early, within 2 weeks, after high fat feeding to promote differentiation of adipose resident CD4+ Th1 cells, which are a major source of IFNγ (Deng et al. 2013). IFNγ is a primary cytokine that stimulates the MHCII pathway in adipocytes. This novel adipocyte/T-cell dialog promotes an escalating cycle of inflammation as obesity progresses. Genetic ablation of the antigen presenting capability of the adipocyte impairs the rise in VAT CD4+ Th1 and prevents the loss of VAT Tregs in obese mice (Deng et al. 2017). Moreover, loss of VAT Tregs in these mice negated the improvement in insulin sensitivity, highlighting the relationship of Th1/Tregs and the importance on VAT Tregs to systemic metabolism. A similar process may occur in humans, in that adipocyte MHCII upregulation occurs in obesity (Deng et al. 2013), and expression of AT Th1 markers and IFNy also increases, while expression of FOXP3 decreases (Dastych et al. 2008; Feuerer et al. 2009; Freigang et al. 2013; Schwartz et al. 1995). Only three studies have performed T-cell flow analyses of human AT (Gyllenhammer et al. 2016; McLaughlin et al. 2014; Winer et al. 2009). McLaughlin et al. identified Th1, Th2, Th17, and CD8 T cells by flow analyses in VAT and SAT of ten overweight/obese subjects but did not assess Tregs. Gyllenhammer et al. measured Tregs in SAT and VAT of 44 obese subjects and found a negative association with fasting glucose and a positive association with islet cell function measured as HOMA-β; however, lean subjects were not studied (Gyllenhammer et al. 2016). Recently, Vijay et al. (2020) performed single-cell analyses of human AT and identified a population of CD8+ T cells that express metallothionines and adipocyte progenitor cells unique to VAT, but did not report potential effects of Tregs. Thus, further studies of human AT are necessary to define the adipocyte/Treg relationship in humans. As mentioned above, leptin secretion by adipocytes increases soon after exposure to nutrient excess and also mediates the adipocyte/T-cell exchange; leptin levels are chronically elevated in obesity, initiating and contributing to the proinflammatory microenvironment in VAT.

AT macrophages can also present antigen to enhance differentiation of CD4+ Th1 cells in AT (Lumeng et al. 2007b). Specific knockout of AT macrophage antigen-presenting capability in obese mice using FITC+ glucan-coated particles that are exclusively taken up by AT and not systemic macrophages prevents the increase in AT CD4+ Th1 cells and the development of glucose intolerance (Blaszczak et al. 2019b). Thus, both adipocytes and macrophages contribute to the increase in AT CD4+ Th1 cells and mediate the resulting loss of AT Tregs in obesity. Although macrophages were the first cell type reported to produce pro-inflammatory cytokines in AT, taking on a classically activated M1-like phenotype in obese animals and humans (Lumeng et al. 2007a; Weisberg et al. 2003; Xu and Geczy 2000), their increase in AT occurs later than the adipocyte and T cell changes during HFD. In mice, adipocyte and T-cell changes occur early, with a week of 2 after HFD, and prior to detectable macrophage changes, which occur after 6–12 weeks of HFD (Deng et al. 2013, 2017; Liu and Nikolajczyk 2019). Similarly, human subjects become insulin resistant after a 2.7 kg weight gain induced by 4 weeks of overfeeding, but there were no major SAT macrophage changes, suggesting AT macrophage changes occur later in the development of obesity (Tam et al. 2010). MCP-1 is elevated in obesity, largely produced by enlarging adipocytes, and leads to the infiltration of macrophages into obese WAT in mice and humans (Kanda et al. 2006; Weisberg et al. 2006). Genetic loss of inflammatory factors, including PPARγ, or the M1 phenotype in macrophages (Hevener et al. 2007; Lumeng et al. 2007a; Weisberg et al. 2003; Xu and Geczy 2000) attenuates insulin resistance in obese mice, suggesting an important role of macrophages. However, all of these maneuvers modify both AT and systemic macrophages, so the exact role of AT macrophages is unclear. Moreover, although M1 and M2 phenotypes are well delineated in AT macrophages in mice, this distinction of pro- and anti-inflammatory macrophages is not clear in humans (Blaszczak et al. 2019a).

In mice, AT ILC2 cells decrease in obesity, contributing to a critical loss of Treg support and contributing to the drop in AT Tregs (Hams et al. 2013). IFNγ antagonizes ILC2s and may contribute to the ILC2 loss (Molofsky et al. 2015). In humans, we did not see a drop in AT ILC2s in obesity (unpublished observations).

8.5 Regulation of Treg Differentiation and Function: Key Factors in VAT

Although the precise mechanisms for enrichment of Tregs are incompletely understood, several factors are known to be important. In mice, Onodera et al. identified that VAT Tregs had the highest degree of proliferation compared to other VAT immune cells, suggesting that local differentiation and proliferation is a major mechanism to expand VAT Tregs (Onodera et al. 2015). Antigen presentation plays a key role in Treg differentiation. Antigen-presenting cells tightly regulate Treg differentiation by presenting MHCII-bound antigen to the T-cell receptor of a naïve T cell (Ramalingam et al. 2012). The co-activators B7 (Cluster of differentiation (CD) 80/B7-1 and CD86/B7-2) on the adipocyte then associate with MHCII to engage CD28 and cytotoxic T-lymphocyte-associated protein 4 (CTLA-4) on the T-cell surface. Treg differentiation is enhanced by low levels of antigen and increased by B7/CTLA-4 binding (Zheng et al. 2006), but inhibited by high antigen concentrations that increase B7/CD28 binding (Benson et al. 2007). This intricate competition between CTLA-4 and CD28 for B7 binding determines normal Treg differentiation (Yamaguchi et al. 2013). Accordingly, B7 deficiency markedly decreases Tregs, leading to increased inflammation in blood and spleen (Bar-On et al. 2011; Salomon et al. 2000) and in VAT, which is associated with insulin resistance (Zhong et al. 2014). Loss of antigen presentation in both adipocytes and macrophages simultaneously leads to a marked reduction in VAT Tregs, further emphasizing the importance of antigen presentation, while loss in only one of these compartments is associated with increased VAT Tregs (Blaszczak et al. 2019b).

The cytokines IL-10 and transforming growth factor beta (TGFβ), as well as all-trans retinoic acid (atRA), are potent enhancers of Treg differentiation and FOXP3 stability, while IFNγ, IL-1β, and IL-6 are negative regulators (Murai et al. 2009; Priceman et al. 2013; Wei et al. 2007) (Lu et al. 2011; Zheng et al. 2002, 2004). In the lean state, the stability of Tregs is enhanced by IL-10 (Fujisaka et al. 2009), a potent anti-inflammatory cytokine derived from adipocytes, macrophages, and Tregs. atRA may also play an important role in maintaining an anti-inflammatory environment by co-stimulating Treg differentiation, specifically, through its action on retinoic acid receptor alpha (RAR-a), and preventing conversion to Th1/Th17 cells (Lu et al. 2014). After storage in the liver, adipocytes are the second major site of RA storage, which could impact VAT Treg levels (Brun et al. 2013).

Peroxisome proliferator-activated receptor gamma (PPARγ), working in concert with FOXP3, is a requirement for the accumulation and function of VAT Tregs in lean mice, whereas this signature is abrogated by serine 273 phosphorylation of PPARγ, which commonly occurs in the presence of inflammatory factors, like TNFα (Cipolletta 2014; Cipolletta et al. 2012a). T-cell receptor activation is necessary to induce PPARγ expression in Tregs and induces IRF4 and basic leucine zipper transcription factor, ATF-like (BATF) to transcriptionally upregulate PPARγ to promote VAT Treg accumulation (Kolodin et al. 2015). Administration of a PPARγ ligand increases VAT Tregs, which is implicated as a mechanism contributing to the insulin-sensitizing effects of PPARγ activation (Cipolletta et al. 2012a).

IL-33, working through its receptor, ST2, which is highly expressed on VAT Tregs in both humans and mice, is important in the amplification of VAT Tregs (Vasanthakumar et al. 2015). Injection of IL-33 into obese mice increases VAT Tregs and improves insulin resistance with little effect on lymphoid Tregs (Han et al. 2015; Miller 2010). In contrast, blocking this effect with an ST2 antibody decreases VAT Treg abundance and worsens insulin sensitivity with little change in splenic Tregs (Deng et al. 2017), and mice deficient in IL-33 or ST2 have decreased VAT Tregs and are insulin resistant, even on a normal diet (Vasanthakumar et al. 2015). Local sources of IL-33 include endothelial cells, VAT mesenchymal stem cells, types 1, 2, and 3, and NK cells (Chang et al. 2017; Kohlgruber et al. 2018; Molofsky et al. 2015; Spallanzani et al. 2019). IL-33 also enriches ILC2s, which support Tregs (Molofsky et al. 2015).

Co-stimulatory and co-inhibitory receptors are a class of proteins that have been shown to either enhance or inhibit Treg function (Lucca and Dominguez-Villar 2020), and particular interest has been promoted by the success of checkpoint inhibitor blockade in the treatment of cancer. While no mechanistic studies have been completed with AT Tregs, involvement of co-receptors in AT Treg function is supported by gene expression profiles showing differential expression of such co-receptors as lymphocyte-activation gene 3 (LAG3), programmed cell death 1 (PD-1), CD137, OX40, and glucocorticoid-induced TNFR family-related gene (GITR) (Cipolletta et al. 2015a; Feuerer et al. 2009; Wu et al. 2019).

Lipid species are in high abundance in AT and have been shown to modulate immune responses, with some studies showing a direct effect on Tregs. Mice deficient in acid sphingomyelinase (ASM), an enzyme involved in conversion of sphingomyelin to ceramide, are reported to have increased percentages of Tregs. Additionally, exogenous addition of ceramide C6 to cultures resulted in reduced induction and proliferation of Tregs in both ASM-deficient and wild-type mice (Zhou et al. 2016). Short-chain fatty acids, lipid species produced by commensal bacteria, have been shown to enhance Treg generation (Lu et al. 2014). It is likely that additional lipid species are capable of modulating Treg responses as shown by their general effect on metabolic disorders. Retinoic acid, a vitamin A metabolite, promotes the differentiation and stabilization of Tregs, and AT is a significant depot for retinoic acid (Brun et al. 2013; Lu et al. 2014). Other lipid species which associate with insulin resistance include ceramides (Sokolowska and Blachnio-Zabielska 2019) and hydroxy fatty acids (Yore et al. 2014). Each of these lipid species have been identified from isolated AT (Sokolowska and Blachnio-Zabielska 2019; Yore et al. 2014) (unpublished data). In addition, lipids, such as short-chain fatty acids, are capable of entering into the PPARγ pocket (Al-Lahham et al. 2010; Varga et al. 2011), potentially leading to expansion of VAT Tregs.

8.6 A Distinct Phenotype of AT Tregs

VAT resident Tregs are distinct from lymphoid organ Tregs. They acquire a unique phenotype in a multi-step, multi-site manner owing to T-cell receptor (TCR) specificity (Li et al. 2018). Although they share some classic signature genes with lymphoid organ Tregs that code for CD25, GITR, CTLA-4, and FOXP3, the VAT Treg-specific phenotype is associated with the upregulation of genes that likely are an adaptation to the environment. In mice, these genes include cytokines and chemokines along with their receptors (CCR2, CCR3, IL-10, IL-9R, IL-33R, or ST2) (Feuerer et al. 2009; Han et al. 2015; Kolodin et al. 2015; Vasanthakumar et al. 2015), and transcription factors (PPARγ, BATF, IRF4, BLIMP1, ID2) (Cipolletta et al. 2012b; Frias et al. 2019; Kolodin et al. 2015; Vasanthakumar et al. 2015). Of note factors involved in lipid metabolism (low-density lipoprotein receptor [LDLR], diglyceride acyltransferase [DGAT], CD36) and circadian clock regulators (Nuclear Receptor Subfamily 1 Group D Member 1 [NR1D1], RAR-related orphan receptor alpha [RORα]) (Cipolletta et al. 2012b; DiSpirito et al. 2018) are also associated with this phenotype. In contrast, some genes involved in lymphoid organ processing (CCR7, Sphingosine-1-Phosphate Receptor 1 [S1PR1], CD62L) and a few other transcription factors (T-cell factor 1 [TCF1], lymphoid enhancer-binding factor-1 [LEF1], ID3) (Sullivan et al. 2019; Yang et al. 2019) are downregulated in VAT Tregs. There is some degree of overlap between the transcriptional profile of a VAT Treg and an activated Treg, except for the genes involved in lipid metabolism (Li et al. 2018). This AT Treg phenotype is well described in mice, but very little is known about human AT Tregs.

A recent study by Wu et al. assessed the immune landscape of human omental AT and its changes in obesity and T2D (Wu et al. 2019). In agreement with previous studies (Wentworth et al. 2010; Zeyda et al. 2007), they reported a rise in pro-inflammatory M1 macrophage ($CD14^{+}CD11c^{+}CD206^{+}$) frequencies in omental AT of aging subjects and T2D patients. They also found a decrease in anti-inflammatory M2 macrophages with age in contrast to studies in mice that showed accumulation of M2 macrophages in spleen, lymph node, and bone marrow of older mice (Jackaman et al. 2013; Kelly et al. 2007). In addition, they found that proinflammatory cytokines like CCL2, CXCL8,

IL-1B, and IL-6 positively correlated with increasing BMI. Using nanostring to compare candidate gene expression in human omental AT Tregs vs. thymic Tregs from two different cohorts, Wu et al. found comparable expression in important transcription factors like FOXP3, GATA Binding Protein 3 (GATA3), TBET, and RAR-related orphan receptor C (RORC) with both sources of Tregs expressing high levels of IKZF2, CD25, and a few markers of suppressive potential, CTLA4, GITR, and ICOS. Distinct from thymic Tregs, omental Tregs expressed higher levels of CD39 and lower levels of TGFβ transcripts. Human AT Tregs also revealed upregulation of genes that comprise the unique signature of mouse AT Tregs: PPARγ, PRDM1 (BLIMP1), and CXCL2 (Cipolletta et al. 2015b; Han et al. 2015; Vasanthakumar et al. 2015). ST2 plays a critical role in maintenance and functionality of AT Tregs in mice (Vasanthakumar et al. 2015), but surprisingly, in this study ST2 was undetectable in omental AT Tregs. Furthermore, some of the highly expressed genes in mouse visceral AT Tregs, amphiregulin (AREG), annexin A1 (ANXA1), CXCR6, killer cell lectin-like receptor subfamily G member 1 (KLRG1), perilipin 2 (PLIN2), and DGAT (Cipolletta et al. 2015b) were found to be either reduced or similar to thymic/Tconv Tregs. Thus far, very little is known about the human VAT Treg signature, which requires further extensive studies.

Conflicts of Interest None

References

Alcala M, Calderon-Dominguez M, Bustos E, Ramos P, Casals N, Serra D, Viana M, Herrero L (2017) Increased inflammation, oxidative stress and mitochondrial respiration in brown adipose tissue from obese mice. Sci Rep 7:16082

Al-Lahham SH, Peppelenbosch MP, Roelofsen H, Vonk RJ, Venema K (2010) Biological effects of propionic acid in humans; metabolism, potential applications and underlying mechanisms. Biochim Biophys Acta 1801:1175–1183

Alwarawrah Y, Kiernan K, MacIver NJ (2018) Changes in nutritional status impact immune cell metabolism and function. Front Immunol 9:1055

Baker NA, Muir LA, Washabaugh AR, Neeley CK, Chen SY, Flesher CG, Vorwald J, Finks JF, Ghaferi AA, Mulholland MW et al (2017) Diabetes-specific regulation of adipocyte metabolism by the adipose tissue extracellular matrix. J Clin Endocrinol Metab 102:1032–1043

Bar-On L, Birnberg T, Kim KW, Jung S (2011) Dendritic cell-restricted CD80/86 deficiency results in peripheral regulatory T-cell reduction but is not associated with lymphocyte hyperactivation. Eur J Immunol 41:291–298

Benezech C, Jackson-Jones LH (2019) ILC2 orchestration of local immune function in adipose tissue. Front Immunol 10:171

Benson MJ, Pino-Lagos K, Rosemblatt M, Noelle RJ (2007) All-trans retinoic acid mediates enhanced T reg cell growth, differentiation, and gut homing in the face of high levels of co-stimulation. J Exp Med 204:1765–1774

Biswas SK, Mantovani A (2010) Macrophage plasticity and interaction with lymphocyte subsets: cancer as a paradigm. Nat Immunol 11:889–896

Blaszczak AM, Jalilvand A, Liu J, Wright VP, Suzo A, Needleman B, Noria S, Lafuse W, Hsueh WA, Bradley D (2019a) Human visceral adipose tissue macrophages are not adequately defined by standard methods of characterization. J Diabetes Res 2019:8124563

Blaszczak AM, Wright VP, Anandani K, Liu J, Jalilvand A, Bergin S, Nicoloro SM, Czech MP, Lafuse W, Deng T et al (2019b) Loss of antigen presentation in adipose tissue macrophages or in adipocytes, but not both, improves glucose metabolism. J Immunol 202:2451–2459

Boden G (2006) Fatty acid-induced inflammation and insulin resistance in skeletal muscle and liver. Curr Diab Rep 6:177–181

Brestoff JR, Artis D (2015) Immune regulation of metabolic homeostasis in health and disease. Cell 161:146–160

Brun PJ, Yang KJ, Lee SA, Yuen JJ, Blaner WS (2013) Retinoids: potent regulators of metabolism. Biofactors 39:151–163

Chang AY, Dao T, Gejman RS, Jarvis CA, Scott A, Dubrovsky L, Mathias MD, Korontsvit T, Zakhaleva V, Curcio M et al (2017) A therapeutic T cell receptor mimic antibody targets tumor-associated PRAME peptide/HLA-I antigens. J Clin Invest 127:3557

Cheng KK, Iglesias MA, Lam KS, Wang Y, Sweeney G, Zhu W, Vanhoutte PM, Kraegen EW, Xu A (2009) APPL1 potentiates insulin-mediated inhibition of hepatic glucose production and alleviates diabetes via Akt activation in mice. Cell Metab 9:417–427

Choi M, Doss HM, Kim KS (2020) Multifaceted physiological roles of adiponectin in inflammation and diseases. Int J Mol Sci 21:1219

Cipolletta D (2014) Adipose tissue-resident regulatory T cells: phenotypic specialization, functions and therapeutic potential. Immunology 142:517–525

Cipolletta D, Feuerer M, Li A, Kamei N, Lee J, Shoelson SE, Benoist C, Mathis D (2012a) PPAR-gamma is a major driver of the accumulation and phenotype of adipose tissue Treg cells. Nature 486:549–553

Cipolletta D, Feuerer M, Li A, Kamei N, Lee J, Shoelson SE, Benoist C, Mathis D (2012b) PPAR-γ is a major driver of the accumulation and phenotype of adipose tissue T reg cells. Nature 486:549–553

Cipolletta D, Cohen P, Spiegelman BM, Benoist C, Mathis D (2015a) Appearance and disappearance of the mRNA signature characteristic of Treg cells in visceral adipose tissue: age, diet, and PPARgamma effects. Proc Natl Acad Sci U S A 112:482–487

Cipolletta D, Cohen P, Spiegelman BM, Benoist C, Mathis D (2015b) Appearance and disappearance of the mRNA signature characteristic of Treg cells in visceral adipose tissue: age, diet, and PPARγ effects. Proc Natl Acad Sci 112:482–487

Crewe C, Joffin N, Rutkowski JM, Kim M, Zhang F, Towler DA, Gordillo R, Scherer PE (2018) An endothelial-to-adipocyte extracellular vesicle Axis governed by metabolic state. Cell 175(695–708):e613

Dastych M, Dastych M Jr, Novotna H, Cihalova J (2008) Lactulose/mannitol test and specificity, sensitivity, and area under curve of intestinal permeability parameters in patients with liver cirrhosis and Crohn's disease. Dig Dis Sci 53:2789–2792

De Rosa V, Procaccini C, Cali G, Pirozzi G, Fontana S, Zappacosta S, La Cava A, Matarese G (2007) A key role of leptin in the control of regulatory T cell proliferation. Immunity 26:241–255

DeFronzo RA (2004) Pathogenesis of type 2 diabetes mellitus. Med Clin North Am 88:787–835, ix

DeFronzo RA, Jacot E, Jequier E, Maeder E, Wahren J, Felber JP (1981) The effect of insulin on the disposal of intravenous glucose. Results from indirect calorimetry and hepatic and femoral venous catheterization. Diabetes 30:1000–1007

Deng T, Lyon CJ, Minze LJ, Lin J, Zou J, Liu JZ, Ren Y, Yin Z, Hamilton DJ, Reardon PR et al (2013) Class II major histocompatibility complex plays an essential role in obesity-induced adipose inflammation. Cell Metab 17:411–422

Deng T, Liu J, Deng Y, Minze L, Xiao X, Wright V, Yu R, Li XC, Blaszczak A, Bergin S et al (2017) Adipocyte adaptive immunity mediates diet-induced adipose inflammation and insulin resistance by decreasing adipose Treg cells. Nat Commun 8:15725

Deshmukh AS, Peijs L, Beaudry JL, Jespersen NZ, Nielsen CH, Ma T, Brunner AD, Larsen TJ, Bayarri-Olmos R, Prabhakar BS et al (2019) Proteomics-based comparative mapping of the secretomes of human Brown and White adipocytes reveals EPDR1 as a novel Batokine. Cell Metab 30:963–975.e967

DiSpirito JR, Zemmour D, Ramanan D, Cho J, Zilionis R, Klein AM, Benoist C, Mathis D (2018) Molecular diversification of regulatory T cells in nonlymphoid tissues. Sci Immunol 3:eaat5861

El Karim IA, Linden GJ, Irwin CR, Lundy FT (2009) Neuropeptides regulate expression of angiogenic growth factors in human dental pulp fibroblasts. J Endod 35:829–833

Ferrannini E, Barrett EJ, Bevilacqua S, DeFronzo RA (1983) Effect of fatty acids on glucose production and utilization in man. J Clin Invest 72:1737–1747

Feuerer M, Herrero L, Cipolletta D, Naaz A, Wong J, Nayer A, Lee J, Goldfine AB, Benoist C, Shoelson S (2009) Lean, but not obese, fat is enriched for a unique population of regulatory T cells that affect metabolic parameters. Nat Med 15:930

Fitzgibbons TP, Kogan S, Aouadi M, Hendricks GM, Straubhaar J, Czech MP (2011) Similarity of mouse perivascular and brown adipose tissues and their resistance to diet-induced inflammation. Am J Physiol Heart Circ Physiol 301:H1425–H1437

Flaherty SE III, Grijalva A, Xu X, Ables E, Nomani A, Ferrante AW Jr (2019) A lipase-independent pathway of lipid release and immune modulation by adipocytes. Science 363:989–993

Freigang S, Ampenberger F, Weiss A, Kanneganti TD, Iwakura Y, Hersberger M, Kopf M (2013) Fatty acid-induced mitochondrial uncoupling elicits inflammasome-independent IL-1alpha and sterile vascular inflammation in atherosclerosis. Nat Immunol 14:1045–1053

Frias AB, Hyzny EJ, Buechel HM, Beppu LY, Xie B, Jurczak MJ, D'Cruz LM (2019) The transcriptional regulator Id2 is critical for adipose-resident regulatory T cell differentiation, survival, and function. J Immunol 203:658–664

Fujisaka S, Usui I, Bukhari A, Ikutani M, Oya T, Kanatani Y, Tsuneyama K, Nagai Y, Takatsu K, Urakaze M et al (2009) Regulatory mechanisms for adipose tissue M1 and M2 macrophages in diet-induced obese mice. Diabetes 58:2574–2582

Gainsford T, Willson TA, Metcalf D, Handman E, McFarlane C, Ng A, Nicola NA, Alexander WS, Hilton DJ (1996) Leptin can induce proliferation, differentiation, and functional activation of hemopoietic cells. Proc Natl Acad Sci U S A 93:14564–14568

Gesta S, Tseng YH, Kahn CR (2007) Developmental origin of fat: tracking obesity to its source. Cell 131:242–256

Gyllenhammer LE, Lam J, Alderete TL, Allayee H, Akbari O, Katkhouda N, Goran MI (2016) Lower omental T-regulatory cell count is associated with higher fasting glucose and lower beta-cell function in adults with obesity. Obesity (Silver Spring) 24:1274–1282

Hams E, Locksley RM, McKenzie AN, Fallon PG (2013) Cutting edge: IL-25 elicits innate lymphoid type 2 and type II NKT cells that regulate obesity in mice. J Immunol 191:5349–5353

Han JM, Wu D, Denroche HC, Yao Y, Verchere CB, Levings MK (2015) IL-33 reverses an obesity-induced deficit in visceral adipose tissue ST2+ T regulatory cells and ameliorates adipose tissue inflammation and insulin resistance. J Immunol 194:4777–4783

Heilbronn LK, Campbell LV (2008) Adipose tissue macrophages, low grade inflammation and insulin resistance in human obesity. Curr Pharm Des 14:1225–1230

Hevener AL, Olefsky JM, Reichart D, Nguyen MT, Bandyopadyhay G, Leung HY, Watt MJ, Benner C, Febbraio MA, Nguyen AK et al (2007) Macrophage PPAR gamma is required for normal skeletal muscle and hepatic insulin sensitivity and full antidiabetic effects of thiazolidinediones. J Clin Invest 117:1658–1669

Holland WL, Bikman BT, Wang LP, Yuguang G, Sargent KM, Bulchand S, Knotts TA, Shui G, Clegg DJ, Wenk MR et al (2011) Lipid-induced insulin resistance mediated by the proinflammatory receptor TLR4 requires saturated fatty acid-induced ceramide biosynthesis in mice. J Clin Invest 121:1858–1870

Hotamisligil GS, Shargill NS, Spiegelman BM (1993) Adipose expression of tumor necrosis factor-alpha: direct role in obesity-linked insulin resistance. Science 259:87–91

Hotamisligil GS, Arner P, Caro JF, Atkinson RL, Spiegelman BM (1995) Increased adipose tissue expression of tumor necrosis factor-alpha in human obesity and insulin resistance. J Clin Invest 95:2409–2415

Huang Q, Niu Z, Tan J, Yang J, Liu Y, Ma H, Lee VW, Sun S, Song X, Guo M et al (2015) IL-25 elicits innate lymphoid cells and multipotent progenitor type 2 cells that reduce renal ischemic/reperfusion injury. J Am Soc Nephrol 26(9):2199–2211

Isomaa B, Almgren P, Tuomi T, Forsen B, Lahti K, Nissen M, Taskinen MR, Groop L (2001) Cardiovascular morbidity and mortality associated with the metabolic syndrome. Diabetes Care 24:683–689

Jackaman C, Radley-Crabb HG, Soffe Z, Shavlakadze T, Grounds MD, Nelson DJ (2013) Targeting macrophages rescues age-related immune deficiencies in C57 BL/6J geriatric mice. Aging Cell 12:345–357

Kanda H, Tateya S, Tamori Y, Kotani K, Hiasa K, Kitazawa R, Kitazawa S, Miyachi H, Maeda S, Egashira K et al (2006) MCP-1 contributes to macrophage infiltration into adipose tissue, insulin resistance, and hepatic steatosis in obesity. J Clin Invest 116:1494–1505

Kelley DE, Mokan M, Simoneau JA, Mandarino LJ (1993) Interaction between glucose and free fatty acid metabolism in human skeletal muscle. J Clin Invest 92:91–98

Kelly J, Khan AA, Yin J, Ferguson TA, Apte RS (2007) Senescence regulates macrophage activation and angiogenic fate at sites of tissue injury in mice. J Clin Invest 117:3421–3426

Khan T, Muise ES, Iyengar P, Wang ZV, Chandalia M, Abate N, Zhang BB, Bonaldo P, Chua S, Scherer PE (2009) Metabolic dysregulation and adipose tissue fibrosis: role of collagen VI. Mol Cell Biol 29:1575–1591

Kita S, Maeda N, Shimomura I (2019) Interorgan communication by exosomes, adipose tissue, and adiponectin in metabolic syndrome. J Clin Invest 129:4041–4049

Kohlgruber AC, Gal-Oz ST, LaMarche NM, Shimazaki M, Duquette D, Koay HF, Nguyen HN, Mina AI, Paras T, Tavakkoli A et al (2018) Gammadelta T cells producing interleukin-17A regulate adipose regulatory T cell homeostasis and thermogenesis. Nat Immunol 19:464–474

Kolodin D, van Panhuys N, Li C, Magnuson AM, Cipolletta D, Miller CM, Wagers A, Germain RN, Benoist C, Mathis D (2015) Antigen-and cytokine-driven accumulation of regulatory T cells in visceral adipose tissue of lean mice. Cell Metab 21:543–557

Kong X, Yao T, Zhou P, Kazak L, Tenen D, Lyubetskaya A, Dawes BA, Tsai L, Kahn BB, Spiegelman BM et al (2018) Brown adipose tissue controls skeletal muscle function via the secretion of myostatin. Cell Metab 28:631–643.e633

La Cava A (2017) Leptin in inflammation and autoimmunity. Cytokine 98:51–58

Lam QL, Liu S, Cao X, Lu L (2006) Involvement of leptin signaling in the survival and maturation of bone marrow-derived dendritic cells. Eur J Immunol 36:3118–3130

Large V, Reynisdottir S, Langin D, Fredby K, Klannemark M, Holm C, Arner P (1999) Decreased expression and function of adipocyte hormone-sensitive lipase in subcutaneous fat cells of obese subjects. J Lipid Res 40:2059–2066

Lee MW, Odegaard JI, Mukundan L, Qiu Y, Molofsky AB, Nussbaum JC, Yun K, Locksley RM, Chawla A (2015) Activated type 2 innate lymphoid cells regulate beige fat biogenesis. Cell 160:74–87

Li C, DiSpirito JR, Zemmour D, Spallanzani RG, Kuswanto W, Benoist C, Mathis D (2018) TCR transgenic mice reveal stepwise, multi-site acquisition of the distinctive fat-Treg phenotype. Cell 174:285–299.e212

Liu R, Nikolajczyk BS (2019) Tissue immune cells fuel obesity-associated inflammation in adipose tissue and beyond. Front Immunol 10:1587

Loffreda S, Yang SQ, Lin HZ, Karp CL, Brengman ML, Wang DJ, Klein AS, Bulkley GB, Bao C, Noble PW et al (1998) Leptin regulates proinflammatory immune responses. FASEB J 12:57–65

Lu L, Ma J, Li Z, Lan Q, Chen M, Liu Y, Xia Z, Wang J, Han Y, Shi W et al (2011) All-trans retinoic acid promotes TGF-beta-induced Tregs via histone modification but not DNA demethylation on Foxp3 gene locus. PLoS One 6:e24590

Lu L, Lan Q, Li Z, Zhou X, Gu J, Li Q, Wang J, Chen M, Liu Y, Shen Y et al (2014) Critical role of all-trans retinoic acid in stabilizing human natural regulatory T cells under inflammatory conditions. Proc Natl Acad Sci U S A 111:E3432–E3440

Lucca LE, Dominguez-Villar M (2020) Modulation of regulatory T cell function and stability by co-inhibitory receptors. Nat Rev Immunol 20:680

Lumeng CN, Bodzin JL, Saltiel AR (2007a) Obesity induces a phenotypic switch in adipose tissue macrophage polarization. J Clin Invest 117:175–184
Lumeng CN, Deyoung SM, Bodzin JL, Saltiel AR (2007b) Increased inflammatory properties of adipose tissue macrophages recruited during diet-induced obesity. Diabetes 56:16–23
Luo Y, Liu M (2016) Adiponectin: a versatile player of innate immunity. J Mol Cell Biol 8:120–128
MacDougald OA, Burant CF (2007) The rapidly expanding family of adipokines. Cell Metab 6:159–161
Maeda Y, Nishikawa H, Sugiyama D, Ha D, Hamaguchi M, Saito T, Nishioka M, Wing JB, Adeegbe D, Katayama I et al (2014) Detection of self-reactive CD8(+) T cells with an anergic phenotype in healthy individuals. Science 346:1536–1540
Mancuso P, McNish RW, Peters-Golden M, Brock TG (2001) Evaluation of phagocytosis and arachidonate metabolism by alveolar macrophages and recruited neutrophils from F344xBN rats of different ages. Mech Ageing Dev 122:1899–1913
van Marken Lichtenbelt WD, Vanhommerig JW, Smulders NM, Drossaerts JM, Kemerink GJ, Bouvy ND, Schrauwen P, Teule GJ (2009) Cold-activated brown adipose tissue in healthy men. N Engl J Med 360:1500–1508
McLaughlin T, Liu LF, Lamendola C, Shen L, Morton J, Rivas H, Winer D, Tolentino L, Choi O, Zhang H et al (2014) T-cell profile in adipose tissue is associated with insulin resistance and systemic inflammation in humans. Arterioscler Thromb Vasc Biol 34:2637–2643
Miller GD (2010) Improved nutrient intake in older obese adults undergoing a structured diet and exercise intentional weight loss program. J Nutr Health Aging 14:461–466
Molofsky AB, Nussbaum JC, Liang HE, Van Dyken SJ, Cheng LE, Mohapatra A, Chawla A, Locksley RM (2013) Innate lymphoid type 2 cells sustain visceral adipose tissue eosinophils and alternatively activated macrophages. J Exp Med 210:535–549
Molofsky AB, Savage AK, Locksley RM (2015) Interleukin-33 in tissue homeostasis, injury, and inflammation. Immunity 42:1005–1019
Moon HS, Dalamaga M, Kim SY, Polyzos SA, Hamnvik OP, Magkos F, Paruthi J, Mantzoros CS (2013) Leptin's role in lipodystrophic and nonlipodystrophic insulin-resistant and diabetic individuals. Endocr Rev 34:377–412
Muller S, Kulenkampff E, Wolfrum C (2016) Adipose tissue stem cells. Handb Exp Pharmacol 233:251–263
Murai M, Turovskaya O, Kim G, Madan R, Karp CL, Cheroutre H, Kronenberg M (2009) Interleukin 10 acts on regulatory T cells to maintain expression of the transcription factor Foxp3 and suppressive function in mice with colitis. Nat Immunol 10:1178–1184
Myers MG Jr (2004) Leptin receptor signaling and the regulation of mammalian physiology. Recent Prog Horm Res 59:287–304
Naylor C, Petri WA Jr (2016) Leptin regulation of immune responses. Trends Mol Med 22:88–98
Nussbaum JC, Van Dyken SJ, von Moltke J, Cheng LE, Mohapatra A, Molofsky AB, Thornton EE, Krummel MF, Chawla A, Liang HE et al (2013) Type 2 innate lymphoid cells control eosinophil homeostasis. Nature 502:245–248
Ogden CL, Carroll MD, Kit BK, Flegal KM (2014) Prevalence of childhood and adult obesity in the United States, 2011-2012. JAMA 311:806–814
Onodera T, Fukuhara A, Jang MH, Shin J, Aoi K, Kikuta J, Otsuki M, Ishii M, Shimomura I (2015) Adipose tissue macrophages induce PPARgamma-high FOXP3(+) regulatory T cells. Sci Rep 5:16801
Osborn O, Olefsky JM (2012) The cellular and signaling networks linking the immune system and metabolism in disease. Nat Med 18:363–374
Ottonello L, Gnerre P, Bertolotto M, Mancini M, Dapino P, Russo R, Garibotto G, Barreca T, Dallegri F (2004) Leptin as a uremic toxin interferes with neutrophil chemotaxis. J Am Soc Nephrol 15:2366–2372
Pan Y, Hui X, Hoo RLC, Ye D, Chan CYC, Feng T, Wang Y, Lam KSL, Xu A (2019) Adipocyte-secreted exosomal microRNA-34a inhibits M2 macrophage polarization to promote obesity-induced adipose inflammation. J Clin Invest 129:834–849
Perez-Perez A, Vilarino-Garcia T, Fernandez-Riejos P, Martin-Gonzalez J, Segura-Egea JJ, Sanchez-Margalet V (2017) Role of leptin as a link between metabolism and the immune system. Cytokine Growth Factor Rev 35:71–84
Priceman SJ, Kujawski M, Shen S, Cherryholmes GA, Lee H, Zhang C, Kruper L, Mortimer J, Jove R, Riggs AD et al (2013) Regulation of adipose tissue T cell subsets by Stat3 is crucial for diet-induced obesity and insulin resistance. Proc Natl Acad Sci U S A 110:13079–13084
Quesada-Lopez T, Cereijo R, Turatsinze JV, Planavila A, Cairo M, Gavalda-Navarro A, Peyrou M, Moure R, Iglesias R, Giralt M et al (2016) The lipid sensor GPR120 promotes brown fat activation and FGF21 release from adipocytes. Nat Commun 7:13479
Ramalingam R, Larmonier CB, Thurston RD, Midura-Kiela MT, Zheng SG, Ghishan FK, Kiela PR (2012) Dendritic cell-specific disruption of TGF-beta receptor II leads to altered regulatory T cell phenotype and spontaneous multiorgan autoimmunity. J Immunol 189:3878–3893
Reynisdottir S, Ellerfeldt K, Wahrenberg H, Lithell H, Arner P (1994) Multiple lipolysis defects in the insulin resistance (metabolic) syndrome. J Clin Invest 93:2590–2599
Ricardo-Gonzalez RR, Red Eagle A, Odegaard JI, Jouihan H, Morel CR, Heredia JE, Mukundan L, Wu D, Locksley RM, Chawla A (2010) IL-4/STAT6 immune axis regulates peripheral nutrient metabolism and insulin sensitivity. Proc Natl Acad Sci U S A 107:22617–22622
Rotter V, Nagaev I, Smith U (2003) Interleukin-6 (IL-6) induces insulin resistance in 3T3-L1 adipocytes and is, like IL-8 and tumor necrosis factor-alpha,

overexpressed in human fat cells from insulin-resistant subjects. J Biol Chem 278:45777–45784

Sabio G, Das M, Mora A, Zhang Z, Jun JY, Ko HJ, Barrett T, Kim JK, Davis RJ (2008) A stress signaling pathway in adipose tissue regulates hepatic insulin resistance. Science 322:1539–1543

Saito M, Okamatsu-Ogura Y, Matsushita M, Watanabe K, Yoneshiro T, Nio-Kobayashi J, Iwanaga T, Miyagawa M, Kameya T, Nakada K et al (2009) High incidence of metabolically active brown adipose tissue in healthy adult humans: effects of cold exposure and adiposity. Diabetes 58:1526–1531

Salomon B, Lenschow DJ, Rhee L, Ashourian N, Singh B, Sharpe A, Bluestone JA (2000) B7/CD28 costimulation is essential for the homeostasis of the CD4+CD25+ immunoregulatory T cells that control autoimmune diabetes. Immunity 12:431–440

Samms RJ, Smith DP, Cheng CC, Antonellis PP, Perfield JW 2nd, Kharitonenkov A, Gimeno RE, Adams AC (2015) Discrete aspects of FGF21 in vivo pharmacology do not require UCP1. Cell Rep 11:991–999

Sano H, Kane S, Sano E, Miinea CP, Asara JM, Lane WS, Garner CW, Lienhard GE (2003) Insulin-stimulated phosphorylation of a Rab GTPase-activating protein regulates GLUT4 translocation. J Biol Chem 278:14599–14602

Scheele C, Wolfrum C (2020) Brown adipose crosstalk in tissue plasticity and human metabolism. Endocr Rev 41:53

Scheja L, Heeren J (2019) The endocrine function of adipose tissues in health and cardiometabolic disease. Nat Rev Endocrinol 15:507–524

Schwartz MA, Schaller MD, Ginsberg MH (1995) Integrins: emerging paradigms of signal transduction. Annu Rev Cell Dev Biol 11:549–599

Shulman GI (2000) Cellular mechanisms of insulin resistance. J Clin Invest 106:171–176

Sokolowska E, Blachnio-Zabielska A (2019) The role of ceramides in insulin resistance. Front Endocrinol 10:577

Spallanzani RG, Zemmour D, Xiao T, Jayewickreme T, Li C, Bryce PJ, Benoist C, Mathis D (2019) Distinct immunocyte-promoting and adipocyte-generating stromal components coordinate adipose tissue immune and metabolic tenors. Sci immunol 4:eaaw3658

Stanford KI, Lynes MD, Takahashi H, Baer LA, Arts PJ, May FJ, Lehnig AC, Middelbeek RJW, Richard JJ, So K et al (2018) 12,13-diHOME: an exercise-induced lipokine that increases skeletal muscle fatty acid uptake. Cell Metab 27:1357

Steculorum SM, Ruud J, Karakasilioti I, Backes H, Engstrom Ruud L, Timper K, Hess ME, Tsaousidou E, Mauer J, Vogt MC et al (2016) AgRP neurons control systemic insulin sensitivity via myostatin expression in Brown adipose tissue. Cell 165:125–138

Stern JH, Rutkowski JM, Scherer PE (2016) Adiponectin, leptin, and fatty acids in the maintenance of metabolic homeostasis through adipose tissue crosstalk. Cell Metab 23:770–784

Sullivan JM, Höllbacher B, Campbell DJ (2019) Cutting edge: dynamic expression of Id3 defines the stepwise differentiation of tissue-resident regulatory T cells. J Immunol 202:31–36

Tam CS, Viardot A, Clement K, Tordjman J, Tonks K, Greenfield JR, Campbell LV, Samocha-Bonet D, Heilbronn LK (2010) Short-term overfeeding may induce peripheral insulin resistance without altering subcutaneous adipose tissue macrophages in humans. Diabetes 59:2164–2170

Tchkonia T, Thomou T, Zhu Y, Karagiannides I, Pothoulakis C, Jensen MD, Kirkland JL (2013) Mechanisms and metabolic implications of regional differences among fat depots. Cell Metab 17:644–656

Thomou T, Mori MA, Dreyfuss JM, Konishi M, Sakaguchi M, Wolfrum C, Rao TN, Winnay JN, Garcia-Martin R, Grinspoon SK et al (2017) Adipose-derived circulating miRNAs regulate gene expression in other tissues. Nature 542:450–455

Varga T, Czimmerer Z, Nagy L (2011) PPARs are a unique set of fatty acid regulated transcription factors controlling both lipid metabolism and inflammation. Biochim Biophys Acta 1812:1007–1022

Vasanthakumar A, Moro K, Xin A, Liao Y, Gloury R, Kawamoto S, Fagarasan S, Mielke LA, Afshar-Sterle S, Masters SL (2015) The transcriptional regulators IRF4, BATF and IL-33 orchestrate development and maintenance of adipose tissue–resident regulatory T cells. Nat Immunol 16:276

Veniant MM, Sivits G, Helmering J, Komorowski R, Lee J, Fan W, Moyer C, Lloyd DJ (2015) Pharmacologic effects of FGF21 are independent of the "Browning" of white adipose tissue. Cell Metab 21:731–738

Vijay J et al (2020) Single-cell analysis of human adipose tissue identifies depot- and disease-specific cell types. Nat Metab 2:97–109

Wang R, Yang F, Qing L, Huang R, Liu Q, Li X (2019) Decreased serum neuregulin 4 levels associated with non-alcoholic fatty liver disease in children with obesity. Clin Obes 9:e12289

Wei J, Duramad O, Perng OA, Reiner SL, Liu YJ, Qin FX (2007) Antagonistic nature of T helper 1/2 developmental programs in opposing peripheral induction of Foxp3+ regulatory T cells. Proc Natl Acad Sci U S A 104:18169–18174

Weisberg SP, McCann D, Desai M, Rosenbaum M, Leibel RL, Ferrante AW Jr (2003) Obesity is associated with macrophage accumulation in adipose tissue. J Clin Invest 112:1796–1808

Weisberg SP, Hunter D, Huber R, Lemieux J, Slaymaker S, Vaddi K, Charo I, Leibel RL, Ferrante AW Jr (2006) CCR2 modulates inflammatory and metabolic effects of high-fat feeding. J Clin Invest 116:115–124

Wentworth JM, Naselli G, Brown WA, Doyle L, Phipson B, Smyth GK, Wabitsch M, O'Brien PE,

Harrison LC (2010) Pro-inflammatory CD11c+ CD206 + adipose tissue macrophages are associated with insulin resistance in human obesity. Diabetes 59:1648–1656

Winer S, Chan Y, Paltser G, Truong D, Tsui H, Bahrami J, Dorfman R, Wang Y, Zielenski J, Mastronardi F et al (2009) Normalization of obesity-associated insulin resistance through immunotherapy. Nat Med 15:921–929

Wolf AM, Wolf D, Rumpold H, Enrich B, Tilg H (2004) Adiponectin induces the anti-inflammatory cytokines IL-10 and IL-1RA in human leukocytes. Biochem Biophys Res Commun 323:630–635

Wu D, Han JM, Yu X, Lam AJ, Hoeppli RE, Pesenacker AM, Huang Q, Chen V, Speake C, Yorke E (2019) Characterization of regulatory T cells in obese omental adipose tissue in humans. Eur J Immunol 49:336–347

Xu K, Geczy CL (2000) IFN-gamma and TNF regulate macrophage expression of the chemotactic S100 protein S100A8. J Immunol 164:4916–4923

Xuan D, Han Q, Tu Q, Zhang L, Yu L, Murry D, Tu T, Tang Y, Lian JB, Stein GS et al (2016) Epigenetic modulation in periodontitis: interaction of adiponectin and JMJD3-IRF4 Axis in macrophages. J Cell Physiol 231:1090–1096

Yamaguchi T, Kishi A, Osaki M, Morikawa H, Prieto-Martin P, Wing K, Saito T, Sakaguchi S (2013) Construction of self-recognizing regulatory T cells from conventional T cells by controlling CTLA-4 and IL-2 expression. Proc Natl Acad Sci U S A 110:E2116–E2125

Yang B-H, Wang K, Wan S, Liang Y, Yuan X, Dong Y, Cho S, Xu W, Jepsen K, Feng G-S (2019) TCF1 and LEF1 control Treg competitive survival and Tfr development to prevent autoimmune diseases. Cell Rep 27:3629–3645.e3626

Yore MM, Syed I, Moraes-Vieira PM, Zhang T, Herman MA, Homan EA, Patel RT, Lee J, Chen S, Peroni OD et al (2014) Discovery of a class of endogenous mammalian lipids with anti-diabetic and anti-inflammatory effects. Cell 159:318–332

Yu QN, Guo YB, Li X, Li CL, Tan WP, Fan XL, Qin ZL, Chen D, Wen WP, Zheng SG et al (2018) ILC2 frequency and activity are inhibited by glucocorticoid treatment via STAT pathway in patients with asthma. Allergy 73:1860–1870

Zarkesh-Esfahani H, Pockley G, Metcalfe RA, Bidlingmaier M, Wu Z, Ajami A, Weetman AP, Strasburger CJ, Ross RJ (2001) High-dose leptin activates human leukocytes via receptor expression on monocytes. J Immunol 167:4593–4599

Zeyda M, Farmer D, Todoric J, Aszmann O, Speiser M, Györi G, Zlabinger G, Stulnig T (2007) Human adipose tissue macrophages are of an anti-inflammatory phenotype but capable of excessive pro-inflammatory mediator production. Int J Obes 31:1420–1428

Zheng SG, Gray JD, Ohtsuka K, Yamagiwa S, Horwitz DA (2002) Generation ex vivo of TGF-beta-producing regulatory T cells from CD4+CD25- precursors. J Immunol 169:4183–4189

Zheng SG, Wang JH, Gray JD, Soucier H, Horwitz DA (2004) Natural and induced CD4+CD25+ cells educate CD4+CD25- cells to develop suppressive activity: the role of IL-2, TGF-beta, and IL-10. J Immunol 172:5213–5221

Zheng SG, Wang JH, Stohl W, Kim KS, Gray JD, Horwitz DA (2006) TGF-beta requires CTLA-4 early after T cell activation to induce FoxP3 and generate adaptive CD4+CD25+ regulatory cells. J Immunol 176:3321–3329

Zhong J, Rao X, Braunstein Z, Taylor A, Narula V, Hazey J, Mikami D, Needleman B, Rutsky J, Sun Q et al (2014) T-cell costimulation protects obesity-induced adipose inflammation and insulin resistance. Diabetes 63:1289–1302

Zhou Y, Salker MS, Walker B, Munzer P, Borst O, Gawaz M, Gulbins E, Singh Y, Lang F (2016) Acid sphingomyelinase (ASM) is a negative regulator of regulatory T cell (Treg) development. Cell Physiol Biochem 39:985–995

Intestinal Regulatory T Cells

9

Vanessa R. Figliuolo da Paz, Deepa R. Jamwal, and Pawel R. Kiela

Abstract

Mucosal surfaces are distinctive sites exposed to environmental, dietary, and microbial antigens. Particularly in the gut, the host continuously actively adapts via complex interactions between the microbiota and dietary compounds and immune and other tissue cells. Regulatory T cells (Tregs) are critical for tuning the intestinal immune response to self- and non-self-antigens in the intestine. Its importance in intestinal homeostasis is illustrated by the onset of overt inflammation caused by deficiency in Treg generation, function, or stability in the gut. A substantial imbalance in Tregs has been observed in intestinal tissue during pathogenic conditions, when a tightly regulated and equilibrated system becomes dysregulated and leads to unimpeded and chronic immune responses. In this chapter, we compile and critically discuss the current knowledge on the key factors that promote Treg-mediated tolerance in the gut, such as those involved in intestinal Treg differentiation, specificity and suppressive function, and their immunophenotype during health and disease. We also discuss the current state of knowledge on Treg dysregulation in human intestine during pathological states such as inflammatory bowel disease (IBD), necrotizing enterocolitis (NEC), graft-versus-host disease (GVHD), and colorectal cancer (CRC), and how that knowledge is guiding development of Treg-targeted therapies to treat or prevent intestinal disorders.

Keywords

Treg · Immune tolerance · Inflammation · Microbiota · Immunotherapy

9.1 Introduction to Intestinal Tregs

Intestine is a unique site that is constantly exposed to wide array of luminal microbial and dietary antigens. Consequently, while safeguarding the host against the harmful pathogens, the intestinal mucosal immune system has to remain resilient and tolerant to the innocuous food antigens and commensal microbes to be able to endure the immunological insults without getting unduly provoked. The intestine employs many defense mechanisms to maintain its immunosuppressive tone, such as physical barrier

Vanessa R. Figliuolo da Paz and Deepa R. Jamwal contributed equally with all other contributors.

V. R. Figliuolo da Paz · D. R. Jamwal
Department of Pediatrics, University of Arizona, Tucson, AZ, USA

P. R. Kiela (✉)
Department of Pediatrics, University of Arizona, Tucson, AZ, USA

Department of Immunobiology, University of Arizona, Tucson, AZ, USA
e-mail: pkiela@peds.arizona.edu

S.-G. Zheng (ed.), *T Regulatory Cells in Human Health and Diseases*, Advances in Experimental Medicine and Biology 1278, https://doi.org/10.1007/978-981-15-6407-9_9

(mucus and epithelial layers) or the production of secretory IgA. Tregs are another instrumental players in keeping a check on the activation of local T cells in the intestine (Powrie and Mason 1990; Powrie et al. 1993; Gambineri et al. 2003; Josefowicz et al. 2012).

To better understand the roles of Tregs in the gut, it is important to place it in the context of the intestinal tissue architecture. Histologically, along its entire length, intestine is made up of four layers: mucosa (further consists of the epithelial layer, lamina propria, and muscularis mucosae), submucosa, muscularis propria, and serosa. Lamina propria (LP) compartment consists of a variety of immune cells such as plasma cells, macrophages, dendritic cells, mast cells, and B and T lymphocytes, including Tregs, the subject of this overview. Small intestine (SI) comprises of duodenum, jejunum, and ileum and is characterized by the highly absorptive surface created by finger-like projections called the villi, further enlarged by cellular microvilli located at the apical, lumen-facing membrane of the enterocytes. The small intestinal epithelium consists of invagination forming the crypts of Lieberkuhn. At the base of the crypts reside the intestinal stem cells, which give rise to transient proliferative cells that differentiate and mature into Paneth cells, goblet cells, enteroendocrine cells, and tuft cells as they travel up through the transition zone, to ultimately shed into the lumen at the apex of the villi. Peyer's patches are organized lymphoid structures unique to the small intestine. They harbor immune cells and are separated from the lumen by a follicle-associated epithelium which contains specialized cells called microfold cells (M cells) capable of sampling the antigens directly from the lumen and delivering it to antigen-presenting cells juxtapositioned at their basolateral side (Fig. 9.1a). The large intestine consists of cecum, colon, and rectum and harbors the bulk of the commensals. Its major function is to absorb water and vitamins while converting undigested food into feces. The most notable histological differences are lack of villi, Paneth cells (except for their metaplastic occurrence in chronic inflammatory bowel diseases), and Peyer's patches (Fig. 9.1b).

Most of the information available regarding the development of intestinal Tregs is based on research on mice. In other organs, Tregs comprise approximately 10% of total $CD4^+$ T-cell population However, in the gut, they represent a much higher fraction with >30% of $CD4^+$ T cells in the colonic lamina propria (cLP) and about 20% in the small intestinal lamina propria (siLP) (Weiss et al. 2012; Round and Mazmanian 2010; Atarashi et al. 2011; Geuking et al. 2011; Stefka et al. 2014). The presence of higher proportion of Tregs in the colon, the site of the highest microbial biomass, intuitively suggests that development and maturation of intestinal Tregs is strongly influenced by microbial antigens. Indeed, germ-free (GF) mice and long-term broad spectrum antibiotic-treated specific pathogen-free (SPF) mice have a dramatic reduction in their colonic Tregs compared to their respective controls (Atarashi et al. 2011; Geuking et al. 2011; Weiss et al. 2012). However, the comparable numbers of Tregs present in the siLP of GF and SPF mice indicate that in this intestinal segment, bacterial antigens contribute to the Treg population to a lesser degree (Kim et al. 2016). The same study showed that once the weaned GF mice were switched to antigen-free diet, siLP (but not cLP) Tregs declined, thus revealing a significant role of dietary antigens in regulation of small intestinal Treg population (Kim et al. 2016).

Human $CD4^+$ Tregs are characterized by expression of the transcription factor forkhead box P3 protein (FOXP3, scurfin), high surface expression of CD25, and low or no expression of CD127. Tregs express high levels of CD25 (the IL-2 receptor α-chain) due to their high dependence on interleukin 2 (IL-2) for their development (Zheng et al. 2007; Davidson et al. 2007) and maintenance and peripheral homeostasis (D'Cruz and Klein 2005; Fontenot et al. 2005). The groundbreaking study led by Sagakuchi demonstrated that depletion of $CD25^+$ subpopulation of $CD4^+$ T cells resulted in the manifestation of multiorgan autoimmunity, including intestinal inflammation (Sakaguchi et al. 1995). A few years later, Fiona Powrie and colleagues successfully developed the $CD4^+CD45RB^{hi}$

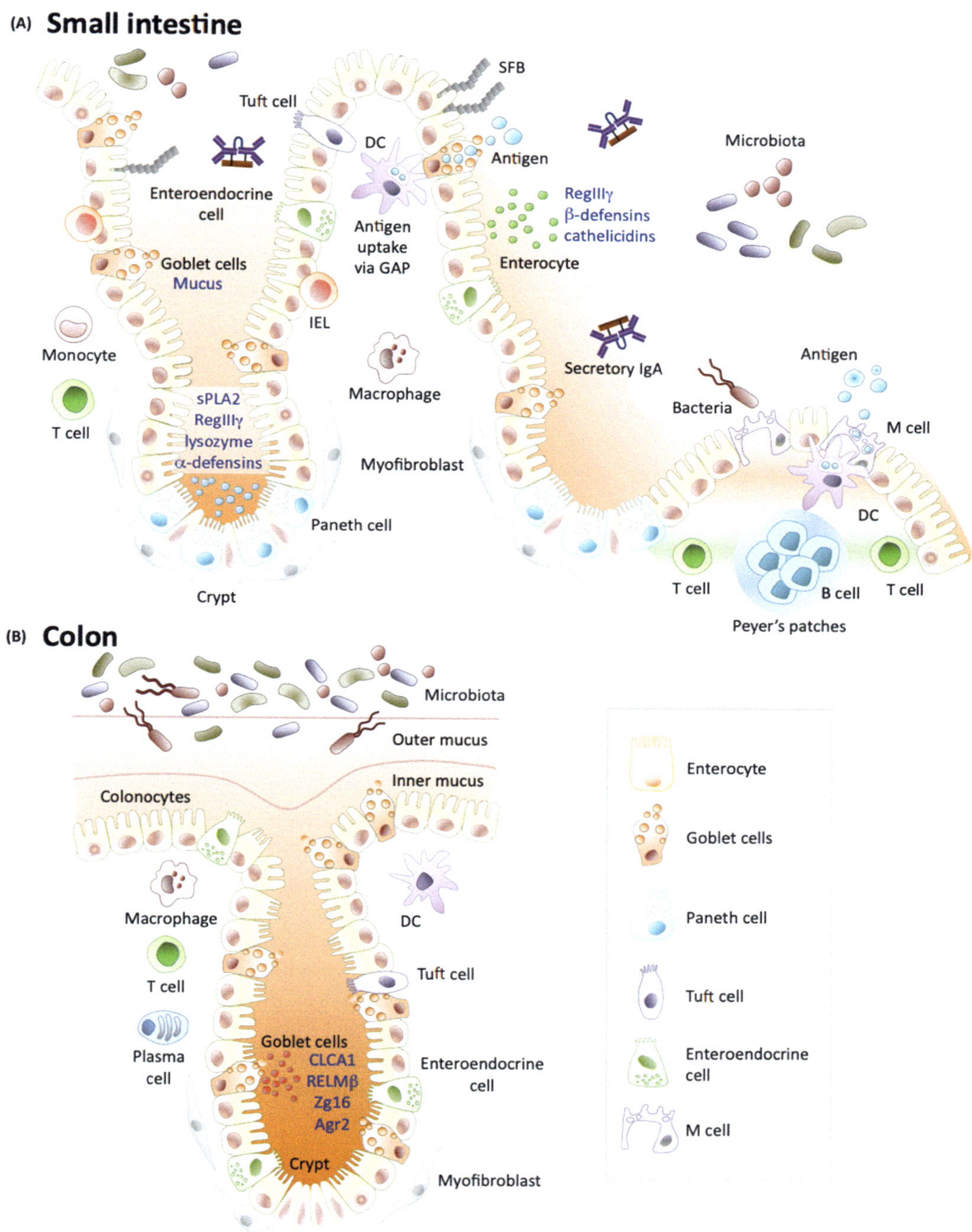

Fig. 9.1 Features of the small intestinal and colonic mucosa. (**a**) Small intestinal epithelium. Enterocyte-derived antimicrobial peptides: RegIIIγ, β-defensins, cathelicidin. Paneth cell-derived antimicrobial peptides [lysozyme, α-defensins, secretory phospholipase 2 (sPLA2)]. M cells and goblet cells facilitate luminal antigen transfer to the underlying dendritic cells. (**b**) Colonic epithelium. Goblet cells produce two-layered mucus layer through Muc2 mucin secretion along with other proteins contributing to the colonic epithelial barrier such as anterior gradient protein 2 homolog (Agr2), small lectin-like protein ZG16 (zymogen granulae protein 16), secreted goblet cell-derived protein Clca1 (chloride channel regulator, calcium-activated-1), or Other abbreviations: *DC* dendritic cell, *IEL* intraepithelial lymphocyte. (Figure adapted from Allaire et al. with permission) (Allaire et al. 2018)

T-cell transfer model of IBD. The consequent discovery that the symptoms of this model could be suppressed via the co-transfer of $CD4^+CD25^+$ T cells, was a huge leap in the field (Mottet et al. 2003). Later, FoxP3 was discovered to be a vital transcription factor for the induction and maintenance of $CD4^+CD25^+$ Tregs. In "scurfy" mice, an X-linked recessive mutation led to scaly skin, lymphoproliferation, hypergammaglobulinemia, lymphadenomegaly, anemia, runting, and early death (Godfrey et al. 1991). In 2001, seminal genetic studies in the "scurfy" mice led to the discovery of *Foxp3* gene (Brunkow et al. 2001). Other contemporary studies revealed that human patients with IPEX syndrome (immunodysregulation polyendocrinopathy enteropathy *X*-linked syndrome) harbored mutations in *Foxp3* gene and its 3′UTR region (Bennett et al. 2001; Wildin et al. 2001). Conspicuously, the FoxP3-deficient mice and patients with IPEX syndrome present with gastrointestinal manifestations in the form of autoimmune enteropathy (Powrie et al. 1993; Gambineri et al. 2003), thus highlighting the role of Tregs in the maintenance of intestinal homeostasis.

While $CD4^+$ Tregs are most widely studied in health and disease, $CD8^+$ Tregs were identified in 1970 by Gershon and Kondo, long before the characterization of $CD4^+$ Tregs (Gershon and Kondo 1970). $CD8^+$ Tregs are a heterogenous population that are not very well characterized and their role in the maintenance of intestinal homeostasis is poorly understood. In this chapter, we will primarily focus on $CD4^+$ Tregs, while the reader is directed to published review articles on $CD8^+$ Tregs and original articles therein (Flippe et al. 2019; Yu et al. 2018; Tsai et al. 2011; Liu et al. 2014; Zhang et al. 2019).

9.2 Intestinal Tregs Subsets

Perhaps as an adaptation to the highly specialized and dynamic intestinal environment, Treg population is characterized by a significant heterogeneity. The current knowledge about the different Treg subsets is mostly based on mice as just like any other biomedical research, studying Treg biology in humans is very challenging. While Foxp3 is considered to be the signature marker associated with murine Tregs, it is not so straightforward in human Tregs. Moreover, unlike in mice, where a FoxP3 reporter mouse lines are available, it is a technical challenge to isolate human Tregs based on FoxP3 which is expressed intracellularly, which precludes isolation of live cells for further functional studies. Even though the presence of Foxp3 is indispensable for the development and function of human Tregs, it is not necessarily synonymous with immunosuppressive functions. There have been many instances when expression of FoxP3 was shown in activated human T cells that lacked the regulatory activity (Allan et al. 2007; Morgan et al. 2005; Wang et al. 2007; Gavin et al. 2006; Tran et al. 2007). Human Tregs are also known to express cytotoxic T lymphocyte-associated antigen 4 (CTLA4) and glucocorticoid-induced TNF receptor (GITR), but just like CD25, they are not very specific to Tregs and can be expressed by effector T cells (T_{eff}) (Read et al. 2000; Takahashi et al. 2000; McHugh et al. 2002). The CD127 and CD27 markers have been proposed to increase the specificity of Treg identification. CD127 expression is much lower in $CD25^+Foxp3^+$ Tregs than in T_{eff} (Seddiki et al. 2006), and this marker is useful to identify Tregs in non-inflammatory conditions. Therefore, most of the human studies are conducted on $CD4^+CD25^{bright}CD127^{lo}$ cells isolated from peripheral blood.

In the murine gut, based on transcription factor and surface molecule expression, $CD4^+$ Tregs can be subdivided into the following subsets: $GATA3^+Helios^+(Nrp1^+)$, retinoic acid receptor-related orphan receptor γt (RORγt)-expressing $ROR\gamma t^+Helios^-$ cells, and $ROR\gamma t^-$ $Nrp1^-(Helios^-)$ Tregs.

9.2.1 Thymic vs. Peripheral Tregs

Based on their origin, Tregs are classified into thymic Tregs (tTregs) which develop in the thymus and into peripheral Treg which are induced from naïve T cells in the periphery (pTreg). Murine tTreg differentiation into $Foxp3^+$ Treg is a two-step process: First, future Tregs recognize MHCII-presented self-antigen on thymic epithelial cells via a high-affinity T-cell receptor (TCR) (Stritesky et al. 2012; Xing and Hogquist 2012). In the second step, γ_c cytokines (mainly IL2) and CD28 co-stimulation lead to an induction of FoxP3 expression (Xing and Hogquist 2012; Fontenot et al. 2005; Salomon et al. 2000; Hori et al. 2003; Tai et al. 2005). In contrast, future pTregs exit the thymus as naive $CD4^+$ T cells and differentiate into $Foxp3^+$ Treg cells in the periphery, upon subsequent recognition of their cognate non self-antigen (Shevach and Thornton 2014; Chen et al. 2003; Yadav et al. 2013) and in the presence of a permissive cytokine microenvironment (Zheng et al. 2002, 2004). The role of cytokines and other factors in the development of Tregs is discussed in Sect. 9.4 of this chapter. One of the big challenges that remains in the Treg field is the paucity of markers (especially cell surface proteins) to distinguish the pTregs from the tTregs. So far, Helios and Neuropilin 1 (Nrp1) have shown to be the two most promising markers expressed by tTregs (Thornton et al. 2010; Yadav et al. 2012; Weiss et al. 2012). However, both of these markers are not exclusive to tTregs. The finding that inflammatory milieu upregulates the expression of both Helios and Nrp1, further complicates their utilization as markers (Gottschalk et al. 2012). While Helios is present in 70% of Tregs in peripheral lymphoid tissues (Thornton et al. 2010), the presence of Nrp1 in humans is still questionable. In humans, both of these markers have not proven to be very reliable. A fraction of human tTregs does not express Helios (Himmel et al. 2013), and Nrp1 expression can be induced in activated humans T_{eff} (Milpied et al. 2009). Intestinal Tregs are thought to have a dual origin as they could either arise in thymus or develop in the periphery, in response to local microbiota and food antigens. However, a vast majority of Tregs present in the small intestine and colon are $Helios^-Nrp1^-$ (Ohnmacht et al. 2015). In fact, it is the frequency of $FoxP3^+Helios^-Nrp1^-$ Tregs which is reduced in the colon of GF and antibiotic-treated SPF mice (Atarashi et al. 2011; Geuking et al. 2011; Weiss et al. 2012).

Development of $FoxP3^+$ pTregs and tTregs is also regulated by *Foxp3* intronic enhancers, termed conserved noncoding sequence 1, 2, and 3(CNS1-3). While CNS3 is instrumental in the initial induction of FoxP3 in tTregs, CNS1 harbors a TGFβ-NFAT response element and a binding site for retinoic acid receptor (RAR) and retinoic X receptor (RXR) heterodimer, which is activated by retinoic acid (Zheng et al. 2010; Xu et al. 2010; Tone et al. 2008). CNS1-deficient mice lack pTregs, while their tTregs differentiation remains unaffected (Zheng et al. 2010). Another study reported that CNS1 deficiency in colonic TCR-transgenic T cells only affected initial, but not late, pTreg cell selection (Nutsch et al. 2016). Nonetheless, CNS1-deficient mice display a dysbiotic microbiota with exacerbated Th2 type pathologies in the lung and colon (Josefowicz et al. 2012; Campbell et al. 2018). pTregs have also been shown to suppress Th1-type responses to food antigens in the gut (Kim et al. 2016). Cumulatively, these studies highlight the key role of pTregs in maintaining tolerance toward luminal antigens.

While pTregs remain the predominating Treg population in colon, it has been suggested that tTregs and pTregs supplement the function of each other, partly by expanding their TCR diversity (Haribhai et al. 2011). Moreover, in the absence of pTregs, tTregs can migrate to intestinal LP and expand locally in response to the cross-reactive microbial antigens (Cebula et al. 2013). Using single-cell and high-throughput sequencing on TCR^{mini} mice with "limited but diversified TCR," this study demonstrated that TCR repertoires of colonic tTregs and pTregs have significant overlap (Cebula et al. 2013). A study with mice with MHCII expression restricted to the cortical thymic epithelial cells and thus cannot provide peripheral MHCII-TCR interaction necessary for the induction of pTregs demonstrated that majority of intestinal Tregs were of thymic origin (Korn et al. 2014). The study revealed that there exists a unique "Treg niche" in the intestine, where tTregs migrate during early life and are maintained without the need for TCR stimulation. However, broad spectrum antibiotic treatment abrogated LP tTregs, suggesting that continuous microbial stimuli might be vital for the maintenance of Tregs in the intestinal compartment (Korn et al. 2014). In summary, both pTregs and tTregs are likely to contribute to the maintenance of intestinal homeostasis, although it is challenging to determine which of the two populations plays a major role.

9.2.2 GATA3$^+$Helios$^+$(Nrp1$^+$) Tregs

GATA-binding protein 3 (GATA3), the canonical Th2-specific transcription factor, is expressed by up to 65% of intestinal Tregs and can be induced in CD4$^+$Foxp3$^+$ Tregs by TCR engagement (Wohlfert et al. 2011). In the same report, human Tregs (CD4$^+$CD25^{bright}CD127^{lo}) isolated from peripheral blood of healthy donors were also able to dramatically upregulate GATA3 expression upon TCR engagement (Wohlfert et al. 2011). However, this subset has not yet been demonstrated in the human gut. GATA3 has been shown to be important for immunosuppressive function and accumulation of Tregs to the extent that GATA3-deficient Tregs were unable to suppress naïve T-cell transfer-induced colitis in mice (Wang et al. 2011). In the mouse gut, GATA3$^+$Helios$^+$ tTregs have been also found to express IL33 receptor ST2 and amphiregulin (Schiering et al. 2014). IL33 is produced by intestinal epithelial cells under inflammatory conditions. In fact, high levels of IL33 have been reported in inflamed lesions of inflammatory bowel disease patients (Kobori et al. 2010; Pastorelli et al. 2010; Seidelin et al. 2010). In the same line of thought, in steady state, ST2$^+$ FoxP3$^+$Tregs showed increased production of IL10 and TGFβ and displayed preferential accumulation in the intestinal LP due to high expression of gut-homing receptors CCR9 and α4β7 (Siede et al. 2016). Furthermore, ST2$^+$ Tregs display an activated phenotype and express high levels of OX40 (Schiering et al. 2014), a secondary co-stimulatory immune checkpoint molecule which is also important for the accumulation of Tregs in colon and for Treg-mediated suppression of naïve T-cell transfer-induced colitis (Griseri et al. 2010). It is thus plausible that GATA3$^+$ tTregs might be home to the sites of damage in the gut to curb excessive inflammation and mediate tissue repair. However, this hypothesis still needs to be verified.

9.2.3 RORγt$^+$Helios$^-$ pTregs

Another Treg subset that is abundantly located in the gut expresses the Th17 master transcription factor, retinoic acid receptor-related orphan receptor γt (RORγt). The absence of Helios (Sefik et al. 2015; Ohnmacht et al. 2015) and Nrp1 (Yang et al. 2016) on most of the RORγt$^+$ Tregs suggests their peripheral origin. Owing to their similar expression of RORγt, the RORγt$^+$Foxp3$^+$ Tregs were once considered to be precursors for unstable Tregs that could differentiate into Th17 cells (Zhou et al. 2008; Lochner et al. 2008). However, it was later revealed that, in steady state, RORγt$^+$Helios$^-$ FoxP3$^+$ Tregs comprise about 65% and ~35% of total Treg population in the colon and small intestine, respectively (Sefik et al. 2015; Ohnmacht et al. 2015; Yang et al. 2016). Comparable frequencies of Rorγt$^+$ Tregs were detected by staining cells from healthy or inflamed (Crohn's) colon biopsies (Sefik et al. 2015) and among healthy human PBMCs (Duhen et al. 2012), confirming the presence of this subset in humans. Interestingly, there were higher number of circulating IL-17-producing RORγt$^+$Foxp3$^+$CD4$^+$ lymphocytes in IBD patients, and the suppressive ability of these Tregs were reduced by approximately 60% in patients with IBD compared with healthy controls (Ueno et al. 2013).

The plasticity between Th17 cells and RORγt$^+$ Tregs is not very well understood. Development of Th17 cells is dependent on IL6, IL1β, and IL23. However, while IL6 and downstream STAT3 signaling are important for the development of intestinal RORγt$^+$ Tregs, IL1β and IL23 have not been shown to play a role under homeostatic conditions (Sefik et al. 2015; Xu et al. 2010; Yang et al. 2016). Making use of the transcriptomic and epigenetic profiling, Yang et al. (2016) revealed that RORγt$^+$ Foxp3$^+$T cells display higher similarity to Tregs than to Th17 cells and that they represent a lineage-stable population with potent suppressive activity. Using fixed TCRβ transgenic mice, it was shown that TCR repertoire of RORγt^{+} Foxp3^{+} Tregs is largely unique compared with other colonic T-cell subsets (Solomon and Hsieh 2016). However, they also observed a subset of Foxp3^{+}RORγt^{+} TCRs that are shared with Th17 cells. The same study suggested that the development of mucosal RORγt^{+} Foxp3^{+}T cells is dependent on CX3CR1^{+} antigen-presenting cells and can be derived from naïve CD4^{+} T cells in the periphery via a RORγt^{-} Treg intermediate before co-expressing RORγt. Furthermore, RORγt^{+} Foxp3^{+}Treg subset is readily lost in GF mice or upon antibiotic treatment, signifying the indispensable role of microbiota in their maintenance (Ohnmacht et al. 2015; Sefik et al. 2015). Interestingly, the differentiation of RORγt^{+} Tregs is also dependent upon constant antigen exposure, as mice deficient in MHCII have decreased frequency of this subset (Ohnmacht et al. 2015). The role of microbiota in the development of RORγt^{+} Tregs would be further discussed later in this chapter. Additionally, c-Maf, a transcription factor which has already been known to be a positive regulator of Th17 cells, has also been found to be highly expressed by RORγt^{+} Foxp3^{+}Tregs (Xu et al. 2018). Not only has c-Maf been found to be required for the development of RORγt^{+} Tregs, but is also important for their suppressive ability, as ablation of c-Maf in Tregs leads to spontaneous inflammation in mice that is not observed in mice with RORγt-deficient Tregs (Sefik et al. 2015; Xu et al. 2018; Yang et al. 2016).

9.2.4 RORγt^{-} Nrp1^{-}(Helios^{-}) pTregs

Another subset of intestinal Tregs are the CD4^{+}Foxp3^{+}RORγt^{-} T cells. They are believed to be of peripheral origin due to absence of Helios or Nrp1. These cells represent about 50% of SI LP Tregs and 15% of colonic Tregs (Kim et al. 2016). Since these Tregs are more frequently present in SI, they are induced by dietary antigens, so much, so that they are not affected by the absence of microbiota but disappear in mice fed "antigen-free" diet (Kim et al. 2016). In fact, the absence of this Treg subset leads to increased susceptibility to food allergy in mice (Kim et al. 2016).

9.2.5 Effector Tregs and IL10^{+} Tregs

In humans, based on the expression levels of Foxp3 and protein tyrosine phosphatase isoform A (CD45RA), human peripheral blood Foxp3 +CD4^{+} T cells can be classified into Foxp3^{lo}CD45RA^{+} naive Tregs which after antigenic stimulation differentiate into Foxp3^{hi}CD45RA^{-} Tregs that represent activated, antigen primed, terminally differentiated and highly suppressive effector Tregs or eTregs. In contrast, Foxp3^{lo}CD45RA^{-} cells do not possess suppressive activity and can even secrete pro-inflammatory cytokines (Miyara et al. 2009). To understand these cells better, some studies have been carried out with mouse CD44^{hi}CD62L^{low}KLRG1^{+}FOXP3^{+} cells, thought to represent the murine counterpart of the human eTregs (Liston and Gray 2014). The CD44^{hi} effector Tregs are most abundant in the colonic LP and depend on constant TCR engagement and subsequent IRF4 expression for their maintenance (Levine et al. 2014). On the other hand, TCR and IRF4 are dispensable for the development of CD44^{lo}CD62L^{hi} naive Tregs as tamoxifen-induced Treg-specific deletion of TCRα or IRF4 does not adversely affect their numbers (Levine et al. 2014). Transcription factors, IRF4 and Blimp-1, jointly regulate the differentiation, function, and homeostasis of eTregs (Levine et al. 2014; Cretney et al. 2011).

Another important Treg subset described in humans is the FOXP3^{-} type 1 (Tr1) cells that develop exclusively in periphery and secrete copious amounts of IL10 and TGFβ1, intermediate levels of IFNγ and no IL4 (Levings and Roncarolo 2005; Roncarolo et al. 2018). In fact, IL10-producing Tr1 cells were originally identified and isolated from patients who have severe combined immunodeficiency (SCID) and had undergone successful HLA-mismatched bone marrow transplantation. In these studies, antigen-specific Tr1 cells were generated in vitro from both mice and human naïve $CD4^+$ T cells (Bacchetta et al. 1994; Groux et al. 1997). The in vitro generated murine Tr1 cells were able to prevent establishment of colitis in SCID mice after $CD4^+CD45RB^{hi}$ adoptive T-cell transfer in IL10- and TGFβ-mediated fashion (Groux et al. 1997). Hence, Tr1 cells appear to significantly contribute to maintaining the intestinal tolerance.

9.3 Antigen Specificity of Intestinal Tregs

One of major challenges in the field has been to understand the antigen specificity of intestinal Tregs. For example, are pTregs generated against self-antigens or antigens derived from commensal bacteria? Although some earlier studies have shown that commensal bacteria are not necessary for colonic Treg generation (Min et al. 2007; Round and Mazmanian 2010), generation of Tregs in response to foreign antigens has been established with TCR transgenic models of oral tolerance (Curotto de Lafaille et al. 2008; Sun et al. 2007). Earlier in this chapter, we have described studies with germ-free and antibiotic-treated mice that clearly showed the importance of the gut commensals in driving colonic Treg numbers. Some human gut commensals such as *Clostridium* clusters IV and XIVa (Atarashi et al. 2011) and *Bacteroides fragilis* (Round and Mazmanian 2010) have been found to increase the frequency or function of colonic Tregs via protease-resistant capsular polysaccharide (PSA) and a TLR2-dependent mechanism (Round et al. 2011). PSA also induced expression of Foxp3 along with CD39 in naïve human CD4 T cells in vitro while promoting IL-10 secretion (Telesford et al. 2015).

To address TCR specificity, labs developed TCR transgenic mouse lines that respond to antigens derived from commensal bacteria, e.g., flagellin (Cong et al. 2009). However, this approach does not allow for the comparison of the normal in vivo frequency of those TCRs in the Tregs with the effector T cells. The study by Lathrop et al. (2011) analyzed the colonic TCR repertoire and concluded that TCR usage in the colon is very different from that in any other organ and that there was very little overlap in TCRs in Tregs and the effector/memory T cells. This study confirmed that the colonic Treg population is strongly shaped and fine-tuned by the local antigenic milieu and that the effector T cells and Tregs are not necessarily activated by the same antigens. Furthermore, it could be safely inferred that colonic Tregs recognize and suppress the immune response evoked against the antigens from commensal bacteria. In fact, "Limited mice," which express only a limited TCR repertoire, display shrunken $Helios^-$ Tregs and eventually develop loss of tolerance to the gut microbiota and spontaneous development of Th17-driven colitis (Nishio et al. 2015).

9.4 Key Factors Required for the Development and Maintenance of Tregs in the Gut

The intestinal Tregs are very unique and highly specialized due to their role in preventing inflammatory responses to commensal flora, diet, and other innocuous antigens. There are many different mechanisms that are engaged in the induction and maintenance of Tregs in the intestine. While most of these intricate pathways have been elucidated in mice, some attempts have been made to validate them in humans.

9.4.1 Host Factors: Cytokines

As already described in the Introduction to Intestinal Tregs, development of Treg precursors in the thymus requires not only the strong TCR stimulation but also co-stimulation through CD28 and a presence of common-γ-chain cytokines such as IL2 and IL15 (Chinen et al. 2016; Setoguchi et al. 2005). The transcription factor STAT5, which is activated downstream of the IL2 receptor α (CD25), binds to CNS2 in the *Foxp3* gene and thus stimulates its expression. In fact, many mouse models mimicking IL2 deficiency have demonstrated IL2 to be a key cytokine for the development and the maintenance of tTregs in the periphery (Papiernik et al. 1998; Almeida et al. 2002; Malek et al. 2002; Setoguchi et al. 2005). Accordingly, CNS2-deficient Tregs are sensitive to low levels of IL2 and eventually lose FoxP3 expression in response to strong TCR activation or pro-inflammatory cytokines, a typical intestinal milieu (Feng et al. 2014; Li et al. 2014). On the other hand, Treg pool can be increased following treatment with agonistic complexes of IL2 or antibody to IL2 (Oldenhove et al. 2009; Wohlfert et al. 2011).

The role of IL2 is central to the development of Tregs as IL2-deficient mice have been known to develop colitis (Sadlack et al. 1993). In general, IL2 potently stimulates T-cell growth and population expansion. However, the function of IL2 signaling for Treg differentiation and maintenance of immune system homeostasis appears to overcome its effects on the effector T cells, since mice deficient in IL-2 or components of the IL-2 receptor, IL-2Ra (CD25) or IL-2Rb (CD122), suffer from an aggressive lymphoproliferative, autoimmune syndrome. Similarly, patients with IL-2Ra deficiency develop immunodeficiency and early onset IBD live-disease (Sharfe et al. 1997). It has been reported that while CD25 activation by IL2 is required for FoxP3$^+$ pTreg cell differentiation, survival, and expansion, it remains dispensable for the induction and differentiation of FoxP3$^+$ tTreg cells (D'Cruz and Klein 2005; Fontenot et al. 2005). However, a later report showed that IL2 seems to be critical for the maintenance of intestinal GATA3$^+$ tTregs, as this subset is severely reduced in the small intestine of IL2$^{-/-}$ mice (Wohlfert et al. 2011). Furthermore, blocking IL2 in a mouse model of oral tolerance to ovalbumin impaired the conversion of adoptively transferred naive OT-II CD4$^+$ T cells into pTreg cells in the mesenteric lymph nodes (MLNs) and Peyer's Patches (PP) (Edwards et al. 2016). However, the role of IL2 in the induction of RORγt$^+$ and RORγt$^-$ pTregs has not been evaluated.

TGFβ gives an initial survival advantage to the Treg precursors in the thymus, though it fails to promote Foxp3 expression and is therefore not important for the induction of tTregs (Li et al. 2006; Marie et al. 2005, 2006; Ouyang et al. 2010; Tone et al. 2008; Konkel and Chen 2011). On the other hand, TGFβ has an unambiguously essential role in the differentiation of pTregs in the intestine, where it is present in copious amounts (Chen et al. 2003; Konkel and Chen 2011). TGFβ1-deficient mice develop lethal multi-organ lymphoproliferative disease, which mainly affects the gut. Notably, this pathology has considerable similarity with the one observed in the Foxp3-deficient mice (Kulkarni et al. 1993; Shull et al. 1992). TGFβ1-activated Smad3 binds to Foxp3 intronic enhancer CNS1 (Josefowicz et al. 2012; Tone et al. 2008; Xu et al. 2010). However, mice with an ablation of the Smad3-binding site in CNS1 did not result in reduction in the frequency of intestinal Tregs (Schlenner et al. 2012). Similarly, the frequency of colonic Tregs is not compromised in mice with the deletion of the cytokine receptor TGFβR1 specifically in Tregs (Konkel and Chen 2011). However, specific Treg subsets were not investigated in that study, and it is possible that while pTregs contracted in the absence of TGFβ signaling, the tTregs expanded to compensate. On the other hand, colitis in mice with TGF-βRII-deficient DCs is accompanied by reduced frequencies of mucosal and MLN Foxp3$^+$ Tregs (Jamwal et al. 2020; Ramalingam et al. 2012).

TGFβ1 is secreted as an inactive precursor molecule combined with latency-associated peptide (LAP) and needs to be modified by integrins, such as αvβ8 and αvβ6 (Travis et al. 2007). Importantly, the expression of integrin β8 is highly upregulated in intestinal CD103$^+$ DCs (Paidassi et al. 2011), and consequently CD103$^+$ DCs deficient in αvβ8 have a compromised capacity for Treg induction in the MLNs (Worthington et al. 2011). Additionally, LAP could be cleaved by pro-protein convertase enzymes, such as Furin. Notably, Furin-deficient Tregs exhibit defective regulatory function in the naïve T-cell transfer-induced colitis (Pesu et al. 2008). Furthermore, in a mouse model of oral tolerance, a complex of GARP (glycoprotein A repetitions predominant) and latent TGFβ1 on

Tregs has been found to be a major source of TGFβ1 needed for the induction of pTreg cells in the MLNs and PPs (Edwards et al. 2016). Overall, IL2 and TGFβ1 seem to be central in Treg homeostasis, so much, so that naïve T cells after in vitro stimulation in the presence of TGFβ and IL2 are induced to differentiate into inducible Tregs (iTregs) (Zheng et al. 2002, 2004).

Although the plasticity of Tregs under inflammatory pressure is well recognized, there are lingering controversies, mainly stemming from varying definitions, protocols, and starting cell populations. One aspect of Treg plasticity is related to reversible FoxP3 expression. The in vitro generated iTregs are generally recognized to have unstable expression of Foxp3 via a mechanism involving histone demethylation at the CNS2 element, resulting in quick reversion to $Foxp3^-$ effector T cells (Kanamori et al. 2016). Another aspect of Treg plasticity is the described gain in IL17 production. IL6 is thought to contribute to the latter, although different Treg lineages seem to respond differently. nTregs can be induced to become Th17-like when exposed to IL6 in a TGFβ-dependent manner, but iTregs were resistant to this conversion under the same protocol (Zheng et al. 2008). Xu et al. also showed that $CD4^+CD25^+FoxP3^+$ mouse splenocytes (constituting primarily nTregs) stimulated with IL6 turned into Th17 cells, even without the presence of exogenous TGFβ (Xu et al. 2007). Similarly, nTregs sorted from mouse thymus or spleens were converted to Th17 cells with CD3/IL6 stimulation, a conversion inhibited by atRA (Zhou et al. 2010). Human peripheral blood $CD4^+CD25^{hi}CD127^{lo}Foxp3^+$ Tregs stimulated with high-salt medium lose their immunosuppressive function via gaining IFNγ expression, though only a very minor increase of IL17 expression was observed in this cell population. In a more recent mouse study, Luo et al. showed that high-salt diet in vivo and elevated NaCl concentration in vitro did not negatively affected pTregs or iTregs, respectively, but impaired the immunosuppressive function of tTregs and reduced FoxP3 expression (Luo et al. 2019). Thus, none of the Treg population should be considered as entirely stable or terminally differentiated. Their function and phenotype can be affected differently by various stimuli dependent on their lineage, tissue, and/or pathophysiological milieu.

Recently, IL33 has also been deemed instrumental in the induction of Tregs in the gut. IL33 is passively released by stromal or epithelial cells during cell necrosis or tissue damage, defining its role as an alarmin that alerts the immune system during trauma (Pichery et al. 2012). As mentioned earlier, intestinal $GATA3^+$ tTregs, but not $ROR\gamma t^+$ pTregs, have high expression of the IL33 receptor ST2. Likewise, in vivo administration of IL33 increases the number of both splenic Tregs and colonic $ROR\gamma t^-$ Tregs, but not $ROR\gamma t^+$ pTregs (Schiering et al. 2014; Sefik et al. 2015). Overall, IL33 seems to positively regulate tTregs but not pTregs in vivo. However, IL33 in combination with TGFβ1 has also been successfully able to induce the differentiation of iTregs (Schiering et al. 2014).

9.4.2 Host Factors: TNFRSF–NF-κB Axis

Survival at a barrier site such as intestine requires adaptation to the changing environmental milieu. Due to continuous exposure to microorganisms, Treg cells display an activated phenotype characterized by the core tumor necrosis factor receptor superfamily (TNFRSF) signature including expression of GITR, OX40, and TNFR2 (Miragaia et al. 2019; Vasanthakumar et al. 2017). The stimulation of TNFRSF members further activates the NF-κB signaling pathway and provides vital survival signals for Tregs (Miragaia et al. 2019; Salomon et al. 2018; Vasanthakumar et al. 2017). OX40 is important for survival, accumulation, and function of Treg cells in the colon (Griseri et al. 2010). Activation of RelA, a NFκB transcription factor, is associated with the adaptation of both tTregs and pTregs in the colon (Vasanthakumar et al. 2017). Overall, the TNFRSF–NF-κB axis may be a central signaling component in Treg maintenance in the gut.

9.4.3 Environmental Factors: Commensal Microbiota

The intestine, especially the colon, harbors abundant bacteria reaching 10^{12} cells per gram of intestinal content. Bacteria remain in a complex relationship with the host immune system, and among their many functions, they profoundly affect mucosal T cells, including their differentiation into different subsets of effector or regulatory cells (Larmonier et al. 2015). As mentioned in the previous sections, the pTregs are induced, maintained, and "fine-tuned" directly by the commensal bacteria or by the metabolites produced by bacteria. While both RORγt$^+$ and RORγt$^-$ pTregs are affected by general presence or absence of microbiota, apparently, only the RORγt$^+$ subset seems to be affected by qualitative changes in the colonic microbiome (Kim et al. 2016; Ohnmacht et al. 2015; Sefik et al. 2015; Ye et al. 2017). Some bacterial genera including *Clostridium, Bacteroides, Bifidobacterium, Lactobacillus*, and *Helicobacter* have been shown to promote pTregs in colon (Geuking et al. 2011; Atarashi et al. 2011, 2013; Chai et al. 2017; Kullberg et al. 2002; Sefik et al. 2015). Most of clostridia are gut commensals, with a few exceptions, such as *Clostridium perfringens, Clostridium difficile*, and *Clostridium tetani*, which are pathogenic and toxigenic. Oral administration of GF mice with 46 strains of commensal *Clostridia* showed strong promotion of differentiation, proliferation, and recruitment of Helios$^-$RORγt$^+$ Tregs in the colon (Atarashi et al. 2011). Similar experiments executed with 17 strains of *Clostridia* isolated from a healthy human adult also demonstrated a strong capacity to induce RORγt$^+$Helios$^-$ pTreg cells rather than GATA3$^+$ tTregs in the colon of mice and rats (Atarashi et al. 2013). *Clostridia*-induced increase in frequency of colonic Tregs is also accompanied by an increase in the expression of IL10, CTLA4, and ICOS on pTregs (Atarashi et al. 2011, 2013) and increased intestinal production of sIgA which coats commensals and reduces T-cell activation by microbial antigens, such as flagellin (Cong et al. 2009). *Faecalibacterium prausnitzii* (cluster IV) is another member of the class of *Clostridia* and one of the most abundant residents in healthy humans (Godefroy et al. 2018; Sarrabayrouse et al. 2014). *F. prausnitzii* promotes the accumulation of Tr1 cells (IL10-producing CD4$^+$CD8$\alpha\alpha^+$ T cells) in the colon and the blood of healthy humans (Sarrabayrouse et al. 2014). Interestingly, relative abundance of *F. prausnitzii* is decreased in IBD patients (Machiels et al. 2014; Sokol et al. 2008, 2009). *F. prausnitzii* is capable of inducing IL10 secretion by healthy human DCs (Alameddine et al. 2019) which may also assist with reducing the overall inflammatory tone in the gut.

Apart from the Clostridia, another resident member of the human gut microbiome capable of impacting Treg homeostasis is *Bacteroides fragilis*. *B. fragilis* enhances the accumulation of IL10-producing Foxp3$^+$ Tregs in the colon (Round et al. 2011). Monoassociations of mice with other Bacteroides spp., e.g. *B. caccae, B. thetaiotaomicron, B. uniformis, B. finegoldii, or B. ovatus,* have been shown to induce the accumulation of FOXP3$^+$ cells, particularly Nrp1$^-$RORγt$^+$ pTreg cells in the mouse colon (Faith et al. 2014; Sefik et al. 2015). De novo generation of Tregs, with the capacity to suppress intestinal inflammation, was also observed in GF mice colonized with *Bifidobacterium bifidum* (Verma et al. 2018). *Clostridia, B. fragilis,* and *B. bifidum* have also been shown to increase the expression of CTLA4 on Tregs (Atarashi et al. 2011, 2013; Round et al. 2011; Verma et al. 2018). While it is not yet established if Clostridium and Bacteroides species function in a TCR-dependent manner, a pathobiont *Helicobacter hepaticus* has been shown to induce antigen-specific RORγt$^+$ Tregs in the colon (Chai et al. 2017; Xu et al. 2018).

Microbes also stimulate the regulatory arm of the mucosal immune system via their components. For example, *B. fragilis* releases outer membrane vesicles (OMVs) containing polysaccharide A (PSA), the most abundant capsular polysaccharide expressed by *B. fragilis.* OMVs are taken up by cLP DCs and further elicit the conversion of CD4$^+$ T cells into IL10-

producing Foxp3$^+$ Tregs via TLR2 (Round et al. 2011). Similarly, cell surface β-glucan/galactan (CSGG) polysaccharides from *B. bifidum* are important effector components able to induce Tregs via a DC-dependent and partially TLR2-dependent mechanism (Verma et al. 2018).

9.4.4 Environmental Factors: Dietary and Microbial Metabolites

The specific mechanism by which symbionts stimulate the accumulation of Tregs is not fully understood. It is believed that *Clostridia* spp. can function synergistically to stimulate IECs leading to Treg induction (Atarashi et al. 2013). One proposed mechanism of action is the cooperative production of short-chain fatty acids (SCFAs), such as butyrate, propionate, and acetate, through fermentation of dietary fiber by the commensals including *Clostridium* and *Bacteroides* spp. in the colon (Arpaia et al. 2013; Atarashi et al. 2013; Furusawa et al. 2013; Smith et al. 2013). Colonic SCFAs are absorbed by passive diffusion and by active H^+-coupled transport via MCT1 (SLC16A1) and MCT4 (SLC16A3), two electroneutral monocarboxylate transporters, or by electrogenic Na^+-coupled monocarboxylate transporters SMCT1 (SLC5A8) and SMCT2 (SLC5A12). Oral administration of butyrate, propionate, and acetate either individually or in combination led to an increase in the number of colonic Treg cells in SPF mice (Smith et al. 2013). Although in monoassociation studies, Sefik et al. could not find a correlation between luminal SCFA concentrations and the frequencies of RORγt$^+$ pTregs (Sefik et al. 2015), this study did not take the combinatorial effects of SCFA producers into account. Upon reaching the colonic LP, SCFAs affect DCs and T cells directly using two mechanisms: via inhibition of histone deacetylases (HDAC) and via signaling through certain G protein–coupled receptors. Butyrate is known to participate in Treg differentiation by facilitating histone H3 acetylation in the promoter region and CNS1 and 3 of *Foxp3* gene (Furusawa et al. 2013). Recognition of butyrate by G protein–coupled receptors such as GPR43and GPR15 expressed by colonic Treg cells and by GPR109A expressed by DCs and macrophages (Arpaia et al. 2013; Atarashi et al. 2013; Furusawa et al. 2013; Singh et al. 2014; Smith et al. 2013) also promotes Treg differentiation. Interestingly, SCFA treatment increased Helios$^+$ Treg numbers in GF mice, indicating that SCFAs also promote the expansion of the tTregs present in the colonic LP (Smith et al. 2013).

Aryl hydrocarbon receptor (AhR) is a nuclear sensor expressed by Tregs that helps them sense and react to compounds that act as AhR ligands. Intestinal pTregs have higher expression of AhR than Tregs in any other tissue (Ye et al. 2017). Dietary tryptophan, an essential amino acid, is metabolized by IDO (indoleamine 2,3-dioxygenase) and TDO (tryptophan 2,3-dioxygenase) expressed in IECs and DCs, into several AhR ligands including kynurenine. Kynurenine enhances Foxp3$^+$ Treg generation in vitro in the presence of TGFβ1 (Mezrich et al. 2010). Gut microbes like *Lactobacillus* spp. can also metabolize tryptophan into many AhR ligands (Gao et al. 2018). AhR is not only important for the induction of Tregs, but it also contributes to their regulatory function. Ahr-expressing Tregs showed enhanced in vivo suppressive activity compared with Ahr-deficient Tregs in a T-cell transfer model of colitis (Ye et al. 2017).

Another dietary metabolite that drives the Treg development in intestine is the all-*trans* retinoic acid (atRA), a bioactive form of vitamin A. Dietary vitamin A, specifically retinol, is absorbed by the intestinal epithelial cells via passive diffusion. Although epithelial cells are capable of synthesizing atRA from dietary vitamin A (D'Ambrosio et al. 2011), a process regulated by commensal bacteria (Grizotte-Lake et al. 2018), the vast majority of studies to date focused on dendritic cells, macrophages, and eosinophils as sources of atRA for iTreg induction. Aldehyde dehydrogenases ALDH1A1, ALDH1A2, and ALDH1A3 are the rate limiting enzymes in the synthesis of atRA (Bazewicz et al. 2019). Of those three, CD103$^+$DCs express primarily the ALDH1A2 isoform (Coombes et al. 2007).

$CD103^+$ DCs are chiefly equipped to promote Tregs in the gut by the activation of latent TGFβ1 via integrin αVβ8. Thus, $CD103^+$ DC-derived RA, along with TGFβ, converts naive $FoxP3^-CD4^+$ T cells into $FoxP3^+$ Tregs in both human and mouse (Coombes et al. 2007; Kang et al. 2007; Mucida et al. 2007; Sun et al. 2007). These Tregs also display an enhanced upregulation of the mucosal tissue-homing receptors C-C chemokine receptor type 9 (CCR9) and integrin β7 along with being potent suppressors of effector T cells in a colitis model (Coombes et al. 2007; Kang et al. 2007; Mucida et al. 2007; Sun et al. 2007). These RA-induced human $FoxP3^+$ cells were also chemotactically responsive to CCR9 ligand CCL25, a chemokine specifically expressed by intestinal epithelial cells (Kang et al. 2007). Furthermore, CNS1 non-coding element in the *FoxP3* gene, which is required for the generation of pTregs, contains binding sites for the heterodimer of retinoic acid receptor (RAR) and retinoid X receptor (RXR). Upon binding of RA to RAR and RXR, CNS1 undergoes histone acetylation leading to the prolonged and stable expression of FoxP3 in Tregs (Xu et al. 2010). atRA was also shown to promote iTreg development and maintenance via an ERK1/2-dependent mechanism and increased histone methylation and acetylation within the CNS2 element of *Foxp3* gene locus (Lu et al. 2011).

Additionally, it is also reported that RA is able to suppress the IL6-induced conversion of $Foxp3^+$ Tregs into inflammatory Th17 cells, a mechanism likely mediated by the retinoic acid receptor α (Elias et al. 2008). Based on experiments with mice fed with a vitamin A–deficient diet or treated with an inhibitor of the RA receptor, vitamin A seems to affect the development of $ROR\gamma t^+$ pTregs cells and not $Helios^+$ tTreg cells in the gut (Ohnmacht et al. 2015). While $ROR\gamma t^+$ Tregs were restored upon feeding RA, it did not "rescue" the defective generation of $ROR\gamma t^-$ Tregs (Kim et al. 2016). Notably, an RA-reactive human Treg cell subset expressing natural killer cell receptor CD161 and RORγt has been described recently (Povoleri et al. 2018).

While vitamin A plays the most prominent role in the development of Tregs, there are other vitamins that have been reported to affect this lymphocyte population. Dietary vitamin D3 is metabolized to 1,25-dihyroxyvitamin D3 which binds to the nuclear vitamin D receptor element (VDRE) within the +1714 to +2554 nt non-coding CNS region of the human *Foxp* gene (Kang et al. 2012). FR4, the receptor for vitamin B9 (folic acid), is highly expressed by Tregs (Yamaguchi et al. 2007) and is known to promote survival of colonic Tregs by upregulating the anti-apoptotic factor BCL2 (Kinoshita et al. 2012). Consistent with this observation, mice fed with a vitamin B9–deficient diet displayed a marked reduction in colonic pTregs and a higher susceptibility to intestinal inflammation (Kinoshita et al. 2012).

9.5 Suppressive Function Mediated by Treg in the Gut

Tregs can exert suppressive function towards a variety of immune cells, such as NK, $CD8^+$, and $CD4^+$ T lymphocytes and antigen-presenting cells (APC). Inhibition of T conventional cells can occur indirectly, i.e., when inhibition involves influencing the activation status of APC, or directly, i.e., in absence of APC. Tregs can suppress intestinal immune responses to maintain homeostasis and exert their functions through various mechanisms, mostly determined on the basis of in vitro assays. The best-known suppressive mechanisms to be discussed here are (1) cytokine-mediated inhibition, (2) expression of inhibitory surface molecules, and (3) cytolysis of target cells. Species-specific mechanisms may complicate data interpretation, and findings identified in murine models may not correspond to those found in human Tregs.

9.5.1 Secretion of Inhibitory Cytokines

$FoxP3^+$ Tregs and Tr1 cells produce high levels of IL10, which can target several cell types in the gut, among them the effector immune cells, to inhibit their expansion and to promote immunosuppression. IL10 also acts in an autocrine manner as a proliferative factor for Tregs, although its requirement for $FoxP3^+$ Tregs and Tr1 maintenance in vivo is unlikely, since both Treg subsets develop normally in the absence of IL10 or IL10R (Maynard et al. 2007; Chaudhry et al. 2011). Even though IL10 can be secreted by many cells of the innate and adaptive immune systems, the importance of IL10 originating in Tregs for the gut homeostasis is illustrated by the development of exacerbated intestinal inflammation when $FoxP3^+$ Tregs lacked the ability to produce IL10, likely due to their inability to suppress pathogenic Th17 cell responses (Asseman et al. 1999; Chaudhry et al. 2011). The high concentration of IL10 produced by mucosal Tr1 enables the suppression of IL1β and TNFα released from human myeloid cells (Cook et al. 2019). In the gut homeostasis, IL10 gene has been identified as susceptibility locus in ulcerative colitis (UC) (Franke et al. 2008) and mutations on IL10RA and IL10RB receptors result in severe and early onset IBD (Glocker et al. 2009; Moran et al. 2013). In the murine colon, $ROR\gamma^+$ $FoxP3^+$ Tregs have been reported to be the main producers of IL10 (Ohnmacht et al. 2015).

The involvement of Treg-derived TGFβ in immune tolerance, inhibition of acute inflammation, and promotion of wound-healing process renders it a pivotal cytokine in the immune regulation in the gut. Soluble and membrane bound TGFβ are produced in high amounts by Tr1 and $FoxP3^+$ Tregs in the gut (Powrie et al. 1996; Levings et al. 2002). TGFβ also contributes to the differentiation of Tregs, converting naïve $CD4^+$ T cells into $FoxP3^+$ Tregs (Chen et al. 2003; Zheng et al. 2002), but other mediators are required for Tregs to fully differentiate since the induction of FoxP3 expression by TGFβ alone does not necessarily confer immunosuppressive abilities to Tregs (Tran et al. 2007). Production of TGFβ by Treg is, however, necessary to prevent intestinal inflammation (Powrie et al. 1996) as much as expression of TGFbRII by the T conventional cells (Gorelik and Flavell 2000), indicating the essential role of Treg-derived TGFβ in controlling pathogenic T-cell response intestinal inflammation.

Tr1 expanded in vitro, but not $FoxP3^+$ Tregs, can also secrete IL22 to promote intestinal barrier function (Cook et al. 2019). Even though IL35 has been described as an important immunoregulatory cytokine secreted by murine Tregs (Collison et al. 2007), human $FoxP3^+$ Tregs in resting or activated state appear not to express IL35 (Bardel et al. 2008).

9.5.2 Expression of Cell Surface Inhibitors of Inflammation

Human Tregs express cell surface molecules that enable the suppression of immune function in target cells via a contact-dependent (e.g., CTLA4, LAG3, and PD1) or contact-independent manner (e.g., IL2R and CD39). The contact-dependent suppression involves binding of the inhibitory receptors expressed on the Treg surface to stimulatory and co-stimulatory receptors expressed on the surface of target cells, ultimately suppressing immunostimulatory intracellular cascades in target cells. The successful antigen presentation by APC requires the antigen–TCR binding in the context of major histocompatibility complex II (MHCII), in coordination with the triggering of CD28 receptors on T cell by co-stimulatory receptors from the B7 family on APC. B7.1 (CD80) and B7.2 (CD86) can bind to a variety of molecule members of the CD28 family: CD28 and ICOS act as positive regulators of T-cell function while CTLA4 and PD1 act as inhibitors. B7.1 and B7.2 are also expressed on T_{eff} cells and can be targeted for inhibition through contact-dependent mechanisms.

9.5.2.1 Contact-Dependent Inhibition

Cytotoxic T-lymphocyte-associated protein 4 (CTLA4, CD152) is highly expressed on $FoxP3^+$ Treg and Tr1 cells relative to the conventional $CD4^+$ T cells. It is a key regulatory molecule, since null or Treg-specific knockout of CTLA4 result in autoimmune disease (Takahashi et al. 2000; Wing et al. 2008). In vitro suppression of T_{eff} cell proliferation by Tregs (in presence of

APC) was shown to be dependent on CTLA4 in murine and human cells (Tang et al. 2004; Birebent et al. 2004) and was mediated by CD80/CD86 downregulation on APC (Cederbom et al. 2000). However, production of inflammatory cytokines does not seem to be affected by Tregs' CTLA4. Expression of lymphocyte activation gene 3 (LAG3) in human Tregs also contributes to their suppressive function (Wang et al. 2018). LAG3 is a CD4 homolog expressed on FoxP3$^+$ and Tr1 T cells and interacts directly with MHCII complex acting as a decoy receptor to inhibit its interaction with the TCR, downregulate cell proliferation and secretion of inflammatory cytokines IFNγ, IL2, and TNFα by APC and B cells (Hannier et al. 1998). Deficiency in LAG3 renders Tregs inefficient to inhibit T-cell proliferation and unable to prevent chronic intestinal inflammation (Huang et al. 2004). PD1 (CD279) is also expressed on both Treg subsets and binds to PD-L1 (B7-H1), PD-L2 (B7-DC), and B7.1. PD1 signaling inhibits TCR-signaling-induced cell cycle progression and TCR expression on CD8$^+$ T cells (Karwacz et al. 2011), and blockade of the PD1/PD-L1 pathway can abrogate Treg-mediated immunoregulation (Kitazawa et al. 2007).

9.5.2.2 Contact-Independent Inhibition

In humans, suppressive Tregs express high levels of CD25. CD25 is a subunit of the high-affinity heterotrimeric IL2 receptor and is also transiently expressed on activated T cells. CD25 expression is a part of the suppressive mechanism of Tregs, supposedly through the scavenging of IL2 and reduction of its effects on T cells (Barthlott et al. 2005). Though the proximity between cells is necessary for inhibition mediated by CD25, contact is not required. Exogenous IL2 abrogates Treg-mediated suppression of T_{eff} in human and mice (Oberle et al. 2007; Pandiyan et al. 2007). Recent reports link the IL2 scavenging by Tregs to induction of apoptosis on T_{eff} cells (Pandiyan et al. 2007), but further studies are required to prove the existence of such regulatory mechanism in humans.

Expression of CD39 on the surface of FoxP3$^+$ Tregs also provides an important immunoregulatory mechanism. Catalysis of ATP to AMP by CD39 (ecto-nucleoside triphosphate diphosphohydrolase 1, E-NTPDase1), and its subsequent hydrolysis into adenosine by CD73 (ecto-5′-nucleotidase, NT5E) regulates the extracellular levels of ATP and adenosine and their effects on immune cells (Su et al. 2019). Since human Tregs do not express high levels of CD73 (in contrast to their murine counterparts), the AMP conversion to adenosine in humans is thought to result from the expression of CD73 on neighboring cells (Dwyer et al. 2010). ATP binding to purinergic receptors such as P2X7 and P2Y2 expressed on the surface of immune cells induces the secretion of inflammatory cytokines and a DC-mediated priming of intestinal T_{eff} cells toward the Th17 phenotype (Pandolfi et al. 2016; Atarashi et al. 2008). On the other hand, activation of A2A receptors by adenosine induces immune anergy and promotes intestinal homeostasis (Ye and Rajendran 2009). Fifty percent of the human Tregs express CD39 (Borsellino et al. 2007), and this population is highly suppressive through the generation of adenosine (Mandapathil et al. 2010). CD39 polymorphisms in humans are associated with IBD susceptibility (Friedman et al. 2009). ATP/adenosine balance by CD39$^+$ Tregs is of relevance to the intestinal inflammation, where high concentrations of ATP produced by the microbiota reach the LP immune cells as a consequence of intestinal epithelial barrier injury (Atarashi et al. 2008).

Inducible co-stimulator (ICOS) has a dual and balancing role to sustain T-cell activation and effector functions and to promote Treg differentiation and suppression. ICOS expression on human Tregs is potentially involved in contact-dependent and -independent immunosuppression in peripheral tissues. High expression of ICOS on natural FoxP3$^+$ Tregs present in thymus, PBMC, and peripheral tissues endows Tregs with the capacity to produce high levels of IL10 and TGFβ to suppress DC and T-cell function, respectively, whereas ICOS$^-$FoxP3$^+$ Tregs play a suppressive function mainly through the secretion of TGFβ1 (Ito et al. 2008). ICOS contact-dependent CD4$^+$ T suppression is mediated by the upregulation of CTLA4 expression on human

Tregs (Zheng et al. 2013). Some of the findings in the literature support the role of the microenvironment in the differentiation of ICOS$^+$ Tregs. These cells are the main suppressive subset in human melanoma (Strauss et al. 2008). In the murine gut, colonic LP ICOS$^+$ Tregs have a strong immunosuppressive profile (Nakanishi et al. 2018) and are stimulated locally by human microbiota (Atarashi et al. 2013).

9.5.3 Cytolysis

A less studied mechanism of Treg-mediated immune suppression in human is the secretion of cytolytic enzymes and killing of target cells. Human CD25$^+$ Tregs generated in vitro from CD4$^+$ T cells stimulated with antibodies to CD3/CD46 and with IL2, express granzyme B and induce cell death in autologous cells in a perforin/granzyme-dependent manner (Grossman et al. 2004a). Although CD25^{high} Tregs from human peripheral blood activated in vitro did not express granzyme B, they expressed granzyme A and presented a cytolytic activity toward various autologous immune cells in a perforin-dependent but FasL-independent fashion (Grossman et al. 2004b). The target cells tested included activated CD4$^+$ and CD8$^+$ T cells, CD14$^+$ monocytes, and dendritic cells. However, iTreg-mediated suppression of B cells seems to be independent of cytolysis (Xu et al. 2016). Two studies reported the presence of FoxP3$^+$ Tregs expressing granzyme B in the colon of patients with colorectal cancer (Sun et al. 2020) and in the lungs of RSV-infected mice (Loebbermann et al. 2012). Very few to none of these cells could be found in lymphoid organs or circulating Tregs, suggesting that granzyme B expression is acquired locally by Tregs. These findings, summarized in Fig. 9.2, indicate the participation of perforin/granzyme-dependent cytotoxicity in

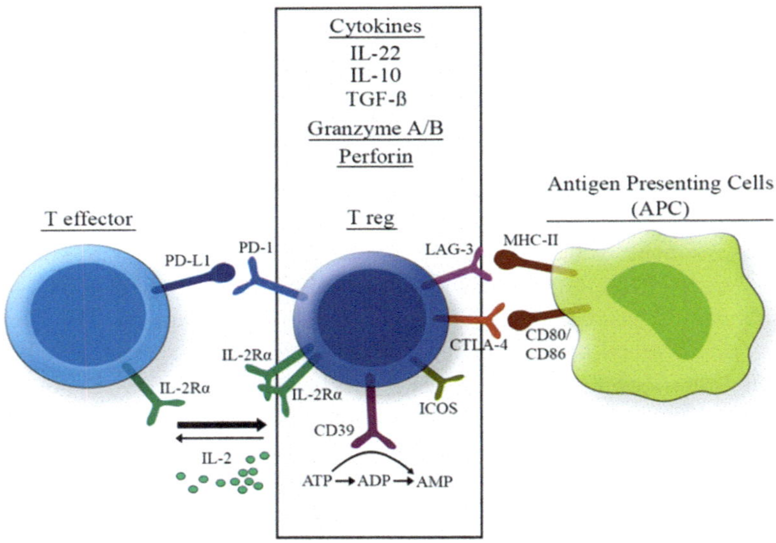

Fig. 9.2 Mechanisms of immunosuppression in human regulatory T cells. Tregs suppress inflammatory cells through secretion of anti-inflammatory cytokines, expression of inhibitory surface molecules and cytolysis of target cells. Secretion of IL22, IL10 and TGFβ, as well as cytolytic Granzyme A/B and Perforin leads to suppression of effector functions of a variety of immune cells. ATP hydrolysis by CD39 also limits the level of such inflammatory signaling molecule in extracellular medium. Expression of LAG3 and CTLA4 constitute negative regulators of APC activation, through MHC competition and inhibition of CD80/86 signaling respectively. High expression of IL2Rα scavenges the extracellular IL2 available for T_{eff} cells, while PD-1 expression inhibits TCR expression and signaling in effector T cells. ICOS induces secretion of TGFβ1 and CTLA4 expression

Treg-mediated suppression in the mucosal surfaces.

9.6 Treg During Intestinal Diseases

Tregs are central regulators of the intestinal immune response and inflammation in the gut. Through the described mechanisms, Tregs suppress immune activation and control the intensity of inflammatory responses in the gut. A breakdown of these tolerogenic mechanisms is usually observed under pathological conditions involving exacerbated inflammation in the intestinal mucosa, e.g., in inflammatory bowel disease, celiac disease, or necrotizing enterocolitis (NEC). On the other hand, the function of Tregs can be detrimental under certain circumstances in which the immune response is necessary to eliminate infectious agents or oncogenically transformed cells. In this topic, we will focus on intestinal diseases where a dysfunction of regulatory T cells has been implicated.

9.6.1 Inflammatory Bowel Disease

IBD is a group of chronic inflammatory diseases resulting from exacerbated immune response in the gut mucosa. There is a consensus that the inappropriate immune response driven against commensal intestinal microbes is the key to IBD etiology and is affected by both the environmental factors and genetic susceptibility of the host. Ulcerative colitis (UC) and Crohn's disease (CD) are the two major types of IBD, distinguished on the basis of histological and endoscopic findings, distribution, and clinical complications (Molodecky et al. 2012). Several lines of evidence support the role of T cells in IBD; however, the adaptive immune profile during CD and UC differs. CD is noted for the involvement of Th1-related cytokines TNFα, IFNγ, and IL12, whereas UC is driven by Th2 and Th9 cytokines such as IL5, IL13, and IL9 (Muzes et al. 2012). Th17-related cytokines, IL17A, and IL23 are increased in intestinal tissue in both UC and CD (Nemeth et al. 2017) and contribute to the disease pathogenesis in these patients.

Multiple human genome-wide association studies (GWAS) have identified a remarkable proportion of single-nucleotide polymorphisms (SNPs) in genes that affect T-cell function in patients with IBD. Among them, polymorphisms in genes linked to the IL10 signaling pathway are associated with severe early onset of IBD (Farh et al. 2015; Glocker et al. 2009). IL10 signaling is essential to maintain homeostasis and mucosal barrier function and to the control of effector and suppressor cells. As described earlier in this chapter, Tr1 and FoxP3$^+$ Tregs are the two major producers of IL10 in the gut. The pivotal role of FoxP3$^+$ Tregs in maintaining intestinal homeostasis is exemplified by the presence of severe intestinal inflammation in patients with FoxP3 mutations, who suffer with global failure in Treg development (Bennett et al. 2001). Interestingly, patients with IBD do not have a quantitative deficiency in frequency of FoxP3$^+$ Tregs in the gut. On the contrary, an increase in this population has been observed in the mucosa and regional lymph nodes of CD and UC patients (Reikvam et al. 2011; Yu et al. 2007). Ex vivo, CD4$^+$CD25$^+$ cells isolated from UC patients showed suppressive activity, and the authors suggested that this activity may either be insufficient or impaired in vivo, in the face of overwhelming mucosal inflammation (Yu et al. 2007).

Helios+ and Helios− FoxP3$^+$ cells in the mucosa of IBD patients were found to express similar levels of immunoinhibitory receptors CD25, CD39, CTLA4, TIGIT, and PD-1 compared to control patients (Lord et al. 2015). Analyses of circulating Tregs in IBD can be controversial, and the frequency of FoxP3$^+$ Tregs and Tr1 cells has been found either increased (Vitale et al. 2019) or decreased (Khalili et al. 2018). In murine adoptive naïve T-cell transfer model of colitis, a reliable and representative model of IBD (te Velde et al. 2007), the administration of a T-cell population containing IL10-producing Tregs, protected from intestinal inflammation (Asseman et al. 1999). CTLA4 was found to be mutated in mucosal immune cells in early onset CD (Zeissig et al. 2015). Impaired dimerization and binding of the mutated CTLA4 to CD80 could explain the loss of Treg regulatory mechanisms in IBD. Chao et al. (2018) showed that lack of CTLA4 in $CD4^+$ T cells in mice impairs differentiation of follicular Tregs and triggers the accumulation of autoantibodies in intestinal epithelial cells. This defect exacerbated the immune response to MCMV (mouse cytomegalovirus), resulting in acute intestinal injury and death.

Another mechanism driving impaired immunosuppression in IBD may be related to $FoxP3^+$ Treg survival in the inflamed intestine as a consequence of the hyperinflammatory environment. It has been reported that Tregs from the inflamed tissue of IBD patients show evidence of enhanced apoptosis, which can be reversed by anti-TFNα treatment (Veltkamp et al. 2011). In line with the evidence of altered Treg survival during IBD, polymorphisms in genes encoding the IL2RA and IL2RB subunits, indispensable for Treg survival in peripheral tissues, were identified in patients with CD and UC (Bouzid et al. 2013). ST2/IL33 axis has also been shown to enhance Treg stability and/or survival. Under homeostasis, IL33 binds to suppressor of tumorigenicity 2 (ST2) highly expressed on colonic Tregs, stimulates FoxP3 and GATA3 expression, and promotes Treg function through enhancing TGFβ1-mediated differentiation, also an inducer of survival pathway in T cells (Schiering et al. 2014). IL33 signaling in DCs may indirectly contribute to this process, by stimulating DC production of IL2 (Matta et al. 2014). Increased levels of sST2, an extracellular ST2 decoy receptor for IL33, in the mucosa and serum of IBD patients is positively associated with the disease severity, suggesting that this axis maybe be disrupted during chronic inflammation in patients with IBD (Boga et al. 2016). In sepsis-surviving patient, IL33 contributes to long-term immunosuppression by expanding the Tregs (Nascimento et al. 2017). However, in UC patients, increased IL33 mRNA expression was reversed in remission induced with infliximab, a TNFα inhibitor, which could be interpreted that IL33 may have a dual role in IBD patients (Gundersen et al. 2016; Chen et al. 2020; Pastorelli et al. 2013).

In addition to a reduced survival observed in Tregs from patients with IBD, alterations in Treg function and phenotype have been reported. Zhou et al. showed that $FoxP3^+$ Tregs have reduced FoxP3 expression and acquire inflammatory phenotype when exposed to an inflammatory milieu (Zhou et al. 2009). Loss of FoxP3 interaction with the epigenetic enzyme Enhancer of Zeste Homolog 2 (EZH2) observed in mucosal Tregs of IBD patients may result from Treg exposure to IL6 (Bamidele et al. 2019) and may lead to an impaired repression of inflammatory genes, normally targeted by FoxP3/EZH2 complex in $CD4^+$ T cells (Sarmento et al. 2017). Sanctuary et al. (2019) described a mechanism for TNFα-induced loss of Treg immunosuppressive function. According to the authors, TNFα induces miR-106a microRNA in both humans and mice, which through NFκB promoter binding suppresses post-transcriptional regulation of IL10 release by Tregs.

In homeostatic conditions, intestinal $FoxP3^+$ Tregs express RORγT. While this population is highly immunosuppressive, it may also display considerable plasticity and gain the ability to produce IL17A in inflammation associated with CD and UC. IL17A producing Tregs from IBD patients showed defective suppressive function, and differences in cytokine requirements for that transition were observed in patient peripheral blood cells from CD and UC (Ueno et al. 2013). In CD patients, IL1β/TGFβ/IL23 are more effective in inducing $IL17A^+$ $FoxP3^+$ Tregs, while cells from UC patients required IL21 and TGFβ (Ueno et al. 2013). Patients with IBD also demonstrated increased Treg/Th1 crossover populations in $T\text{-}bet^+FoxP3^+$ Treg (Li et al. 2017), but the pathways involved in acquisition of T-bet in human intestinal $FoxP3^+$ Tregs and their role during IBD have not been investigated. To complicate matters, another study showed that despite producing IL17, $IL17^+CD161^+$ Tregs enriched in the CD mucosa remained highly suppressive and were enriched for wound healing genes (e.g., PDGFA and CFS2), including soluble mediators (e.g., IL17A and IL22) known for accelerating epithelial barrier healing (Povoleri et al. 2018).

The mucosal inflammatory milieu of IBD may also impair the responsiveness of the effector cells to Treg-mediated suppression. Inflammatory cytokines and inducers of NFκb in T lymphocytes, such as TNFα, have anti-apoptotic effects in these cells (Dudley et al. 1999) and could lead to pathogenic T-cell resistance to mechanisms of cell death induced by Tregs during IBD. Such resistance could promote the persistence and perpetuation of pathogenic $CD4^+$ T cells in the gut tissue. Mechanisms of resistance to TGFβ have also been identified in the LP effector cells during IBD. High levels of Smad7 produced by mucosal T lymphocytes in patients with IBD, likely as a consequence of increased TNFα, IL1β, and IFNγ expression (Ulloa et al. 1999), turn pathogenic T_{eff} cells insensitive to regulation by TGFβ (Laudisi et al. 2016), despite the increased levels of TGFβ in IBD tissues (Babyatsky et al. 1996).

Most of the knowledge on Treg dysfunction in IBD comes from studies on $FoxP3^+$ Treg subset, and the characteristics and function of Tr1 in patients with IBD have yet to be further investigated. A recent report has identified $IFN\gamma^+$ IL10-producing Tr1 as $CCR5^+PD\text{-}1^+$ cells in the human intestine but did not indicate altered frequencies of these population in CD or UC. However, Tr1 showed reduced IL10 expression when isolated from the IBD mucosa, a decrease that could be reproduced by treatment with of IL-23 and IL1β (Alfen et al. 2018).

9.6.2 Celiac Disease

Celiac disease is a T-cell-mediated disease triggered by ingestion of gliadin, the main component of the gluten, in genetically predisposed individuals. Gliadin-derived peptides activate T helper (Th) cells to produce Th1 and Th17 signature cytokines IFNγ (Nilsen et al. 1998) and IL17A (Fernandez et al. 2011) that contribute to the expansion and maintenance of the inflammatory response in the gut of individuals with celiac disease. A break of tolerance to dietary gliadin, and the known role of Tregs in immune tolerance to dietary components, prompted several groups to investigate the Treg population patients with celiac disease. The findings offered some parallels and similarities with IBD. Like in IBD, the numbers of $FoxP3^+$ Tregs are increased in the intestinal mucosa of celiac patients with active disease (Cianci et al. 2012), and its local expansion after gliadin stimulation (Zanzi et al. 2011) suggested that it is an insufficient attempt at reigning in the T_{eff} cells during this disease. Celiac intestinal mucosa also harbors Tr1 cells, able to produce IL10 and TGFβ and to suppress proliferation of pathogenic T cells in vitro (Gianfrani et al. 2006). In fact, upregulation of these two immunosuppressive cytokines by Tregs was observed along with the exacerbated proinflammatory immune response found in celiac disease (Lahat et al. 1999). In vitro studies support the regulatory role of IL10 in the mucosal T cells from celiac patients by inhibiting IFNγ production by gliadin-specific T_{eff} cells (Salvati et al. 2005).

The negative regulation of the immune response exerted by Tregs is not enough to control the ongoing immune response in celiac patients. Circulating $FoxP3^+$ Tregs from children with celiac disease show high expression of CD101 and CD129 (Akesson et al. 2015) and 80% of the circulating $FoxP3^+$ Tregs from celiac patients express high levels of CD39 (Cook et al. 2017), all indicators of their suppressor activity. $FoxP3^+$ Tregs from the intestine of celiac patients showed no impaired function in vitro (Zanzi et al. 2011), but Cook et al. showed that circulating gluten-specific $FoxP3^+CD39^+$ Tregs from celiac disease patients can have reduced suppressive function in response to a polyclonal stimulus (Cook et al. 2017). These results suggest that the intestinal and the inflammatory environment established during the course of the disease may locally impair crucial regulatory mechanisms afforded by Tregs. The overexpression of intestinal IL15 in celiac disease could contribute to this phenomenon. IL15 impairs TGFβ signaling in biopsies from active celiac disease (Benahmed et al. 2007), dampens Treg suppressive activity (Zanzi et al. 2011), and makes T effector cells refractory to the regulatory effects of Tregs through the activation of phosphatidylinositol

3-kinase (PI3K) pathway (Ben Ahmed et al. 2009). Additionally, microbiome-derived metabolites have shown to modulate alternative splicing and the expression of FoxP3 isoforms in intestinal biopsies from patients with active celiac disease, where increased expression of FoxP3 shorter D2 isoform over full length has been observed (Serena et al. 2017). The alternatively spliced isoform FoxP3 D2, thought to be induced by butyrate produced by bacteria expanding during the active celiac disease in presence of IFNγ, is less effective in downregulating the Th17 differentiation. In this scenario, the switch to FoxP3 isoform D2 in patients with celiac disease could promote local Th17 immune response (Serena et al. 2017). Therefore, the microenvironment in the intestinal tissue of patients with celiac disease might orchestrate the differentiation and/or functionality of Tregs, thus restricting their activity.

9.6.3 Necrotizing Enterocolitis

Necrotizing enterocolitis (NEC) is an immunopathology that most commonly affects infants born prematurely. The disease is characterized by dysregulated inflammatory host responses to luminal bacteria with increased levels of inflammatory mediators, such as TNFα and platelet-activating factor (PAF) in the small intestinal tissue and serum (Caplan et al. 1990). The inflammatory injury in NEC causes a variable spectrum of damage to the intestinal tract, from mucosal injury to full-thickness necrosis. Multiple factors are involved in the precipitation of NEC, including formula feeding, ischemic/hypoxic injury, and abnormal bacterial colonization.

The population of FoxP3$^+$ Tregs develop early during gestation. They are detected as early as 23 weeks of gestational age in large and small bowel. Treg numbers are stable during the later phases of development and do not change with increasing intrauterine development or postnatal age (Weitkamp et al. 2009). The immaturity of the immune system in preterm babies is thought to be a critical pathogenic factor in NEC. Several studies have found important correlations between preterm birth and growth restriction with Treg function in the intestinal tissue. The FoxP3$^+$ Treg frequency in the circulating blood of preterm infants inversely correlated with gestational age at birth, indicating a contraction in this population after birth in preterm children (Qazi et al. 2020). Treg suppressive index, which reflects the expression level of FoxP3, is also decreased in preterm infants (Mukhopadhyay et al. 2014). Their homing markers may also be affected in preterm children (Qazi et al. 2020) and children with NEC (Weitkamp et al. 2013). In infants with growth restrictions, Mukhopadhyay et al. found a markedly reduced proportion of circulating Tregs in the cord blood (Mukhopadhyay et al. 2014). In this vulnerable group of infants, circulating FoxP3$^+$ Tregs also showed a reduced expression of the gut homing integrin α4β7, which could also be an indicator of reduced capacity of Tregs to reach the intestinal mucosa (Qazi et al. 2020). Therefore, lower frequency, impaired homing, and/or impaired function of Tregs observed in preterm children may be implicated, or at least contribute to the pathogenesis of NEC.

In NEC, the inflammatory microenvironment may contribute to the impairment of Treg differentiation and expansion, which could perpetuate the inflammation. Flow cytometry analysis has shown lower frequency of FoxP3$^+$ Treg cells in ileum (Weitkamp et al. 2013) and peripheral blood mononuclear cells of infants with NEC (Pang et al. 2018a). The frequency of intestinal FoxP3$^+$ Tregs was recovered in tissue from healed postoperative NEC (Weitkamp et al. 2013), suggesting that Tregs from children with NEC have no intrinsic failure in expansion and accumulation in intestinal tissue. Studies of Tregs in infants with NEC need to be interpreted with caution due the diversity of markers used to identify these cells in different investigations. Detecting FoxP3 expression in Tregs is a useful tool to characterize the frequency of Tregs in the inflamed tissue, but it requires cell fixation and does not permit downstream functional analysis. Because of that, isolation of human Tregs for in vitro assays is usually based on high expression of IL2Rα (CD25). Human CD4$^+$CD25$^{+/hi}$ Tregs from NEC infants showed reduced expression of

CTLA4, LAG3, and Helios, reduced capability to suppress CD4 T_{eff} proliferation in vitro, and remarkably failed to suppress IL17A expression in these cells (Pang et al. 2018a). An imbalance caused by an increased ratio of Th17/Treg cells in the LP was shown to contribute to excessive inflammatory response in NEC (Egan et al. 2016), and the restoration of this balance after Treg induction by retinoic acid administration decreased NEC severity (Nino et al. 2017). IL6, which inhibits TGFβ-induced Treg differentiation and favors Th17 lineage, was found to be increased in the small intestine affected by NEC and could participate in the contraction of Tregs and in perpetuating the disease (Weitkamp et al. 2013). Indeed, IL6 neutralization restored FoxP3$^+$ Tregs and ameliorated experimental NEC in mice (Ma et al. 2019). Monocytes in NEC patients were suggested to play a role in Th17/Treg imbalance by promoting the differentiation of RORγt-expressing Th17 cells via production of IL6 (Pang et al. 2018b). The plasticity of Treg phenotype in hyperinflammatory environment, as we discussed earlier in the chapter, could also contribute to Treg dysfunction in NEC patients. IL17$^+$ Tregs were increased in peripheral blood of NEC patients, and this population could be promoted by in vitro Treg exposure to IL6 (Ma et al. 2019).

9.6.4 Graft-Versus-Host Disease (GVHD)

GVHD is a syndrome commonly associated with allogeneic hematopoietic stem cell transplantation (HCT), caused by the recognition of the host as foreign by the immune cells that remain in the donated tissue (graft). The naïve donor T cells migrate to target organs such as the colon, small bowel, liver, and skin where they cause tissue damage (Wysocki et al. 2005). These tissues are potentially attractive sites for the donor T cells due the release of danger-associated molecular patterns (DAMPs), the presence and epithelial translocation of pathogen-associated molecular patterns (PAMPs), and inflammatory cytokines induced during irradiation and/or chemotherapy. Host's activated APC activate donor T cells, leading to their proliferation and differentiation toward Th1 and Th17 cells, which cause inflammation and damage at mucosal tissues, including the gut. In human allogeneic transplant, localization of Foxp3$^+$ Tregs in the gut has shown negative correlation with the severity of intestinal inflammation in GVHD (Rieger et al. 2006). Increased frequencies of Tregs with gut-homing phenotype ($\alpha 4\beta 7^+$) is associated with reduced gut GVHD in patients subjected to neutrophil engraftment (Engelhardt et al. 2011, 2012), indicating that Tregs inhibit immune responses at mucosal sites during GVHD. However, the involvement of Tregs in homeostasis during this syndrome may differ at different sites of the gastrointestinal tract, since no correlation between Treg frequency and the histological or clinical severity of gastrointestinal GVHD was observed in gastric antral biopsies in allogeneic hematopoietic stem cell transplantation (allo-HCT)recipients (Lord et al. 2011). Both tTreg and pTreg cells are important to mediate immunosuppression in the intestine during GVHD, and their transfer prior or along with the donor T cells has shown to diminish the incidence of GVHD and improved survival (Edinger et al. 2003). The literature about the use of Treg induction or infusion as means of GVHD prevention will be discussed in the next section of this chapter.

9.6.5 Tregs in Colorectal Cancer

Cancers are not just masses of transformed cells, but are complex aggregates with non-transformed cells, cytokines, chemokines, growth factors, and inflammatory and matrix remodeling enzymes. Colorectal cancer (CRC) remains one of the leading causes of death worldwide. Besides genetic predisposition and environmental factors, IBD-associated inflammation has been linked to CRC in an "inflammation-dysplasia-carcinoma sequence" (Zisman and Rubin 2008). The tumor microenvironment is thus a result of complex host–tumor interactions on multiple molecular and cellular levels. Immune cells contribute to this microenvironment with the primary goal to efficiently recognize and destroy neoplastic cells in a process known as tumor-immunosurveillance (Dunn et al. 2004). Tumor infiltrating lymphocytes (TILs) like $CD8^+$ cytotoxic T lymphocytes (CTLs) are the major key players in tumor-immunosurveillance. Infiltration of $CD8^+$ cells in solid tumors including colorectal carcinoma has been correlated with favorable prognosis (Ohtani 2007). In the inflamed colon, Tregs accumulate locally in an effort to limit the inflammation and mitigate the related tissue damage (Holmen et al. 2006). Furthermore, it has been demonstrated that CRC could secrete CCL5 which recruits Tregs to tumors through CCL5/CCR5 signaling (Chang et al. 2012). In fact, $CCR5^{hi}Foxp^{3+}$ Tregs are more suppressive than $CCR5^{lo}FoxP3^+$ Tregs in human CRC (Ward et al. 2015). Tumors can also secrete angiogenic factors such as vascular endothelial growth factor (VEGF) that help recruit Tregs to tumors. Blocking VEGF pathways inhibited the recruitment of $FoxP3^+$ Tregs in a mouse model of CRC and impeded the proliferation of human Tregs in vitro (Terme et al. 2013). The high frequency of Tregs might be important for controlling cancer-driving inflammation, but at the same time, they may promote tumor progression by obstructing specific immune responses and contributing to escape of tumor cells from the immune surveillance.

As of now, the state of the field seems to be divided regarding the role of Tregs in the outcome of CRC. Most of the TILs in CRC patients are $CD3^+$ and using $CD3^+/FoxP3^+$ T-cell ratio as an appropriate prognostic marker in CRC, a low intraepithelial $CD3^+/FOXP3^+$ T-cell ratio has been associated with shorter patient survival (Sinicrope et al. 2009). Another study used intra-tumoral $CD8^+$ T-cell/$FoxP3^+$ cell ratio as a predictive marker for overall survival and found $FoxP3^+$ Tregs to be associated with lymph node metastases (Suzuki et al. 2010). Similarly, a preliminary study showed that peripheral blood derived $CD4^+CD25^+$ T-cell lines derived from CRC patient and inhibited the proliferation of autologous HLA-A1 restricted CTLs via TGFβ, thus allowing for tumor cells to escape the immunosurveillance (Somasundaram et al. 2002). Moreover, patients with less abundant Foxp3 Tregs displayed improved T-cell responses (Galon et al. 2006).

There have been reports of antigen-specific $CD3^+$ T-cell responses in CRC, but spontaneous tumor clearance is a very rare occurrence (Campi et al. 2003; Nagorsen et al. 2005; Sinicrope et al. 2009). Apparently, Treg suppression in CRC is tumor-associated antigen (TAA) specific and not systemic (Betts et al. 2012; Clarke et al. 2006). Another detailed analysis using various CRC-specific TAAs such as carcinoembryonic antigen (CEA) peptides, telomerase, Her-2/neu, and MUC-1 showed that they can lead to Treg activation (Bonertz et al. 2009), which may explain the ineffective immune response against the tumor. Moreover, depleting Tregs in PBMCs from CRC patients markedly enhanced the IFNγ and TNFα production in T cells, which were stimulated with a CEA peptide (Yaqub et al. 2008). TAA-specific Tr1 cells have also been identified using a p53 peptide (Bueter et al. 2006). Highly suppressive TIM3-expressing $FoxP3^+Helios^+$ Tregs have also been demonstrated at different stages of CRC (Toor et al. 2019). Tumor-infiltrating highly suppressive Tregs that express interleukin-1 receptor 2 (IL1R2), PD-1 Ligand 1, PD-1 Ligand 2, TNFRSF9, CCR1and CCR8 have also been reported (De Simone et al. 2016; Fujimoto et al. 2018). Interestingly, high expression in Treg cell signature genes within whole tumor transcriptome, such as LAYN (a transmembrane protein with homology to c-type lectin), and of MAGEH1 (a member of the melanoma antigen gene family) or CCR8, correlated with poor prognosis (De Simone et al. 2016). Another study with CRC patients showed increased levels of Tr1 markers LAG3 and CD49b, as well as IL17 which were inversely associated with patient survival (Chen and Chen 2014). TAA-specific Tregs suppress APC and T-cell function via IL-10 and NK cell function by producing TGFβ, thus incapacitating both adaptive and innate immunity against cancer (Zou 2006). Additionally, tumor-associated Tregs have been shown to express very high levels of CD39 and inhibit the in vitro T-cell transendothelial migration via adenosine synthesis (Sundstrom et al. 2016).

The field is not universally in agreement about the role of Tregs in CRC progression, as the picture becomes more nuanced. Some studies showed favorable outcomes associated with high infiltration with FoxP3$^+$ Tregs at different stages of CRC (Frey et al. 2010; Lee et al. 2018; Salama et al. 2009). It is also worth noting that effector Tregs (eTregs) are very frequently found in CRC patients. In an interesting study, Saito and colleagues revealed the heterogeneity of FoxP3 expression in tumor infiltrating Tregs (Saito et al. 2016). The authors showed that CRC tumors with favorable outcomes were actually more infiltrated with Foxp3^{lo}CD45RA$^+$ non-Tregs and upregulated inflammatory genes like *Il12a*, *Il12b*, *Tgfb1*, and *Tnf*. On the other hand, CRC with higher infiltration of FOXP3^{hi}CD45RA$^-$ eTregs resulted in poor prognosis and lower disease-free survival (Saito et al. 2016). Complementing this study, Lin et al. demonstrated that eTregs, but not Foxp3^{lo}CD45RA$^+$ non-Tregs, were associated with late-stage CRC (Lin et al. 2013). Another study showed that FoxP3$^+$ eTregs that express Blimp1, a critical regulator of CD4 T cell exhaustion, have been identified in human CRC and were associated with long-term disease-free survival (Ward-Hartstonge et al. 2017). FoxP3$^+$ Tregs might also be involved in regulating the immune response directed against high-level microsatellite instability (MSI) CRC at the primary tumor site. Significantly elevated FoxP3$^+$ Tregs were associated with CRC patients with known MSI status (Michel et al. 2008).

The discrepancies in the role of Tregs may partly be explained by Treg plasticity. Inflammation in the gut and tumor microenvironment drive certain Tregs to display plasticity and an acquisition of a pro-inflammatory phenotype. RORγt$^+$FOXP3$^+$ Tregs with pathogenic properties have been identified in CRC patients and were enriched in patients with late-stage cancer (Blatner et al. 2012). Interestingly, RORγt$^+$FOXP3$^+$ Tregs in CRC patients produced both IL10 and IL17 (Blatner et al. 2012), indicating that the balance between protective and pathogenic Tregs could potentially influence the disease outcome. Murine Tregs have also been shown to lose FoxP3 expression and gain the ability to produce IL17, in response to an inflammatory bacterial insult in the gut (Yurchenko et al. 2012). Populations of IL17-producing Tregs have been identified in mouse polyps and human CRC (Blatner et al. 2010). Interestingly, this Treg reprogramming does not represent a classical Th17 conversion as these IL17-producing Tregs did not lose their suppressive function (Blatner et al. 2010). It has been suggested that these Tregs might be transitional and have reserved the ability to differentiate into FoxP3$^+$ Tregs or TH17 cells depending on the local milieu of tumor microenvironment (Du et al. 2014). In a mouse model of CRC, TGFβ and prostaglandin E2 mediated the differentiation of Th17 cells into suppressive IL17$^+$ and IL17$^-$FoxP3$^+$ Tregs (Downs-Canner et al. 2017). Indeed, IL17-producing FoxP3$^+$ Tregs have been identified in human CRC (Girardin et al. 2013; Kryczek et al. 2011) with the ability to suppress the proliferation of CD8$^+$ TILs stimulated with CRC antigens (Ma and Dong 2011).

Besides CD4$^+$Foxp3$^+$ Tregs, there have been some unconventional Tregs that have been identified in the context of CRC. Populations of FoxP3$^+$ and FoxP3$^-$ T cells that express latency-associated protein (LAP) have been identified in CRC (Mahalingam et al. 2014; Scurr et al. 2014; Zhong et al. 2017). The tumor-infiltrating LAP$^+$FOXP3$^-$ cells had much greater suppressive capacity than LAP$^-$FoxP3$^+$ Tregs isolated from the peripheral blood (Scurr et al. 2014) and were associated with poor prognosis. CD8$^+$FoxP3$^+$ Tregs have also been identified in CRC tissues (Taylor et al. 2016). Another study demonstrated tumor infiltrating CD8$^+$CD25$^+$FoxP3$^+$ cells to be able to express CTLA4 and GITR and suppress CD4$^+$ T-cell proliferation and IFNγ production ex vivo (Chaput et al. 2009).

9.7 Treg-Related Therapeutic Approaches

Based on the observed impairment of Treg immunosuppressive abilities in patients with inflammatory disorders of the gut, therapeutic approaches aiming to correct such deficiencies may have the potential to protect the gut tissue from excessive inflammation and successfully restore intestinal homeostasis. Studies on Treg immunotherapy with infusion of autologous Treg or mesenchymal stem cells (MSCs), or enhancement of Treg numbers and function via administration of antibodies, cytokines, or dietary interventions, have shown preliminary efficacy. In this section, we discuss the available evidence for Treg-targeted therapies to modulate intestinal disorders.

9.7.1 Treg Infusion

Purity, homing ability, antigen-specificity, and survival of Treg cells all need to be carefully considered during the design of therapies involving Treg infusion. Varying in vitro protocols are used to obtain sufficient numbers of Tregs to be used in immunotherapy. The expansion of polyclonal FoxP3$^+$ Treg isolated from peripheral blood and the differentiation of Tr1 from human CD4$^+$ T in the presence of IL10 and allogenic monocytes are the most common methods employed to obtain Tregs. The use of specific markers to properly identify and isolate Treg is critical to prevent contamination with other cells. Also, to appropriately reach the desired site of action, infused Tregs should express homing receptor(s) that guide their migration to the gut where the cells can act directly at the inflamed site or to the draining lymph nodes to modulate T-cell priming. Engineering or selecting the right antigen specificity of the infused Tregs can influence the expansion of these cells in the gut, where TCR engagement by microbial and/or dietary antigens would promote Treg expansion, retention, and function after transfer.

T cells expanded from the population of blood CD25$^+$CD127^{low}CD45RA$^+$ (natural) Tregs may represent a promising T-cell transfer therapy approach in CD, as they do not convert to Th17, home to the small intestine and suppress the activation of immune cells in the intestinal tissue of CD patients (Canavan et al. 2016). Phase I and II clinical trials to test the safety and efficacy of Treg immunotherapy using cells obtained with this protocol are ongoing (NCT03185000). The upregulation of α4β7 and CCR9, gut homing molecules, and the induction of a transcriptional profile consistent with gut-homing can be enhanced by addition of atRA, usually produced by CD103$^+$ cells in the gut, during in vitro expansion of human pediatric thymic and adult blood Tregs (Hoeppli et al. 2019). Additionally, the expansion of Tregs from patients with CD in the presence of RAR568, specific agonist for RA receptor A was even more efficient than atRA in inducing α4β7 (Goldberg et al. 2019). New approaches aim to generate antigen-specific Tregs, which would decrease the risk of a non-specific inhibition, allow the use of lower numbers of Tregs and increase the chance of Treg expansion in the gut. As proof of principle, in a phase I/IIa clinical study in 20 patients with refractory CD, a reduction of 40% in Crohn's Disease Activity Index (CDAI) was observed after treatment with single-injection of Tr1 ovalbumin-specific Tregs in response to ovalbumin ingestion (Desreumaux et al. 2012).

In patients with GVHD, tTreg infusion was shown to be well tolerated with no toxicities reported and improved clinical outcomes if injected alone or in conjunction with other anti-GVHD therapies (Duggleby et al. 2018). More recently, a study in mice showed that treatment with selective inhibitor of bromodomain and extra-terminal (BET) proteins, which reduce pro-inflammatory cytokine production, can be successfully combined with Treg expansion therapy for the treatment of GVHD (Copsel et al. 2018).

The field of NEC is perhaps less advanced both in the available knowledge of Treg function and their potential use as preventative therapy. One study has reported that administration of Tregs in a mouse model of NEC has attenuated the severity of NEC by limiting the immune response, also indicating the potential of Tregs in controlling human NEC (Dingle et al. 2013). Additionally, in very low birth weight (VLBW) neonates fed with lactoferrin (LF), the major protein in mammalian milk, prevention of NEC has correlated with increased levels of circulating FoxP3$^+$ Tregs (Akin et al. 2014).

9.7.2 Promoting the Number and Function of Endogenous Tregs

Indirect interventions to affect the gut Treg population represent viable approaches aimed to maximize Treg immunosuppression in vivo. While majority of the available data comes from preclinical models, some therapies have found their way into ongoing clinical trials and are discussed in this section.

9.7.2.1 Targeting Cytokines and Inflammatory Pathways

Treatments to improve Treg population based on the administration of cytokines known to promote their survival and expansion or inhibition of pro-inflammatory cytokines known to promote Treg dysfunction/cell death have been explored in several studies.

Administration of low doses of IL2 preferentially activates Treg cells due to their abundant expression of CD25, avoiding its unspecific action on other immune cells expressing lower levels of CD25. Low dose of IL2 promotes proliferation, thymic export, and resistance to apoptosis among FoxP3$^+$ Tregs (Matsuoka et al. 2013). In a phase I study, patients with active chronic GVHD (cGVHD) refractory to glucocorticoid treatment treated daily with low-dose IL2, augmented Tregs and demonstrated clinical improvement (Koreth et al. 2011). In a phase II trial, 61% of the patients had a clinical response at multiple cGVHD sites, including gastrointestinal tract. During 2 years of extended IL2 therapy, both the clinical response and Treg cell immune response persisted (Koreth et al. 2016). Currently, a clinical trial phase II is recruiting patients to study the induction of Tregs by low-dose IL2 in CD and UC (NCT01988506). Other soluble factor beside IL2 can be used in therapies aimed at Treg expansion. In a mouse model of cGVHD, granulocyte-macrophage colony-stimulating factor (GM-CSF) therapy increased the number CD4$^+$CD8$^-$ DCs, expanded Tregs, and protected against the development of skin GVHD (Hotta et al. 2019). The potential of GM-CSF therapy for the recovery of intestinal homeostasis is yet to be addressed in future studies.

AMG714, a neutralizing anti-IL15 monoclonal antibody, in a clinical phase IIa study in patients with refractory coeliac disease was safe, and improved the clinical symptoms even though it did not prevent mucosal injury after gluten challenge, particularly diarrhea in patients who underwent gluten challenge (Lahdeaho et al. 2019; Cellier et al. 2019). Although not directly studied in clinical trials, IL15 blockade would be expected to restore TGFβ signaling in the LP immune cells and promote the local Treg population.

Restoring TGFβ sensitivity has also been an active area of research and pharmacologic developments. The oral administration of SMAD7 antisense oligonucleotide, GED-0301 (Morgensen; Celgene), was shown safe and effective in double-blind, placebo-controlled trials (Monteleone et al. 2015, 2016; Feagan et al. 2018), although a more recent clinical trial with CD patients failed to show a benefit over placebo (Sands et al. 2019).

Several studies in mice had shown that therapy that promotes IL33 signaling in FoxP3$^+$ Tregs with either rIL33 or sST2-blocking antibody showed the potential to be effective in models of exacerbated intestinal inflammation. The response to such treatments may be contextual, as the intestinal inflammation could be ameliorated if rIL33 was administered during chronic phases of intestinal inflammation (Grobeta et al. 2012), but was exacerbated in acute settings (Oboki et al. 2010). The latter may be attributed to the role of IL33 as alarmin, instigating a pro-inflammatory response in the gut. On the other hand, in chronic stages of the intestinal inflammation, rIL33 could promote tissue homeostasis and healing. In a model of CD, the protection promoted by rIL-33 was mediated by the induction of $FoxP3^+$ Treg expansion and $CD103^+$ DCs in the gut (Duan et al. 2012). The same way, $ST2^+$ Treg expansion during rIL33 pre-conditioning therapy or blocking sST2 during peri-transplant period, decreased GVHD severity and mortality (Matta et al. 2016) (Zhang et al. 2015). One study in patients with celiac disease showed even higher levels of IL33 and sST2 in serum and mucosal samples of celiac patients than in patients with Crohn's disease, which suggests that the blocking sST2 therapy approach could also have a favorable outcome in celiac disease patients (Lopez-Casado et al. 2017).

Inhibition of inflammatory pathways may also improve Treg survival in the gut. For example, the anti-TNFα therapy has shown to reverse $FoxP3^+$ Treg apoptosis in IBD patients (Li et al. 2015). Blocking antibody against OX40L-OX40, promoted Treg responses, such as increased IL10-producing $CD4^+$ T cells and the frequency of $CD25^+Foxp3^+CD127^{lo}CTLA4^+CCR4^+$ Tregs, reduced T-cell infiltration in various sites of inflammation including the gut, and reduced the severity of the GVHD in immunodeficient mice transplanted with human peripheral blood mononuclear cells (Tripathi et al. 2019).

9.7.2.2 Mesenchymal Stem Cell (MSC) Infusion

Mesenchymal Stem Cells (MSC) are multipotent non-hematopoietic cells present in different tissues, which due to their immunomodulatory characteristics are considered as new therapeutic agents in the cell-based therapy of IBD, NEC, and GVHD. Different tissues were used as the source of MSCs in research studies, including bone marrow, amniotic fluid, umbilical cord, and placenta, where these cells are typically identified based on the expression of markers CD73, CD90, and CD105, and lack of CD34, CD45, CD11b, CD14, CD19, or CD79a (Drucker et al. 2018; Chen et al. 2013). Recently published studies emphasized the crucial importance of MSC for the attenuation of colonic inflammation mainly through the promotion of M2 macrophage lineage (Giri et al. 2020), induction of intestinal epithelial repair (Gong et al. 2016), and the expansion of colonic Treg population (Vicente-Suarez et al. 2015). The effects of MSCs on macrophage-mediated mucosal repair involves polarization of $IL10^+$ tissue macrophages via CCL2 and CXCL12 (Giri et al. 2020). In Treg generation, MSCs can either act directly by delivering sphingosine 1-phosphate (S1P) to $CD4^+$ T cells and promoting FoxP3-expressing and IL10-producing Tregs (Li et al. 2019). Indirectly, MSCs can promote a regulatory phenotype in DCs, associated with high production of IDO1 (Zhang et al. 2018), suppression of expansion of inflammatory Th1 and Th17 cells, and favoring of conversion to $CD4^+CD25^+FoxP3^+$ Tregs (Harrell et al. 2018; Zhang et al. 2017).

Cell-free MSC-based therapy have also been postulated as potentially safer. Conditioned medium containing regulatory soluble mediators and exosomes secreted by MSCs mimicked the ability of the MSCs to increase total number of IL10- and TGFβ-producing Tregs in injured tissues and peripheral lymph organs of mice with autoimmune and chronic inflammatory diseases, including IBD (Heidari et al. 2018).

Thus far, MSC infusion in humans has been reported to be safe over long-term use. The efficacy and diversity of the MSC-based therapy protocols being tested in human intestinal disorders have been extensively revised elsewhere (Gregoire et al. 2017; Moheb-Alian et al. 2016; Gao et al. 2016). The improvement in the intestinal disease outcome after MSC therapy is related to improvement in mucosal frequency of $FoxP3^+$ Tregs. Allogeneic bone marrow-derived mesenchymal stromal cell (bmMSC) therapy in perianal Crohn's disease fistulas resulted in closure at week 24 and remained closed after 4 years (Barnhoorn et al. 2020). Ciccocioppo et al. showed an increased percentage of circulating and mucosal Tregs in CD patients after MSC therapy (Ciccocioppo et al. 2011). The use of MSC as adjuvant therapy combined with conventional pharmacological approaches has also been considered. The combination of MSC with anti-cytokine therapy to treat CD patients with perianal lesions showed significant improvement in comparison to the control group receiving cell therapy only (Knyazev et al. 2018). Neutralizing TNFα and IFNγ during MSC-based therapy may increase their efficacy by removing the ability of these two cytokines to impair the pro-repair effects of MSCs (Hu et al. 2019) and increase Treg generation/viability in the inflamed intestine (Visperas et al. 2014) (Veltkamp et al. 2011).

Several trials in safety and efficacy of MSC therapy are ongoing in both IBD and GHVD (clinicaltrial.gov). The results of this novel therapeutic approach and in-depth information of how its findings relate to Treg improvement in the intestinal lamina propria are awaited by the scientific community.

9.7.2.3 Induction of Tregs by Microbiota

In humans and mice, deviation in gut microbial composition from homeostatic condition, referred to as dysbiosis, is associated with intestinal immune disorders, including NEC (Denning and Prince 2018), celiac disease (Girbovan et al. 2017), and IBD (Ni et al. 2017). In patients with GVHD, the conditioning regimen and medications given to patients undergoing allo-HCT reduce the microbial diversity and disturb the homeostatic crosstalk between the microbiome and the host immune system (Hidaka et al. 2018). However, the direct causal relationship between dysbiosis and human intestinal disorders is not straightforward. While it may be possible that dysbiosis arises as a consequence of the inflammatory milieu in the gut, it is also well established that transfer of dysbiosis community can increase susceptibility to and severity of experimental colitis (Harrison et al. 2018; Gerassy-Vainberg et al. 2018; Garrett et al. 2007). Despite that, research into fecal microbiome transplant (FMT) or next-generation probiotics and preclinical studies with mouse models of intestinal inflammation show promise.

In humans, FMT from healthy individuals to subjects with active CD have successfully increased intestinal microbial diversity and mucosal $CD4^{+}CD25^{+}CD127^{low}$ Tregs (Vaughn et al. 2016). However, clinical studies of FMT in active IBD, especially CD, brought mixed results, with notable cases of disease exacerbation attributed to the introduction of novel antigens to an already overreactive immune system and/or increasing bacterial load (Sarrabayrouse et al. 2020; Browne and Kelly 2017).

Administration of selected bacterial species or micro-communities derived from the human gut microbiota offers a promising therapeutic alternative to FMT for the prevention and/or treatment of dysregulated intestinal immune responses. Bacterial species from strains clusters IV, XIVa, and XVIII of Clostridia (Narushima et al. 2014), *Bacteroides fragilis* (Round and Mazmanian 2010) and *Akkermansia muciniphila* (Zhai et al. 2019) have been reported to promote Tregs in murine colon, as was described in more detail in Sect. 9.4.3. Short-chain fatty acids (SCFAs) such as butyrate and propionate produced by Clostridia through fermentation of mainly undigested dietary carbohydrates are the main mediators of Treg induction by this bacterial population. This aspect of Treg regulation was also discussed in more detail in Sect. 9.4.3. With the notable exception of NEC, where the relative abundance of Clostridia clusters IV, XIVa is increased as compared to healthy infants (Brower-Sinning et al. 2014), these bacterial groups are generally decreased in the intestine under inflammatory conditions such as IBD and celiac disease (Narushima et al. 2014; Sanchez et al. 2010). Their reduction may contribute to intestinal immune dysregulation, and replenishment may offer a therapeutic strategy.

Clostridium *F. prausnitzii* can induce an immunoregulatory phenotype in human dendritic cells, characterized by the expression of IL10, IL27, CD39, IDO1, and PDL-1 and inhibition of inflammatory mediator production in response to TLR4 stimulation, and ultimately promotes colonic Treg development (Alameddine et al. 2019).

B. fragilis colonization in mice was found to reduce colitis and inflammation-associated colon cancer in mice in a mechanism dependent on the production on PSA and TLR2 signaling (Lee et al. 2018), and promoted circulating Tregs, alleviating intestinal inflammation (Tan et al. 2019). PSA produced by *B. fragilis*-derived PSA (but no other TRL2 ligands) stimulated FoxP3$^+$ Tregs while suppressing intestinal Th17 generation. IL17 suppression by PSA is a critical mechanism by which *B. fragilis* associates with its host and promotes mucosal colonic niche and immune tolerance (Round et al. 2011). The first study in *B. fragilis* PSA induction of human Tregs showed that PSA presentation by total APC induced human peripheral CD4$^+$ T differentiation into IL10-producing Tr1 cells (Kreisman and Cobb 2011). Later, Telesford et al. showed that PSA promoted frequency and function in human IL10-producing CD39$^+$Foxp3$^+$ Tregs in the presence of DCs through HLA-DR, CD86, CD40, and PD-L1 and enhanced Treg suppressive function (Telesford et al. 2015). Therefore, the presence and specific phenotype of the APC is required to prime CD4$^+$ T cells with PSA to promote differentiation into one of the two Treg subtypes. IBD patients have a significantly lower relative abundance of *B. fragilis* actively producing PSA, thus not just overall decrease of *B. fragilis* but also a reduction in PSA production by this species may contribute to the breakdown of the intestinal homeostasis in IBD patients (Blandford et al. 2019).

Akkermansia muciniphila, a mucin-degrading and abundant member of a healthy microbiota, is reduced in IBD (Lopez-Siles et al. 2018; Earley et al. 2019). It is primarily localized in the transverse and descending colon, where its abundance depends on the mucus thickness and is affected by the immune status of the host (Van den Abbeele et al. 2010). *A. muciniphila* administration in mice modulated the intestinal inflammation and ameliorated acute colitis (Bian et al. 2019). Its beneficial effect in chronic colitis was associated with an induction of Treg differentiation (Zhai et al. 2019). The role of *A. muciniphila* in preventing intestinal disorders has to be addressed with caution because in contrast to the findings in colitis, expansion of *A. muciniphila* in mice with GVHD was associated with loss of the colonic mucus layer and compromised intestinal barrier function (Shono et al. 2016).

There is an accumulating evidence that restoration of intestinal flora by classical probiotic bacteria also generates FoxP3 Tregs in the gut, making the probiotic consumption an interesting approach to induce this cell population in patients with intestinal disorders. Commonly probiotic mixture contains *Bifidobacterium* spp., *Lactobacillus* spp., and *Streptococcus thermophilus*. The beneficial impact of probiotics on intestinal homeostasis has been shown in many studies (Abraham and Quigley 2017; Cristofori et al. 2018) (Patel and Underwood 2018; Andermann et al. 2016). For instance, *L. fermentum*, *L. acidophilus*, and *L. casei* Lbs2 protect from experimental colitis, through the induction of Treg responses (Jang et al. 2019; Park et al. 2018; Thakur et al. 2016). Treg promotion by Lactobacilli is usually mediated by induction of a regulatory phenotype in the intestinal DCs (Mikulic et al. 2017). Furthermore, Bifidobacteria and *S. thermophilus* stimulated significant concentrations of TGFβ in human PBMC (Donkor et al. 2012). Oral consumption of *Bifidobacterium infantis* induces FoxP3 expression and enhances IL10 and TGFβ production in the mouse colon (Zhou et al. 2019). Although limited clinical trials suggest a promising efficacy of classical probiotic strains and their combinations in ulcerative colitis, there remains a lot of skepticism in the IBD field regarding the efficacy of such approaches in Crohn's disease (Guandalini and Sansotta 2019).

9.7.2.4 Dietary Vitamins

Vitamins are organic dietary compounds with significant impact on the gut mucosal immune function. As we described in more detail in Sect. 9.4.4., vitamins A and D provide important signaling cues for the generation and survival of intestinal Foxp3$^+$ Tregs (Huang et al. 2018; Fakhoury et al. 2020). Retinoic acid (RA), product of vit A degradation by RALDH2 expressed by CD103$^+$ DCs in the intestine, has a large spectrum of modulatory effects on Tregs. Such pivotal role of vitamin A metabolites on intestinal Foxp3$^+$ Tregs and homeostasis is illustrated by the findings that vitamin A deficiency exacerbates experimental colitis (Reifen et al. 2002) and supplementation of vitamin A results in decreased inflammatory markers in NEC patients (Xiao et al. 2018). Therefore, administration of vitamin A/retinoic acid in clinical patients with outcomes related to intestinal inflammation may stimulate FoxP3$^+$ Tregs. A clinical trial on the efficacy of high doses of vitamin A to prevent gastrointestinal GVHD out of the Ohio State University Comprehensive Cancer Center is in early stages **(NCT03719092)**, and the outcome of this therapeutic approach may inform other intestinal disorders going forward.

Along with vitamin A, vitamin D is another potent immunomodulator that affects T cells directly and indirectly (Cantorna et al. 2015, 2019). We have already discussed the effects of vitamin D on Treg biology earlier in this chapter. In IBD patients, VDR polymorphisms have been associated with higher disease risk and increased morbidity (Eloranta et al. 2011). Vitamin D deficiency is commonly observed in IBD patients, and association between vitamin D status and the extent of inflammation has been reported by several studies (Jorgensen et al. 2013; Gubatan et al. 2019; Fletcher et al. 2019; Burrelli Scotti et al. 2019; Law et al. 2019). In a double-blind randomized controlled trial, where 90 mild-to-moderate UC patients were administered one single muscular injection of vitamin D (7.5 mg) and reduced serum concentrations of TNFα, IFNγ, and IL12p70 levels, but it did not affect serum IL4 and IL10 levels. However, no other clinical parameters were evaluated or included in the report (Sharifi et al. 2019). In a murine model of GVHD, vitamin D promoted FoxP3$^+$ Treg generation and slowed the disease progression (Ni et al. 2019). A severe vitamin D deficiency has also been documented in patients with celiac disease (Sulimani 2019) and maternal vitamin D deficiency has been proposed to be a new risk factor for NEC in preterm infants (Cetinkaya et al. 2017). Importantly, vitamin D has pleiotropic effects on the cells of the innate and adaptive immune system and affects intestinal barrier function and gut microbial ecology, thus any beneficial effects of this vitamin/hormone cannot be solely attributed to promotion of Tregs and their immunosuppressive functions.

Finally, dietary plant sterols capable of reducing intestinal inflammation in mice also promote Treg numbers. These compounds are consumed in habitual diets derived from vegetable oils, grain products, nuts, seeds, fruits, and vegetables (te Velde et al. 2015). Fiber-enriched diets were associated with a higher colonic in situ production of total SCFA including butyrate, which promotes the gut Tregs (Rodriguez-Cabezas et al. 2002). Dietary interventions and application of prebiotics that enhance butyrate production, such as 2′-fucosyllactose, may help induce remission in UC without involving additional immune suppression (phase IV clinical trial **NCT02345733**) or may contribute to the maintenance of remission in pediatric and young adult IBD patients undergoing anti-cytokine therapy (**NCT03847467**). Collectively, the available data suggests somewhat instinctively that the supplementation with vitamins (or correction of deficiencies) and intake of a balanced diet rich in fruits, grains, and vegetables should be a part of the comprehensive therapy of intestinal inflammatory disorders, and a part of the associated mechanism may be through promotion of intestinal Tregs and immunoregulatory environment in the gut.

9.8 Conclusions

Intestinal Tregs is a complex and heterogenous population of immunosuppressive cells whose development and function are affected (positively and negatively) by an array of factors, both host-intrinsic and extrinsic. They play crucial roles in maintaining the low mucosal inflammatory tone in the gut under continuous antigenic challenge, contribute to the maintenance of intestinal barrier function, keep the immune response to pathogens in check, and promote the resolution of inflammation and mucosal restitution. As such, they represent attractive target of research and therapeutic developments. Although much is known already about their phenotypic and functional features in mice, translating these results to humans has been challenging. A better understanding of human Tregs at mucosal sites, especially in the gut, will help guide better therapeutic options, primary or adjuvant, to treat chronic inflammatory conditions of the gut.

Acknowledgment We thank Dr. Claudio Bernardazzi M. Azevedo for his assistance with the preparation of figures.

References

Abraham BP, Quigley EMM (2017) Probiotics in inflammatory bowel disease. Gastroenterol Clin N Am 46 (4):769–782. https://doi.org/10.1016/j.gtc.2017.08.003

Akesson K, Tompa A, Ryden A, Faresjo M (2015) Low expression of CD39(+)/CD45RA(+) on regulatory T cells (Treg) cells in type 1 diabetic children in contrast to high expression of CD101(+) /CD129(+) on Treg cells in children with coeliac disease. Clin Exp Immunol 180(1):70–82. https://doi.org/10.1111/cei.12559

Akin IM, Atasay B, Dogu F, Okulu E, Arsan S, Karatas HD, Ikinciogullari A, Turmen T (2014) Oral lactoferrin to prevent nosocomial sepsis and necrotizing enterocolitis of premature neonates and effect on T-regulatory cells. Am J Perinatol 31(12):1111–1120. https://doi.org/10.1055/s-0034-1371704

Alameddine J, Godefroy E, Papargyris L, Sarrabayrouse G, Tabiasco J, Bridonneau C, Yazdanbakhsh K, Sokol H, Altare F, Jotereau F (2019) Faecalibacterium prausnitzii skews human DC to prime IL10-producing T cells through TLR2/6/JNK signaling and IL-10, IL-27, CD39, and IDO-1 induction. Front Immunol 10:143. https://doi.org/10.3389/fimmu.2019.00143

Alfen JS, Larghi P, Facciotti F, Gagliani N, Bosotti R, Paroni M, Maglie S, Gruarin P, Vasco CM, Ranzani V, Frusteri C, Iseppon A, Moro M, Crosti MC, Gatti S, Pagani M, Caprioli F, Abrignani S, Flavell RA, Geginat J (2018) Intestinal IFN-gamma-producing type 1 regulatory T cells coexpress CCR5 and programmed cell death protein 1 and downregulate IL-10 in the inflamed guts of patients with inflammatory bowel disease. J Allergy Clin Immunol 142 (5):1537–1547.e1538. https://doi.org/10.1016/j.jaci.2017.12.984

Allaire JM, Crowley SM, Law HT, Chang SY, Ko HJ, Vallance BA (2018) The intestinal epithelium: central coordinator of mucosal immunity. Trends Immunol 39 (9):677–696. https://doi.org/10.1016/j.it.2018.04.002

Allan SE, Crome SQ, Crellin NK, Passerini L, Steiner TS, Bacchetta R, Roncarolo MG, Levings MK (2007) Activation-induced FOXP3 in human T effector cells does not suppress proliferation or cytokine production. Int Immunol 19(4):345–354. https://doi.org/10.1093/intimm/dxm014

Almeida AR, Legrand N, Papiernik M, Freitas AA (2002) Homeostasis of peripheral CD4+ T cells: IL-2R alpha and IL-2 shape a population of regulatory cells that controls CD4+ T cell numbers. J Immunol 169 (9):4850–4860. https://doi.org/10.4049/jimmunol.169.9.4850

Andermann TM, Rezvani A, Bhatt AS (2016) Microbiota manipulation with prebiotics and probiotics in patients undergoing stem cell transplantation. Curr Hematol Malig Rep 11(1):19–28. https://doi.org/10.1007/s11899-016-0302-9

Arpaia N, Campbell C, Fan X, Dikiy S, van der Veeken J, deRoos P, Liu H, Cross JR, Pfeffer K, Coffer PJ, Rudensky AY (2013) Metabolites produced by commensal bacteria promote peripheral regulatory T-cell generation. Nature 504(7480):451–455. https://doi.org/10.1038/nature12726

Asseman C, Mauze S, Leach MW, Coffman RL, Powrie F (1999) An essential role for interleukin 10 in the function of regulatory T cells that inhibit intestinal inflammation. J Exp Med 190(7):995–1004. https://doi.org/10.1084/jem.190.7.995

Atarashi K, Nishimura J, Shima T, Umesaki Y, Yamamoto M, Onoue M, Yagita H, Ishii N, Evans R, Honda K, Takeda K (2008) ATP drives lamina propria T(H)17 cell differentiation. Nature 455 (7214):808–812. https://doi.org/10.1038/nature07240

Atarashi K, Tanoue T, Shima T, Imaoka A, Kuwahara T, Momose Y, Cheng G, Yamasaki S, Saito T, Ohba Y, Taniguchi T, Takeda K, Hori S, Ivanov II, Umesaki Y, Itoh K, Honda K (2011) Induction of colonic regulatory T cells by indigenous Clostridium species. Science 331(6015):337–341. https://doi.org/10.1126/science.1198469

Atarashi K, Tanoue T, Oshima K, Suda W, Nagano Y, Nishikawa H, Fukuda S, Saito T, Narushima S, Hase K, Kim S, Fritz JV, Wilmes P, Ueha S, Matsushima K, Ohno H, Olle B, Sakaguchi S, Taniguchi T, Morita H, Hattori M, Honda K (2013) Treg induction by a rationally selected mixture of Clostridia strains from the human microbiota. Nature 500(7461):232–236. https://doi.org/10.1038/nature12331

Babyatsky MW, Rossiter G, Podolsky DK (1996) Expression of transforming growth factors alpha and beta in colonic mucosa in inflammatory bowel disease. Gastroenterology 110(4):975–984. https://doi.org/10.1053/gast.1996.v110.pm8613031

Bacchetta R, Bigler M, Touraine JL, Parkman R, Tovo PA, Abrams J, de Waal Malefyt R, de Vries JE, Roncarolo MG (1994) High levels of interleukin 10 production in vivo are associated with tolerance in SCID patients transplanted with HLA mismatched hematopoietic stem cells. J Exp Med 179(2):493–502. https://doi.org/10.1084/jem.179.2.493

Bamidele AO, Svingen PA, Sagstetter MR, Sarmento OF, Gonzalez M, Braga Neto MB, Kugathasan S, Lomberk G, Urrutia RA, Faubion WA Jr (2019) Disruption of FOXP3-EZH2 interaction represents a pathobiological mechanism in intestinal inflammation. Cell Mol Gastroenterol Hepatol 7(1):55–71. https://doi.org/10.1016/j.jcmgh.2018.08.009

Bardel E, Larousserie F, Charlot-Rabiega P, Coulomb-L'Hermine A, Devergne O (2008) Human CD4+ CD25+ Foxp3+ regulatory T cells do not constitutively express IL-35. J Immunol 181(10):6898–6905. https://doi.org/10.4049/jimmunol.181.10.6898

Barnhoorn MC, Wasser M, Roelofs H, Maljaars PWJ, Molendijk I, Bonsing BA, Oosten LEM, Dijkstra G, van der Woude CJ, Roelen DL, Zwaginga JJ, Verspaget HW, Fibbe WE, Hommes DW, Peeters K, van der Meulen-de Jong AE (2020) Long-term evaluation of allogeneic bone marrow-derived mesenchymal stromal cell therapy for Crohn's disease perianal fistulas. J Crohns Colitis 14(1):64–70. https://doi.org/10.1093/ecco-jcc/jjz116

Barthlott T, Moncrieffe H, Veldhoen M, Atkins CJ, Christensen J, O'Garra A, Stockinger B (2005) CD25 + CD4+ T cells compete with naive CD4+ T cells for IL-2 and exploit it for the induction of IL-10 production. Int Immunol 17(3):279–288. https://doi.org/10.1093/intimm/dxh207

Bazewicz CG, Dinavahi SS, Schell TD, Robertson GP (2019) Aldehyde dehydrogenase in regulatory T-cell development, immunity and cancer. Immunology 156 (1):47–55. https://doi.org/10.1111/imm.13016

Ben Ahmed M, Belhadj Hmida N, Moes N, Buyse S, Abdeladhim M, Louzir H, Cerf-Bensussan N (2009) IL-15 renders conventional lymphocytes resistant to suppressive functions of regulatory T cells through activation of the phosphatidylinositol 3-kinase pathway. J Immunol 182(11):6763–6770. https://doi.org/10.4049/jimmunol.0801792

Benahmed M, Meresse B, Arnulf B, Barbe U, Mention JJ, Verkarre V, Allez M, Cellier C, Hermine O, Cerf-Bensussan N (2007) Inhibition of TGF-beta signaling by IL-15: a new role for IL-15 in the loss of immune homeostasis in celiac disease. Gastroenterology 132 (3):994–1008. https://doi.org/10.1053/j.gastro.2006.12.025

Bennett CL, Christie J, Ramsdell F, Brunkow ME, Ferguson PJ, Whitesell L, Kelly TE, Saulsbury FT, Chance PF, Ochs HD (2001) The immune dysregulation, polyendocrinopathy, enteropathy, X-linked syndrome (IPEX) is caused by mutations of FOXP3. Nat Genet 27(1):20–21. https://doi.org/10.1038/83713

Betts G, Jones E, Junaid S, El-Shanawany T, Scurr M, Mizen P, Kumar M, Jones S, Rees B, Williams G, Gallimore A, Godkin A (2012) Suppression of tumour-specific CD4(+) T cells by regulatory T cells is associated with progression of human colorectal cancer. Gut 61(8):1163–1171. https://doi.org/10.1136/gutjnl-2011-300970

Bian X, Wu W, Yang L, Lv L, Wang Q, Li Y, Ye J, Fang D, Wu J, Jiang X, Shi D, Li L (2019) Administration of Akkermansia muciniphila ameliorates dextran sulfate sodium-induced ulcerative colitis in mice. Front Microbiol 10:2259. https://doi.org/10.3389/fmicb.2019.02259

Birebent B, Lorho R, Lechartier H, de Guibert S, Alizadeh M, Vu N, Beauplet A, Robillard N, Semana G (2004) Suppressive properties of human CD4+CD25 + regulatory T cells are dependent on CTLA-4 expression. Eur J Immunol 34(12):3485–3496. https://doi.org/10.1002/eji.200324632

Blandford LE, Johnston EL, Sanderson JD, Wade WG, Lax AJ (2019) Promoter orientation of the immunomodulatory Bacteroides fragilis capsular polysaccharide A (PSA) is off in individuals with inflammatory bowel disease (IBD). Gut Microbes 10(5):569–577. https://doi.org/10.1080/19490976.2018.1560755

Blatner NR, Bonertz A, Beckhove P, Cheon EC, Krantz SB, Strouch M, Weitz J, Koch M, Halverson AL, Bentrem DJ, Khazaie K (2010) In colorectal cancer mast cells contribute to systemic regulatory T-cell dysfunction. Proc Natl Acad Sci U S A 107 (14):6430–6435. https://doi.org/10.1073/pnas.0913683107

Blatner NR, Mulcahy MF, Dennis KL, Scholtens D, Bentrem DJ, Phillips JD, Ham S, Sandall BP, Khan MW, Mahvi DM, Halverson AL, Stryker SJ, Boller AM, Singal A, Sneed RK, Sarraj B, Ansari MJ, Oft M, Iwakura Y, Zhou L, Bonertz A, Beckhove P, Gounari F, Khazaie K (2012) Expression of RORgammat marks a pathogenic regulatory T cell subset in human colon cancer. Sci Transl Med 4 (164):164ra159. https://doi.org/10.1126/scitranslmed.3004566

Boga S, Alkim H, Koksal AR, Ozagari AA, Bayram M, Tekin Neijmann S, Sen I, Alkim C (2016) Serum ST2 in inflammatory bowel disease: a potential biomarker for disease activity. J Investig Med 64(5):1016–1024. https://doi.org/10.1136/jim-2016-000062

Bonertz A, Weitz J, Pietsch DH, Rahbari NN, Schlude C, Ge Y, Juenger S, Vlodavsky I, Khazaie K, Jaeger D, Reissfelder C, Antolovic D, Aigner M, Koch M, Beckhove P (2009) Antigen-specific Tregs control T cell responses against a limited repertoire of tumor antigens in patients with colorectal carcinoma. J Clin Invest 119(11):3311–3321. https://doi.org/10.1172/JCI39608

Borsellino G, Kleinewietfeld M, Di Mitri D, Sternjak A, Diamantini A, Giometto R, Hopner S, Centonze D, Bernardi G, Dell'Acqua ML, Rossini PM, Battistini L, Rotzschke O, Falk K (2007) Expression of ectonucleotidase CD39 by Foxp3+ Treg cells: hydrolysis of extracellular ATP and immune suppression. Blood 110(4):1225–1232. https://doi.org/10.1182/blood-2006-12-064527

Bouzid D, Amouri A, Fourati H, Marques I, Abida O, Tahri N, Goncalves CP, Masmoudi H (2013) Polymorphisms in the IL2RA and IL2RB genes in inflammatory bowel disease risk. Genet Test Mol Biomarkers 17(11):833–839. https://doi.org/10.1089/gtmb.2013.0291

Brower-Sinning R, Zhong D, Good M, Firek B, Baker R, Sodhi CP, Hackam DJ, Morowitz MJ (2014) Mucosa-associated bacterial diversity in necrotizing enterocolitis. PLoS One 9(9):e105046. https://doi.org/10.1371/journal.pone.0105046

Browne AS, Kelly CR (2017) Fecal transplant in inflammatory bowel disease. Gastroenterol Clin N Am 46 (4):825–837. https://doi.org/10.1016/j.gtc.2017.08.005

Brunkow ME, Jeffery EW, Hjerrild KA, Paeper B, Clark LB, Yasayko SA, Wilkinson JE, Galas D, Ziegler SF, Ramsdell F (2001) Disruption of a new forkhead/winged-helix protein, scurfin, results in the fatal lymphoproliferative disorder of the scurfy mouse. Nat Genet 27(1):68–73. https://doi.org/10.1038/83784

Bueter M, Gasser M, Schramm N, Lebedeva T, Tocco G, Gerstlauer C, Grimm M, Nichiporuk E, Thalheimer A, Thiede A, Meyer D, Benichou G, Waaga-Gasser AM (2006) T-cell response to p53 tumor-associated antigen in patients with colorectal carcinoma. Int J Oncol 28 (2):431–438

Burrelli Scotti G, Afferri MT, De Carolis A, Vaiarello V, Fassino V, Ferrone F, Minisola S, Nieddu L, Vernia P (2019) Factors affecting vitamin D deficiency in active inflammatory bowel diseases. Dig Liver Dis 51 (5):657–662. https://doi.org/10.1016/j.dld.2018.11.036

Campbell C, Dikiy S, Bhattarai SK, Chinen T, Matheis F, Calafiore M, Hoyos B, Hanash A, Mucida D, Bucci V, Rudensky AY (2018) Extrathymically generated regulatory T cells establish a niche for intestinal border-dwelling Bacteria and affect physiologic metabolite balance. Immunity 48(6):1245–1257. e1249. https://doi.org/10.1016/j.immuni.2018.04.013

Campi G, Crosti M, Consogno G, Facchinetti V, Conti-Fine BM, Longhi R, Casorati G, Dellabona P, Protti MP (2003) CD4(+) T cells from healthy subjects and colon cancer patients recognize a carcinoembryonic antigen-specific immunodominant epitope. Cancer Res 63(23):8481–8486

Canavan JB, Scotta C, Vossenkamper A, Goldberg R, Elder MJ, Shoval I, Marks E, Stolarczyk E, Lo JW, Powell N, Fazekasova H, Irving PM, Sanderson JD, Howard JK, Yagel S, Afzali B, MacDonald TT, Hernandez-Fuentes MP, Shpigel NY, Lombardi G, Lord GM (2016) Developing in vitro expanded CD45RA+ regulatory T cells as an adoptive cell therapy for Crohn's disease. Gut 65(4):584–594. https://doi.org/10.1136/gutjnl-2014-306919

Cantorna MT, Snyder L, Lin YD, Yang L (2015) Vitamin D and 1,25(OH)2D regulation of T cells. Nutrients 7 (4):3011–3021. https://doi.org/10.3390/nu7043011

Cantorna MT, Snyder L, Arora J (2019) Vitamin A and vitamin D regulate the microbial complexity, barrier function, and the mucosal immune responses to ensure intestinal homeostasis. Crit Rev Biochem Mol Biol 54 (2):184–192. https://doi.org/10.1080/10409238.2019.1611734

Caplan MS, Sun XM, Hseuh W, Hageman JR (1990) Role of platelet activating factor and tumor necrosis factor-alpha in neonatal necrotizing enterocolitis. J Pediatr 116(6):960–964. https://doi.org/10.1016/s0022-3476(05)80661-4

Cebula A, Seweryn M, Rempala GA, Pabla SS, McIndoe RA, Denning TL, Bry L, Kraj P, Kisielow P, Ignatowicz L (2013) Thymus-derived regulatory T cells contribute to tolerance to commensal microbiota. Nature 497(7448):258–262. https://doi.org/10.1038/nature12079

Cederbom L, Hall H, Ivars F (2000) CD4+CD25+ regulatory T cells down-regulate co-stimulatory molecules on antigen-presenting cells. Eur J Immunol 30 (6):1538–1543. https://doi.org/10.1002/1521-4141(200006)30:6<1538::AID-IMMU1538>3.0.CO;2-X

Cellier C, Bouma G, van Gils T, Khater S, Malamut G, Crespo L, Collin P, Green PHR, Crowe SE, Tsuji W, Butz E, Cerf-Bensussan N, Macintyre E, Parnes JR, Leon F, Hermine O, Mulder CJ, Investigators R-ISG (2019) Safety and efficacy of AMG 714 in patients with type 2 refractory coeliac disease: a phase 2a, randomised, double-blind, placebo-controlled, parallel-group study. Lancet Gastroenterol Hepatol 4 (12):960–970. https://doi.org/10.1016/S2468-1253(19)30265-1

Cetinkaya M, Erener-Ercan T, Kalayci-Oral T, Babayigit A, Cebeci B, Semerci SY, Buyukkale G (2017) Maternal/neonatal vitamin D deficiency: a new risk factor for necrotizing enterocolitis in preterm infants? J Perinatol 37(6):673–678. https://doi.org/10.1038/jp.2017.18

Chai JN, Peng Y, Rengarajan S, Solomon BD, Ai TL, Shen Z, Perry JSA, Knoop KA, Tanoue T, Narushima S, Honda K, Elson CO, Newberry RD, Stappenbeck TS, Kau AL, Peterson DA, Fox JG, Hsieh CS (2017) Helicobacter species are potent drivers of colonic T cell responses in homeostasis and inflammation. Sci Immunol 2(13):eaal5068. https://doi.org/10.1126/sciimmunol.aal5068

Chang LY, Lin YC, Mahalingam J, Huang CT, Chen TW, Kang CW, Peng HM, Chu YY, Chiang JM, Dutta A, Day YJ, Chen TC, Yeh CT, Lin CY (2012) Tumor-derived chemokine CCL5 enhances TGF-beta-mediated killing of CD8(+) T cells in colon cancer by T-regulatory cells. Cancer Res 72(5):1092–1102. https://doi.org/10.1158/0008-5472.CAN-11-2493

Chao G, Li X, Ji Y, Zhu Y, Li N, Zhang N, Feng Z, Niu M (2018) CTLA-4 regulates T follicular regulatory cell differentiation and participates in intestinal damage caused by spontaneous autoimmunity. Biochem Biophys Res Commun 505(3):865–871. https://doi.org/10.1016/j.bbrc.2018.09.182

Chaput N, Louafi S, Bardier A, Charlotte F, Vaillant JC, Menegaux F, Rosenzwajg M, Lemoine F, Klatzmann D, Taieb J (2009) Identification of CD8 +CD25+Foxp3+ suppressive T cells in colorectal cancer tissue. Gut 58(4):520–529. https://doi.org/10.1136/gut.2008.158824

Chaudhry A, Samstein RM, Treuting P, Liang Y, Pils MC, Heinrich JM, Jack RS, Wunderlich FT, Bruning JC, Muller W, Rudensky AY (2011) Interleukin-10 signaling in regulatory T cells is required for suppression of Th17 cell-mediated inflammation. Immunity 34 (4):566–578. https://doi.org/10.1016/j.immuni.2011.03.018

Chen J, Chen Z (2014) The effect of immune microenvironment on the progression and prognosis of colorectal cancer. Med Oncol 31(8):82. https://doi.org/10.1007/s12032-014-0082-9

Chen W, Jin W, Hardegen N, Lei KJ, Li L, Marinos N, McGrady G, Wahl SM (2003) Conversion of peripheral CD4+CD25- naive T cells to CD4+CD25+ regulatory T cells by TGF-beta induction of transcription factor Foxp3. J Exp Med 198(12):1875–1886. https://doi.org/10.1084/jem.20030152

Chen M, Su W, Lin X, Guo Z, Wang J, Zhang Q, Brand D, Ryffel B, Huang J, Liu Z, He X, Le AD, Zheng SG (2013) Adoptive transfer of human gingiva-derived mesenchymal stem cells ameliorates collagen-induced arthritis via suppression of Th1 and Th17 cells and enhancement of regulatory T cell differentiation. Arthritis Rheum 65(5):1181–1193. https://doi.org/10.1002/art.37894

Chen J, He Y, Tu L, Duan L (2020) Dual immune functions of IL-33 in inflammatory bowel disease. Histol Histopathol 35(2):137–146. https://doi.org/10.14670/HH-18-149

Chinen T, Kannan AK, Levine AG, Fan X, Klein U, Zheng Y, Gasteiger G, Feng Y, Fontenot JD, Rudensky AY (2016) An essential role for the IL-2 receptor in Treg cell function. Nat Immunol 17 (11):1322–1333. https://doi.org/10.1038/ni.3540

Cianci R, Cammarota G, Frisullo G, Pagliari D, Ianiro G, Martini M, Frosali S, Plantone D, Damato V, Casciano F, Landolfi R, Paola Batocchi A, Pandolfi F (2012) Tissue-infiltrating lymphocytes analysis reveals large modifications of the duodenal "immunological niche" in coeliac disease after gluten-free diet. Clin Transl Gastroenterol 3:e28. https://doi.org/10.1038/ctg.2012.22

Ciccocioppo R, Bernardo ME, Sgarella A, Maccario R, Avanzini MA, Ubezio C, Minelli A, Alvisi C, Vanoli A, Calliada F, Dionigi P, Perotti C, Locatelli F, Corazza GR (2011) Autologous bone marrow-derived mesenchymal stromal cells in the treatment of fistulising Crohn's disease. Gut 60 (6):788–798. https://doi.org/10.1136/gut.2010.214841

Clarke SL, Betts GJ, Plant A, Wright KL, El-Shanawany TM, Harrop R, Torkington J, Rees BI, Williams GT, Gallimore AM, Godkin AJ (2006) CD4+CD25+FOXP3+ regulatory T cells suppress anti-tumor immune responses in patients with colorectal cancer. PLoS One 1:e129. https://doi.org/10.1371/journal.pone.0000129

Collison LW, Workman CJ, Kuo TT, Boyd K, Wang Y, Vignali KM, Cross R, Sehy D, Blumberg RS, Vignali DA (2007) The inhibitory cytokine IL-35 contributes to regulatory T-cell function. Nature 450 (7169):566–569. https://doi.org/10.1038/nature06306

Cong Y, Feng T, Fujihashi K, Schoeb TR, Elson CO (2009) A dominant, coordinated T regulatory cell-IgA response to the intestinal microbiota. Proc Natl Acad Sci U S A 106(46):19256–19261. https://doi.org/10.1073/pnas.0812681106

Cook L, Munier CML, Seddiki N, van Bockel D, Ontiveros N, Hardy MY, Gillies JK, Levings MK, Reid HH, Petersen J, Rossjohn J, Anderson RP, Zaunders JJ, Tye-Din JA, Kelleher AD (2017) Circulating gluten-specific FOXP3(+)CD39(+) regulatory T cells have impaired suppressive function in patients with celiac disease. J Allergy Clin Immunol 140(6):1592–1603.e1598. https://doi.org/10.1016/j.jaci.2017.02.015

Cook L, Stahl M, Han X, Nazli A, MacDonald KN, Wong MQ, Tsai K, Dizzell S, Jacobson K, Bressler B, Kaushic C, Vallance BA, Steiner TS, Levings MK (2019) Suppressive and gut-reparative functions of human type 1 T regulatory cells. Gastroenterology 157(6):1584–1598. https://doi.org/10.1053/j.gastro.2019.09.002

Coombes JL, Siddiqui KR, Arancibia-Carcamo CV, Hall J, Sun CM, Belkaid Y, Powrie F (2007) A functionally specialized population of mucosal CD103+ DCs induces Foxp3+ regulatory T cells via a TGF-beta and retinoic acid-dependent mechanism. J Exp Med 204(8):1757–1764. https://doi.org/10.1084/jem.20070590

Copsel SN, Lightbourn CO, Barreras H, Lohse I, Wolf D, Bader CS, Manov J, Kale BJ, Shah D, Brothers SP, Perez VL, Komanduri KV, Wahlestedt C, Levy RB (2018) BET bromodomain inhibitors which permit Treg function enable a combinatorial strategy to suppress GVHD in pre-clinical allogeneic HSCT. Front Immunol 9:3104. https://doi.org/10.3389/fimmu.2018.03104

Cretney E, Xin A, Shi W, Minnich M, Masson F, Miasari M, Belz GT, Smyth GK, Busslinger M, Nutt SL, Kallies A (2011) The transcription factors Blimp-1 and IRF4 jointly control the differentiation and function of effector regulatory T cells. Nat Immunol 12 (4):304–311. https://doi.org/10.1038/ni.2006

Cristofori F, Indrio F, Miniello VL, De Angelis M, Francavilla R (2018) Probiotics in celiac disease. Nutrients 10(12):1824. https://doi.org/10.3390/nu10121824

Curotto de Lafaille MA, Kutchukhidze N, Shen S, Ding Y, Yee H, Lafaille JJ (2008) Adaptive Foxp3+ regulatory T cell-dependent and -independent control of allergic inflammation. Immunity 29(1):114–126. https://doi.org/10.1016/j.immuni.2008.05.010

D'Ambrosio DN, Clugston RD, Blaner WS (2011) Vitamin a metabolism: an update. Nutrients 3(1):63–103. https://doi.org/10.3390/nu3010063

Davidson TS, DiPaolo RJ, Andersson J, Shevach EM (2007) Cutting edge: IL-2 is essential for TGF-beta-mediated induction of Foxp3+ T regulatory cells. J

Immunol 178(7):4022–4026. https://doi.org/10.4049/jimmunol.178.7.4022
D'Cruz LM, Klein L (2005) Development and function of agonist-induced CD25+Foxp3+ regulatory T cells in the absence of interleukin 2 signaling. Nat Immunol 6 (11):1152–1159. https://doi.org/10.1038/ni1264
De Simone M, Arrigoni A, Rossetti G, Gruarin P, Ranzani V, Politano C, Bonnal RJP, Provasi E, Sarnicola ML, Panzeri I, Moro M, Crosti M, Mazzara S, Vaira V, Bosari S, Palleschi A, Santambrogio L, Bovo G, Zucchini N, Totis M, Gianotti L, Cesana G, Perego RA, Maroni N, Pisani Ceretti A, Opocher E, De Francesco R, Geginat J, Stunnenberg HG, Abrignani S, Pagani M (2016) Transcriptional landscape of human tissue lymphocytes unveils uniqueness of tumor-infiltrating T regulatory cells. Immunity 45(5):1135–1147. https://doi.org/10.1016/j.immuni.2016.10.021
Denning NL, Prince JM (2018) Neonatal intestinal dysbiosis in necrotizing enterocolitis. Mol Med 24 (1):4. https://doi.org/10.1186/s10020-018-0002-0
Desreumaux P, Foussat A, Allez M, Beaugerie L, Hebuterne X, Bouhnik Y, Nachury M, Brun V, Bastian H, Belmonte N, Ticchioni M, Duchange A, Morel-Mandrino P, Neveu V, Clerget-Chossat N, Forte M, Colombel JF (2012) Safety and efficacy of antigen-specific regulatory T-cell therapy for patients with refractory Crohn's disease. Gastroenterology 143 (5):1207–1217. e1202. https://doi.org/10.1053/j.gastro.2012.07.116
Dingle BM, Liu Y, Fatheree NY, Min J, Rhoads JM, Tran DQ (2013) FoxP3(+) regulatory T cells attenuate experimental necrotizing enterocolitis. PLoS One 8 (12):e82963. https://doi.org/10.1371/journal.pone.0082963
Donkor ON, Ravikumar M, Proudfoot O, Day SL, Apostolopoulos V, Paukovics G, Vasiljevic T, Nutt SL, Gill H (2012) Cytokine profile and induction of T helper type 17 and regulatory T cells by human peripheral mononuclear cells after microbial exposure. Clin Exp Immunol 167(2):282–295. https://doi.org/10.1111/j.1365-2249.2011.04496.x
Downs-Canner S, Berkey S, Delgoffe GM, Edwards RP, Curiel T, Odunsi K, Bartlett DL, Obermajer N (2017) Suppressive IL-17A(+)Foxp3(+) and ex-Th17 IL-17A (neg)Foxp3(+) Treg cells are a source of tumour-associated Treg cells. Nat Commun 8:14649. https://doi.org/10.1038/ncomms14649
Drucker NA, McCulloh CJ, Li B, Pierro A, Besner GE, Markel TA (2018) Stem cell therapy in necrotizing enterocolitis: current state and future directions. Semin Pediatr Surg 27(1):57–64. https://doi.org/10.1053/j.sempedsurg.2017.11.011
Du R, Zhao H, Yan F, Li H (2014) IL-17+Foxp3+ T cells: an intermediate differentiation stage between Th17 cells and regulatory T cells. J Leukoc Biol 96(1):39–48. https://doi.org/10.1189/jlb.1RU0114-010RR
Duan L, Chen J, Zhang H, Yang H, Zhu P, Xiong A, Xia Q, Zheng F, Tan Z, Gong F, Fang M (2012) Interleukin-33 ameliorates experimental colitis through promoting Th2/Foxp3(+) regulatory T-cell responses in mice. Mol Med 18:753–761. https://doi.org/10.2119/molmed.2011.00428
Dudley E, Hornung F, Zheng L, Scherer D, Ballard D, Lenardo M (1999) NF-kappaB regulates Fas/APO-1/CD95- and TCR- mediated apoptosis of T lymphocytes. Eur J Immunol 29(3):878–886. https://doi.org/10.1002/(SICI)1521-4141(199903)29:03<878::AID-IMMU878>3.0.CO;2-9
Duggleby R, Danby RD, Madrigal JA, Saudemont A (2018) Clinical grade regulatory CD4(+) T cells (Tregs): moving toward cellular-based immunomodulatory therapies. Front Immunol 9:252. https://doi.org/10.3389/fimmu.2018.00252
Duhen T, Duhen R, Lanzavecchia A, Sallusto F, Campbell DJ (2012) Functionally distinct subsets of human FOXP3+ Treg cells that phenotypically mirror effector Th cells. Blood 119(19):4430–4440. https://doi.org/10.1182/blood-2011-11-392324
Dunn GP, Old LJ, Schreiber RD (2004) The immunobiology of cancer immunosurveillance and immunoediting. Immunity 21(2):137–148. https://doi.org/10.1016/j.immuni.2004.07.017
Dwyer KM, Hanidziar D, Putheti P, Hill PA, Pommey S, McRae JL, Winterhalter A, Doherty G, Deaglio S, Koulmanda M, Gao W, Robson SC, Strom TB (2010) Expression of CD39 by human peripheral blood CD4+ CD25+ T cells denotes a regulatory memory phenotype. Am J Transplant 10(11):2410–2420. https://doi.org/10.1111/j.1600-6143.2010.03291.x
Earley H, Lennon G, Balfe A, Coffey JC, Winter DC, O'Connell PR (2019) The abundance of Akkermansia muciniphila and its relationship with sulphated colonic mucins in health and ulcerative colitis. Sci Rep 9 (1):15683. https://doi.org/10.1038/s41598-019-51878-3
Edinger M, Hoffmann P, Ermann J, Drago K, Fathman CG, Strober S, Negrin RS (2003) CD4+CD25+ regulatory T cells preserve graft-versus-tumor activity while inhibiting graft-versus-host disease after bone marrow transplantation. Nat Med 9(9):1144–1150. https://doi.org/10.1038/nm915
Edwards JP, Hand TW, Morais da Fonseca D, Glass DD, Belkaid Y, Shevach EM (2016) The GARP/latent TGF-beta1 complex on Treg cells modulates the induction of peripherally derived Treg cells during oral tolerance. Eur J Immunol 46(6):1480–1489. https://doi.org/10.1002/eji.201546204
Egan CE, Sodhi CP, Good M, Lin J, Jia H, Yamaguchi Y, Lu P, Ma C, Branca MF, Weyandt S, Fulton WB, Nino DF, Prindle T Jr, Ozolek JA, Hackam DJ (2016) Toll-like receptor 4-mediated lymphocyte influx induces neonatal necrotizing enterocolitis. J Clin Invest 126 (2):495–508. https://doi.org/10.1172/JCI83356
Elias KM, Laurence A, Davidson TS, Stephens G, Kanno Y, Shevach EM, O'Shea JJ (2008) Retinoic acid inhibits Th17 polarization and enhances FoxP3 expression through a Stat-3/Stat-5 independent signaling pathway. Blood 111(3):1013–1020. https://doi.org/10.1182/blood-2007-06-096438
Eloranta JJ, Wenger C, Mwinyi J, Hiller C, Gubler C, Vavricka SR, Fried M, Kullak-Ublick GA, Swiss

IBDCSG (2011) Association of a common vitamin D-binding protein polymorphism with inflammatory bowel disease. Pharmacogenet Genomics 21 (9):559–564. https://doi.org/10.1097/FPC.0b013e328348f70c

Engelhardt BG, Jagasia M, Savani BN, Bratcher NL, Greer JP, Jiang A, Kassim AA, Lu P, Schuening F, Yoder SM, Rock MT, Crowe JEJ (2011) Regulatory T cell expression of CLA or alpha(4)beta(7) and skin or gut acute GVHD outcomes. Bone Marrow Transplant 46(3):436–442. https://doi.org/10.1038/bmt.2010.127

Engelhardt BG, Sengsayadeth SM, Jagasia M, Savani BN, Kassim AA, Lu P, Shyr Y, Yoder SM, Rock MT, Crowe JE Jr (2012) Tissue-specific regulatory T cells: biomarker for acute graft-vs-host disease and survival. Exp Hematol 40(12):974–982. e971. https://doi.org/10.1016/j.exphem.2012.08.002

Faith JJ, Ahern PP, Ridaura VK, Cheng J, Gordon JI (2014) Identifying gut microbe-host phenotype relationships using combinatorial communities in gnotobiotic mice. Sci Transl Med 6(220):220ra211. https://doi.org/10.1126/scitranslmed.3008051

Fakhoury HMA, Kvietys PR, AlKattan W, Anouti FA, Elahi MA, Karras SN, Grant WB (2020) Vitamin D and intestinal homeostasis: barrier, microbiota, and immune modulation. J Steroid Biochem Mol Biol 200:105663. https://doi.org/10.1016/j.jsbmb.2020.105663

Farh KK, Marson A, Zhu J, Kleinewietfeld M, Housley WJ, Beik S, Shoresh N, Whitton H, Ryan RJ, Shishkin AA, Hatan M, Carrasco-Alfonso MJ, Mayer D, Luckey CJ, Patsopoulos NA, De Jager PL, Kuchroo VK, Epstein CB, Daly MJ, Hafler DA, Bernstein BE (2015) Genetic and epigenetic fine mapping of causal autoimmune disease variants. Nature 518 (7539):337–343. https://doi.org/10.1038/nature13835

Feagan BG, Sands BE, Rossiter G, Li X, Usiskin K, Zhan X, Colombel JF (2018) Effects of Mongersen (GED-0301) on endoscopic and clinical outcomes in patients with active Crohn's disease. Gastroenterology 154(1):61–64.e66. https://doi.org/10.1053/j.gastro.2017.08.035

Feng Y, Arvey A, Chinen T, van der Veeken J, Gasteiger G, Rudensky AY (2014) Control of the inheritance of regulatory T cell identity by a cis element in the Foxp3 locus. Cell 158(4):749–763. https://doi.org/10.1016/j.cell.2014.07.031

Fernandez S, Molina IJ, Romero P, Gonzalez R, Pena J, Sanchez F, Reynoso FR, Perez-Navero JL, Estevez O, Ortega C, Santamaria M (2011) Characterization of gliadin-specific Th17 cells from the mucosa of celiac disease patients. Am J Gastroenterol 106(3):528–538. https://doi.org/10.1038/ajg.2010.465

Fletcher J, Cooper SC, Ghosh S, Hewison M (2019) The role of vitamin D in inflammatory bowel disease: mechanism to management. Nutrients 11(5):1019. https://doi.org/10.3390/nu11051019

Flippe L, Bezie S, Anegon I, Guillonneau C (2019) Future prospects for CD8(+) regulatory T cells in immune tolerance. Immunol Rev 292(1):209–224. https://doi.org/10.1111/imr.12812

Fontenot JD, Rasmussen JP, Gavin MA, Rudensky AY (2005) A function for interleukin 2 in Foxp3-expressing regulatory T cells. Nat Immunol 6 (11):1142–1151. https://doi.org/10.1038/ni1263

Franke A, Balschun T, Karlsen TH, Sventoraityte J, Nikolaus S, Mayr G, Domingues FS, Albrecht M, Nothnagel M, Ellinghaus D, Sina C, Onnie CM, Weersma RK, Stokkers PC, Wijmenga C, Gazouli M, Strachan D, McArdle WL, Vermeire S, Rutgeerts P, Rosenstiel P, Krawczak M, Vatn MH, INSTEN Study Group, Mathew CG, Schreiber S (2008) Sequence variants in IL10, ARPC2 and multiple other loci contribute to ulcerative colitis susceptibility. Nat Genet 40 (11):1319–1323. https://doi.org/10.1038/ng.221

Frey DM, Droeser RA, Viehl CT, Zlobec I, Lugli A, Zingg U, Oertli D, Kettelhack C, Terracciano L, Tornillo L (2010) High frequency of tumor-infiltrating FOXP3(+) regulatory T cells predicts improved survival in mismatch repair-proficient colorectal cancer patients. Int J Cancer 126(11):2635–2643. https://doi.org/10.1002/ijc.24989

Friedman DJ, Kunzli BM, YI AR, Sevigny J, Berberat PO, Enjyoji K, Csizmadia E, Friess H, Robson SC (2009) From the cover: CD39 deletion exacerbates experimental murine colitis and human polymorphisms increase susceptibility to inflammatory bowel disease. Proc Natl Acad Sci U S A 106(39):16788–16793. https://doi.org/10.1073/pnas.0902869106

Fujimoto H, Saito Y, Ohuchida K, Kawakami E, Fujiki S, Watanabe T, Ono R, Kaneko A, Takagi S, Najima Y, Hijikata A, Cui L, Ueki T, Oda Y, Hori S, Ohara O, Nakamura M, Saito T, Ishikawa F (2018) Deregulated mucosal immune surveillance through gut-associated regulatory T cells and PD-1(+) T cells in human colorectal cancer. J Immunol 200(9):3291–3303. https://doi.org/10.4049/jimmunol.1701222

Furusawa Y, Obata Y, Fukuda S, Endo TA, Nakato G, Takahashi D, Nakanishi Y, Uetake C, Kato K, Kato T, Takahashi M, Fukuda NN, Murakami S, Miyauchi E, Hino S, Atarashi K, Onawa S, Fujimura Y, Lockett T, Clarke JM, Topping DL, Tomita M, Hori S, Ohara O, Morita T, Koseki H, Kikuchi J, Honda K, Hase K, Ohno H (2013) Commensal microbe-derived butyrate induces the differentiation of colonic regulatory T cells. Nature 504(7480):446–450. https://doi.org/10.1038/nature12721

Galon J, Costes A, Sanchez-Cabo F, Kirilovsky A, Mlecnik B, Lagorce-Pages C, Tosolini M, Camus M, Berger A, Wind P, Zinzindohoue F, Bruneval P, Cugnenc PH, Trajanoski Z, Fridman WH, Pages F (2006) Type, density, and location of immune cells within human colorectal tumors predict clinical outcome. Science 313(5795):1960–1964. https://doi.org/10.1126/science.1129139

Gambineri E, Torgerson TR, Ochs HD (2003) Immune dysregulation, polyendocrinopathy, enteropathy, and X-linked inheritance (IPEX), a syndrome of systemic autoimmunity caused by mutations of FOXP3, a critical regulator of T-cell homeostasis. Curr Opin Rheumatol 15(4):430–435. https://doi.org/10.1097/00002281-200307000-00010

Gao F, Chiu SM, Motan DA, Zhang Z, Chen L, Ji HL, Tse HF, Fu QL, Lian Q (2016) Mesenchymal stem cells and immunomodulation: current status and future prospects. Cell Death Dis 7:e2062. https://doi.org/10.1038/cddis.2015.327

Gao J, Xu K, Liu H, Liu G, Bai M, Peng C, Li T, Yin Y (2018) Impact of the gut microbiota on intestinal immunity mediated by tryptophan metabolism. Front Cell Infect Microbiol 8:13. https://doi.org/10.3389/fcimb.2018.00013

Garrett WS, Lord GM, Punit S, Lugo-Villarino G, Mazmanian SK, Ito S, Glickman JN, Glimcher LH (2007) Communicable ulcerative colitis induced by T-bet deficiency in the innate immune system. Cell 131(1):33–45. https://doi.org/10.1016/j.cell.2007.08.017

Gavin MA, Torgerson TR, Houston E, DeRoos P, Ho WY, Stray-Pedersen A, Ocheltree EL, Greenberg PD, Ochs HD, Rudensky AY (2006) Single-cell analysis of normal and FOXP3-mutant human T cells: FOXP3 expression without regulatory T cell development. Proc Natl Acad Sci U S A 103(17):6659–6664. https://doi.org/10.1073/pnas.0509484103

Gerassy-Vainberg S, Blatt A, Danin-Poleg Y, Gershovich K, Sabo E, Nevelsky A, Daniel S, Dahan A, Ziv O, Dheer R, Abreu MT, Koren O, Kashi Y, Chowers Y (2018) Radiation induces proinflammatory dysbiosis: transmission of inflammatory susceptibility by host cytokine induction. Gut 67 (1):97–107. https://doi.org/10.1136/gutjnl-2017-313789

Gershon RK, Kondo K (1970) Cell interactions in the induction of tolerance: the role of thymic lymphocytes. Immunology 18(5):723–737

Geuking MB, Cahenzli J, Lawson MA, Ng DC, Slack E, Hapfelmeier S, McCoy KD, Macpherson AJ (2011) Intestinal bacterial colonization induces mutualistic regulatory T cell responses. Immunity 34(5):794–806. https://doi.org/10.1016/j.immuni.2011.03.021

Gianfrani C, Levings MK, Sartirana C, Mazzarella G, Barba G, Zanzi D, Camarca A, Iaquinto G, Giardullo N, Auricchio S, Troncone R, Roncarolo MG (2006) Gliadin-specific type 1 regulatory T cells from the intestinal mucosa of treated celiac patients inhibit pathogenic T cells. J Immunol 177 (6):4178–4186. https://doi.org/10.4049/jimmunol.177.6.4178

Girardin A, McCall J, Black MA, Edwards F, Phillips V, Taylor ES, Reeve AE, Kemp RA (2013) Inflammatory and regulatory T cells contribute to a unique immune microenvironment in tumor tissue of colorectal cancer patients. Int J Cancer 132(8):1842–1850. https://doi.org/10.1002/ijc.27855

Girbovan A, Sur G, Samasca G, Lupan I (2017) Dysbiosis a risk factor for celiac disease. Med Microbiol Immunol 206(2):83–91. https://doi.org/10.1007/s00430-017-0496-z

Giri J, Das R, Nylen E, Chinnadurai R, Galipeau J (2020) CCL2 and CXCL12 derived from mesenchymal stromal cells cooperatively polarize IL-10+ tissue macrophages to mitigate gut injury. Cell Rep 30 (6):1923–1934. e1924. https://doi.org/10.1016/j.celrep.2020.01.047

Glocker EO, Kotlarz D, Boztug K, Gertz EM, Schaffer AA, Noyan F, Perro M, Diestelhorst J, Allroth A, Murugan D, Hatscher N, Pfeifer D, Sykora KW, Sauer M, Kreipe H, Lacher M, Nustede R, Woellner C, Baumann U, Salzer U, Koletzko S, Shah N, Segal AW, Sauerbrey A, Buderus S, Snapper SB, Grimbacher B, Klein C (2009) Inflammatory bowel disease and mutations affecting the interleukin-10 receptor. N Engl J Med 361(21):2033–2045. https://doi.org/10.1056/NEJMoa0907206

Godefroy E, Alameddine J, Montassier E, Mathe J, Desfrancois-Noel J, Marec N, Bossard C, Jarry A, Bridonneau C, Le Roy A, Sarrabayrouse G, Kerdreux E, Bourreille A, Sokol H, Jotereau F, Altare F (2018) Expression of CCR6 and CXCR6 by gut-derived CD4(+)/CD8alpha(+) T-regulatory cells, which are decreased in blood samples from patients with inflammatory bowel diseases. Gastroenterology 155(4):1205–1217. https://doi.org/10.1053/j.gastro.2018.06.078

Godfrey VL, Wilkinson JE, Rinchik EM, Russell LB (1991) Fatal lymphoreticular disease in the scurfy (sf) mouse requires T cells that mature in a sf thymic environment: potential model for thymic education. Proc Natl Acad Sci U S A 88(13):5528–5532. https://doi.org/10.1073/pnas.88.13.5528

Goldberg R, Scotta C, Cooper D, Nissim-Eliraz E, Nir E, Tasker S, Irving PM, Sanderson J, Lavender P, Ibrahim F, Corcoran J, Prevost T, Shpigel NY, Marelli-Berg F, Lombardi G, Lord GM (2019) Correction of defective T-regulatory cells from patients with Crohn's disease by ex vivo ligation of retinoic acid receptor-alpha. Gastroenterology 156(6):1775–1787. https://doi.org/10.1053/j.gastro.2019.01.025

Gong W, Guo M, Han Z, Wang Y, Yang P, Xu C, Wang Q, Du L, Li Q, Zhao H, Fan F, Liu Q (2016) Mesenchymal stem cells stimulate intestinal stem cells to repair radiation-induced intestinal injury. Cell Death Dis 7(9):e2387. https://doi.org/10.1038/cddis.2016.276

Gorelik L, Flavell RA (2000) Abrogation of TGFbeta signaling in T cells leads to spontaneous T cell differentiation and autoimmune disease. Immunity 12 (2):171–181. https://doi.org/10.1016/s1074-7613(00)80170-3

Gottschalk RA, Corse E, Allison JP (2012) Expression of Helios in peripherally induced Foxp3+ regulatory T cells. J Immunol 188(3):976–980. https://doi.org/10.4049/jimmunol.1102964

Gregoire C, Lechanteur C, Briquet A, Baudoux E, Baron F, Louis E, Beguin Y (2017) Review article:

mesenchymal stromal cell therapy for inflammatory bowel diseases. Aliment Pharmacol Ther 45 (2):205–221. https://doi.org/10.1111/apt.13864

Griseri T, Asquith M, Thompson C, Powrie F (2010) OX40 is required for regulatory T cell-mediated control of colitis. J Exp Med 207(4):699–709. https://doi.org/10.1084/jem.20091618

Grizotte-Lake M, Zhong G, Duncan K, Kirkwood J, Iyer N, Smolenski I, Isoherranen N, Vaishnava S (2018) Commensals suppress intestinal epithelial cell retinoic acid synthesis to regulate interleukin-22 activity and prevent microbial dysbiosis. Immunity 49 (6):1103–1115. e1106. https://doi.org/10.1016/j.immuni.2018.11.018

Grobeta P, Doser K, Falk W, Obermeier F, Hofmann C (2012) IL-33 attenuates development and perpetuation of chronic intestinal inflammation. Inflamm Bowel Dis 18(10):1900–1909. https://doi.org/10.1002/ibd.22900

Grossman WJ, Verbsky JW, Barchet W, Colonna M, Atkinson JP, Ley TJ (2004a) Human T regulatory cells can use the perforin pathway to cause autologous target cell death. Immunity 21(4):589–601. https://doi.org/10.1016/j.immuni.2004.09.002

Grossman WJ, Verbsky JW, Tollefsen BL, Kemper C, Atkinson JP, Ley TJ (2004b) Differential expression of granzymes A and B in human cytotoxic lymphocyte subsets and T regulatory cells. Blood 104 (9):2840–2848. https://doi.org/10.1182/blood-2004-03-0859

Groux H, O'Garra A, Bigler M, Rouleau M, Antonenko S, de Vries JE, Roncarolo MG (1997) A CD4+ T-cell subset inhibits antigen-specific T-cell responses and prevents colitis. Nature 389(6652):737–742. https://doi.org/10.1038/39614

Guandalini S, Sansotta N (2019) Probiotics in the treatment of inflammatory bowel disease. Adv Exp Med Biol 1125:101–107. https://doi.org/10.1007/5584_2018_319

Gubatan J, Chou ND, Nielsen OH, Moss AC (2019) Systematic review with meta-analysis: association of vitamin D status with clinical outcomes in adult patients with inflammatory bowel disease. Aliment Pharmacol Ther 50(11–12):1146–1158. https://doi.org/10.1111/apt.15506

Gundersen MD, Goll R, Hol J, Olsen T, Rismo R, Sorbye SW, Sundnes O, Haraldsen G, Florholmen J (2016) Loss of interleukin 33 expression in colonic crypts - a potential marker for disease remission in ulcerative colitis. Sci Rep 6:35403. https://doi.org/10.1038/srep35403

Hannier S, Tournier M, Bismuth G, Triebel F (1998) CD3/TCR complex-associated lymphocyte activation gene-3 molecules inhibit CD3/TCR signaling. J Immunol 161(8):4058–4065

Haribhai D, Williams JB, Jia S, Nickerson D, Schmitt EG, Edwards B, Ziegelbauer J, Yassai M, Li SH, Relland LM, Wise PM, Chen A, Zheng YQ, Simpson PM, Gorski J, Salzman NH, Hessner MJ, Chatila TA, Williams CB (2011) A requisite role for induced regulatory T cells in tolerance based on expanding antigen receptor diversity. Immunity 35(1):109–122. https://doi.org/10.1016/j.immuni.2011.03.029

Harrell CR, Simovic Markovic B, Fellabaum C, Arsenijevic A, Djonov V, Arsenijevic N, Volarevic V (2018) Therapeutic potential of mesenchymal stem cell-derived exosomes in the treatment of eye diseases. Adv Exp Med Biol 1089:47–57. https://doi.org/10.1007/5584_2018_219

Harrison CA, Laubitz D, Ohland CL, Midura-Kiela MT, Patil K, Besselsen DG, Jamwal DR, Jobin C, Ghishan FK, Kiela PR (2018) Microbial dysbiosis associated with impaired intestinal Na(+)/H(+) exchange accelerates and exacerbates colitis in ex-germ free mice. Mucosal Immunol 11(5):1329–1341. https://doi.org/10.1038/s41385-018-0035-2

Heidari M, Pouya S, Baghaei K, Aghdaei HA, Namaki S, Zali MR, Hashemi SM (2018) The immunomodulatory effects of adipose-derived mesenchymal stem cells and mesenchymal stem cells-conditioned medium in chronic colitis. J Cell Physiol 233(11):8754–8766. https://doi.org/10.1002/jcp.26765

Hidaka D, Hayase E, Shiratori S, Hasegawa Y, Ishio T, Tateno T, Okada K, Goto H, Sugita J, Onozawa M, Nakagawa M, Kahata K, Endo T, Hashimoto D, Teshima T (2018) The association between the incidence of intestinal graft-vs-host disease and antibiotic use after allogeneic hematopoietic stem cell transplantation. Clin Transpl 32(9):e13361. https://doi.org/10.1111/ctr.13361

Himmel ME, MacDonald KG, Garcia RV, Steiner TS, Levings MK (2013) Helios+ and Helios- cells coexist within the natural FOXP3+ T regulatory cell subset in humans. J Immunol 190(5):2001–2008. https://doi.org/10.4049/jimmunol.1201379

Hoeppli RE, MacDonald KN, Leclair P, Fung VCW, Mojibian M, Gillies J, Rahavi SMR, Campbell AIM, Gandhi SK, Pesenacker AM, Reid G, Lim CJ, Levings MK (2019) Tailoring the homing capacity of human Tregs for directed migration to sites of Th1-inflammation or intestinal regions. Am J Transplant 19(1):62–76. https://doi.org/10.1111/ajt.14936

Holmen N, Lundgren A, Lundin S, Bergin AM, Rudin A, Sjovall H, Ohman L (2006) Functional CD4+CD25high regulatory T cells are enriched in the colonic mucosa of patients with active ulcerative colitis and increase with disease activity. Inflamm Bowel Dis 12(6):447–456. https://doi.org/10.1097/00054725-200606000-00003

Hori S, Nomura T, Sakaguchi S (2003) Control of regulatory T cell development by the transcription factor Foxp3. Science 299(5609):1057–1061. https://doi.org/10.1126/science.1079490

Hotta M, Yoshimura H, Satake A, Tsubokura Y, Ito T, Nomura S (2019) GM-CSF therapy inhibits chronic graft-versus-host disease via expansion of regulatory T cells. Eur J Immunol 49(1):179–191. https://doi.org/10.1002/eji.201847684

Hu S, Yuan J, Xu J, Li X, Zhang G, Ma Q, Zhang B, Hu T, Song G (2019) TNF-alpha and IFN-gamma synergistically inhibit the repairing ability of mesenchymal stem cells on mice colitis and colon cancer. Am J Transl Res 11(9):6207–6220

Huang CT, Workman CJ, Flies D, Pan X, Marson AL, Zhou G, Hipkiss EL, Ravi S, Kowalski J, Levitsky HI, Powell JD, Pardoll DM, Drake CG, Vignali DA (2004) Role of LAG-3 in regulatory T cells. Immunity 21 (4):503–513. https://doi.org/10.1016/j.immuni.2004.08.010

Huang Z, Liu Y, Qi G, Brand D, Zheng SG (2018) Role of vitamin a in the immune system. J Clin Med 7(9):258. https://doi.org/10.3390/jcm7090258

Ito T, Hanabuchi S, Wang YH, Park WR, Arima K, Bover L, Qin FX, Gilliet M, Liu YJ (2008) Two functional subsets of FOXP3+ regulatory T cells in human thymus and periphery. Immunity 28 (6):870–880. https://doi.org/10.1016/j.immuni.2008.03.018

Jamwal DR, Marati RV, Harrison CA, Midura-Kiela MT, Figliuolo Paz VR, Besselsen DG, Ghishan FK, Kiela PR (2020) Total CD3 T cells are necessary and sufficient to induce colitis in Immunodeficient mice with dendritic cell-specific deletion of TGFbR2: a novel IBD model to study CD4 and CD8 T-cell interaction. Inflamm Bowel Dis 26(2):229–241. https://doi.org/10.1093/ibd/izz191

Jang YJ, Kim WK, Han DH, Lee K, Ko G (2019) Lactobacillus fermentum species ameliorate dextran sulfate sodium-induced colitis by regulating the immune response and altering gut microbiota. Gut Microbes 10(6):696–711. https://doi.org/10.1080/19490976.2019.1589281

Jorgensen SP, Hvas CL, Agnholt J, Christensen LA, Heickendorff L, Dahlerup JF (2013) Active Crohn's disease is associated with low vitamin D levels. J Crohns Colitis 7(10):e407–e413. https://doi.org/10.1016/j.crohns.2013.01.012

Josefowicz SZ, Niec RE, Kim HY, Treuting P, Chinen T, Zheng Y, Umetsu DT, Rudensky AY (2012) Extrathymically generated regulatory T cells control mucosal TH2 inflammation. Nature 482 (7385):395–399. https://doi.org/10.1038/nature10772

Kanamori M, Nakatsukasa H, Okada M, Lu Q, Yoshimura A (2016) Induced regulatory T cells: their development, stability, and applications. Trends Immunol 37 (11):803–811. https://doi.org/10.1016/j.it.2016.08.012

Kang SG, Lim HW, Andrisani OM, Broxmeyer HE, Kim CH (2007) Vitamin a metabolites induce gut-homing FoxP3+ regulatory T cells. J Immunol 179 (6):3724–3733. https://doi.org/10.4049/jimmunol.179.6.3724

Kang SW, Kim SH, Lee N, Lee WW, Hwang KA, Shin MS, Lee SH, Kim WU, Kang I (2012) 1,25-Dihyroxyvitamin D3 promotes FOXP3 expression via binding to vitamin D response elements in its conserved noncoding sequence region. J Immunol 188(11):5276–5282. https://doi.org/10.4049/jimmunol.1101211

Karwacz K, Bricogne C, MacDonald D, Arce F, Bennett CL, Collins M, Escors D (2011) PD-L1 co-stimulation contributes to ligand-induced T cell receptor down-modulation on CD8+ T cells. EMBO Mol Med 3 (10):581–592. https://doi.org/10.1002/emmm.201100165

Khalili A, Ebrahimpour S, Maleki I, Abediankenari S, Afrouzi MM (2018) CD4+CD25+CD127low FoxP3+ regulatory T cells in Crohn's disease. Rom J Intern Med 56(3):158–166. https://doi.org/10.2478/rjim-2018-0006

Kim KS, Hong SW, Han D, Yi J, Jung J, Yang BG, Lee JY, Lee M, Surh CD (2016) Dietary antigens limit mucosal immunity by inducing regulatory T cells in the small intestine. Science 351(6275):858–863. https://doi.org/10.1126/science.aac5560

Kinoshita M, Kayama H, Kusu T, Yamaguchi T, Kunisawa J, Kiyono H, Sakaguchi S, Takeda K (2012) Dietary folic acid promotes survival of Foxp3 + regulatory T cells in the colon. J Immunol 189 (6):2869–2878. https://doi.org/10.4049/jimmunol.1200420

Kitazawa Y, Fujino M, Wang Q, Kimura H, Azuma M, Kubo M, Abe R, Li XK (2007) Involvement of the programmed death-1/programmed death-1 ligand pathway in CD4+CD25+ regulatory T-cell activity to suppress alloimmune responses. Transplantation 83 (6):774–782. https://doi.org/10.1097/01.tp.0000256293.90270.e8

Knyazev OV, Fadeeva NA, Kagramanova AV, Belyakov NI, Orlova NV, Lishchinskaya AA, Konoplyannikov AG, Parfenov AI (2018) Stem cell therapy for perianal Crohn's disease. Ter Arkh 90(3):60–66. https://doi.org/10.26442/terarkh201890360-66

Kobori A, Yagi Y, Imaeda H, Ban H, Bamba S, Tsujikawa T, Saito Y, Fujiyama Y, Andoh A (2010) Interleukin-33 expression is specifically enhanced in inflamed mucosa of ulcerative colitis. J Gastroenterol 45(10):999–1007. https://doi.org/10.1007/s00535-010-0245-1

Konkel JE, Chen W (2011) Balancing acts: the role of TGF-beta in the mucosal immune system. Trends Mol Med 17(11):668–676. https://doi.org/10.1016/j.molmed.2011.07.002

Koreth J, Matsuoka K, Kim HT, McDonough SM, Bindra B, Alyea EP 3rd, Armand P, Cutler C, Ho VT, Treister NS, Bienfang DC, Prasad S, Tzachanis D, Joyce RM, Avigan DE, Antin JH, Ritz J, Soiffer RJ (2011) Interleukin-2 and regulatory T cells in graft-versus-host disease. N Engl J Med 365(22):2055–2066. https://doi.org/10.1056/NEJMoa1108188

Koreth J, Kim HT, Jones KT, Lange PB, Reynolds CG, Chammas MJ, Dusenbury K, Whangbo J, Nikiforow S, Alyea EP III, Armand P, Cutler CS, Ho VT, Chen YB, Avigan D, Blazar BR, Antin JH, Ritz J, Soiffer RJ (2016) Efficacy, durability, and response predictors of low-dose interleukin-2 therapy for chronic graft-versus-host disease. Blood 128(1):130–137. https://doi.org/10.1182/blood-2016-02-702852

Korn LL, Hubbeling HG, Porrett PM, Yang Q, Barnett LG, Laufer TM (2014) Regulatory T cells occupy an isolated niche in the intestine that is antigen independent. Cell Rep 9(5):1567–1573. https://doi.org/10.1016/j.celrep.2014.11.006

Kreisman LS, Cobb BA (2011) Glycoantigens induce human peripheral Tr1 cell differentiation with gut-homing specialization. J Biol Chem 286 (11):8810–8818. https://doi.org/10.1074/jbc.M110.206011

Kryczek I, Wu K, Zhao E, Wei S, Vatan L, Szeliga W, Huang E, Greenson J, Chang A, Rolinski J, Radwan P, Fang J, Wang G, Zou W (2011) IL-17+ regulatory T cells in the microenvironments of chronic inflammation and cancer. J Immunol 186(7):4388–4395. https://doi.org/10.4049/jimmunol.1003251

Kulkarni AB, Huh CG, Becker D, Geiser A, Lyght M, Flanders KC, Roberts AB, Sporn MB, Ward JM, Karlsson S (1993) Transforming growth factor beta 1 null mutation in mice causes excessive inflammatory response and early death. Proc Natl Acad Sci U S A 90 (2):770–774. https://doi.org/10.1073/pnas.90.2.770

Kullberg MC, Jankovic D, Gorelick PL, Caspar P, Letterio JJ, Cheever AW, Sher A (2002) Bacteria-triggered CD4(+) T regulatory cells suppress ? Helicobacter hepaticus-induced colitis. J Exp Med 196 (4):505–515. https://doi.org/10.1084/jem.20020556

Lahat N, Shapiro S, Karban A, Gerstein R, Kinarty A, Lerner A (1999) Cytokine profile in coeliac disease. Scand J Immunol 49(4):441–446. https://doi.org/10.1046/j.1365-3083.1999.00523.x

Lahdeaho ML, Scheinin M, Vuotikka P, Taavela J, Popp A, Laukkarinen J, Koffert J, Koivurova OP, Pesu M, Kivela L, Lovro Z, Keisala J, Isola J, Parnes JR, Leon F, Maki M (2019) Safety and efficacy of AMG 714 in adults with coeliac disease exposed to gluten challenge: a phase 2a, randomised, double-blind, placebo-controlled study. Lancet Gastroenterol Hepatol 4(12):948–959. https://doi.org/10.1016/S2468-1253(19)30264-X

Larmonier CB, Shehab KW, Ghishan FK, Kiela PR (2015) T lymphocyte dynamics in inflammatory bowel diseases: role of the microbiome. Biomed Res Int 2015:504638. https://doi.org/10.1155/2015/504638

Lathrop SK, Bloom SM, Rao SM, Nutsch K, Lio CW, Santacruz N, Peterson DA, Stappenbeck TS, Hsieh CS (2011) Peripheral education of the immune system by colonic commensal microbiota. Nature 478 (7368):250–254. https://doi.org/10.1038/nature10434

Laudisi F, Dinallo V, Di Fusco D, Monteleone G (2016) Smad7 and its potential as therapeutic target in inflammatory bowel diseases. Curr Drug Metab 17(3):303–306. https://doi.org/10.2174/1389200217666151210130103

Law AD, Dutta U, Kochhar R, Vaishnavi C, Kumar S, Noor T, Bhadada S, Singh K (2019) Vitamin D deficiency in adult patients with ulcerative colitis: prevalence and relationship with disease severity, extent, and duration. Indian J Gastroenterol 38(1):6–14. https://doi.org/10.1007/s12664-019-00932-z

Lee YK, Mehrabian P, Boyajian S, Wu WL, Selicha J, Vonderfecht S, Mazmanian SK (2018) The protective role of Bacteroides fragilis in a murine model of colitis-associated colorectal cancer. mSphere 3(6):e00587-18. https://doi.org/10.1128/mSphere.00587-18

Levine AG, Arvey A, Jin W, Rudensky AY (2014) Continuous requirement for the TCR in regulatory T cell function. Nat Immunol 15(11):1070–1078. https://doi.org/10.1038/ni.3004

Levings MK, Roncarolo MG (2005) Phenotypic and functional differences between human CD4+CD25+ and type 1 regulatory T cells. Curr Top Microbiol Immunol 293:303–326. https://doi.org/10.1007/3-540-27702-1_14

Levings MK, Sangregorio R, Sartirana C, Moschin AL, Battaglia M, Orban PC, Roncarolo MG (2002) Human CD25+CD4+ T suppressor cell clones produce transforming growth factor beta, but not interleukin 10, and are distinct from type 1 T regulatory cells. J Exp Med 196(10):1335–1346. https://doi.org/10.1084/jem.20021139

Li MO, Sanjabi S, Flavell RA (2006) Transforming growth factor-beta controls development, homeostasis, and tolerance of T cells by regulatory T cell-dependent and -independent mechanisms. Immunity 25 (3):455–471. https://doi.org/10.1016/j.immuni.2006.07.011

Li X, Liang Y, LeBlanc M, Benner C, Zheng Y (2014) Function of a Foxp3 cis-element in protecting regulatory T cell identity. Cell 158(4):734–748. https://doi.org/10.1016/j.cell.2014.07.030

Li Z, Vermeire S, Bullens D, Ferrante M, Van Steen K, Noman M, Rutgeerts P, Ceuppens JL, Van Assche G (2015) Restoration of Foxp3+ regulatory T-cell subsets and Foxp3- type 1 regulatory-like T cells in inflammatory bowel diseases during anti-tumor necrosis factor therapy. Inflamm Bowel Dis 21(10):2418–2428. https://doi.org/10.1097/MIB.0000000000000509

Li J, Ueno A, Iacucci M, Fort Gasia M, Jijon HB, Panaccione R, Kaplan GG, Beck PL, Luider J, Barkema HW, Qian J, Gui X, Ghosh S (2017) Crossover subsets of CD4(+) T lymphocytes in the intestinal lamina propria of patients with Crohn's disease and ulcerative colitis. Dig Dis Sci 62(9):2357–2368. https://doi.org/10.1007/s10620-017-4596-9

Li Y, Wang F, Guo R, Zhang Y, Chen D, Li X, Tian W, Xie X, Jiang Z (2019) Exosomal sphingosine 1-phosphate secreted by mesenchymal stem cells regulated Treg/Th17 balance in aplastic anemia. IUBMB Life 71(9):1284–1292. https://doi.org/10.1002/iub.2035

Lin YC, Mahalingam J, Chiang JM, Su PJ, Chu YY, Lai HY, Fang JH, Huang CT, Chiu CT, Lin CY (2013) Activated but not resting regulatory T cells accumulated in tumor microenvironment and correlated with tumor progression in patients with colorectal cancer. Int J Cancer 132(6):1341–1350. https://doi.org/10.1002/ijc.27784

Liston A, Gray DH (2014) Homeostatic control of regulatory T cell diversity. Nat Rev Immunol 14 (3):154–165. https://doi.org/10.1038/nri3605

Liu Y, Lan Q, Lu L, Chen M, Xia Z, Ma J, Wang J, Fan H, Shen Y, Ryffel B, Brand D, Quismorio F, Liu Z, Horwitz DA, Xu A, Zheng SG (2014) Phenotypic and functional characteristic of a newly identified CD8 + Foxp3- CD103+ regulatory T cells. J Mol Cell Biol 6 (1):81–92. https://doi.org/10.1093/jmcb/mjt026

Lochner M, Peduto L, Cherrier M, Sawa S, Langa F, Varona R, Riethmacher D, Si-Tahar M, Di Santo JP, Eberl G (2008) In vivo equilibrium of proinflammatory IL-17+ and regulatory IL-10+ Foxp3+ RORgamma t+ T cells. J Exp Med 205(6):1381–1393. https://doi.org/10.1084/jem.20080034

Loebbermann J, Thornton H, Durant L, Sparwasser T, Webster KE, Sprent J, Culley FJ, Johansson C, Openshaw PJ (2012) Regulatory T cells expressing granzyme B play a critical role in controlling lung inflammation during acute viral infection. Mucosal Immunol 5(2):161–172. https://doi.org/10.1038/mi.2011.62

Lopez-Casado MA, Lorite P, Palomeque T, Torres MI (2017) Potential role of the IL-33/ST2 axis in celiac disease. Cell Mol Immunol 14(3):285–292. https://doi.org/10.1038/cmi.2015.85

Lopez-Siles M, Enrich-Capo N, Aldeguer X, Sabat-Mir M, Duncan SH, Garcia-Gil LJ, Martinez-Medina M (2018) Alterations in the abundance and co-occurrence of Akkermansia muciniphila and Faecalibacterium prausnitzii in the colonic mucosa of inflammatory bowel disease subjects. Front Cell Infect Microbiol 8:281. https://doi.org/10.3389/fcimb.2018.00281

Lord JD, Hackman RC, Gooley TA, Wood BL, Moklebust AC, Hockenbery DM, Steinbach G, Ziegler SF, McDonald GB (2011) Blood and gastric FOXP3+ T cells are not decreased in human gastric graft-versus-host disease. Biol Blood Marrow Transplant 17 (4):486–496. https://doi.org/10.1016/j.bbmt.2010.09.015

Lord JD, Shows DM, Chen J, Thirlby RC (2015) Human blood and mucosal regulatory T cells express activation markers and inhibitory receptors in inflammatory bowel disease. PLoS One 10(8):e0136485. https://doi.org/10.1371/journal.pone.0136485

Lu L, Ma J, Li Z, Lan Q, Chen M, Liu Y, Xia Z, Wang J, Han Y, Shi W, Quesniaux V, Ryffel B, Brand D, Li B, Liu Z, Zheng SG (2011) All-trans retinoic acid promotes TGF-beta-induced Tregs via histone modification but not DNA demethylation on Foxp3 gene locus. PLoS One 6(9):e24590. https://doi.org/10.1371/journal.pone.0024590

Luo Y, Xue Y, Wang J, Dang J, Fang Q, Huang G, Olsen N, Zheng SG (2019) Negligible effect of sodium chloride on the development and function of TGF-beta-induced CD4(+) Foxp3(+) regulatory T cells. Cell Rep 26(7):1869–1879. e1863. https://doi.org/10.1016/j.celrep.2019.01.066

Ma C, Dong X (2011) Colorectal cancer-derived Foxp3(+) IL-17(+) T cells suppress tumour-specific CD8+ T cells. Scand J Immunol 74(1):47–51. https://doi.org/10.1111/j.1365-3083.2011.02539.x

Ma F, Li S, Gao X, Zhou J, Zhu X, Wang D, Cai Y, Li F, Yang Q, Gu X, Ge W, Liu H, Xiao X, Hao H (2019) Interleukin-6-mediated CCR9(+) interleukin-17-producing regulatory T cells polarization increases the severity of necrotizing enterocolitis. EBioMedicine 44:71–85. https://doi.org/10.1016/j.ebiom.2019.05.042

Machiels K, Joossens M, Sabino J, De Preter V, Arijs I, Eeckhaut V, Ballet V, Claes K, Van Immerseel F, Verbeke K, Ferrante M, Verhaegen J, Rutgeerts P, Vermeire S (2014) A decrease of the butyrate-producing species Roseburia hominis and Faecalibacterium prausnitzii defines dysbiosis in patients with ulcerative colitis. Gut 63(8):1275–1283. https://doi.org/10.1136/gutjnl-2013-304833

Mahalingam J, Lin CY, Chiang JM, Su PJ, Chu YY, Lai HY, Fang JH, Huang CT, Lin YC (2014) CD4(+) T cells expressing latency-associated peptide and Foxp3 are an activated subgroup of regulatory T cells enriched in patients with colorectal cancer. PLoS One 9(9):e108554. https://doi.org/10.1371/journal.pone.0108554

Malek TR, Yu A, Vincek V, Scibelli P, Kong L (2002) CD4 regulatory T cells prevent lethal autoimmunity in IL-2Rbeta-deficient mice. Implications for the nonredundant function of IL-2. Immunity 17(2):167–178. https://doi.org/10.1016/s1074-7613(02)00367-9

Mandapathil M, Hilldorfer B, Szczepanski MJ, Czystowska M, Szajnik M, Ren J, Lang S, Jackson EK, Gorelik E, Whiteside TL (2010) Generation and accumulation of immunosuppressive adenosine by human CD4+CD25highFOXP3+ regulatory T cells. J Biol Chem 285(10):7176–7186. https://doi.org/10.1074/jbc.M109.047423

Marie JC, Letterio JJ, Gavin M, Rudensky AY (2005) TGF-beta1 maintains suppressor function and Foxp3 expression in CD4+CD25+ regulatory T cells. J Exp Med 201(7):1061–1067. https://doi.org/10.1084/jem.20042276

Marie JC, Liggitt D, Rudensky AY (2006) Cellular mechanisms of fatal early-onset autoimmunity in mice with the T cell-specific targeting of transforming growth factor-beta receptor. Immunity 25(3):441–454. https://doi.org/10.1016/j.immuni.2006.07.012

Matsuoka K, Koreth J, Kim HT, Bascug G, McDonough S, Kawano Y, Murase K, Cutler C, Ho VT, Alyea EP, Armand P, Blazar BR, Antin JH, Soiffer RJ, Ritz J (2013) Low-dose interleukin-2 therapy restores regulatory T cell homeostasis in patients with chronic graft-versus-host disease. Sci Transl Med 5(179):179ra143. https://doi.org/10.1126/scitranslmed.3005265

Matta BM, Lott JM, Mathews LR, Liu Q, Rosborough BR, Blazar BR, Turnquist HR (2014) IL-33 is an unconventional Alarmin that stimulates IL-2 secretion by dendritic cells to selectively expand IL-33R/ST2+ regulatory T cells. J Immunol 193(8):4010–4020. https://doi.org/10.4049/jimmunol.1400481

Matta BM, Reichenbach DK, Zhang X, Mathews L, Koehn BH, Dwyer GK, Lott JM, Uhl FM, Pfeifer D, Feser CJ, Smith MJ, Liu Q, Zeiser R, Blazar BR,

Turnquist HR (2016) Peri-alloHCT IL-33 administration expands recipient T-regulatory cells that protect mice against acute GVHD. Blood 128(3):427–439. https://doi.org/10.1182/blood-2015-12-684142

Maynard CL, Harrington LE, Janowski KM, Oliver JR, Zindl CL, Rudensky AY, Weaver CT (2007) Regulatory T cells expressing interleukin 10 develop from Foxp3+ and Foxp3- precursor cells in the absence of interleukin 10. Nat Immunol 8(9):931–941. https://doi.org/10.1038/ni1504

McHugh RS, Whitters MJ, Piccirillo CA, Young DA, Shevach EM, Collins M, Byrne MC (2002) CD4(+) CD25(+) immunoregulatory T cells: gene expression analysis reveals a functional role for the glucocorticoid-induced TNF receptor. Immunity 16 (2):311–323. https://doi.org/10.1016/s1074-7613(02) 00280-7

Mezrich JD, Fechner JH, Zhang X, Johnson BP, Burlingham WJ, Bradfield CA (2010) An interaction between kynurenine and the aryl hydrocarbon receptor can generate regulatory T cells. J Immunol 185 (6):3190–3198. https://doi.org/10.4049/jimmunol.0903670

Michel S, Benner A, Tariverdian M, Wentzensen N, Hoefler P, Pommerencke T, Grabe N, von Knebel Doeberitz M, Kloor M (2008) High density of FOXP3-positive T cells infiltrating colorectal cancers with microsatellite instability. Br J Cancer 99 (11):1867–1873. https://doi.org/10.1038/sj.bjc.6604756

Mikulic J, Longet S, Favre L, Benyacoub J, Corthesy B (2017) Secretory IgA in complex with Lactobacillus rhamnosus potentiates mucosal dendritic cell-mediated Treg cell differentiation via TLR regulatory proteins, RALDH2 and secretion of IL-10 and TGF-beta. Cell Mol Immunol 14(6):546–556. https://doi.org/10.1038/cmi.2015.110

Milpied P, Renand A, Bruneau J, Mendes-da-Cruz DA, Jacquelin S, Asnafi V, Rubio MT, MacIntyre E, Lepelletier Y, Hermine O (2009) Neuropilin-1 is not a marker of human Foxp3+ Treg. Eur J Immunol 39 (6):1466–1471. https://doi.org/10.1002/eji.200839040

Min B, Thornton A, Caucheteux SM, Younes SA, Oh K, Hu-Li J, Paul WE (2007) Gut flora antigens are not important in the maintenance of regulatory T cell heterogeneity and homeostasis. Eur J Immunol 37 (7):1916–1923. https://doi.org/10.1002/eji.200737236

Miragaia RJ, Gomes T, Chomka A, Jardine L, Riedel A, Hegazy AN, Whibley N, Tucci A, Chen X, Lindeman I, Emerton G, Krausgruber T, Shields J, Haniffa M, Powrie F, Teichmann SA (2019) Single-cell transcriptomics of regulatory T cells reveals trajectories of tissue adaptation. Immunity 50 (2):493–504. e497. https://doi.org/10.1016/j.immuni.2019.01.001

Miyara M, Yoshioka Y, Kitoh A, Shima T, Wing K, Niwa A, Parizot C, Taflin C, Heike T, Valeyre D, Mathian A, Nakahata T, Yamaguchi T, Nomura T, Ono M, Amoura Z, Gorochov G, Sakaguchi S (2009) Functional delineation and differentiation dynamics of human CD4+ T cells expressing the FoxP3 transcription factor. Immunity 30(6):899–911. https://doi.org/10.1016/j.immuni.2009.03.019

Moheb-Alian A, Forouzesh F, Rostami-Nejad M, Rostami K (2016) Mesenchymal stem cells as potential therapeutic approaches in celiac disease. Gastroenterol Hepatol Bed Bench 9(suppl 1):S1–S7

Molodecky NA, Soon IS, Rabi DM, Ghali WA, Ferris M, Chernoff G, Benchimol EI, Panaccione R, Ghosh S, Barkema HW, Kaplan GG (2012) Increasing incidence and prevalence of the inflammatory bowel diseases with time, based on systematic review. Gastroenterology 142(1):46–54.e42, quiz e30. https://doi.org/10.1053/j.gastro.2011.10.001

Monteleone G, Neurath MF, Ardizzone S, Di Sabatino A, Fantini MC, Castiglione F, Scribano ML, Armuzzi A, Caprioli F, Sturniolo GC, Rogai F, Vecchi M, Atreya R, Bossa F, Onali S, Fichera M, Corazza GR, Biancone L, Savarino V, Pica R, Orlando A, Pallone F (2015) Mongersen, an oral SMAD7 antisense oligonucleotide, and Crohn's disease. N Engl J Med 372 (12):1104–1113. https://doi.org/10.1056/NEJMoa1407250

Monteleone G, Di Sabatino A, Ardizzone S, Pallone F, Usiskin K, Zhan X, Rossiter G, Neurath MF (2016) Impact of patient characteristics on the clinical efficacy of mongersen (GED-0301), an oral Smad7 antisense oligonucleotide, in active Crohn's disease. Aliment Pharmacol Ther 43(6):717–724. https://doi.org/10.1111/apt.13526

Moran CJ, Walters TD, Guo CH, Kugathasan S, Klein C, Turner D, Wolters VM, Bandsma RH, Mouzaki M, Zachos M, Langer JC, Cutz E, Benseler SM, Roifman CM, Silverberg MS, Griffiths AM, Snapper SB, Muise AM (2013) IL-10R polymorphisms are associated with very-early-onset ulcerative colitis. Inflamm Bowel Dis 19(1):115–123. https://doi.org/10.1002/ibd.22974

Morgan ME, van Bilsen JH, Bakker AM, Heemskerk B, Schilham MW, Hartgers FC, Elferink BG, van der Zanden L, de Vries RR, Huizinga TW, Ottenhoff TH, Toes RE (2005) Expression of FOXP3 mRNA is not confined to CD4+CD25+ T regulatory cells in humans. Hum Immunol 66(1):13–20. https://doi.org/10.1016/j.humimm.2004.05.016

Mottet C, Uhlig HH, Powrie F (2003) Cutting edge: cure of colitis by CD4+CD25+ regulatory T cells. J Immunol 170(8):3939–3943. https://doi.org/10.4049/jimmunol.170.8.3939

Mucida D, Park Y, Kim G, Turovskaya O, Scott I, Kronenberg M, Cheroutre H (2007) Reciprocal TH17 and regulatory T cell differentiation mediated by retinoic acid. Science 317(5835):256–260. https://doi.org/10.1126/science.1145697

Mukhopadhyay D, Weaver L, Tobin R, Henderson S, Beeram M, Newell-Rogers MK, Perger L (2014) Intrauterine growth restriction and prematurity influence regulatory T cell development in newborns. J Pediatr

Surg 49(5):727–732. https://doi.org/10.1016/j.jpedsurg.2014.02.055

Muzes G, Molnar B, Tulassay Z, Sipos F (2012) Changes of the cytokine profile in inflammatory bowel diseases. World J Gastroenterol 18(41):5848–5861. https://doi.org/10.3748/wjg.v18.i41.5848

Nagorsen D, Scheibenbogen C, Letsch A, Germer CT, Buhr HJ, Hegewisch-Becker S, Rivoltini L, Thiel E, Keilholz U (2005) T cell responses against tumor associated antigens and prognosis in colorectal cancer patients. J Transl Med 3(1):3. https://doi.org/10.1186/1479-5876-3-3

Nakanishi Y, Ikebuchi R, Chtanova T, Kusumoto Y, Okuyama H, Moriya T, Honda T, Kabashima K, Watanabe T, Sakai Y, Tomura M (2018) Regulatory T cells with superior immunosuppressive capacity emigrate from the inflamed colon to draining lymph nodes. Mucosal Immunol 11(2):437–448. https://doi.org/10.1038/mi.2017.64

Narushima S, Sugiura Y, Oshima K, Atarashi K, Hattori M, Suematsu M, Honda K (2014) Characterization of the 17 strains of regulatory T cell-inducing human-derived Clostridia. Gut Microbes 5 (3):333–339. https://doi.org/10.4161/gmic.28572

Nascimento DC, Melo PH, Pineros AR, Ferreira RG, Colon DF, Donate PB, Castanheira FV, Gozzi A, Czaikoski PG, Niedbala W, Borges MC, Zamboni DS, Liew FY, Cunha FQ, Alves-Filho JC (2017) IL-33 contributes to sepsis-induced long-term immunosuppression by expanding the regulatory T cell population. Nat Commun 8:14919. https://doi.org/10.1038/ncomms14919

Nemeth ZH, Bogdanovski DA, Barratt-Stopper P, Paglinco SR, Antonioli L, Rolandelli RH (2017) Crohn's disease and ulcerative colitis show unique cytokine profiles. Cureus 9(4):e1177. https://doi.org/10.7759/cureus.1177

Ni J, Wu GD, Albenberg L, Tomov VT (2017) Gut microbiota and IBD: causation or correlation? Nat Rev Gastroenterol Hepatol 14(10):573–584. https://doi.org/10.1038/nrgastro.2017.88

Ni C, Gan X, Li X, Sun H, Chen Z, Lu H (2019) Vitamin D alleviates acute graft-versus-host disease through promoting the generation of Foxp3(+) T cells. Ann Transl Med 7(23):748. https://doi.org/10.21037/atm.2019.11.102

Nilsen EM, Jahnsen FL, Lundin KE, Johansen FE, Fausa O, Sollid LM, Jahnsen J, Scott H, Brandtzaeg P (1998) Gluten induces an intestinal cytokine response strongly dominated by interferon gamma in patients with celiac disease. Gastroenterology 115 (3):551–563. https://doi.org/10.1016/s0016-5085(98)70134-9

Nino DF, Sodhi CP, Egan CE, Zhou Q, Lin J, Lu P, Yamaguchi Y, Jia H, Martin LY, Good M, Fulton WB, Prindle T Jr, Ozolek JA, Hackam DJ (2017) Retinoic acid improves incidence and severity of necrotizing enterocolitis by lymphocyte balance restitution and repopulation of LGR5+ intestinal stem cells. Shock 47(1):22–32. https://doi.org/10.1097/SHK.0000000000000713

Nishio J, Baba M, Atarashi K, Tanoue T, Negishi H, Yanai H, Habu S, Hori S, Honda K, Taniguchi T (2015) Requirement of full TCR repertoire for regulatory T cells to maintain intestinal homeostasis. Proc Natl Acad Sci U S A 112(41):12770–12775. https://doi.org/10.1073/pnas.1516617112

Nutsch K, Chai JN, Ai TL, Russler-Germain E, Feehley T, Nagler CR, Hsieh CS (2016) Rapid and efficient generation of regulatory T cells to commensal antigens in the periphery. Cell Rep 17(1):206–220. https://doi.org/10.1016/j.celrep.2016.08.092

Oberle N, Eberhardt N, Falk CS, Krammer PH, Suri-Payer E (2007) Rapid suppression of cytokine transcription in human CD4+CD25 T cells by CD4+Foxp3+ regulatory T cells: independence of IL-2 consumption, TGF-beta, and various inhibitors of TCR signaling. J Immunol 179(6):3578–3587. https://doi.org/10.4049/jimmunol.179.6.3578

Oboki K, Ohno T, Kajiwara N, Arae K, Morita H, Ishii A, Nambu A, Abe T, Kiyonari H, Matsumoto K, Sudo K, Okumura K, Saito H, Nakae S (2010) IL-33 is a crucial amplifier of innate rather than acquired immunity. Proc Natl Acad Sci U S A 107(43):18581–18586. https://doi.org/10.1073/pnas.1003059107

Ohnmacht C, Park JH, Cording S, Wing JB, Atarashi K, Obata Y, Gaboriau-Routhiau V, Marques R, Dulauroy S, Fedoseeva M, Busslinger M, Cerf-Bensussan N, Boneca IG, Voehringer D, Hase K, Honda K, Sakaguchi S, Eberl G (2015) MUCOSAL IMMUNOLOGY. The microbiota regulates type 2 immunity through RORgammat(+) T cells. Science 349(6251):989–993. https://doi.org/10.1126/science.aac4263

Ohtani H (2007) Focus on TILs: prognostic significance of tumor infiltrating lymphocytes in human colorectal cancer. Cancer Immun 7:4

Oldenhove G, Bouladoux N, Wohlfert EA, Hall JA, Chou D, Dos Santos L, O'Brien S, Blank R, Lamb E, Natarajan S, Kastenmayer R, Hunter C, Grigg ME, Belkaid Y (2009) Decrease of Foxp3+ Treg cell number and acquisition of effector cell phenotype during lethal infection. Immunity 31(5):772–786. https://doi.org/10.1016/j.immuni.2009.10.001

Ouyang W, Beckett O, Ma Q, Li MO (2010) Transforming growth factor-beta signaling curbs thymic negative selection promoting regulatory T cell development. Immunity 32(5):642–653. https://doi.org/10.1016/j.immuni.2010.04.012

Paidassi H, Acharya M, Zhang A, Mukhopadhyay S, Kwon M, Chow C, Stuart LM, Savill J, Lacy-Hulbert A (2011) Preferential expression of integrin alphavbeta8 promotes generation of regulatory T cells by mouse CD103+ dendritic cells. Gastroenterology 141(5):1813–1820. https://doi.org/10.1053/j.gastro.2011.06.076

Pandiyan P, Zheng L, Ishihara S, Reed J, Lenardo MJ (2007) CD4+CD25+Foxp3+ regulatory T cells induce cytokine deprivation-mediated apoptosis of effector CD4+ T cells. Nat Immunol 8(12):1353–1362. https://doi.org/10.1038/ni1536

Pandolfi JB, Ferraro AA, Sananez I, Gancedo MC, Baz P, Billordo LA, Fainboim L, Arruvito L (2016) ATP-induced inflammation drives tissue-resident Th17 cells in metabolically unhealthy obesity. J Immunol 196(8):3287–3296. https://doi.org/10.4049/jimmunol.1502506

Pang Y, Du X, Xu X, Wang M, Li Z (2018a) Impairment of regulatory T cells in patients with neonatal necrotizing enterocolitis. Int Immunopharmacol 63:19–25. https://doi.org/10.1016/j.intimp.2018.07.029

Pang Y, Du X, Xu X, Wang M, Li Z (2018b) Monocyte activation and inflammation can exacerbate Treg/Th17 imbalance in infants with neonatal necrotizing enterocolitis. Int Immunopharmacol 59:354–360. https://doi.org/10.1016/j.intimp.2018.04.026

Papiernik M, de Moraes ML, Pontoux C, Vasseur F, Penit C (1998) Regulatory CD4 T cells: expression of IL-2R alpha chain, resistance to clonal deletion and IL-2 dependency. Int Immunol 10(4):371–378. https://doi.org/10.1093/intimm/10.4.371

Park JS, Choi JW, Jhun J, Kwon JY, Lee BI, Yang CW, Park SH, Cho ML (2018) Lactobacillus acidophilus improves intestinal inflammation in an acute colitis mouse model by regulation of Th17 and Treg cell balance and fibrosis development. J Med Food 21 (3):215–224. https://doi.org/10.1089/jmf.2017.3990

Pastorelli L, Garg RR, Hoang SB, Spina L, Mattioli B, Scarpa M, Fiocchi C, Vecchi M, Pizarro TT (2010) Epithelial-derived IL-33 and its receptor ST2 are dysregulated in ulcerative colitis and in experimental Th1/Th2 driven enteritis. Proc Natl Acad Sci U S A 107(17):8017–8022. https://doi.org/10.1073/pnas.0912678107

Pastorelli L, De Salvo C, Vecchi M, Pizarro TT (2013) The role of IL-33 in gut mucosal inflammation. Mediat Inflamm 2013:608187. https://doi.org/10.1155/2013/608187

Patel RM, Underwood MA (2018) Probiotics and necrotizing enterocolitis. Semin Pediatr Surg 27(1):39–46. https://doi.org/10.1053/j.sempedsurg.2017.11.008

Pesu M, Watford WT, Wei L, Xu L, Fuss I, Strober W, Andersson J, Shevach EM, Quezado M, Bouladoux N, Roebroek A, Belkaid Y, Creemers J, O'Shea JJ (2008) T-cell-expressed proprotein convertase furin is essential for maintenance of peripheral immune tolerance. Nature 455(7210):246–250. https://doi.org/10.1038/nature07210

Pichery M, Mirey E, Mercier P, Lefrancais E, Dujardin A, Ortega N, Girard JP (2012) Endogenous IL-33 is highly expressed in mouse epithelial barrier tissues, lymphoid organs, brain, embryos, and inflamed tissues: in situ analysis using a novel Il-33-LacZ gene trap reporter strain. J Immunol 188(7):3488–3495. https://doi.org/10.4049/jimmunol.1101977

Povoleri GAM, Nova-Lamperti E, Scotta C, Fanelli G, Chen YC, Becker PD, Boardman D, Costantini B, Romano M, Pavlidis P, McGregor R, Pantazi E, Chauss D, Sun HW, Shih HY, Cousins DJ, Cooper N, Powell N, Kemper C, Pirooznia M, Laurence A, Kordasti S, Kazemian M, Lombardi G, Afzali B (2018) Human retinoic acid-regulated CD161 (+) regulatory T cells support wound repair in intestinal mucosa. Nat Immunol 19(12):1403–1414. https://doi.org/10.1038/s41590-018-0230-z

Powrie F, Mason D (1990) OX-22high CD4+ T cells induce wasting disease with multiple organ pathology: prevention by the OX-22low subset. J Exp Med 172 (6):1701–1708. https://doi.org/10.1084/jem.172.6.1701

Powrie F, Leach MW, Mauze S, Caddle LB, Coffman RL (1993) Phenotypically distinct subsets of CD4+ T cells induce or protect from chronic intestinal inflammation in C. B-17 scid mice. Int Immunol 5(11):1461–1471. https://doi.org/10.1093/intimm/5.11.1461

Powrie F, Carlino J, Leach MW, Mauze S, Coffman RL (1996) A critical role for transforming growth factor-beta but not interleukin 4 in the suppression of T helper type 1-mediated colitis by CD45RB(low) CD4+ T cells. J Exp Med 183(6):2669–2674. https://doi.org/10.1084/jem.183.6.2669

Qazi KR, Bach Jensen G, van der Heiden M, Bjorkander S, Holmlund U, Haileselassie Y, Kokkinou E, Marchini G, Jenmalm MC, Abrahamsson T, Sverremark-Ekstrom E (2020) Extremely preterm infants have significant alterations in their conventional T cell compartment during the first weeks of life. J Immunol 204(1):68–77. https://doi.org/10.4049/jimmunol.1900941

Ramalingam R, Larmonier CB, Thurston RD, Midura-Kiela MT, Zheng SG, Ghishan FK, Kiela PR (2012) Dendritic cell-specific disruption of TGF-beta receptor II leads to altered regulatory T cell phenotype and spontaneous multiorgan autoimmunity. J Immunol 189(8):3878–3893. https://doi.org/10.4049/jimmunol.1201029

Read S, Malmstrom V, Powrie F (2000) Cytotoxic T lymphocyte-associated antigen 4 plays an essential role in the function of CD25(+)CD4(+) regulatory cells that control intestinal inflammation. J Exp Med 192(2):295–302. https://doi.org/10.1084/jem.192.2.295

Reifen R, Nur T, Ghebermeskel K, Zaiger G, Urizky R, Pines M (2002) Vitamin A deficiency exacerbates inflammation in a rat model of colitis through activation of nuclear factor-kappaB and collagen formation. J Nutr 132(9):2743–2747. https://doi.org/10.1093/jn/132.9.2743

Reikvam DH, Perminow G, Lyckander LG, Gran JM, Brandtzaeg P, Vatn M, Carlsen HS (2011) Increase of regulatory T cells in ileal mucosa of untreated pediatric Crohn's disease patients. Scand J Gastroenterol 46 (5):550–560. https://doi.org/10.3109/00365521.2011.551887

Rieger K, Loddenkemper C, Maul J, Fietz T, Wolff D, Terpe H, Steiner B, Berg E, Miehlke S, Bornhauser M,

Schneider T, Zeitz M, Stein H, Thiel E, Duchmann R, Uharek L (2006) Mucosal FOXP3+ regulatory T cells are numerically deficient in acute and chronic GvHD. Blood 107(4):1717–1723. https://doi.org/10.1182/blood-2005-06-2529

Rodriguez-Cabezas ME, Galvez J, Lorente MD, Concha A, Camuesco D, Azzouz S, Osuna A, Redondo L, Zarzuelo A (2002) Dietary fiber down-regulates colonic tumor necrosis factor alpha and nitric oxide production in trinitrobenzenesulfonic acid-induced colitic rats. J Nutr 132(11):3263–3271. https://doi.org/10.1093/jn/132.11.3263

Roncarolo MG, Gregori S, Bacchetta R, Battaglia M, Gagliani N (2018) The biology of T regulatory type 1 cells and their therapeutic application in immune-mediated diseases. Immunity 49(6):1004–1019. https://doi.org/10.1016/j.immuni.2018.12.001

Round JL, Mazmanian SK (2010) Inducible Foxp3+ regulatory T-cell development by a commensal bacterium of the intestinal microbiota. Proc Natl Acad Sci U S A 107(27):12204–12209. https://doi.org/10.1073/pnas.0909122107

Round JL, Lee SM, Li J, Tran G, Jabri B, Chatila TA, Mazmanian SK (2011) The Toll-like receptor 2 pathway establishes colonization by a commensal of the human microbiota. Science 332(6032):974–977. https://doi.org/10.1126/science.1206095

Sadlack B, Merz H, Schorle H, Schimpl A, Feller AC, Horak I (1993) Ulcerative colitis-like disease in mice with a disrupted interleukin-2 gene. Cell 75 (2):253–261. https://doi.org/10.1016/0092-8674(93)80067-o

Saito T, Nishikawa H, Wada H, Nagano Y, Sugiyama D, Atarashi K, Maeda Y, Hamaguchi M, Ohkura N, Sato E, Nagase H, Nishimura J, Yamamoto H, Takiguchi S, Tanoue T, Suda W, Morita H, Hattori M, Honda K, Mori M, Doki Y, Sakaguchi S (2016) Two FOXP3(+)CD4(+) T cell subpopulations distinctly control the prognosis of colorectal cancers. Nat Med 22(6):679–684. https://doi.org/10.1038/nm.4086

Sakaguchi S, Sakaguchi N, Asano M, Itoh M, Toda M (1995) Immunologic self-tolerance maintained by activated T cells expressing IL-2 receptor alpha-chains (CD25). Breakdown of a single mechanism of self-tolerance causes various autoimmune diseases. J Immunol 155(3):1151–1164

Salama P, Phillips M, Grieu F, Morris M, Zeps N, Joseph D, Platell C, Iacopetta B (2009) Tumor-infiltrating FOXP3+ T regulatory cells show strong prognostic significance in colorectal cancer. J Clin Oncol 27(2):186–192. https://doi.org/10.1200/JCO.2008.18.7229

Salomon B, Lenschow DJ, Rhee L, Ashourian N, Singh B, Sharpe A, Bluestone JA (2000) B7/CD28 costimulation is essential for the homeostasis of the CD4+CD25+ immunoregulatory T cells that control autoimmune diabetes. Immunity 12(4):431–440. https://doi.org/10.1016/s1074-7613(00)80195-8

Salomon BL, Leclerc M, Tosello J, Ronin E, Piaggio E, Cohen JL (2018) Tumor necrosis factor alpha and regulatory T cells in oncoimmunology. Front Immunol 9:444. https://doi.org/10.3389/fimmu.2018.00444

Salvati VM, Mazzarella G, Gianfrani C, Levings MK, Stefanile R, De Giulio B, Iaquinto G, Giardullo N, Auricchio S, Roncarolo MG, Troncone R (2005) Recombinant human interleukin 10 suppresses gliadin dependent T cell activation in ex vivo cultured coeliac intestinal mucosa. Gut 54(1):46–53. https://doi.org/10.1136/gut.2003.023150

Sanchez E, Donat E, Ribes-Koninckx C, Calabuig M, Sanz Y (2010) Intestinal bacteroides species associated with coeliac disease. J Clin Pathol 63(12):1105–1111. https://doi.org/10.1136/jcp.2010.076950

Sanctuary MR, Huang RH, Jones AA, Luck ME, Aherne CM, Jedlicka P, de Zoeten EF, Collins CB (2019) miR-106a deficiency attenuates inflammation in murine IBD models. Mucosal Immunol 12 (1):200–211. https://doi.org/10.1038/s41385-018-0091-7

Sands BE, Feagan BG, Sandborn WJ, Schreiber S, Peyrin-Biroulet L, Colombel JF, Rossiter G, Usiskin K, Ather S, Zhan X, D'Haens G (2019) Mongersen (GED-0301) for active Crohn's disease: results of a phase 3 study. Am J Gastroenterol 115:738–745. https://doi.org/10.14309/ajg.0000000000000493

Sarmento OF, Svingen PA, Xiong Y, Sun Z, Bamidele AO, Mathison AJ, Smyrk TC, Nair AA, Gonzalez MM, Sagstetter MR, Baheti S, McGovern DP, Friton JJ, Papadakis KA, Gautam G, Xavier RJ, Urrutia RA, Faubion WA (2017) The role of the histone methyltransferase enhancer of Zeste homolog 2 (EZH2) in the Pathobiological mechanisms underlying inflammatory bowel disease (IBD). J Biol Chem 292(2):706–722. https://doi.org/10.1074/jbc.M116.749663

Sarrabayrouse G, Bossard C, Chauvin JM, Jarry A, Meurette G, Quevrain E, Bridonneau C, Preisser L, Asehnoune K, Labarriere N, Altare F, Sokol H, Jotereau F (2014) CD4CD8alphaalpha lymphocytes, a novel human regulatory T cell subset induced by colonic bacteria and deficient in patients with inflammatory bowel disease. PLoS Biol 12(4):e1001833. https://doi.org/10.1371/journal.pbio.1001833

Sarrabayrouse G, Landolfi S, Pozuelo M, Willamil J, Varela E, Clark A, Campos D, Herrera C, Santiago A, Machiels K, Vermeire S, Marti M, Espin E, Manichanh C (2020) Mucosal microbial load in Crohn's disease: a potential predictor of response to faecal microbiota transplantation. EBioMedicine 51:102611. https://doi.org/10.1016/j.ebiom.2019.102611

Schiering C, Krausgruber T, Chomka A, Frohlich A, Adelmann K, Wohlfert EA, Pott J, Griseri T, Bollrath J, Hegazy AN, Harrison OJ, Owens BMJ, Lohning M, Belkaid Y, Fallon PG, Powrie F (2014) The alarmin IL-33 promotes regulatory T-cell function

in the intestine. Nature 513(7519):564–568. https://doi.org/10.1038/nature13577
Schlenner SM, Weigmann B, Ruan Q, Chen Y, von Boehmer H (2012) Smad3 binding to the foxp3 enhancer is dispensable for the development of regulatory T cells with the exception of the gut. J Exp Med 209(9):1529–1535. https://doi.org/10.1084/jem.20112646
Scurr M, Ladell K, Besneux M, Christian A, Hockey T, Smart K, Bridgeman H, Hargest R, Phillips S, Davies M, Price D, Gallimore A, Godkin A (2014) Highly prevalent colorectal cancer-infiltrating LAP(+) Foxp3(−) T cells exhibit more potent immunosuppressive activity than Foxp3(+) regulatory T cells. Mucosal Immunol 7(2):428–439. https://doi.org/10.1038/mi.2013.62
Seddiki N, Santner-Nanan B, Martinson J, Zaunders J, Sasson S, Landay A, Solomon M, Selby W, Alexander SI, Nanan R, Kelleher A, Fazekas de St Groth B (2006) Expression of interleukin (IL)-2 and IL-7 receptors discriminates between human regulatory and activated T cells. J Exp Med 203(7):1693–1700. https://doi.org/10.1084/jem.20060468
Sefik E, Geva-Zatorsky N, Oh S, Konnikova L, Zemmour D, McGuire AM, Burzyn D, Ortiz-Lopez A, Lobera M, Yang J, Ghosh S, Earl A, Snapper SB, Jupp R, Kasper D, Mathis D, Benoist C (2015) MUCOSAL IMMUNOLOGY. Individual intestinal symbionts induce a distinct population of RORgamma(+) regulatory T cells. Science 349 (6251):993–997. https://doi.org/10.1126/science.aaa9420
Seidelin JB, Bjerrum JT, Coskun M, Widjaya B, Vainer B, Nielsen OH (2010) IL-33 is upregulated in colonocytes of ulcerative colitis. Immunol Lett 128(1):80–85. https://doi.org/10.1016/j.imlet.2009.11.001
Serena G, Yan S, Camhi S, Patel S, Lima RS, Sapone A, Leonard MM, Mukherjee R, Nath BJ, Lammers KM, Fasano A (2017) Proinflammatory cytokine interferon-gamma and microbiome-derived metabolites dictate epigenetic switch between forkhead box protein 3 isoforms in coeliac disease. Clin Exp Immunol 187 (3):490–506. https://doi.org/10.1111/cei.12911
Setoguchi R, Hori S, Takahashi T, Sakaguchi S (2005) Homeostatic maintenance of natural Foxp3(+) CD25 (+) CD4(+) regulatory T cells by interleukin (IL)-2 and induction of autoimmune disease by IL-2 neutralization. J Exp Med 201(5):723–735. https://doi.org/10.1084/jem.20041982
Sharfe N, Dadi HK, Shahar M, Roifman CM (1997) Human immune disorder arising from mutation of the alpha chain of the interleukin-2 receptor. Proc Natl Acad Sci U S A 94(7):3168–3171. https://doi.org/10.1073/pnas.94.7.3168
Sharifi A, Vahedi H, Nedjat S, Rafiei H, Hosseinzadeh-Attar MJ (2019) Effect of single-dose injection of vitamin D on immune cytokines in ulcerative colitis patients: a randomized placebo-controlled trial. APMIS 127(10):681–687. https://doi.org/10.1111/apm.12982
Shevach EM, Thornton AM (2014) tTregs, pTregs, and iTregs: similarities and differences. Immunol Rev 259 (1):88–102. https://doi.org/10.1111/imr.12160
Shono Y, Docampo MD, Peled JU, Perobelli SM, Velardi E, Tsai JJ, Slingerland AE, Smith OM, Young LF, Gupta J, Lieberman SR, Jay HV, Ahr KF, Porosnicu Rodriguez KA, Xu K, Calarfiore M, Poeck H, Caballero S, Devlin SM, Rapaport F, Dudakov JA, Hanash AM, Gyurkocza B, Murphy GF, Gomes C, Liu C, Moss EL, Falconer SB, Bhatt AS, Taur Y, Pamer EG, van den Brink MRM, Jenq RR (2016) Increased GVHD-related mortality with broad-spectrum antibiotic use after allogeneic hematopoietic stem cell transplantation in human patients and mice. Sci Transl Med 8(339):339ra371. https://doi.org/10.1126/scitranslmed.aaf2311
Shull MM, Ormsby I, Kier AB, Pawlowski S, Diebold RJ, Yin M, Allen R, Sidman C, Proetzel G, Calvin D et al (1992) Targeted disruption of the mouse transforming growth factor-beta 1 gene results in multifocal inflammatory disease. Nature 359(6397):693–699. https://doi.org/10.1038/359693a0
Siede J, Frohlich A, Datsi A, Hegazy AN, Varga DV, Holecska V, Saito H, Nakae S, Lohning M (2016) IL-33 receptor-expressing regulatory T cells are highly activated, Th2 biased and suppress CD4 T cell proliferation through IL-10 and TGFbeta release. PLoS One 11(8):e0161507. https://doi.org/10.1371/journal.pone.0161507
Singh N, Gurav A, Sivaprakasam S, Brady E, Padia R, Shi H, Thangaraju M, Prasad PD, Manicassamy S, Munn DH, Lee JR, Offermanns S, Ganapathy V (2014) Activation of Gpr109a, receptor for niacin and the commensal metabolite butyrate, suppresses colonic inflammation and carcinogenesis. Immunity 40 (1):128–139. https://doi.org/10.1016/j.immuni.2013.12.007
Sinicrope FA, Rego RL, Ansell SM, Knutson KL, Foster NR, Sargent DJ (2009) Intraepithelial effector (CD3+)/regulatory (FoxP3+) T-cell ratio predicts a clinical outcome of human colon carcinoma. Gastroenterology 137(4):1270–1279. https://doi.org/10.1053/j.gastro.2009.06.053
Smith PM, Howitt MR, Panikov N, Michaud M, Gallini CA, Bohlooly YM, Glickman JN, Garrett WS (2013) The microbial metabolites, short-chain fatty acids, regulate colonic Treg cell homeostasis. Science 341 (6145):569–573. https://doi.org/10.1126/science.1241165
Sokol H, Pigneur B, Watterlot L, Lakhdari O, Bermudez-Humaran LG, Gratadoux JJ, Blugeon S, Bridonneau C, Furet JP, Corthier G, Grangette C, Vasquez N, Pochart P, Trugnan G, Thomas G, Blottiere HM, Dore J, Marteau P, Seksik P, Langella P (2008) Faecalibacterium prausnitzii is an anti-inflammatory commensal bacterium identified by gut microbiota analysis of Crohn disease patients. Proc Natl Acad

Sci U S A 105(43):16731–16736. https://doi.org/10.1073/pnas.0804812105

Sokol H, Seksik P, Furet JP, Firmesse O, Nion-Larmurier I, Beaugerie L, Cosnes J, Corthier G, Marteau P, Dore J (2009) Low counts of Faecalibacterium prausnitzii in colitis microbiota. Inflamm Bowel Dis 15 (8):1183–1189. https://doi.org/10.1002/ibd.20903

Solomon BD, Hsieh CS (2016) Antigen-specific development of mucosal Foxp3+RORgammat+ T cells from regulatory T cell precursors. J Immunol 197 (9):3512–3519. https://doi.org/10.4049/jimmunol.1601217

Somasundaram R, Jacob L, Swoboda R, Caputo L, Song H, Basak S, Monos D, Peritt D, Marincola F, Cai D, Birebent B, Bloome E, Kim J, Berencsi K, Mastrangelo M, Herlyn D (2002) Inhibition of cytolytic T lymphocyte proliferation by autologous CD4+/CD25+ regulatory T cells in a colorectal carcinoma patient is mediated by transforming growth factor-beta. Cancer Res 62(18):5267–5272

Stefka AT, Feehley T, Tripathi P, Qiu J, McCoy K, Mazmanian SK, Tjota MY, Seo GY, Cao S, Theriault BR, Antonopoulos DA, Zhou L, Chang EB, Fu YX, Nagler CR (2014) Commensal bacteria protect against food allergen sensitization. Proc Natl Acad Sci U S A 111(36):13145–13150. https://doi.org/10.1073/pnas.1412008111

Strauss L, Bergmann C, Szczepanski MJ, Lang S, Kirkwood JM, Whiteside TL (2008) Expression of ICOS on human melanoma-infiltrating CD4+CD25highFoxp3+ T regulatory cells: implications and impact on tumor-mediated immune suppression. J Immunol 180(5):2967–2980. https://doi.org/10.4049/jimmunol.180.5.2967

Stritesky GL, Jameson SC, Hogquist KA (2012) Selection of self-reactive T cells in the thymus. Annu Rev Immunol 30:95–114. https://doi.org/10.1146/annurev-immunol-020711-075035

Su W, Chen X, Zhu W, Yu J, Li W, Li Y, Li Z, Olsen N, Liang D, Zheng SG (2019) The cAMP-adenosine feedback loop maintains the suppressive function of regulatory T cells. J Immunol 203(6):1436–1446. https://doi.org/10.4049/jimmunol.1801306

Sulimani RA (2019) Celiac disease and severe vitamin D deficiency: the case for anti-tissue transglutaminase antibody screening. Arch Osteoporos 14(1):30. https://doi.org/10.1007/s11657-018-0554-1

Sun CM, Hall JA, Blank RB, Bouladoux N, Oukka M, Mora JR, Belkaid Y (2007) Small intestine lamina propria dendritic cells promote de novo generation of Foxp3 T reg cells via retinoic acid. J Exp Med 204 (8):1775–1785. https://doi.org/10.1084/jem.20070602

Sun B, Liu M, Cui M, Li T (2020) Granzyme B-expressing Treg cells are enriched in colorectal cancer and present the potential to eliminate autologous T conventional cells. Immunol Lett 217:7–14. https://doi.org/10.1016/j.imlet.2019.10.007

Sundstrom P, Stenstad H, Langenes V, Ahlmanner F, Theander L, Ndah TG, Fredin K, Borjesson L, Gustavsson B, Bastid J, Quiding-Jarbrink M (2016) Regulatory T cells from colon cancer patients inhibit effector T-cell migration through an adenosine-dependent mechanism. Cancer Immunol Res 4 (3):183–193. https://doi.org/10.1158/2326-6066.CIR-15-0050

Suzuki H, Chikazawa N, Tasaka T, Wada J, Yamasaki A, Kitaura Y, Sozaki M, Tanaka M, Onishi H, Morisaki T, Katano M (2010) Intratumoral CD8(+) T/FOXP3 (+) cell ratio is a predictive marker for survival in patients with colorectal cancer. Cancer Immunol Immunother 59(5):653–661. https://doi.org/10.1007/s00262-009-0781-9

Tai X, Cowan M, Feigenbaum L, Singer A (2005) CD28 costimulation of developing thymocytes induces Foxp3 expression and regulatory T cell differentiation independently of interleukin 2. Nat Immunol 6 (2):152–162. https://doi.org/10.1038/ni1160

Takahashi T, Tagami T, Yamazaki S, Uede T, Shimizu J, Sakaguchi N, Mak TW, Sakaguchi S (2000) Immunologic self-tolerance maintained by CD25(+)CD4(+) regulatory T cells constitutively expressing cytotoxic T lymphocyte-associated antigen 4. J Exp Med 192 (2):303–310. https://doi.org/10.1084/jem.192.2.303

Tan H, Zhao J, Zhang H, Zhai Q, Chen W (2019) Novel strains of Bacteroides fragilis and Bacteroides ovatus alleviate the LPS-induced inflammation in mice. Appl Microbiol Biotechnol 103(5):2353–2365. https://doi.org/10.1007/s00253-019-09617-1

Tang Q, Boden EK, Henriksen KJ, Bour-Jordan H, Bi M, Bluestone JA (2004) Distinct roles of CTLA-4 and TGF-beta in CD4+CD25+ regulatory T cell function. Eur J Immunol 34(11):2996–3005. https://doi.org/10.1002/eji.200425143

Taylor ES, McCall JL, Girardin A, Munro FM, Black MA, Kemp RA (2016) Functional impairment of infiltrating T cells in human colorectal cancer. Onco Targets Ther 5(11):e1234573. https://doi.org/10.1080/2162402X.2016.1234573

Telesford KM, Yan W, Ochoa-Reparaz J, Pant A, Kircher C, Christy MA, Begum-Haque S, Kasper DL, Kasper LH (2015) A commensal symbiotic factor derived from Bacteroides fragilis promotes human CD39(+)Foxp3(+) T cells and Treg function. Gut Microbes 6(4):234–242. https://doi.org/10.1080/19490976.2015.1056973

Terme M, Pernot S, Marcheteau E, Sandoval F, Benhamouda N, Colussi O, Dubreuil O, Carpentier AF, Tartour E, Taieb J (2013) VEGFA-VEGFR pathway blockade inhibits tumor-induced regulatory T-cell proliferation in colorectal cancer. Cancer Res 73 (2):539–549. https://doi.org/10.1158/0008-5472.CAN-12-2325

Thakur BK, Saha P, Banik G, Saha DR, Grover S, Batish VK, Das S (2016) Live and heat-killed probiotic Lactobacillus casei Lbs2 protects from experimental colitis through toll-like receptor 2-dependent induction of T-regulatory response. Int Immunopharmacol

36:39–50. https://doi.org/10.1016/j.intimp.2016.03.033

Thornton AM, Korty PE, Tran DQ, Wohlfert EA, Murray PE, Belkaid Y, Shevach EM (2010) Expression of Helios, an Ikaros transcription factor family member, differentiates thymic-derived from peripherally induced Foxp3+ T regulatory cells. J Immunol 184(7):3433–3441. https://doi.org/10.4049/jimmunol.0904028

Tone Y, Furuuchi K, Kojima Y, Tykocinski ML, Greene MI, Tone M (2008) Smad3 and NFAT cooperate to induce Foxp3 expression through its enhancer. Nat Immunol 9(2):194–202. https://doi.org/10.1038/ni1549

Toor SM, Murshed K, Al-Dhaheri M, Khawar M, Abu Nada M, Elkord E (2019) Immune checkpoints in circulating and tumor-infiltrating CD4(+) T cell subsets in colorectal cancer patients. Front Immunol 10:2936. https://doi.org/10.3389/fimmu.2019.02936

Tran DQ, Ramsey H, Shevach EM (2007) Induction of FOXP3 expression in naive human CD4+FOXP3 T cells by T-cell receptor stimulation is transforming growth factor-beta dependent but does not confer a regulatory phenotype. Blood 110(8):2983–2990. https://doi.org/10.1182/blood-2007-06-094656

Travis MA, Reizis B, Melton AC, Masteller E, Tang Q, Proctor JM, Wang Y, Bernstein X, Huang X, Reichardt LF, Bluestone JA, Sheppard D (2007) Loss of integrin alpha(v)beta8 on dendritic cells causes autoimmunity and colitis in mice. Nature 449(7160):361–365. https://doi.org/10.1038/nature06110

Tripathi T, Yin W, Xue Y, Zurawski S, Fujita H, Hanabuchi S, Liu YJ, Oh S, Joo H (2019) Central roles of OX40L-OX40 interaction in the induction and progression of human T cell-driven acute graft-versus-host disease. Immunohorizons 3(3):110–120. https://doi.org/10.4049/immunohorizons.1900001

Tsai S, Clemente-Casares X, Santamaria P (2011) CD8(+) Tregs in autoimmunity: learning "self"-control from experience. Cell Mol Life Sci 68(23):3781–3795. https://doi.org/10.1007/s00018-011-0738-y

Ueno A, Jijon H, Chan R, Ford K, Hirota C, Kaplan GG, Beck PL, Iacucci M, Fort Gasia M, Barkema HW, Panaccione R, Ghosh S (2013) Increased prevalence of circulating novel IL-17 secreting Foxp3 expressing CD4+ T cells and defective suppressive function of circulating Foxp3+ regulatory cells support plasticity between Th17 and regulatory T cells in inflammatory bowel disease patients. Inflamm Bowel Dis 19(12):2522–2534. https://doi.org/10.1097/MIB.0b013e3182a85709

Ulloa L, Doody J, Massague J (1999) Inhibition of transforming growth factor-beta/SMAD signalling by the interferon-gamma/STAT pathway. Nature 397(6721):710–713. https://doi.org/10.1038/17826

Van den Abbeele P, Grootaert C, Marzorati M, Possemiers S, Verstraete W, Gerard P, Rabot S, Bruneau A, El Aidy S, Derrien M, Zoetendal E, Kleerebezem M, Smidt H, Van de Wiele T (2010) Microbial community development in a dynamic gut model is reproducible, colon region specific, and selective for Bacteroidetes and Clostridium cluster IX. Appl Environ Microbiol 76(15):5237–5246. https://doi.org/10.1128/AEM.00759-10

Vasanthakumar A, Liao Y, Teh P, Pascutti MF, Oja AE, Garnham AL, Gloury R, Tempany JC, Sidwell T, Cuadrado E, Tuijnenburg P, Kuijpers TW, Lalaoui N, Mielke LA, Bryant VL, Hodgkin PD, Silke J, Smyth GK, Nolte MA, Shi W, Kallies A (2017) The TNF receptor superfamily-NF-kappaB Axis is critical to maintain effector regulatory T cells in lymphoid and non-lymphoid tissues. Cell Rep 20(12):2906–2920. https://doi.org/10.1016/j.celrep.2017.08.068

Vaughn BP, Vatanen T, Allegretti JR, Bai A, Xavier RJ, Korzenik J, Gevers D, Ting A, Robson SC, Moss AC (2016) Increased intestinal microbial diversity following fecal microbiota transplant for active Crohn's disease. Inflamm Bowel Dis 22(9):2182–2190. https://doi.org/10.1097/MIB.0000000000000893

te Velde AA, de Kort F, Sterrenburg E, Pronk I, ten Kate FJ, Hommes DW, van Deventer SJ (2007) Comparative analysis of colonic gene expression of three experimental colitis models mimicking inflammatory bowel disease. Inflamm Bowel Dis 13(3):325–330. https://doi.org/10.1002/ibd.20079

te Velde AA, Brull F, Heinsbroek SE, Meijer SL, Lutjohann D, Vreugdenhil A, Plat J (2015) Effects of dietary plant sterols and Stanol esters with low- and high-fat diets in chronic and acute models for experimental colitis. Nutrients 7(10):8518–8531. https://doi.org/10.3390/nu7105412

Veltkamp C, Anstaett M, Wahl K, Moller S, Gangl S, Bachmann O, Hardtke-Wolenski M, Langer F, Stremmel W, Manns MP, Schulze-Osthoff K, Bantel H (2011) Apoptosis of regulatory T lymphocytes is increased in chronic inflammatory bowel disease and reversed by anti-TNFalpha treatment. Gut 60(10):1345–1353. https://doi.org/10.1136/gut.2010.217117

Verma R, Lee C, Jeun EJ, Yi J, Kim KS, Ghosh A, Byun S, Lee CG, Kang HJ, Kim GC, Jun CD, Jan G, Suh CH, Jung JY, Sprent J, Rudra D, De Castro C, Molinaro A, Surh CD, Im SH (2018) Cell surface polysaccharides of Bifidobacterium bifidum induce the generation of Foxp3(+) regulatory T cells. Sci Immunol 3(28):eaat6975. https://doi.org/10.1126/sciimmunol.aat6975

Vicente-Suarez I, Larange A, Reardon C, Matho M, Feau S, Chodaczek G, Park Y, Obata Y, Gold R, Wang-Zhu Y, Lena C, Zajonc DM, Schoenberger SP, Kronenberg M, Cheroutre H (2015) Unique lamina propria stromal cells imprint the functional phenotype of mucosal dendritic cells. Mucosal Immunol 8(1):141–151. https://doi.org/10.1038/mi.2014.51

Visperas A, Shen B, Min B (2014) gammadelta T cells restrain extrathymic development of Foxp3+−inducible regulatory T cells via IFN-gamma. Eur J

Immunol 44(8):2448–2456. https://doi.org/10.1002/eji.201344331
Vitale A, Strisciuglio C, Vitale S, Santopaolo M, Bruzzese D, Micillo T, Scarpato E, Miele E, Staiano A, Troncone R, Matarese G, Gianfrani C (2019) Increased frequency of regulatory T cells in pediatric inflammatory bowel disease at diagnosis: a compensative role? Pediatr Res 87:853. https://doi.org/10.1038/s41390-019-0662-7
Wang J, Ioan-Facsinay A, van der Voort EI, Huizinga TW, Toes RE (2007) Transient expression of FOXP3 in human activated nonregulatory CD4+ T cells. Eur J Immunol 37(1):129–138. https://doi.org/10.1002/eji.200636435
Wang Y, Su MA, Wan YY (2011) An essential role of the transcription factor GATA-3 for the function of regulatory T cells. Immunity 35(3):337–348. https://doi.org/10.1016/j.immuni.2011.08.012
Wang J, Ti Y, Wang Y, Guo G, Jiang H, Chang M, Qian H, Zhao J, Sun G (2018) LAG-3 represents a marker of CD4+ T cells with regulatory activity in patients with bone fracture. Immunol Investig 47(5):492–503. https://doi.org/10.1080/08820139.2018.1458107
Ward ST, Li KK, Hepburn E, Weston CJ, Curbishley SM, Reynolds GM, Hejmadi RK, Bicknell R, Eksteen B, Ismail T, Rot A, Adams DH (2015) The effects of CCR5 inhibition on regulatory T-cell recruitment to colorectal cancer. Br J Cancer 112(2):319–328. https://doi.org/10.1038/bjc.2014.572
Ward-Hartstonge KA, McCall JL, McCulloch TR, Kamps AK, Girardin A, Cretney E, Munro FM, Kemp RA (2017) Inclusion of BLIMP-1(+) effector regulatory T cells improves the Immunoscore in a cohort of New Zealand colorectal cancer patients: a pilot study. Cancer Immunol Immunother 66(4):515–522. https://doi.org/10.1007/s00262-016-1951-1
Weiss JM, Bilate AM, Gobert M, Ding Y, Curotto de Lafaille MA, Parkhurst CN, Xiong H, Dolpady J, Frey AB, Ruocco MG, Yang Y, Floess S, Huehn J, Oh S, Li MO, Niec RE, Rudensky AY, Dustin ML, Littman DR, Lafaille JJ (2012) Neuropilin 1 is expressed on thymus-derived natural regulatory T cells, but not mucosa-generated induced Foxp3+ T reg cells. J Exp Med 209(10):1723–1742, S1721. https://doi.org/10.1084/jem.20120914
Weitkamp JH, Rudzinski E, Koyama T, Correa H, Matta P, Alberty B, Polk DB (2009) Ontogeny of FOXP3(+) regulatory T cells in the postnatal human small intestinal and large intestinal lamina propria. Pediatr Dev Pathol 12(6):443–449. https://doi.org/10.2350/08-09-0533.1
Weitkamp JH, Koyama T, Rock MT, Correa H, Goettel JA, Matta P, Oswald-Richter K, Rosen MJ, Engelhardt BG, Moore DJ, Polk DB (2013) Necrotising enterocolitis is characterised by disrupted immune regulation and diminished mucosal regulatory (FOXP3)/effector (CD4, CD8) T cell ratios. Gut 62(1):73–82. https://doi.org/10.1136/gutjnl-2011-301551
Wildin RS, Ramsdell F, Peake J, Faravelli F, Casanova JL, Buist N, Levy-Lahad E, Mazzella M, Goulet O, Perroni L, Bricarelli FD, Byrne G, McEuen M, Proll S, Appleby M, Brunkow ME (2001) X-linked neonatal diabetes mellitus, enteropathy and endocrinopathy syndrome is the human equivalent of mouse scurfy. Nat Genet 27(1):18–20. https://doi.org/10.1038/83707
Wing K, Onishi Y, Prieto-Martin P, Yamaguchi T, Miyara M, Fehervari Z, Nomura T, Sakaguchi S (2008) CTLA-4 control over Foxp3+ regulatory T cell function. Science 322(5899):271–275. https://doi.org/10.1126/science.1160062
Wohlfert EA, Grainger JR, Bouladoux N, Konkel JE, Oldenhove G, Ribeiro CH, Hall JA, Yagi R, Naik S, Bhairavabhotla R, Paul WE, Bosselut R, Wei G, Zhao K, Oukka M, Zhu J, Belkaid Y (2011) GATA3 controls Foxp3(+) regulatory T cell fate during inflammation in mice. J Clin Invest 121(11):4503–4515. https://doi.org/10.1172/JCI57456
Worthington JJ, Czajkowska BI, Melton AC, Travis MA (2011) Intestinal dendritic cells specialize to activate transforming growth factor-beta and induce Foxp3+ regulatory T cells via integrin alphavbeta8. Gastroenterology 141(5) 1802-1812:1802. https://doi.org/10.1053/j.gastro.2011.06.057
Wysocki CA, Panoskaltsis-Mortari A, Blazar BR, Serody JS (2005) Leukocyte migration and graft-versus-host disease. Blood 105(11):4191–4199. https://doi.org/10.1182/blood-2004-12-4726
Xiao S, Li Q, Hu K, He Y, Ai Q, Hu L, Yu J (2018) Vitamin A and retinoic acid exhibit protective effects on necrotizing enterocolitis by regulating intestinal flora and enhancing the intestinal epithelial barrier. Arch Med Res 49(1):1–9. https://doi.org/10.1016/j.arcmed.2018.04.003
Xing Y, Hogquist KA (2012) T-cell tolerance: central and peripheral. Cold Spring Harb Perspect Biol 4(6):a006957. https://doi.org/10.1101/cshperspect.a006957
Xu L, Kitani A, Fuss I, Strober W (2007) Cutting edge: regulatory T cells induce CD4+CD25-Foxp3- T cells or are self-induced to become Th17 cells in the absence of exogenous TGF-beta. J Immunol 178(11):6725–6729. https://doi.org/10.4049/jimmunol.178.11.6725
Xu L, Kitani A, Stuelten C, McGrady G, Fuss I, Strober W (2010) Positive and negative transcriptional regulation of the Foxp3 gene is mediated by access and binding of the Smad3 protein to enhancer I. Immunity 33(3):313–325. https://doi.org/10.1016/j.immuni.2010.09.001
Xu A, Liu Y, Chen W, Wang J, Xue Y, Huang F, Rong L, Lin J, Liu D, Yan M, Li QZ, Li B, Song J, Olsen N, Zheng SG (2016) TGF-beta-induced regulatory T cells directly suppress B cell responses through a noncytotoxic mechanism. J Immunol 196(9):3631–3641. https://doi.org/10.4049/jimmunol.1501740

Xu M, Pokrovskii M, Ding Y, Yi R, Au C, Harrison OJ, Galan C, Belkaid Y, Bonneau R, Littman DR (2018) c-MAF-dependent regulatory T cells mediate immunological tolerance to a gut pathobiont. Nature 554 (7692):373–377. https://doi.org/10.1038/nature25500

Yadav M, Louvet C, Davini D, Gardner JM, Martinez-Llordella M, Bailey-Bucktrout S, Anthony BA, Sverdrup FM, Head R, Kuster DJ, Ruminski P, Weiss D, Von Schack D, Bluestone JA (2012) Neuropilin-1 distinguishes natural and inducible regulatory T cells among regulatory T cell subsets in vivo. J Exp Med 209(10):1713–1722, S1711–1719. https://doi.org/10.1084/jem.20120822

Yadav M, Stephan S, Bluestone JA (2013) Peripherally induced tregs - role in immune homeostasis and autoimmunity. Front Immunol 4:232. https://doi.org/10.3389/fimmu.2013.00232

Yamaguchi T, Hirota K, Nagahama K, Ohkawa K, Takahashi T, Nomura T, Sakaguchi S (2007) Control of immune responses by antigen-specific regulatory T cells expressing the folate receptor. Immunity 27 (1):145–159. https://doi.org/10.1016/j.immuni.2007.04.017

Yang BH, Hagemann S, Mamareli P, Lauer U, Hoffmann U, Beckstette M, Fohse L, Prinz I, Pezoldt J, Suerbaum S, Sparwasser T, Hamann A, Floess S, Huehn J, Lochner M (2016) Foxp3(+) T cells expressing RORgammat represent a stable regulatory T-cell effector lineage with enhanced suppressive capacity during intestinal inflammation. Mucosal Immunol 9(2):444–457. https://doi.org/10.1038/mi.2015.74

Yaqub S, Henjum K, Mahic M, Jahnsen FL, Aandahl EM, Bjornbeth BA, Tasken K (2008) Regulatory T cells in colorectal cancer patients suppress anti-tumor immune activity in a COX-2 dependent manner. Cancer Immunol Immunother 57(6):813–821. https://doi.org/10.1007/s00262-007-0417-x

Ye JH, Rajendran VM (2009) Adenosine: an immune modulator of inflammatory bowel diseases. World J Gastroenterol 15(36):4491–4498. https://doi.org/10.3748/wjg.15.4491

Ye J, Qiu J, Bostick JW, Ueda A, Schjerven H, Li S, Jobin C, Chen ZE, Zhou L (2017) The aryl hydrocarbon receptor preferentially marks and promotes gut regulatory T cells. Cell Rep 21(8):2277–2290. https://doi.org/10.1016/j.celrep.2017.10.114

Yu QT, Saruta M, Avanesyan A, Fleshner PR, Banham AH, Papadakis KA (2007) Expression and functional characterization of FOXP3+ CD4+ regulatory T cells in ulcerative colitis. Inflamm Bowel Dis 13 (2):191–199. https://doi.org/10.1002/ibd.20053

Yu Y, Ma X, Gong R, Zhu J, Wei L, Yao J (2018) Recent advances in CD8(+) regulatory T cell research. Oncol Lett 15(6):8187–8194. https://doi.org/10.3892/ol.2018.8378

Yurchenko E, Shio MT, Huang TC, Da Silva Martins M, Szyf M, Levings MK, Olivier M, Piccirillo CA (2012) Inflammation-driven reprogramming of CD4+ Foxp3+ regulatory T cells into pathogenic Th1/Th17 T effectors is abrogated by mTOR inhibition in vivo. PLoS One 7(4):e35572. https://doi.org/10.1371/journal.pone.0035572

Zanzi D, Stefanile R, Santagata S, Iaffaldano L, Iaquinto G, Giardullo N, Lania G, Vigliano I, Vera AR, Ferrara K, Auricchio S, Troncone R, Mazzarella G (2011) IL-15 interferes with suppressive activity of intestinal regulatory T cells expanded in celiac disease. Am J Gastroenterol 106(7):1308–1317. https://doi.org/10.1038/ajg.2011.80

Zeissig S, Petersen BS, Tomczak M, Melum E, Huc-Claustre E, Dougan SK, Laerdahl JK, Stade B, Forster M, Schreiber S, Weir D, Leichtner AM, Franke A, Blumberg RS (2015) Early-onset Crohn's disease and autoimmunity associated with a variant in CTLA-4. Gut 64(12):1889–1897. https://doi.org/10.1136/gutjnl-2014-308541

Zhai R, Xue X, Zhang L, Yang X, Zhao L, Zhang C (2019) Strain-specific anti-inflammatory properties of two Akkermansia muciniphila strains on chronic colitis in mice. Front Cell Infect Microbiol 9:239. https://doi.org/10.3389/fcimb.2019.00239

Zhang J, Ramadan AM, Griesenauer B, Li W, Turner MJ, Liu C, Kapur R, Hanenberg H, Blazar BR, Tawara I, Paczesny S (2015) ST2 blockade reduces sST2-producing T cells while maintaining protective mST2-expressing T cells during graft-versus-host disease. Sci Transl Med 7(308):308ra160. https://doi.org/10.1126/scitranslmed.aab0166

Zhang W, Zhou L, Dang J, Zhang X, Wang J, Chen Y, Liang J, Li D, Ma J, Yuan J, Chen W, Zadeh HH, Olsen N, Zheng SG (2017) Human gingiva-derived mesenchymal stem cells ameliorate Streptozoticin-induced T1DM in mice via suppression of T effector cells and up-regulating Treg subsets. Sci Rep 7 (1):15249. https://doi.org/10.1038/s41598-017-14979-5

Zhang Q, Fu L, Liang Y, Guo Z, Wang L, Ma C, Wang H (2018) Exosomes originating from MSCs stimulated with TGF-beta and IFN-gamma promote Treg differentiation. J Cell Physiol 233(9):6832–6840. https://doi.org/10.1002/jcp.26436

Zhang X, Ouyang X, Xu Z, Chen J, Huang Q, Liu Y, Xu T, Wang J, Olsen N, Xu A, Zheng SG (2019) CD8+CD103+ iTregs inhibit chronic graft-versus-host disease with lupus nephritis by the increased expression of CD39. Mol Ther 27(11):1963–1973. https://doi.org/10.1016/j.ymthe.2019.07.014

Zheng SG, Gray JD, Ohtsuka K, Yamagiwa S, Horwitz DA (2002) Generation ex vivo of TGF-beta-producing regulatory T cells from CD4+CD25- precursors. J Immunol 169(8):4183–4189. https://doi.org/10.4049/jimmunol.169.8.4183

Zheng SG, Wang JH, Gray JD, Soucier H, Horwitz DA (2004) Natural and induced CD4+CD25+ cells educate CD4+CD25- cells to develop suppressive activity: the role of IL-2, TGF-beta, and IL-10. J Immunol 172 (9):5213–5221. https://doi.org/10.4049/jimmunol.172.9.5213

Zheng SG, Wang J, Wang P, Gray JD, Horwitz DA (2007) IL-2 is essential for TGF-beta to convert naive CD4+CD25- cells to CD25+Foxp3+ regulatory T cells and for expansion of these cells. J Immunol 178(4):2018–2027. https://doi.org/10.4049/jimmunol.178.4.2018

Zheng SG, Wang J, Horwitz DA (2008) Cutting edge: Foxp3+CD4+CD25+ regulatory T cells induced by IL-2 and TGF-beta are resistant to Th17 conversion by IL-6. J Immunol 180(11):7112–7116. https://doi.org/10.4049/jimmunol.180.11.7112

Zheng Y, Josefowicz S, Chaudhry A, Peng XP, Forbush K, Rudensky AY (2010) Role of conserved non-coding DNA elements in the Foxp3 gene in regulatory T-cell fate. Nature 463(7282):808–812. https://doi.org/10.1038/nature08750

Zheng J, Chan PL, Liu Y, Qin G, Xiang Z, Lam KT, Lewis DB, Lau YL, Tu W (2013) ICOS regulates the generation and function of human CD4+ Treg in a CTLA-4 dependent manner. PLoS One 8(12):e82203. https://doi.org/10.1371/journal.pone.0082203

Zhong W, Jiang ZY, Zhang L, Huang JH, Wang SJ, Liao C, Cai B, Chen LS, Zhang S, Guo Y, Cao YF, Gao F (2017) Role of LAP(+)CD4(+) T cells in the tumor microenvironment of colorectal cancer. World J Gastroenterol 23(3):455–463. https://doi.org/10.3748/wjg.v23.i3.455

Zhou L, Lopes JE, Chong MM, Ivanov II, Min R, Victora GD, Shen Y, Du J, Rubtsov YP, Rudensky AY, Ziegler SF, Littman DR (2008) TGF-beta-induced Foxp3 inhibits T(H)17 cell differentiation by antagonizing RORgammat function. Nature 453(7192):236–240. https://doi.org/10.1038/nature06878

Zhou X, Bailey-Bucktrout SL, Jeker LT, Penaranda C, Martinez-Llordella M, Ashby M, Nakayama M, Rosenthal W, Bluestone JA (2009) Instability of the transcription factor Foxp3 leads to the generation of pathogenic memory T cells in vivo. Nat Immunol 10(9):1000–1007. https://doi.org/10.1038/ni.1774

Zhou X, Kong N, Wang J, Fan H, Zou H, Horwitz D, Brand D, Liu Z, Zheng SG (2010) Cutting edge: all-trans retinoic acid sustains the stability and function of natural regulatory T cells in an inflammatory milieu. J Immunol 185(5):2675–2679. https://doi.org/10.4049/jimmunol.1000598

Zhou L, Liu D, Xie Y, Yao X, Li Y (2019) Bifidobacterium infantis induces protective colonic PD-L1 and Foxp3 regulatory T cells in an acute murine experimental model of inflammatory bowel disease. Gut Liver 13(4):430–439. https://doi.org/10.5009/gnl18316

Zisman TL, Rubin DT (2008) Colorectal cancer and dysplasia in inflammatory bowel disease. World J Gastroenterol 14(17):2662–2669. https://doi.org/10.3748/wjg.14.2662

Zou W (2006) Regulatory T cells, tumour immunity and immunotherapy. Nat Rev Immunol 6(4):295–307. https://doi.org/10.1038/nri1806

The Association of Gut Microbiota and Treg Dysfunction in Autoimmune Diseases

10

Yuying Liu, Dat Q. Tran, John William Lindsey, and Jon Marc Rhoads

Abstract

Autoimmune conditions affect 23 million Americans or 7% of the US population. There are more than 100 autoimmune disorders, affecting every major organ system in humans. This chapter aims to further explain Treg dysfunction autoimmune disorders, including monogenic primary immune deficiency such as immune dysregulation polyendocrinopathy, enteropathy, X-linked inheritance (IPEX) syndrome, and polygenic autoimmune diseases with Treg dysfunction such as multiple sclerosis (MS), inflammatory bowel disease (IBD), and food allergy. These conditions are associated with an abnormal small intestinal and colonic microbiome. Some disorders clearly improve with therapies aimed at microbial modification, including probiotics and fecal microbiota transplantation (FMT). Approaches to prevent and treat these disorders will need to focus on the acquisition and maintenance of a healthy colonic microbiota, in addition to more focused approaches at immune suppression during acute disease exacerbations.

Keywords

Probiotic · Dysbiosis · Immunodeficiency · Multiple sclerosis · Inflammatory bowel disease

10.1 Tregs Are Shaped by the Gut Microbiota

The gut microbiota is essential for the development and maturation of the immune system; reciprocally, the microbial community is also profoundly affected by the complex host immune system. The gut epithelium spatially segregates the lamina propria (LP), rich in lymphocytes, plasma cells, dendritic cells (DCs), macrophages, neutrophils, and scaffolding myofibroblasts, from the luminal contents, particularly the gut microbiome (bacteria, fungi, viruses, archaea, and phage) and a variety of food antigens. One of the most fascinating observations is that the subepithelial immune network can exert pressure against selective microbes by elaborating cytokines, antimicrobial peptides, nutrients, and mucus (Zhou and Sonnenberg 2018). Likewise, immune cells can "welcome" certain organisms. T cells in the LP are the crucial source of anti-inflammatory cytokine IL-10. To exemplify microbial induction of immune tolerance, luminal

Y. Liu · D. Q. Tran · J. M. Rhoads (✉)
Department of Pediatrics, Division of Gastroenterology, The University of Texas Health Science Center at Houston McGovern Medical School, Houston, TX, USA
e-mail: j.marc.rhoads@uth.tmc.edu

J. W. Lindsey
Department of Neurology, Division of Multiple Sclerosis and Neuroimmunology, The University of Texas Health Science Center at Houston McGovern Medical School, Houston, TX, USA

S.-G. Zheng (ed.), *T Regulatory Cells in Human Health and Diseases*, Advances in Experimental Medicine and Biology 1278, https://doi.org/10.1007/978-981-15-6407-9_10

Bacillus fragilis produces polysaccharide antigen A (PSA), which acts on DCs, resulting in TGF-β and IL-10 production, which supports the development of gut anti-inflammatory cells, especially regulatory T cells (Tregs). Consequently, Tregs produce IL-10 that induces in macrophages a tolerogenic phenotype. However, even in the absence of IL-10 or in hosts with IL-10 receptor mutations, an unusual condition in humans with very early onset of inflammatory bowel disease (IBD), different *Lactobacillus* species can effectively prevent or attenuate the colitis that develops (Madsen et al. 1999).

A seminal observation related to luminal microbial impact on the immune system was that LP $CD4^+$ (helper) T cells were shifted toward an inflammatory T_H17 phenotype when mice were colonized with segmented filamentous bacteria (SFB). SFB are interesting coiled microbes which penetrate the mucus gel in the ileum and directly attach like leeches to the epithelial cells (Ivanov et al. 2009). However, *Clostridia* species and other microbes produced short-chain fatty acids (SCFAs) via the metabolism of dietary fiber, resulting in colonic Tregs development and reduced inflammation in this model, via a mechanism involving G-protein-coupled receptor GPCR43 (Smith et al. 2013). In humans with IBD (to be discussed below), certain colonic microbes have been clearly associated with active diseases, such as *Ruminococcus gnavus* (Schirmer et al. 2019) and oral cavity inhabitants (*Veillonella dispar, Aggregatibacter, Lachnospiraceae, and Haemophilus parainfluenzae*) (Schirmer et al. 2018a), whereas other organisms such as *Faecalibacterium prausnitzii,* which produces the SCFA butyrate, are clearly associated with disease remission (Schirmer et al. 2018b).

10.2 Gut Dysbiosis-Related Autoimmune Disorders Involve Tregs

Autoimmune disorders prevalent in humans include disorders of monogenic primary immune deficiency and polygenic autoimmune diseases including diseases of the central nervous system (CNS) (multiple sclerosis, transverse myelitis, and Guillain Barre syndrome); the gastrointestinal tract (inflammatory bowel disease, autoimmune hepatitis, and celiac disease); and autoimmune conditions of the skin (including psoriasis, eczema, and scleroderma).

10.2.1 Monogenic Primary Autoimmune Disorders Related to Tregs Are Associated with Gut Microbial Dysbiosis

Monogenic autoimmune disorders are due to a single-gene defect and are generally quite rare. These mutations cause impairment in one of the principal mechanisms controlling adaptive immunity, such as loss of control of inflammatory effector T cells (T_H1/T_H2) caused by deficiency of $Foxp3^+$Tregs. $Foxp3^+$Treg deficiency can be due to a *Foxp3* gene mutation or deletion, resulting clinically in a condition characterized by immune dysregulation polyendocrinopathy, enteropathy, and X-linked inheritance, called IPEX syndrome and in mice the scurfy (SF) phenotype. A recent review (Chervonsky 2013) stated that monogenic diseases including IPEX (Chinen et al. 2010) and autoimmune polyendocrinopathy-candidiasis-ectodermal dystrophy (APECED, due to mutations in the transcription factor autoimmune regulator (AIRE)) (Gray et al. 2007) are insensitive to commensal microbial regulation, because these diseases occur in germ-free (GF) animals. The authors suggested that autoimmune T cells are activated in the absence of the innate-adaptive connection. A critical role for Treg-mediated control of inflammation was studied using GF mice compared with specific pathogen-free (SPF) mice and demonstrated that Treg development and suppressive function were not dependent on gut microbiota. However, in a Treg-depleted model (Foxp3-DTR), inflammation in the small intestine of SPF mice was more severe than in GF mice, as shown by significantly increased gut lymphocyte infiltration, decreased body weight, and increased percentage of IFN-γ-producing T helper cells, indicating that Treg deficiency-induced intestinal inflammation is indeed influenced by the composition of the gut microbiota (Chinen et al. 2010).

Additional experiments provide evidence that a single-gene immune-related mutation is linked to intestinal dysbiosis. Dysbiosis is defined as atypical microbial populations, which carry proinflammatory characteristics. It is not surprising that the feces of individuals with autoimmune disorders harbor skewed populations of microbes that differ from those of healthy controls. We observed in Foxp3-deficient scurfy mice that development of autoimmunity was accompanied by a progressive gut microbial dysbiosis over the first 22 days of life (He et al. 2017b). We demonstrated reduced bacterial diversity and altered bacterial composition. Altered gut microbiota has also been shown in immunodeficient mice lacking B cells (Ighm$^{-/-}$), T cells (Cd3e$^{-/-}$), or both B and T cells (Rag1$^{-/-}$), whereas administration of Foxp3^{+}Tregs to T-cell-deficient mice restored bacterial diversity (Kawamoto et al. 2014). A detailed study compared Rag1$^{+/+}$ and Rag1$^{-/-}$ mice of the same genetic background and eliminated cage effects. Investigators sought to determine the effect of Rag1 and the adaptive immune system on gut microbiota. They demonstrated that Rag1 status is a source of variation in community structure of gut microbiota (Zhang et al. 2015). A recent study of individuals with APECED also indicated that their microbiota composition is altered. These patients developed early and sustained responses to gut microbial antigens, and abnormal immune recognition of gut commensals was linked to enteropathy, the development of defensin-specific T cells and anti-defensin antibodies, and production of anti-*Saccharomyces cerevisiae* antibodies (ASCA). ASCA level was highly correlated with the depletion of gut-associated Tregs ($r = 0.7$, $P < 0.01$); and it is worthy to note that ASCA has been linked to Crohn's disease in humans, indicating that transcription factor AIRE is an important regulator of intestinal homeostasis (Dobes et al. 2015; Hetemaki et al. 2016). Thus, overall evidence leads us to conclude that host adaptive immunity in diseases with single-gene mutations alters gut microbiota.

Our team has studied the previously mentioned Foxp3-deficient mouse SF mouse. We observed dynamic changes of autoimmunity and gut microbial dysbiosis in this disease. We noticed that microbial beta-diversity in SF mice was successfully shifted by orally administering a probiotic or health-promoting bacteria (*Lactobacillus reuteri* DSM 17938, LR 17938) to produce a distinct microbial signature and a concomitant reduction in disease severity (He et al. 2017b).

Metabolites produced by bacteria may be major mediators of immune tolerance. We discovered that the adenosine metabolite *inosine* is reduced in plasma of SF mice and is restored by LR 17938 treatment. Orally gavaging this nucleoside to SF mice, similar to gavaging the probiotic LR itself, prolonged the survival of SF mice and reduced inflammation in multiple organs such as liver and lung. Inosine and LR 17938 each controlled inflammation by acting as a ligand to the adenosine receptor 2A (A_{2A}), as evidenced by the lack of inhibition of naïve CD4^{+} T-cell differentiation in mice with genetically deleted adenosine receptor A_{2A} ($A_{2A}{}^{-/-}$ mice). Inosine and LR 17938 were ineffective in mice with mutations of three other adenosine receptors (A_1, A_{2B}, and A_3) (He et al. 2017b). We concluded that A_{2A} is required for the protection by this probiotic (He et al. 2017a).

10.2.2 Polygenic Autoimmune Diseases with Treg Dysfunction Are Associated with Gut Microbial Dysbiosis

Polygenic autoimmune diseases generally develop gradually, with incomplete penetration of the disease, meaning that many individuals expected to develop symptoms do not become sick. The conditions eventually occur because there is a "perfect storm" of environmental insults, even in some cases emotional trauma, coupled with aberrant microbial population in the gut which predisposes to systemic inflammation.

The intestine harbors ~10^{14} microbes (more than our own cell population) and approximately 1000 different taxa, accounting for 3% of our total body mass. In almost every autoimmune disorder, an abnormal population of enteric microbes has been identified, called *dysbiosis*. However, it is remarkable that the microbial populations segregate differentially with respect to each other, as well as segregating differently from controls when analyzed by principal components analysis. Investigators recently showed that when looking at stool 16S rDNA sequences of patients with different autoimmune disorders, the microbial beta diversity was distinct for Sjogren syndrome, compared with systemic lupus erythematosus (SLE) and primary phospholipid syndrome (PPLS), and these were distinct from controls not having autoimmune disease (Fig. 10.1).

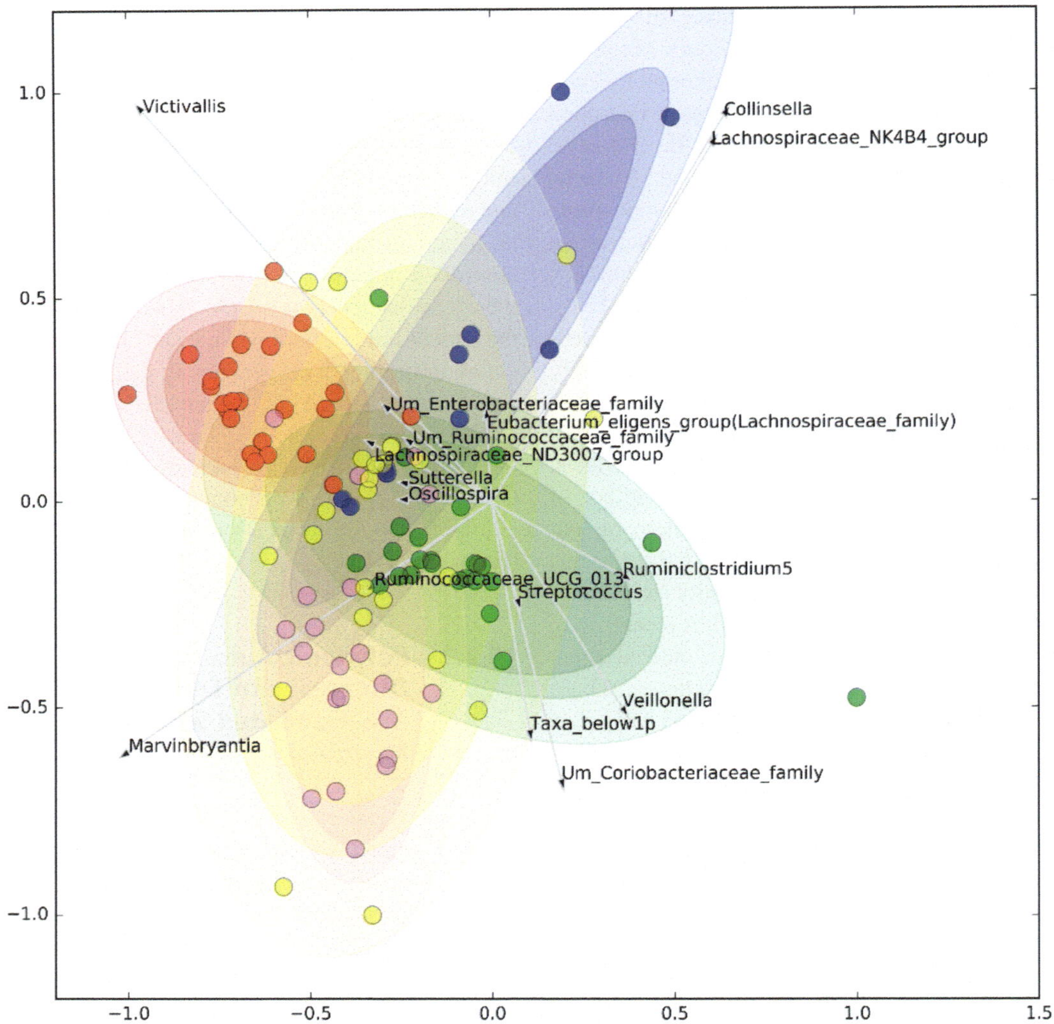

Fig. 10.1 Each specific disorder has microbial signature which segregated the disease via its own microbial signature. Clustering of selected microbiome genera. Graphical representation is optimized to maximize the compactness and separation of clusters. Red = healthy controls, purple = systemic lupus erythematosus, green = Sjogren's syndrome, blue = primary antiphospholipid syndrome, and yellow = undifferentiated connective tissue disease. Reproduced with permission from Bellocchi et al. (2019). The association between autoimmunity and gut microbiota would be discussed in three polygenic autoimmune diseases as examples, multiple sclerosis, inflammatory bowel diseases, and food allergy

10.2.3 Autoimmune Diseases in the Central Nervous System (CNS)

Multiple sclerosis (MS). A polygenic autoimmune disorder associated with microbial dysbiosis is multiple sclerosis (MS). This disease affects >1 million Americans and is characterized by autoimmune cells invading the CNS, producing impaired vision, weakness, loss of sensation, chronic pain, and declining cognition (Wallin et al. 2019). Although there is no cure, the major thrust of treatment is immunosuppression. T cells play a major role in the pathogenesis of MS, with T_H1- and T_H17-associated cytokines being elevated, along with reported major abnormalities in the function of Tregs (Danikowski et al. 2017). Gut bacteria have an important influence on the development of this disease. In MS and its mouse model called experimental autoimmune encephalomyelitis (EAE), antibodies are produced that react to myelin. In the laboratory, to induce EAE, myelin oligodendrocyte (MOG) peptides with Freund's adjuvant are injected in the presence of pertussis toxin to produce the gradual onset of varying degrees of paralysis. MOG peptides presented by antigen-presenting cells subsequently activate T cells, which migrate to the CNS. Unlike conventional mice, germ-free mice are protected from paralysis, whereas those inoculated with a single organism SFB (as discussed above) develop the symptoms (Lee et al. 2011). In these studies, SFBs were shown to induce IL-17A-producing $CD4^+$ T cells (T_H17) in the CNS.

When peripheral blood leukocytes are examined from individuals with MS, many have been found to reduce Foxp3 expression, as well as reduced numbers of Foxp3-negative regulatory T cells or Tr1 cells. About 30% of the brain lesions in MS patients have no Foxp3 expression. Not surprisingly, many but not all treatments for MS are associated with the ability to increase Tregs. Notably, indoleamine 2,3-dioxygenase (IDO), a product of the metabolism of tryptophan (now well characterized as a product of gut microbes), when administered parenterally ameliorates experimental EAE (Yan et al. 2010). Specific chemokines, such as CCR4 and dendritic cell-derived CCL17 are required for Treg recruitment to the CSF, whereas CXCR3 is important for the transmigration of Tregs and Tr1 cells across the blood–brain barrier to combat the neuroinflammation (Danikowski et al. 2017).

We discovered that the probiotic LR 17938 when given enterally to mice with EAE was able to reduce the severity of clinical disease, measured by the extent of paralysis, as well as inflammation in the spinal cord, as measured by hematoxylin and eosin staining and $CD3^+$ T-cell staining. LR 17938 also reduced the number of circulating T_H1/T_H17 cells and their associated cytokines IFN-γ/IL-17. We also showed that the loss of diversity of gut microbiota induced by EAE was largely reversed by LR 17938 treatment (He et al. 2019). Taxonomy-based analysis of gut microbiota showed that three "beneficial" genera *Bifidobacterium, Prevotella*, and *Lactobacillus* were negatively correlated with EAE clinical severity. Notably *Prevotella* was previously shown to be reduced in humans with MS (Schepici et al. 2019). Conversely, genera *Anaeroplasma, Rikenellaceae*, and *Clostridium* were positively correlated with disease severity. Remarkably, LR 17938 treatment coordinately altered the relative abundance of these EAE-associated genera. These studies suggested that LR 17938 might represent a novel adjunctive treatment in future studies to modify the severity of MS.

10.2.4 Gastrointestinal Autoimmune Disorders

Inflammatory bowel disease (IBD). IBD is the most common chronic gastrointestinal illness of children and young adults, affecting 100–200 children per 100,000 in the US (or ~70,000 children) (Rosen et al. 2015). The two most prevalent conditions are Crohn's disease (CD), which can cause ulceration in any part of the gastrointestinal tract, especially the ileum and proximal colon, and ulcerative colitis (UC), which involves circumferential ulceration of the colon only. Symptoms include diarrhea, abdominal pain, fever, weight loss, arthritis, mucositis, perianal disease, bowel obstruction, and fistula development. The most accepted working hypothesis for the pathogenesis of IBD is that a dysfunctional immune system interacts with an abnormal microbiota, with exacerbating environmental influences to induce disease. Contributing factors, such as antibiotics, high animal protein intake, cigarette smoke, and high linoleic acid intake, all may result in an imbalance between pro- and anti-inflammatory cells in the gastrointestinal mucosa (Bernstein 2017).

There are >250 different mutations that have now been linked to CD. Some mutations affect T cells, but others involve defective phagocytic bacterial killing (such as the NOD1 mutation) or may impair more global cellular protective processes such as autophagy (e.g., the ATG16L1 mutation) (Bianco et al. 2015). When one begins to look at genetic mutations in IBD, it becomes evident that IBD represents a common "denominator" of many different abnormalities in immune cell function and microbial homeostasis.

Reproducible findings of T-cell functional abnormalities seen in individuals with IBD have led to the hypothesis that unregulated inflammation must be related to an imbalance between helper T-cell subsets, specifically the balance between Tregs and T_H1 and T_H17 cells (Packey and Sartor 2008). The important roles of both T_H1 and T_H2 cells in UC and T_H1 and T_H17 cells in CD have been described, but the role of Tregs is complex. For example, in pediatric IBD, two types of Tregs are actually elevated. Tr1 cells are more abundant than normal in the peripheral circulation of children with ulcerative colitis and Crohn's disease, whereas in UC, circulating Foxp3 $^{+}$ Treg cells are increased at the time of diagnosis but dwindle during remission (Vitale et al. 2020).

Animal models uncover a complex and somewhat paradoxical role of Tregs in IBD. For example, Tbet (the transcription factor associated with T_H1 cells) can also be expressed on Tregs (which are characterized by Foxp3 expression) in the initial stages of colitis in humans. Knockout of Tbet on Foxp3-positive Tregs reduces colitis severity, implicating a pro-inflammatory role not previously recognized for this group of Treg cells. In a mouse model of colitis induced by dextran sodium sulfate (DSS), there was an influx into the gut epithelium of cells positive for Tbet, interferon-gamma, *and* Foxp3 occurring prior to the accumulation of classic T_H1 cells into the mucosa of the colon that was preventable by a conditional knockout of Tbet on Tregs (Di et al. 2019).

Clearly, T cells and microbes play a central role in the regulation of IBD. Biologic therapies that are widely used include immunosuppression by the anti-tumor necrosis factor (TNF) antibodies infliximab and adalimumab and administration of anti-integrin alpha7beta4 (α7β4) antibodies such as vedolizumab, which prevents homing of leukocytes to the gastrointestinal (GI) tract (Jeong et al. 2019).

In taking an "outside the box" approach to severe IBD, Treg infusion therapy was found to have only limited success in humans with CD. In 2012, investigators participated in an 8-week open-label dose-ranging trial of human Tregs for 20 adults with CD. Tregs were isolated from patients' peripheral blood mononuclear cells, exposed to chicken ovalbumin, and administered intravenously (ova-Tregs); subsequently, a meringue cake diet (rich in ovalbumin) stimulated Treg proliferation in the GI tract. There was an improvement in Crohn's disease activity (CDA) in 40% of patients, but only minimal improvement in circulating inflammatory markers (CRP and fecal calprotectin), without an increase in Treg numbers in the blood; and of the four doses tested, only the lowest dose appeared to be effective (Desreumaux et al. 2012). No significant human trials of Treg isolation, in vitro expansion, and administration have been subsequently published for patients with IBD.

A review pointed out that the persistence of inflammation in patients with IBD in the face of high levels of $Foxp3^+$ Tregs in the affected tissues suggests that the condition is resistant to Tregs, perhaps due to other effector cell activation resistant to Tregs or tissue/extracellular matrix factors unsupportive of their inhibitory action (Lord 2015).

Recently, interest in Tr1 cells which do not express Foxp3 has arisen as a potential therapeutic target (Huang et al. 2017). These cells secrete IL-10 and IL-22, both with anti-inflammatory properties, and Tr1 cells were shown to reduce the proliferation of effector T cells and the secretion of cytokines by myeloid cells. These cells also protected barrier function of cultured T84 cells in vitro (Cook et al. 2019). Extensive research has also shown that commensal microbes produce SCFAs that interact with cell surface G-protein-coupled receptors GPR43 and GPR41 on myeloid cells to subsequently induce regulatory phenotype and IL-10 secretion by Tr1 cells (Sun et al. 2018). Commensals can also act on monocytes (Ly6C^{hi} cells) via their SCFA, resulting in prostaglandin E2 production that reduces the activation level of neutrophils (Grainger et al. 2013).

Food allergy. Given the importance of Tregs in maintaining tolerance to endogenous and exogenous stimuli, it is not surprising that Tregs are localized predominantly in the intestines and skin, two tissues with the largest surface area exposing to the external environment (Harrison and Powrie 2013). It is know that the skin is more sensitizing, while the gut is more tolerizing. In murine model of allergy, the skin is used to create an allergic immune development. If the mice are fed with the allergen, for example, ovalbumin, prior to the skin sensitization, inducing the allergy is more difficult. This observation indicates that the intestinal microenvironment is designed to be more immune tolerant. Previously, the American Academy of Pediatrics (AAP) had recommended avoiding early introduction of highly allergenic foods to infants until after 1 year of age. Ironically, over 10 years later, the incidence of food allergy continues to rise. After a pivotal clinical trial of early introduction of peanuts in infants had demonstrated a significant reduction in the development of peanut allergy, the new AAP guidelines now recommend introduction of highly allergenic foods to infants as early as 4–-6 months of age (Du et al. 2015).

These epidemiologic and clinical studies have demonstrated the importance of the intestines in regulating food allergy development. However, the mechanisms involved in tolerance induction are quite complex. Tregs clearly play a critical role in controlling food allergy development, since patients with IPEX syndrome have a higher incidence of severe food allergy (Torgerson et al. 2007). Clinical studies using oral immunotherapy (OIT) for the prevention or treatment of food allergy have shown an increase in Tregs, particularly allergen-induced Tregs, throughout the therapy (Hardy et al. 2019). Interestingly, food allergy is a problem in developed countries, supporting the hygiene hypothesis (Platts-Mills 2015). The western diet, sedentary lifestyle, cleanliness, and indoor living have altered the microbiomes on the skin, mucosa, and intestines as well as reduction in vitamin D level (Du et al. 2016). In atopic infants living in this modernized environment, skin exposure to allergenic food proteins might induce a more sensitizing immune response, since the immune system might be regulated differently. In contrast, those in developing or underdeveloped countries, the immune system might be more tolerant since its focus would be on controlling the various microbes on the skin, mucosa, and intestines. An analogy would be in the time of war, people concentrate more attention on survival.

It is important to note that there are two major subsets of Tregs: thymic Tregs (tTregs) and peripheral induced Tregs (pTregs) (Horwitz et al. 2008; Zhang et al. 2020). tTregs are developed in the thymus with a stable expression of Foxp3 due to high demethylation of the Foxp3 promoter region called Treg-specific demethylated region (TSDR). pTregs are generated in the peripheral tissues through the induction of Foxp3 expression in CD4$^+$ T cells. These pTregs have greater plasticity and significantly lower demethylation in the TSDR. Current data support that tTregs are important in controlling intrinsic autoimmunity, while pTregs help to regulate extrinsic immune response (Shevach and Thornton 2014). pTregs are essential for maternal–fetal tolerance. In the absence of pTregs, using a specific murine model deficient in conserved noncoding sequence 1 (CNS1) of the Foxp3 enhancer region, there is increased fetal resorption, resulting in a significant reduction in live birth (Samstein et al. 2012). Another important insight using this pTreg-deficient murine model is the importance of pTregs in maintaining intestinal microbial communities by modulating immune defense mechanisms that can lead to dysbiosis (Campbell et al. 2018). The microbiome is vital in contributing to immune homeostasis through its interaction with Tregs. Perturbation in the microbiome can lead to dysbiosis, resulting in aberrant immune responses such as food allergy (Bunyavanich and Berin 2019). Technologies continue to advance rapidly; and so it would be challenging to change people's lifestyle to living that is more rural. Therefore, the focus would be on developing natural therapies and diet that would promote Treg homeostasis and prevent dysbiosis.

10.3 Resetting Microbiota and Tregs in Diseases

Probiotics. Probiotics (as mentioned above) are "live organisms that when administered in adequate doses confer a health benefit to the host". They are provided as fermentable foods, pills, powders, beverages, and liquid drops. Common probiotics are available in pharmacies, groceries, and online. They include but are not limited to *Lactobacilli*, *Bacillus* species, several strains of *Bifidobacteria*, *Streptococcus thermophilus*, *Escherichia coli* strain Nissle 1917, and yeasts, including *Saccharomyces boulardii* and *Saccharomyces cerevisiae*. Many commercially available probiotics contain mixtures of two or more individual species.

Although not discussed earlier in this review, a major medical condition for which probiotics are being studied is neonatal necrotizing enterocolitis (NEC), the leading cause of intestinal failure in premature babies. We have found in mouse and rat models of NEC that probiotic administration in the setting of experimental neonatal gut hypoxia reset the balance between inflammatory effector T cells and their Treg counterparts, reducing both the incidence and severity of NEC (Liu et al. 2013, 2014). Our results revealed that there were low numbers of Tregs in the terminal ileum of newborn rats with NEC. Adoptive transfer of Tregs significantly improved weight loss, improved survival and reduced the incidence of NEC (Dingle et al. 2013). Our findings suggested that while Tregs are present in the newborn intestine, their numbers might be insufficient to dampen the excessive inflammatory state in NEC.

We have not directly tested the impact of probiotics on the microbiota on fecal community composition in the NEC model, but in subsequent investigations in healthy newborn mice, we showed that LR 17938 increased the proportion of Foxp3^{+} Tregs and also increased bacterial diversity (Liu et al. 2019). LR 17938 was associated with increased relative abundance of phylum Firmicutes, families *Lachnospiraceae* and *Ruminococcaceae*, and genera *Clostridium* and *Candidatus arthromitus*. LR 17938 concomitantly decreased the relative abundance of phylum Bacteroidetes, families *Bacteroidaceae* and *Verrucomicrobiaceae*, and genera *Bacteroides*, *Ruminococcus*, *Akkermansia*, and *Sutterella*. Finally, LR 17938 also exerted a major impact on the plasma metabolome, upregulating amino acid metabolites formed via the urea, tricarboxylic acid, and methionine cycles, and simultaneously increasing tryptophan metabolism. Of significance, LR 17938 feeding increased the levels of key tryptophan metabolites indolepropionate and indoleacrylate and purine nucleosides (adenosine and inosine), each of which is known to enhance tolerance to inflammatory stimuli (Liu et al. 2019).

The mechanisms by which probiotics induce Tregs in different diseases have been demonstrated to be variable and strain specific, as summarized in a review article by Dwivedi et al. (2016) and Fig. 10.2. Probiotics and probiotic-modulated beneficial gut microbiota induce intestinal Tregs through direct microbial–immune cell or microbial-intestinal epithelial cell interactions, for example, via stimulation of Toll-like receptors (TLRs 2, 3, 4, or 9) to activate mucosal tolerogenic DCs to produce IL10 and/or Treg-interacting molecules (IL10 or

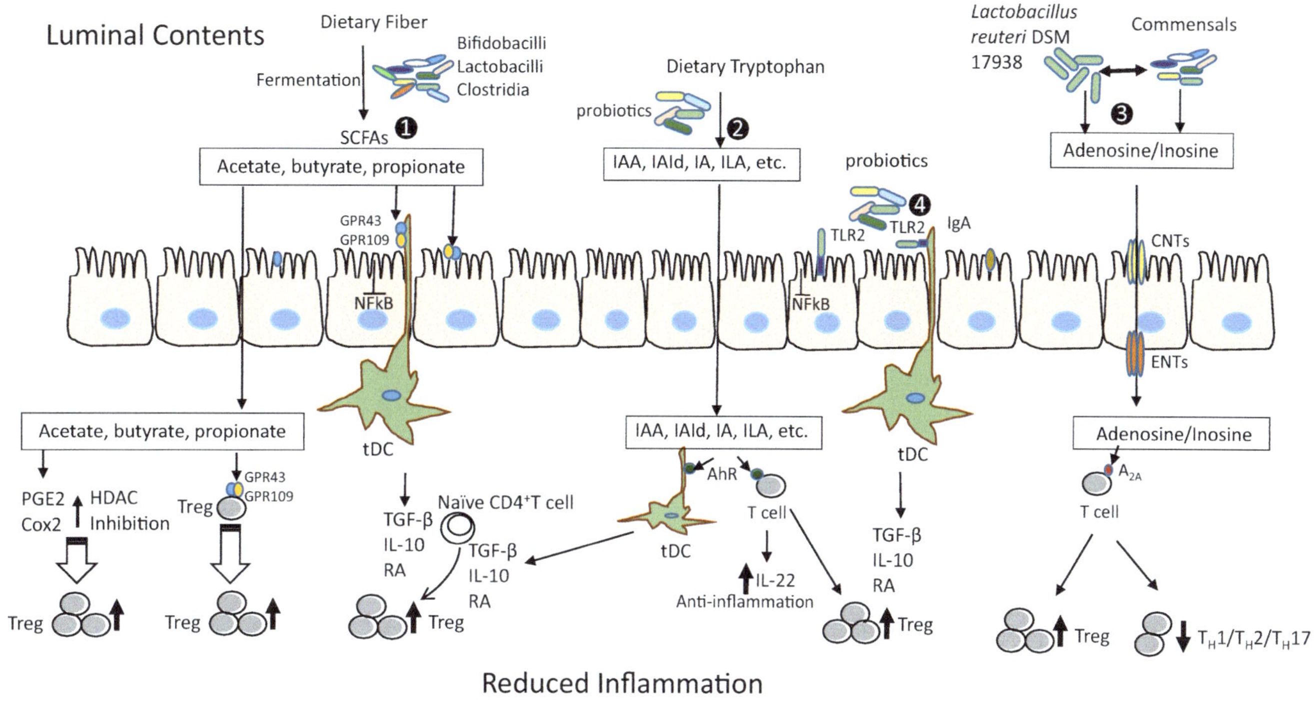

Fig. 10.2 Different mechanisms of Treg induction by probiotics to reduce inflammation. (1) Probiotics metabolize the dietary fibers such as resistant starch, inulin, FOS, β-GOS to release SCFAs: acetate, butyrate and propionate. These SCFAs activate the GPR43 and GPR109A: on intestinal epithelial cells to inhibit the NF-κB pathway to reduce inflammation; on DCs to generate tDCs producing TGFβ, IL-10, and RA; and on Treg cells. This interaction of SCFAs and GPCRs results into induction and expansion of Tregs capable of suppressing the inflammatory state. GPR43 is activated by all three SCFAs; however, GPR109A is activated only by butyrate. SCFA induction also involves inhibition of HDACs which in turn increases the production and suppressive function of FoxP3$^+$Tregs by promoting the acetylation of Foxp3 protein. In addition, SCFAs modulate the production of PGE2 and COX2 (butyrate in particular), promoting expansion of pre-existing Tregs. (2) Microbial tryptophan metabolites including IAA, IAld, IA, and ILA interact with the AHR on DCs to generate tDCs and AHR on T cells to promote Treg function and/or produce anti-inflammatory IL-22. (3) Probiotic *Lactobacillus reuteri* DSM 17938-mediated production of adenosine and its metabolite inosine, by serving as an A_{2A} agonist, inhibits $T_H1/T_H2/T_H17$ cells and promotes Treg-mediated immunoregulation. (4) Probiotics interact with TLR2 expressed on intestinal epithelial cells and immune cells to induce tDCs and subsequently to induce Tregs (Illustration by Yuying Liu). *Treg* regulatory T cell, *FOS* fructooligosaccharides, *GOS* galactooligosaccharide, *SCFA* short-chain fatty acid, *NF-κB* nuclear factor kappa-light-chain-enhancer of activated B cells, *DCs* dendritic cells, *tDCs* tolerogenic dendritic cells, *GPR43 and GPR109A* G-protein-coupled receptors (GPCRs), *TGFβ* transforming growth factor beta, *IL* interleukin, *RA* retinoic acid, *HDACs* histone deacetylases, *PGE2* prostaglandin E2, *COX-2* cyclooxygenase-2, *IAA* indoleacetic acid, *IAld* indolealdehyde, *IA* indoleacrylic acid, *ILA* indolelactic acid, *AHR* aryl hydrocarbon receptor, A_{2A} adenosine receptor 2A, T_H *cell* T helper cell, *TLR2* toll-like receptor 2, *CNT* concentrative nucleoside transporters, *ENT* equilibrative nucleoside transporters

TGFβ) (Dwivedi et al. 2016). In addition, several studies have revealed a complex relationship between Tregs and microbial metabolism, for example, bacterial production of different metabolites (SCFA, tryptophan derivatives, and adenosine/inosine). Probiotics or probiotic-modulated beneficial gut bacteria produce SCFAs (acetate, propionate, and butyrate) through fermentation of dietary fibers, which maintain immune homeostasis. Studies have indicated the significant alteration in the number of butyrate-producing bacteria in colon of patients with inflammatory disorders (Frank et al. 2007; Wang et al. 2012). SCFAs may induce intestinal Treg cells by several mechanisms: (a) by interacting with specific receptors including G-protein-coupled receptors (GPCRs) GPR43 and GPR109A (also known as hydroxycarboxylic acid receptor 2 or HCA2) expressed in colonic epithelium, adipose tissue, and immune cells (Blad et al. 2012; Ganapathy et al. 2013); (b) by inhibiting histone deacetylases (HDACs), resulting in increased *Foxp3* gene expression and the suppressive function of $FoxP3^+$ Tregs (Arpaia et al. 2013; Li et al. 2007; Tao et al. 2007); or (c) by enhancing production of prostaglandin E2 (PGE2), which has been shown to promote the development of Tregs in humans and in mice (Mahic et al. 2006; Sharma et al. 2005). Interestingly, PGE2 is also involved in mediating the suppressive activity of Tregs via IL-10 and reduced production of multiple proinflammatory cytokines (Baratelli et al. 2005). Microbial tryptophan metabolites including indoleacetic acid (IAA), indolealdehyde (IAld), indoleacrylic acid (IA), and indolelactic acid (ILA) act on the aryl hydrocarbon receptor (AHR) found in intestinal immune cells, and thereby mitigate innate and adaptive immune responses in a ligand-specific fashion to promote Treg function and/or produce anti-inflammatory IL22 (Roager and Licht 2018).

A subset of iTreg, expressing ectonucleotidases CD39 and CD73 is able to hydrolyze ATP to 5′-AMP, generate adenosine, and thus mediate suppression of inflammatory T cells while promoting Foxp3 expression of Tregs via interaction with adenosine receptors. These immune cells express predominately adenosine receptor 2A (A_{2A}) (Su et al. 2019). As mentioned above, one of our previous studies showed that probiotic LR 17938-mediated production of adenosine metabolite inosine, as an A_{2A} agonist, inhibited T_H1 and T_H2 cells and their associated cytokines during of Treg-deficiency-induced autoimmunity (He et al. 2017a, b).

Fecal microbiota transplantation (FMT). The use of processed stool cultures from highly screened healthy donors for humans with *C. difficile* infection (CDI) has been shown to be lifesaving. These FMTs are generally given by a series of colonoscopies with infusions into the colon through the biopsy channel. Preclinical work has shown that FMT is beneficial in mouse models of autoimmune diseases, including myocarditis (Hu et al. 2019), myasthenia gravis (Zheng et al. 2019), and experimental autoimmune uveitis (Ye et al. 2018). In humans, only preliminary studies are available for autoimmune diseases, but efficacy has been suggested for a plethora of conditions, including ulcerative colitis (Dang et al. 2020), acute graft versus host disease (Cohen and Maharshak 2017), and a variety of other conditions which include pouchitis, multiple sclerosis, metabolic syndrome, autism, and irritable bowel syndrome (Cohen and Maharshak 2017). In severe cases of inflammatory bowel disease, FMT has been undertaken with partial success (Hirten et al. 2019). In a mouse model of colitis-associated colon cancer (CAC), FMT improved the severity of colitis. $Foxp3^+$ Tregs were dramatically upregulated among splenic, mesenteric lymph node, and lamina propia lymphocytes of CAC mice after FMT (Wang et al. 2019). Ongoing studies are determining the safety of FMT and duration of response in IBD.

FMT has many common side effects, including bloating, discomfort, low-grade fevers, and flatulence (Dailey et al. 2019). More severe delayed effects may also be seen; some fatal cases may have resulted from transmission of antibiotic-resistant microbial sepsis, such as that due to *E. coli* (DeFilipp et al. 2019). Transmission of viruses is often seen (Chehoud et al. 2016), leading to theoretical concerns about performing this procedure during the coronavirus era (Ianiro et al. 2020). One must consider that in some of these diseases, probiotics may be equally effective (Dang et al. 2020). New approaches that may carry less risk included targeted-bacterial therapy and administration of health-promoting bacterial-derived metabolites (postbiotics). Nevertheless, the FMT story testifies to the strong relationship between the gut microbiome and host autoimmunity.

References

Arpaia N, Campbell C, Fan X et al (2013) Metabolites produced by commensal bacteria promote peripheral regulatory T-cell generation. Nature 504 (7480):451–455

Baratelli F, Lin Y, Zhu L et al (2005) Prostaglandin E2 induces FOXP3 gene expression and T regulatory cell function in human CD4+ T cells. J Immunol 175 (3):1483–1490

Bellocchi C, Fernandez-Ochoa A, Montanelli G et al (2019) Identification of a shared microbiomic and metabolomic profile in systemic autoimmune diseases. J Clin Med 8(9):1291

Bernstein CN (2017) Review article: changes in the epidemiology of inflammatory bowel disease-clues for aetiology. Aliment Pharmacol Ther 46(10):911–919

Bianco AM, Girardelli M, Tommasini A (2015) Genetics of inflammatory bowel disease from multifactorial to monogenic forms. World J Gastroenterol 21 (43):12296–12310

Blad CC, Tang C, Offermanns S (2012) G protein-coupled receptors for energy metabolites as new therapeutic targets. Nat Rev Drug Discov 11(8):603–619

Bunyavanich S, Berin MC (2019) Food allergy and the microbiome: current understandings and future directions. J Allergy Clin Immunol 144(6):1468–1477

Campbell C, Dikiy S, Bhattarai SK et al (2018) Extrathymically generated regulatory T cells establish a niche for intestinal border-dwelling Bacteria and affect physiologic metabolite balance. Immunity 48 (6):1245–1257

Chehoud C, Dryga A, Hwang Y et al (2016) Transfer of viral communities between human individuals during fecal microbiota transplantation. MBio 7(2):e00322

Chervonsky AV (2013) Microbiota and autoimmunity. Cold Spring Harb Perspect Biol 5(3):a007294

Chinen T, Volchkov PY, Chervonsky AV et al (2010) A critical role for regulatory T cell-mediated control of inflammation in the absence of commensal microbiota. J Exp Med 207(11):2323–2330

Cohen NA, Maharshak N (2017) Novel indications for fecal microbial transplantation: update and review of the literature. Dig Dis Sci 62(5):1131–1145

Cook L, Stahl M, Han X et al (2019) Suppressive and gut-reparative functions of human type 1 T regulatory cells. Gastroenterology 157(6):1584–1598

Dailey FE, Turse EP, Daglilar E et al (2019) The dirty aspects of fecal microbiota transplantation: a review of its adverse effects and complications. Curr Opin Pharmacol 49:29–33

Dang X, Xu M, Liu D et al (2020) Assessing the efficacy and safety of fecal microbiota transplantation and probiotic VSL#3 for active ulcerative colitis: a systematic review and meta-analysis. PLoS One 15(3):e0228846

Danikowski KM, Jayaraman S, Prabhakar BS (2017) Regulatory T cells in multiple sclerosis and myasthenia gravis. J Neuroinflammation 14(1):117

DeFilipp Z, Bloom PP, Torres SM et al (2019) Drug-resistant E. coli bacteremia transmitted by fecal microbiota transplant. N Engl J Med 381 (21):2043–2050

Desreumaux P, Foussat A, Allez M et al (2012) Safety and efficacy of antigen-specific regulatory T-cell therapy for patients with refractory Crohn's disease. Gastroenterology 143(5):1207–1217

Di GM, Rizzo A, Franze E et al (2019) Tbet expression in regulatory T cells is required to initiate Th1-mediated colitis. Front Immunol 10:2158

Dingle BM, Liu Y, Fatheree NY et al (2013) FoxP3(+) regulatory T cells attenuate experimental necrotizing enterocolitis. PLoS One 8(12):e82963

Dobes J, Neuwirth A, Dobesova M et al (2015) Gastrointestinal autoimmunity associated with loss of central tolerance to enteric alpha-Defensins. Gastroenterology 149(1):139–150

Du TG, Roberts G, Sayre PH et al (2015) Randomized trial of peanut consumption in infants at risk for peanut allergy. N Engl J Med 372(9):803–813

Du TG, Tsakok T, Lack S et al (2016) Prevention of food allergy. J Allergy Clin Immunol 137(4):998–1010

Dwivedi M, Kumar P, Laddha NC et al (2016) Induction of regulatory T cells: a role for probiotics and prebiotics to suppress autoimmunity. Autoimmun Rev 15(4):379–392

Frank DN, St Amand AL, Feldman RA et al (2007) Molecular-phylogenetic characterization of microbial community imbalances in human inflammatory bowel diseases. Proc Natl Acad Sci U S A 104 (34):13780–13785

Ganapathy V, Thangaraju M, Prasad PD et al (2013) Transporters and receptors for short-chain fatty acids as the molecular link between colonic bacteria and the host. Curr Opin Pharmacol 13(6):869–874

Grainger JR, Wohlfert EA, Fuss IJ et al (2013) Inflammatory monocytes regulate pathologic responses to commensals during acute gastrointestinal infection. Nat Med 19(6):713–721

Gray DH, Gavanescu I, Benoist C et al (2007) Danger-free autoimmune disease in Aire-deficient mice. Proc Natl Acad Sci U S A 104(46):18193–18198

Hardy LC, Smeekens JM, Kulis MD (2019) Biomarkers in food allergy immunotherapy. Curr Allergy Asthma Rep 19(12):61

Harrison OJ, Powrie FM (2013) Regulatory T cells and immune tolerance in the intestine. Cold Spring Harb Perspect Biol 5(7):a018341

He B, Hoang TK, Tran DQ et al (2017a) Adenosine A2A receptor deletion blocks the beneficial effects of lactobacillus reuteri in regulatory T-deficient scurfy mice. Front Immunol 8:1680

He B, Hoang TK, Wang T et al (2017b) Resetting microbiota by lactobacillus reuteri inhibits T reg deficiency-induced autoimmunity via adenosine A2A receptors. J Exp Med 214(1):107–123

He B, Hoang TK, Tian X et al (2019) Lactobacillus reuteri reduces the severity of experimental autoimmune encephalomyelitis in mice by modulating gut microbiota. Front Immunol 10:385

Hetemaki I, Jarva H, Kluger N et al (2016) Anticommensal responses are associated with regulatory T cell defect in autoimmune polyendocrinopathy-candidiasis-ectodermal dystrophy patients. J Immunol 196 (7):2955–2964

Hirten RP, Grinspan A, Fu SC et al (2019) Microbial engraftment and efficacy of fecal microbiota transplant for Clostridium Difficile in patients with and without inflammatory bowel disease. Inflamm Bowel Dis 25 (6):969–979

Horwitz DA, Zheng SG, Gray JD (2008) Natural and TGF-beta-induced Foxp3(+)CD4(+) CD25(+) regulatory T cells are not mirror images of each other. Trends Immunol 29(9):429–435

Hu XF, Zhang WY, Wen Q et al (2019) Fecal microbiota transplantation alleviates myocardial damage in myocarditis by restoring the microbiota composition. Pharmacol Res 139:412–421

Huang W, Solouki S, Koylass N et al (2017) ITK signalling via the Ras/IRF4 pathway regulates the development and function of Tr1 cells. Nat Commun 8:15871

Ianiro G, Mullish BH, Kelly CR et al (2020) Screening of faecal microbiota transplant donors during the COVID-19 outbreak: suggestions for urgent updates from an international expert panel. Lancet Gastroenterol Hepatol 5(5):430–432

Ivanov II, Atarashi K, Manel N et al (2009) Induction of intestinal Th17 cells by segmented filamentous bacteria. Cell 139(3):485–498

Jeong DY, Kim S, Son MJ et al (2019) Induction and maintenance treatment of inflammatory bowel disease: a comprehensive review. Autoimmun Rev 18 (5):439–454

Kawamoto S, Maruya M, Kato LM et al (2014) Foxp3(+) T cells regulate immunoglobulin a selection and facilitate diversification of bacterial species responsible for immune homeostasis. Immunity 41(1):152–165

Lee YK, Menezes JS, Umesaki Y et al (2011) Proinflammatory T-cell responses to gut microbiota promote experimental autoimmune encephalomyelitis. Proc Natl Acad Sci U S A 108(Suppl):14615–14622

Li B, Samanta A, Song X et al (2007) FOXP3 interactions with histone acetyltransferase and class II histone deacetylases are required for repression. Proc Natl Acad Sci U S A 104(11):4571–4576

Liu Y, Fatheree NY, Dingle BM et al (2013) Lactobacillus reuteri DSM 17938 changes the frequency of Foxp3+ regulatory T cells in the intestine and mesenteric lymph node in experimental necrotizing enterocolitis. PLoS One 8(2):e56547

Liu Y, Tran DQ, Fatheree NY et al (2014) Lactobacillus reuteri DSM 17938 differentially modulates effector memory T cells and Foxp3+ regulatory T cells in a mouse model of necrotizing enterocolitis. Am J Physiol Gastrointest Liver Physiol 307(2):G177–G186

Liu Y, Tian X, He B et al (2019) Lactobacillus reuteri DSM 17938 feeding of healthy newborn mice regulates immune responses while modulating gut microbiota and boosting beneficial metabolites. Am J Physiol Gastrointest Liver Physiol 317(6):G824–G838

Lord JD (2015) Promises and paradoxes of regulatory T cells in inflammatory bowel disease. World J Gastroenterol 21(40):11236–11245

Madsen KL, Doyle JS, Jewell LD et al (1999) Lactobacillus species prevents colitis in interleukin 10 gene-deficient mice. Gastroenterology 116(5):1107–1114

Mahic M, Yaqub S, Johansson CC et al (2006) FOXP3 +CD4+CD25+ adaptive regulatory T cells express cyclooxygenase-2 and suppress effector T cells by a prostaglandin E2-dependent mechanism. J Immunol 177(1):246–254

Packey CD, Sartor RB (2008) Interplay of commensal and pathogenic bacteria, genetic mutations, and immunoregulatory defects in the pathogenesis of inflammatory bowel diseases. J Intern Med 263(6):597–606

Platts-Mills TA (2015) The allergy epidemics: 1870-2010. J Allergy Clin Immunol 136(1):3–13

Roager HM, Licht TR (2018) Microbial tryptophan catabolites in health and disease. Nat Commun 9 (1):3294

Rosen MJ, Dhawan A, Saeed SA (2015) Inflammatory bowel disease in children and adolescents. JAMA Pediatr 169(11):1053–1060

Samstein RM, Josefowicz SZ, Arvey A et al (2012) Extrathymic generation of regulatory T cells in placental mammals mitigates maternal-fetal conflict. Cell 150 (1):29–38

Schepici G, Silvestro S, Bramanti P et al (2019) The gut microbiota in multiple sclerosis: an overview of clinical trials. Cell Transplant 28(12):1507–1527

Schirmer M, Denson L, Vlamakis H et al (2018a) Compositional and temporal changes in the gut microbiome of pediatric ulcerative colitis patients are linked to disease course. Cell Host Microbe 24(4):600–610

Schirmer M, Franzosa EA, Lloyd-Price J et al (2018b) Dynamics of metatranscription in the inflammatory bowel disease gut microbiome. Nat Microbiol 3 (3):337–346

Schirmer M, Garner A, Vlamakis H et al (2019) Microbial genes and pathways in inflammatory bowel disease. Nat Rev Microbiol 17(8):497–511

Sharma S, Yang SC, Zhu L et al (2005) Tumor cyclooxygenase-2/prostaglandin E2-dependent promotion of FOXP3 expression and CD4+ CD25+ T regulatory cell activities in lung cancer. Cancer Res 65 (12):5211–5220

Shevach EM, Thornton AM (2014) tTregs, pTregs, and iTregs: similarities and differences. Immunol Rev 259 (1):88–102

Smith PM, Howitt MR, Panikov N et al (2013) The microbial metabolites, short-chain fatty acids, regulate colonic Treg cell homeostasis. Science 341 (6145):569–573

Su W, Chen X, Zhu W et al (2019) The cAMP-adenosine feedback loop maintains the suppressive function of regulatory T cells. J Immunol 203(6):1436–1446

Sun M, Wu W, Chen L et al (2018) Microbiota-derived short-chain fatty acids promote Th1 cell IL-10 production to maintain intestinal homeostasis. Nat Commun 9 (1):3555

Tao R, de Zoeten EF, Ozkaynak E et al (2007) Deacetylase inhibition promotes the generation and function of regulatory T cells. Nat Med 13(11):1299–1307

Torgerson TR, Linane A, Moes N et al (2007) Severe food allergy as a variant of IPEX syndrome caused by a deletion in a noncoding region of the FOXP3 gene. Gastroenterology 132(5):1705–1717

Vitale A, Strisciuglio C, Vitale S et al (2020) Increased frequency of regulatory T cells in pediatric inflammatory bowel disease at diagnosis: a compensative role? Pediatr Res 87(5):853–861

Wallin MT, Culpepper WJ, Campbell JD et al (2019) The prevalence of MS in the United States: a population-based estimate using health claims data. Neurology 92 (10):e1029–e1040

Wang T, Cai G, Qiu Y et al (2012) Structural segregation of gut microbiota between colorectal cancer patients and healthy volunteers. ISME J 6(2):320–329

Wang Z, Hua W, Li C et al (2019) Protective role of fecal microbiota transplantation on colitis and colitis-associated colon cancer in mice is associated with Treg cells. Front Microbiol 10:2498

Yan Y, Zhang GX, Gran B et al (2010) IDO upregulates regulatory T cells via tryptophan catabolite and suppresses encephalitogenic T cell responses in experimental autoimmune encephalomyelitis. J Immunol 185(10):5953–5961

Ye Z, Zhang N, Wu C et al (2018) A metagenomic study of the gut microbiome in Behcet's disease. Microbiome 6(1):135

Zhang H, Sparks JB, Karyala SV et al (2015) Host adaptive immunity alters gut microbiota. ISME J 9 (3):770–781

Zhang X, Olsen N, Zheng SG (2020) The progress and prospect of regulatory T cells in autoimmune diseases. J Autoimmun 111:102461

Zheng P, Li Y, Wu J et al (2019) Perturbed microbial ecology in myasthenia gravis: evidence from the gut microbiome and fecal metabolome. Adv Sci (Weinh) 6 (18):1901441

Zhou L, Sonnenberg GF (2018) Essential immunologic orchestrators of intestinal homeostasis. Sci Immunol 3(20):eaao1605

Tregs in Autoimmune Uveitis 11

Zhaohao Huang, Wenli Li, and Wenru Su

Abstract

Uveitis is a chronic disease with relapsing and remitting ocular attack, which requires corticosteroids and systemic immunosuppression to prevent severe vision loss. Classically, uveitis is referred to an autoimmune disease, mediated by pro-inflammatory Th17 cells and immunosuppressive CD4+CD25+FoxP3+ T-regulatory cells (Tregs). More and more evidence indicates that Tregs are involved in development, resolution, and remission of uveitis. Clinically, many researchers have conducted quantitative and functional analyses of peripheral blood from patients with different subtypes of uveitis, in an attempt to find the changing rules of Tregs. Consistently, using the experimental autoimmune uveitis (EAU) model, researchers have explored the development and resolution mechanism of uveitis in many aspects. In addition, many drug and Tregs therapy investigations have yielded encouraging results. In this chapter, we introduced the current understanding of Tregs, summarized the clinical changes in the number and function of patients with uveitis and the immune mechanism of Tregs involved in EAU model, as well as discussed the progress and shortcomings of Tregs-related drug therapy and Tregs therapy. Although the exact mechanism of Tregs-mediated uveitis protection remains to be elucidated, the strategy of Tregs regulation may provide a specific and meaningful way for the prevention and treatment of uveitis.

Keywords

Tregs · Autoimmune uveitis · Experimental autoimmune uveitis · Transforming growth factor β (TGF-β) · Interleukin 10 (IL-10)

11.1 Introduction

Uveitis is a common disease in ophthalmology, which refers to an autoimmune inflammatory disease that occurs in the inner eye and often leads to poor vision. Anatomically, the uvea includes the iris, ciliary body, and choroid. But clinically, uveitis also includes inflammation of other intraocular structure, such as retina and sclera. Because uveitis is a chronic disease, and it often recurs due to unknown factors. The intraocular structure will be destroyed by inflammation gradually, and the final outcome of vision is always unsatisfied (Wildner and Diedrichs-Mohring 2019). Among all blinders, uveitis causes 10–15% of blindness in developed country and 25% in developing country (Rao 2013; Durrani

Z. Huang · W. Li · W. Su (✉)
State Key Laboratory of Ophthalmology, Zhongshan Ophthalmic Center, Sun Yat-sen University, Guangzhou, China
e-mail: huangzhh73@mail2.sysu.edu.cn; suwr3@mail.sysu.edu.cn

S.-G. Zheng (ed.), *T Regulatory Cells in Human Health and Diseases*, Advances in Experimental Medicine and Biology 1278, https://doi.org/10.1007/978-981-15-6407-9_11

et al. 2004). Although the cause of most uveitis is still unclear, ophthalmologists believe that microbial infection and autoimmunity are closely related to it. In developed countries, autoimmune uveitis is more common, while in developing countries, infectious factors are more common (Lee et al. 2014). According to epidemiological survey, non-infectious uveitis accounts for approximately 90% of the uveitis in the developed country (Thorne et al. 2016; Bertrand et al. 2019). Common noninfectious uveitis includes Behcet's disease (BD), Vogt-Koyanagi-Harada (VKH), HLA-B27-related uveitis, spondyloarthritis, and other systemic diseases. Study has shown that around 37% of patients with uveitis are caused by systemic diseases, which are often thought to be due to the body's autoimmune dysfunction (Bertrand et al. 2019). Consistently, accumulating creditably evidence suggests that immunomodulatory drugs such as methotrexate (MTX) and mycophenolate mofetil (MMF) are effective in the treatment of uveitis, which greatly improves the vision outcome (Rathinam et al. 2019).

T cells, especially CD4+ T cells, are considered to play central roles in the occurrence, development, and remission of uveitis (Becker et al. 2000), which have been widely verified in clinical patients and experimental autoimmune uveitis (EAU) (Zhu et al. 2018; Gilbert et al. 2018). Abnormal CD4+ T cells number and function are critical in the immunopathogenesis of uveitis. When CD4+ T cells are activated, they differentiate into different subtypes, such as T helper 1 (Th1), Th2, Th17, Tfh, and regulatory T cells (Tregs), to function as immune regulators and protect tissues (Klein et al. 2019; Zheng et al. 2002, 2008; Lu et al. 2010a; Chen et al. 2012, 2014). These subtypes and the corresponding secreted cytokines are vital factors of immune response and function, and their abnormality will damage the body (Wahren-Herlenius and Dorner 2013). Specifically, the increase of effector T cells (Teffs) and the decrease of Tregs will break the Tregs/Teffs balance, which will promote the disease onset and progress. At present, it is believed that in the pathogenesis of uveitis, Teffs play the role in promoting inflammation, and Tregs play anti-inflammatory role (Silver et al. 2015). Th1 and Th17 cells, by virtue of the expression of IFN-γ and IL-17, respectively, deploy their pro-inflammatory function (Dong 2008; Mullen et al. 2001). Investigations have shown that Tregs display strongly immunosuppressive function and ideal inhibitory effect on Teffs, which help the immune balance of the body (Wahren-Herlenius and Dorner 2013; Su et al. 2012; Gu et al. 2014). In this regard, emerging evidence indicates that patients with uveitis always exhibit lower number and impaired function of Tregs (Gilbert et al. 2018). In line with this concept, similar findings have been confirmed in EAU, which also deepens our understanding of uveitis occurrence, progression, and treatment. Drugs targeting Tregs have shown an optimal therapeutic efficacy in EAU (Matta et al. 2019). Here, we will focus on recent discoveries and advances of Tregs in uveitis, which we wish could provide ideas and directions for future research.

11.2 Foxp3 and Tregs

Tregs are subpopulation of CD4+ T cells, which have strongly immunosuppressive function and play an important role in maintaining the homeostasis of host tissues and the pathogenesis of autoimmune diseases, allergic diseases, and cancer (Sharabi et al. 2018; Horwitz et al. 2008; Zhang et al. 2020; Liu et al. 2020). As early as 1995, Sakaguchi et al. found that CD4+ CD25+ (also known as IL-2 receptor alpha-chains) cells performed the function of self-tolerance to prevent the occurrence of autoimmune diseases (Sakaguchi et al. 1995). It has been widely accepted that transcription factor forkhead box P3 (Foxp3) is constitutively expressed in the Tregs and plays an indispensable role in immune suppressive capability. Study has shown that *Foxp3* gene mutations in humans can lead to a serious autoimmune disease-immune dysregulation, polyendocrinopathy, enteropathy, and X-linked syndrome (IPEX) (Bennett et al. 2001). Likewise, mutated *foxp3* gene in scurfy mice can break tissue homeostasis, which contributes to fatal consequences (Brunkow et al. 2001). By forcing Foxp3 expression, naive conventional T (Tconv) cells could convert into Tregs-cell-like phenotypes with immunosuppressive capacity (Hori et al. 2003; Fontenot et al. 2003). Therefore, Foxp3 is regard as a Tregs-specific transcription factor and a critical regulator of its biological feature. On the flip side, some studies have shown that Foxp3 expression could not be effectively relied to identify Tregs and maintain its immunosuppressive function. Allan et al. found that activated Teffs could transiently express Foxp3 with a lower level compared to Tregs. The transient expression could not induce the Tregs and endow the cells with inhibitory function (Allan et al. 2007). Miyara et al. validated that in human peripheral blood, there was a kind of CD4+ Foxp3+ T cell subtype that had no inhibitory function. On the contrary, it could produce inflammatory factors after being stimulated (Miyara et al. 2009). A pivotal study by Lin suggested that although Foxp3− EGFP+ CD4+ T cells from Foxp3DeltaEGFP mice could not display Tregs immunosuppressive capacity, they can still show Tregs gene characteristics, such as *Ctla4* and *Il2ra* (Lin et al. 2007). Uniformly, Hill et al. demonstrated some Tregs features are not completely ascribed to Foxp3 (Hill et al. 2007). In patients with rheumatoid arthritis, Yang et al. demonstrated that Helios seems to be better than Foxp3 to represent Tregs (Yang et al. 2019a). In general, Foxp3+ T cells are not all functional Tregs, and T cells without Foxp3 expression could also deploy some signature molecules of Tregs. These research realities indicated although Foxp3 is an important regulator of Tregs, it is relatively complex to define the Tregs. It is likely that Foxp3 may not be only inducible transcript to determine the development and function of Tregs.

11.3 Tregs Development and Maintenance

According to the sites of T-cell developmental origin, one often divides T cells into at least two subtypes. The thymus-derived Tregs (tTregs), also called as natural Tregs (nTregs), are considered as a mature T-cell subpopulation. tTregs might be generated from the single positive thymocytes, which could bind to self-antigen peptides via high affinity T-cell receptor (TCR) in the presence of antigen-presenting cells (APC) with specific MHC expression (Dominguez-Villar and Hafler 2018). Likewise, when peripheral Tconv cells are activated by antigens in a specific environment, such as TGF-β and IL-2, they will gradually differentiate into peripherally derived Tregs (pTregs). When aforementioned process happens in vitro, we call these cells induced Tregs (iTregs) (Zheng et al. 2002, 2004b, a, 2006; Kanamori et al. 2016; Kong et al. 2012a).

As we mentioned earlier, the development of tTregs requires recognition of self-antigens by TCR, cytokines such as TGF-β. Using transgenic mice, Itoh et al. documented that abnormal TCR function resulted in fewer CD4+CD8−CD25+ thymocytes, which revealed the cardinal role of TCR in the development of tTregs (Itoh et al. 1999). In addition, Melissa et al. vindicated that the CD4+CD25+ Tregs are sensitive to variations in the expression of self-peptides (Lerman et al. 2004). Rafal et al. utilized high-throughput single-cell analysis to identify that Tregs have higher TCR diversity than naive T cells, even in the absence of self-antigens (Pacholczyk et al. 2006). The diversity of Tregs is conducive to better recognition of self-antigens.

In addition to diversity of Tregs, Tregs development and Foxp3 expression also require the presence of a variety of cytokines, such as TGF-β, IL-2, and IL-7. At present, it is widely believed that Tregs have high expression of CD25, which is also known as IL-2 receptor α chain. The experimental data show that the abnormal expression of IL-2 receptor could reduce the number of T cells and make the mice prone to a variety of autoimmune diseases (Malek et al. 2002; Willerford et al. 1995). In addition, IL-2 is also crucial for iTregs development (Zheng et al. 2007; Davidson et al. 2007). Moreover, low dose of IL-2 administration has demonstrated the therapeutic effect on several autoimmune diseases mainly through enhancing Tregs

(Saadoun et al. 2011; Koreth et al. 2011; Ye et al. 2018). Of note, IL-2 receptor is not indispensable. IL-7 and IL-15 could make up for the lack of IL-2 receptor to some extent and partly restore the function of Tregs (Vang et al. 2008). Furthermore, investigation has shown that TGF-β signaling could participate in thymic negative selection, reducing Tregs apoptosis (Ouyang et al. 2010).

In addition to the expression of Foxp3 and cytokines mentioned above, it has been confirmed that epigenetic regulation plays an important role in the development and function of Tregs. The demethylation status of a Tregs-specific demethylation region (TSDR) in the *foxp3* promoter plays an essential role in Tregs lineage maintenance where the demethylation of the TSDR correlates with stable Tregs phenotype. It has been speculated that the thymocytes will not express Foxp3 until they get Tregs-specific epigenetic changes after receiving certain TCR stimulation (Ohkura et al. 2013). Other studies have shown that CNS1 (Foxp3 conserved noncoding sequence 1) is very important for iTregs in vivo. Specifically, the absence of CNS1 could lead to the decrease of iTregs, but it has no effect on nTregs (Zheng et al. 2010). CNS2 is very important for maintaining the expression of FOXP3, which is helpful for the development and function of Tregs (Ohkura et al. 2013). However, iTregs are well functional in vitro and in vivo although their TSDR possibly is methylated (Reynolds et al. 2014; Lan et al. 2012; Kong et al. 2012b; Xu et al. 2012; Yang et al. 2017). Indeed, all-trans retinoic acid promotes Tregs stability and function (Mucida et al. 2007; Zhou et al. 2010; Lu et al. 2010b, 2014; Ma et al. 2014), interestingly, all-trans retinoic acid promotes TGF-β-induced Tregs via histone modification but not DNA demethylation on *Foxp3* gene locus (Lu et al. 2011). Therefore, more in-depth studies are needed to determine how epigenetic pattern modulates Tregs stability and function.

In the periphery, combined TCR, TGF-β, and IL-2 signals polarize naive CD4+ T cells into pTregs. These pTregs possess similar suppressive capacities as tTregs in vitro and in vivo (Park et al. 2004). Both tTregs and pTregs express *FOXP3*, *CD25*, *CTLA-4*, *GITR*, *CD39*, and *CD73*, along with low levels of *IL-7Rα* (*CD127*) (Liu et al. 2006). The current view believes that tTregs and pTregs have different immunosuppressive functions and play different roles in immune regulation. pTregs may play a more important role in mucosal tolerance, while tTregs may play a role in immune tolerance. Since no specific lineage marker has been found to distinguish these two subtypes of Tregs, it is difficult for us to understand the differences in biological functions between tTregs and pTregs. Researchers have found that Helios can be used as a marker for tTregs (Thornton et al. 2010). However, later studies found that there was also a subpopulation of tTregs that did not express Helios, which challenged Helios as a marker for distinguishing tTregs and pTregs (Himmel et al. 2013). Also of note, Neuropilin1 (NRP1), which was found specifically higher expression in tTregs, also overexpressed in pTregs when exist in pro-inflammatory microenvironment (Yadav et al. 2012; Weiss et al. 2012). So far, there is no specialized marker that clearly distinguishes them.

11.4 Tregs in Clinic Patients with Uveitis

More and more attentions have been paid to the changes of Tregs quantity and function in patients with uveitis. We summarize these studies in Table 11.1. Most of the results showed that Tregs were abnormal in number and/or function. As early as 2008, a study on patients with VKH syndrome found that Tregs in peripheral blood of patients with active uveitis were significantly reduced, and their inhibitory function was incomplete (Chen et al. 2008). Recent study has indicated that Tregs did decrease in VKH patients with active ocular attack, but there is no difference in Tregs frequency between VKH patients who were at inactive phase and normal people (Liang et al. 2019). Moreover, Wang et al. documented that in patients with BD, the frequency of T cells was also significantly lower than that of healthy subjects (Liang et al. 2011). The research of Sugita et al. also proved this point

Table 11.1 Tregs in the patients with uveitis

Study population (*n*)	Marker of Tregs	Source	Changes of Tregs	Functionality and immune association	Clinical association	References
Da: 11 BDi: 8 BDwo: 4 HC: 4	CD4+ $CD25+^{high}$	Peripheral blood	1. BDo: Tregs decreased before ocular attack 2. Decreased level in patients 3. BDi and BDwo: normal level	NA	Decreased Tregs may be a predictive marker of ocular attack in BD patients	Nanke et al. (2008)
Active VKH:30 Inactive VKH:19 HC:26	CD4+ $CD25+^{high}$	Peripheral blood	1. Decreased Tregs frequency in active VKH patients compared with HC and inactive VKH 2. No differences between HC and inactive VKH	Diminished inhibitory function of Tregs in active VKH patients	NA	Chen et al. (2008)
Active NIU:8 Inactive NIU:12	CD4+ Foxp3+	Peripheral blood	Decreased Tregs frequency in active patients	NA	NA	Yeh et al. (2009)
BD: 30 C: 30	CD4+ CD25+	Peripheral blood	Decreased Tregs frequency in BD patients	Decreased IL-10 level in BD patients	NA	Liang et al. (2011)
IP: 1 AS: 3 VKH: 1 ABD: 2 SLE: 1 SpA: 1 JIA: 3 HC: 10	CD3+ CD4+ $CD25+^{high}$ Foxp3+ CD127−	Peripheral blood	Tregs increase after treatment	NA	Clinical efficacy may be mediated through upregulation of Tregs	Calleja et al. (2012)
B27: 10 B51: 10 B35: 9 IU: 35 Controls: 51	CD4+ CD25+ Foxp3+	Peripheral blood	1. Decreased Tregs frequency in active patients 2. No differences between HC and inactive patients	Tregs inhibit the cell proliferation of CD4 +CD25− lymphocytes	Tregs correlate with clinical remission	Ruggieri et al. (2012)
B27: 22 HC:16	CD4+ CD25+ Foxp3+	Peripheral blood	Decreased Tregs frequency in patients	Tregs inhibit the activation and proliferation of auto-reactive T and B cells, immunoglobulin secretion, and the production of inflammatory cytokines	Tregs had a negative correlation with the disease activity score	Zou et al. (2014)
BC: 3 IU: 1 PIC: 1 BD: 6 PP: 2 VKH: 2 SAR: 3 AS: 2 IAU: 1 HC: 18	CD3+ CD4+ Foxp3+	Peripheral blood	No differences between patients and HC	NA	NA	Molins et al. (2015)

(continued)

Table 11.1 (continued)

Study population (*n*)	Marker of Tregs	Source	Changes of Tregs	Functionality and immune association	Clinical association	References
IIU: 9 HC: 28	CD4+ Foxp3+	Peripheral blood	No differences between patients and HC, Th17/Tregs ratio increased in patients	NA	NA	Walscheid et al. (2016)
B27: 20 HC: 20	CD4+ CD25+ Foxp3+	Peripheral blood	Decreased Tregs frequency in patients	Decreased Foxp3 mRNA level in patients	1. Tregs may contribute to the remission of the disease 2. Tregs have a negative correlation with the disease activity score	Zhuang et al. (2017)
TIN: 14 TINU: 19	CD4+ Foxp3+	Kidney biopsy	Decreased Tregs frequency in TINU	Tregs suppress pathological and physiological immune responses	Tregs may contribute to the development of uveitis	Rytkonen et al. (2018)
Inactive 37 Active 13 HC: 10	CD4+ CD25+ Foxp3+	Peripheral blood	1. Tregs and Tregs/TH1 ratio increased in inactive patients 2. No differences between active patients and HC	Tregs in clinical remission patients show stronger suppressive function	The increase of Tregs in patients with uveitis in remission was apparently related to the effect of treatment	Gilbert et al. (2018)
IU:2 Pan:2 AU:10	CD4+ Foxp3+	Peripheral blood	1. No differences before or after treatment 2. Th17/Tregs ratio decreased after therapy	NA	NA	Walscheid et al. (2019)
BD:27 HC:27	CD3+ CD4+ Foxp3+	Peripheral blood	1. Increased Tregs frequency in patients 2. Decreased Tregs frequency after therapy	Favorable effects of IFN-alfa-2a may be associated with the recovery of Tregs function, suppression of Th17 cells, and reduced expression of TLRs on CD4+ T cells and monocytes	Treg dysfunction may play a role in the induction of BD uveitis	Albayrak et al. (2019)
Active VKH: 68 Inactive VKH: 72 HC: 100	CD4+ CD25+ Foxp3+	Peripheral blood	1. Decreased Tregs frequency in active patients compared with HC 2. Decreased Tregs frequency in active patients compared with inactive patients 3. No differences between inactive patients and HC	Tregs perform their anti-inflammatory function mainly by expressing IL-10	Tregs and IL-10 may exert a protective role in uveitis	Liang et al. (2019)

(continued)

Table 11.1 (continued)

Study population (*n*)	Marker of Tregs	Source	Changes of Tregs	Functionality and immune association	Clinical association	References
ID:3 B27: 20 BD: 3 VKH: 2 TINU: 1 HC: 15	CD3+ CD4+ CD25+^{high} CD127 +^{low}	Peripheral blood	1. Total Tregs show no differences between patients and HC	Patients with higher IL-17A levels also showed higher of memory and naïve Tregs	Patients with a lower grade of anterior chamber or vitreous inflammatory cellular reaction showed higher memory Tregs counts than HC	Guedes et al. (2020)
Tubercular uveitis: 17 HC: 18	CD3+ CD4+ CD25+^{high} Foxp3+	Peripheral blood	1. Decreased Tregs frequency in patients compared with HC	The expression of TGF-β and IL-2Rα, but not CTLA4, was reduced in Tregs	Low Tregs frequency contribute to proinflammatory responses in tubercular uveitis	Sharma et al. (2018)

IGP idiopathic granulomatous panuveitis, *JIA* juvenile idiopathic arthritis, *SLE* systemic lupus erythematosus, *IP* idiopathic panuveitis, *SpA* spondyloarthropathy, *BC* birdshot chorioretinopathy, *IU* intermediate uveitis, *PIC* punctuate inner choroidopathy, *BD* Behçet's disease, *PP* pars planitis, *SAR* sarcoidosis, *AS* ankylosing spondylitis, *IAU* idiopathic anterior uveitis, *VKH* Vogt-Koyanagi-Harada, *Pan* panuveitis, *AU* anterior uveitis, *TIN* tubulointerstitial nephritis, *TINU* TIN with uveitis, *IIU* idiopathic intermediate uveitis, *B27* HLA-B27-positive uveitis, *B51* HLA-B51-positive uveitis, *B35* HLA-B35-associated uveitis, *IU* idiopathic uveitis, *BDa* BD patients with ocular attack, *BDi* BD patients with inactive ocular complications, *BDwo* BD patients without ocular complications, *ID* idiopathic disease, *HC* healthy controls

(Sugita et al. 2011). Another study about BD has shown that before ocular attack, the level of Tregs in the patient was reduced, which may be used to predict ocular attack (Nanke et al. 2008). In contrast to the above results, Albayrak et al. validated that peripheral blood Tregs frequency in BD patients was higher than those in normal people (Albayrak et al. 2019). In addition to the research on VKH and BD, HLA-B27-associated uveitis is also the focus of the researchers. Similarly, studies of Zou and Zhuang et al. have demonstrated a significant reduction of Tregs in peripheral blood. Zhuang et al. validated the Foxp3 mRNA expression was drastically diminished. Furthermore, they both revealed that the level of Tregs was significantly negatively correlated with the HLA-B27 associated uveitis activity scores (Zou et al. 2014; Zhuang et al. 2017). Moreover, a rare type of uveitis, tubulointerstitial nephritis (TIN)-associated uveitis, also showed a similar Tregs frequency in the peripheral blood of patients. Additionally, they documented that Tregs were present in kidney biopsy samples, and the density of Tregs were noticeably increased in TIN patients with uveitis, compared to TIN patients without uveitis (Rytkonen et al. 2018). This may suggest that the pathogenesis of these two subtypes may be different, and Tregs may play an important role in the pathogenesis of TIN with uveitis (Rytkonen et al. 2018).

All of the above studies are aimed at a specific type of uveitis. However, due to the limitations of many types of uveitis and regions, there are also many studies aimed at non-infectious uveitis (NIU). In 2009, Yeh et al. reported a decrease in the frequency of Tregs in active patients. In particular, patients with refractory uveitis showing extremely low levels of Tregs and Foxp3 mRNA expression (Yeh et al. 2009). But in recent years, accumulating evidence suggests that there was no difference in Tregs frequency between patients with NIU and healthy controls (Molins et al. 2015; Walscheid et al. 2016; Gilbert et al. 2018; Guedes et al. 2020). On the other hand, Sharma et al. found that Tregs frequency in patients with tubercular uveitis, which is infectious uveitis, was reduced in the peripheral blood (Sharma et al. 2018). However, for these patients, the suppressive capability of Tregs was not weakened, and the frequency of Tregs was negatively correlated with the level of intraocular IFN-γ and IL-17A (Sharma et al. 2018).

When the study is limited to a specific type of uveitis, one often concludes that Tregs frequency is reduced. When the type is extended to NIU, most studies showed that Tregs frequency does not differ. The contradiction of these results cannot be explained only by the marker selection of different Tregs. We should also consider flow cytometry, an experimental method for measuring the frequency of T cells, which depends on the subjective choice of the experimenter. Furthermore, these discrepancies may be caused by disease duration, the choice of medication, the severity of the disease, and the choice of detection time.

Current studies have shown that Th17 cells also play key role in uveitis. Different from the contradictory results of Tregs frequency, Th17/Tregs ratio is significantly increased in patients with uveitis, which has been confirmed by many studies (Liang et al. 2011, 2019; Walscheid et al. 2016; Zhuang et al. 2017). In addition, Tregs-related cytokines such as IL-10 and TGF-β would be significantly reduced in the active phase of the disease (Liang et al. 2011, 2019; Molins et al. 2015). On the contrary, Albayrak et al. vindicated that in Tregs-related cytokines, such as TGF-β, IL-35, and IL-2, there were no significant differences between active patients and healthy controls (Albayrak et al. 2019).

As for the influence of drug treatment on Tregs, some studies have been reported. Sugita et al. showed that Tregs frequency increased significantly in infliximab (IFX, antitumor necrosis factor-alpha monoclonal antibody) treated patients, but did not change in colchicine- and cyclosporine-treated patients (Sugita et al. 2011). It could also partly explain the clinical efficacy of IFX in the treatment of uveitis. Similarly, Calleja et al. reported an increased Tregs frequency in Adalimumab (ADA, anti–tumor necrosis factor-alpha monoclonal antibody)-treated patients, while corticosteroid group failed to increase the frequency (Calleja et al. 2012). Another study on the treatment of NIU with ADA also proved a similar phenomenon (Walscheid et al. 2019; Liang et al. 2019). Liang et al. demonstrated that VKH patients after treatment exhibited higher Tregs frequency and lower TH17/Treg ratio (Liang et al. 2019). In terms of Tregs-related cytokines, studies have shown that patients with systemic therapy have significantly higher levels of IL-10 and IL-35 (Albayrak et al. 2019; Molins et al. 2015). IL-10 and IL-35 are important suppressive cytokines (Huang et al. 2017, 2018a; Collison et al. 2007). Nonetheless, the TGF-β levels did not change before and after treatment (Albayrak et al. 2019). Another study has shown that the levels of TGF-β and IL-10 in serum were positively correlated with Tregs frequency, which contributed to clinical remission (Gilbert et al. 2018).

In the above, we have described the importance of epigenetic regulation for Tregs development and function. At present, corresponding research results have been obtained, for example, Gilbert et al. found that PBMC from patients in clinical remission had significantly lower methylation levels at the Foxp3 TSDR, Foxp3 promoter, and TIGIT loci than those with active disease. Previous duration of oral immunosuppressive treatment is a significant predictor of methylation at Foxp3 TSDR (Gilbert et al. 2018). As far as the current research results are concerned, the clinical research on epigenetic regulation of Tregs is relatively few, which is also the focus of our research in the future. Understanding the epigenetic pattern of Tregs in patients is helpful for our understanding of the Tregs involvement in the pathogenesis of uveitis.

Some scholars are concerned about the relationship between Tregs subsets and the progression and remission of disease. Guedes et al. demonstrated that total Tregs frequency was not diminished in patients. But when they payed close attention to memory Tregs, they found that patients with lighter clinical symptoms had higher levels of memory Tregs. In addition, the results also showed that the expression of IL-10 and TGF-β in the peripheral blood of patients was negatively correlated memory Tregs (Guedes et al. 2020). There are few studies on the role of memory Tregs in uveitis. More researches are needed to reveal its function and role in uveitis.

11.5 The Role of Tregs in the EAU

Experimental autoimmune uveitis (EAU) model, which usually induced by interphotoreceptor retinoid-binding protein (IRBP) in C57BL/6J mice, is often used to explore the pathogenesis of non-infectious uveitis and the therapeutic effect of drugs as its clinical manifestations are similar to those of patients with uveitis (Zhu et al. 2018). The important role of Tregs in EAU has been reported in many literatures. As early as 2010, Sun et al. had proved that Tregs contribute to the remission of EAU. The results showed that the frequency of Tregs in lymph nodes (LN) of EAU mice increased significantly. Tregs showed inhibition of CD4+CD25− T-cell proliferation and IFN-γ expression. Transfer of Tregs mitigated the EAU induction by IRBP (Sun et al. 2010). Similarly, Ke et al. revealed that ocular Tregs obtained from monophasic EAU (mEAU) showed more powerful immune suppressive capability than Tregs obtained from recurrent EAU (rEAU). Transfer of ocular Tregs from mEAU could convert rEAU into mEAU (Ke et al. 2008). Moreover, Silver et al. found IRBP-specific Tregs and Teffs are highly enriched in the inflamed eyes compared with lymph nodes. IRBP-specific Tregs showed inhibition of Teff proliferation and function. Consistently, IRBP-specific Tregs could contribute to resolution and keep remission of EAU (Silver et al. 2015). More in-depth researches showed that the balance between TH17 and Tregs are indeed involved in the development, resolution, and remission of EAU (Jia et al. 2011; Zhang et al. 2016). The level of Th17 increased gradually and decreased after reaching the peak on 13 days. However, after the peak of Tregs on the 18th day, it is kept at a high level. As the balance gradually shifted to Tregs, EAU tended to have a remission (Zhang et al. 2016).

Because of the important role of Tregs in uveitis, increasingly researches are focused on applying retinal pigment epithelial cells (RPE) to induce the Tregs. In 2010, Horie et al. obtained human RPE-induced Tregs by the pretreatment of human RPE cells with TGF-β. In vitro experiments testified that human RPE-induced Tregs, possessing the suppressive capability, could produce TGF-β1 and IL-10 but not IFN-γ. Importantly, these cells also expressed high levels of CTLA-4 and Foxp3, which are central molecules for the function of Tregs (Horie et al. 2010; Li et al. 2014). Two years later, Imai et al. verified the results hosted by Horie. Differently, the results of Imai et al. showed that human RPE-induced Tregs were able to produce IFN-γ. Moreover, human RPE-induced Tregs displayed higher expression of CD25, Foxp3, CTLA-4, and TNFRSF18 and repressed bystander Teffs. Interestingly, due to long-term culture in vitro, these cells would gradually lose Foxp3 expression and thus harmed immune suppressive capability. They demonstrated that the $CD25^{high}Foxp3^{high}CD45RA^{-}$ Tregs were the cardinal effector Treg subtypes with remarkable immunosuppressive property. Kawazoe et al. indicated that retinoic acid (RA, a metabolite of vitamin A) played a central role in RPE-induced Tregs. Results suggested that vitamin A-deficient RPE were unable to produce Tregs due to the lack of TGFβ and RA (Huang et al. 2018b). However, the symptoms of EAU in vitamin A-deficient mice were not more serious than control group. This indicates that there are other mechanisms to alleviate symptoms in EAU, which is also worth exploring in the future (Kawazoe et al. 2012).

11.6 Tregs-Related Drug Therapy in the EAU Model

More and more Tregs-related drugs have made some progress in the study of therapeutic effect in EAU (as shown in Table 11.2). Commodaro et al. reported in 2010 that fingolimod (immunomodulatory drug) could alleviate the clinical symptoms of EAU by promoting the proportion of Tregs in lymph nodes (Commodaro et al. 2010). Sampson et al. illustrated that Tregs response in the draining lymph nodes of EAU was markedly

Table 11.2 Tregs-related drug therapy in the EAU model

Drugs	Type	The expression marker of Tregs	Quantity or proportion of Tregs	Immunomodulatory function of Tregs	Signal pathway mechanism	Outcomes	References
FTY720 (fingolimod)	An immunomodulatory drug capable of preventing T-cell migration to inflammatory sites	CD4+ CD25+ Foxp3+	The frequency of Tregs decreases inside the eye; the frequency of Tregs increase in lymph nodes	NA	FTY720 blocks sphingosine-1 phosphate receptor 1 (S1p1) signaling pathway, leads to T-cell retention in lymphoid organs, and prevents their migration to inflammatory sites	Reduces the frequency of CD4+CD25+ T-cell population in the eye A significant increase in the frequency of CD4 +IFN-γ+ T cells in the lymph nodes Decreasing the absolute cell numbers of CD4+T cells in the spleen	Commodaro et al. (2010)
MR16-1	Anti-mouse IL-6 receptor monoclonal antibody	CD4+ Foxp3+	Increased frequency of Tregs	NA	NA	MR16-1 inhibits TH17 differentiation but not TH1 differentiation in vivo Induction of Tregs	Hohki et al. (2010)
Benzoic acid (Am80)	Retinoic acid receptor-α/β-specific ligand	CD4+ Foxp3+	The number of Tregs in the spleen remained constant; promotion of Tregs conversion in vitro	NA	Am80 may be a promising agent for preventing autoimmune uveoretinal inflammation mediated by the TH1/TH17 pathway	Inhibits TH17 and TH1; IL-17 and IFN-γ; IL-6Rα on CD4+ T cells	Keino et al. (2011)
Anti-CD3 antibody	Anti-CD3 antibody	CD4+ Foxp3+	The frequency of Tregs increased	Protects mice from subsequent EAU induction	NA	Decreases IFN-γ, IL-17, and IL-6 Increases IL-10 and TGF-β Low expression of the transcription factors t-bet and RORγt for TH1 and TH17 differentiation Anti-CD3 mab treatment results in long-term tolerance	Ke et al. (2011)

Aicar	Adenosine monophosphate (AMP) analog	CD4+ CD25+ Foxp3+	Tregs population was not changed	NA	NA	mRNA expression of IL-1b, IL-6, IFN-γ, and TNF-α was elevated in the retina; inhibit IFN-γ, IL-17, and IL-10	Suzuki et al. (2012)
Rapamycin	T-cell inhibitors	CD25+ Foxp3+	The number of Tregs increase in the spleen	Tregs control effector function of T helper cells, including TH1, TH2, and TH17	Regulation of the development of T-cell subsets by rapamycin is through its mammalian target of rapamycin c2 (MTORC2) pathway	Inhibits IFN-γ, IL-2, IL-17, IL-4, and IL-10; increases TGF-β1; inhibits TH17	Yuan et al. (2015)
Galectin-8	A tandem-repeat-type member of the galectin family	CD4+ Foxp3+	Frequency and number of Tregs increase in the retina and dLN	Gal-8 treatment modulates the severity of EAU pathology by enhancing anti-inflammatory Tregs responses	NA	Retina: frequency of Tregs and TH2 was increased; IFN-γ and IL-17A decreased; IL-10 increased; (CTLA)-4 increased; CD103 increased	Sampson et al. (2015)
Vorinostat (HDACi)	Histone deacetylase (HDAC) inhibitors	CD4+ CD25+ Foxp3+	Frequency of Tregs increases	NA	Inhibits NF-κb and STAT signaling pathways to induce anti-inflammatory effects	Inhibits TH1 and TH17; increases TH0 and Tregs; inhibits macrophage, INF-γ, and IL-17a; increases IL-10; inhibits STAT1, STAT3, and p65	Fang et al. (2016)
Sodium butyrate (Nab)	A potent activator for NRF2; a governor of antioxidant signaling	CD4+ CD25+ Foxp3+	The number of Tregs increases in dLNs and spleens	NA	The profound activation of the NRF2/HO-1 signaling pathway, which plays crucial roles in Nab-mediated inhibition on TH17 cell differentiation and EAU	Inhibits TH17; inhibits CXCLl9 (T-cell chemokine) and MCP-1 (monocyte chemokines); inhibits IL-17a, IFN-γ, and TNF-α	Chen et al. (2017b)
Chrysin	A member of the flavonoid family	CD4+ CD25+	Frequency of Tregs increases in the spleens	NA	Chrysin may exert its anti-inflammatory effects on EAU by inhibiting the canonical NF-κb signaling pathway	Inhibits TH1, TH17, and CD4+CD3+CD62l+ TH0 cells; increases Tregs; inhibits IFN-γ, IL-17a, IL-6, IL-1β, and TNF-α; inhibits NF-Kb p65; inhibits macrophage	Meng et al. (2017)

(continued)

Table 11.2 (continued)

Drugs	Type	The expression marker of Tregs	Quantity or proportion of Tregs	Immunomodulatory function of Tregs	Signal pathway mechanism	Outcomes	References
Novel CD28 antagonist mpeg pv1-fab′ (pv1)	CD28 antagonist	CD3+ CD4+ CD25+ Foxp3+	Treg population decreases in peripheral lymphoid organs	NA	The blockade of CD28 signaling pathway can also induce anergy in T lymphocytes	Inhibits CD69, CD25, and PD-1 molecules; inhibits CD4+IFN-γ+ T cells (TH1)	Papotto et al. (2017)
Baicalin	A traditional Chinese medicine	CD4+ Foxp3+	The frequency and number of Tregs increase in the dLN	Tregs play a protective and anti-inflammatory role	NA	Significant decreases in IFN-γ, IL-17a, and TNF-α expression in the retinas and dLN; increases the ratio and number of Tregs and decreases the frequency and number of Teffs in the dLN	Zhu et al. (2018)
TGF-β2	TGF-β2	CD4+ CD25+ Foxp3+	The frequency of Tregs increases	NA	NA	Significantly inhibited the proliferation (in vitro) of CD4+ T cells in response to MAA; increased frequency of Tregs	Matta et al. (2019)
As101	Immunomodulator	CD4+ Foxp3+	The frequency of Tregs increases	NA	Jak/STAT intracellular signaling pathway; PI3k/AKT signaling	Alters effector and regulatory T-cell balance and suppresses EAU; inhibits pathways of TH1 and TH17 effector cell differentiation; enhances Tregs generation and/or expansion	Bing et al. (2019)

Aminooxy-acetic acid (AOA)	A small molecular compound	CD4+ Foxp3+	The frequency of Tregs increases in the spleen	NA	AOA inhibits NF-κB and STAT1 signaling pathways	Decreases the frequency of TH1 and TH17 cells; increases the frequency of Tregs in spleen; a decreased expression of IL-17 and IFN-γ, whereas increased expression of IL-10 in the spleen and retina	Mei et al. (2020)
Berberine	Isoquinoline alkaloid (a traditional Chinese medicine)	CD4+ Foxp3+	Increased frequency of Tregs	NA	NA	IFN-γ and IL-17 were significantly decreased; IL-10 showed a slight increase; BBR treatment alters effector T cell and Tregs balance	Du et al. (2020)

increased after Gal-8 (immunomodulatory drug) treatment. Specifically, Treg immune suppressive capability-related factors including CTLA-4, IL-10, and CD103 were obviously boosted. Indeed, CD103+ Tregs may function independent upon Foxp3 expression (Liu et al. 2014; Zhong et al. 2018; Zhang et al. 2019). Meanwhile, Gal-8 treatment downregulated the TH17/Tregs and TH1/Tregs ratios, decreased the expression of IL-17 and IFN-β, and upregulated the expression of IL-10, which protected the retina from damage caused by inflammation (Sampson et al. 2015). Chen et el. documented that sodium butyrate (NaB, immunomodulatory drug) could not only increase the frequency of Tregs in lymph nodes but also augment the frequency of Tregs in spleen (Chen et al. 2017b). Some types of uveitis, such as BD, often have leakage of fundus vessels, which often leads to poor prognosis of vision. It is very important to reduce the leakage of fundus vessels as early as possible for the recovery of diseases and the maintenance of good vision. As such, the researchers also found that several drugs that could protect vascular tissue and reduce leakage of fundus vessels. Fang et al. reveal the protective effect of vorinostat (histone deacetylase inhibitors) on retina, blood–retinal barrier, and retinal vasculature (Fang et al. 2016). Similarly, Meng et al. demonstrated that chrysin (a member of the flavonoid family with immunomodulatory function) could improve the levels of tight junction proteins ZO-1 and occludin, which maintains the blood–retinal barrier (Meng et al. 2017).

We have discussed the importance of TGF-β for Tregs development and function maintenance. It makes us wonder if TGF-β is an ideal way to treat uveitis. Just last year, Matta et al. obtained gratifying results during the treatment of EAU with human recombinant TGF-β2 (rTGF-β2). Early intravenous injection of rTGF-β2 can effectively avoid the occurrence of EAU. This protective effect is mainly achieved by inhibiting CD4+ cell proliferation and promoting Tregs generation. Moreover, administration of rTGF-β2 at late stage of EAU would lose its protective effect (Matta et al. 2019). This is also consistent with the clinical observation. The earlier the treatment, the better the visual outcome.

Antibody drugs also showed surprising outcomes in the EAU. As early as 2011, Ke et al. proved the efficacy of anti-CD3 antibody in EAU (Ke et al. 2011). On the one hand, anti-CD3 antibody could inhibit the activation of IRBP-specific T cells, the differentiation of Th1/Th17 and the expression of its cytokines. On the other hand, the proportion of Tregs in the spleen was greatly increased and its suppressive capability was also significantly enhanced. Neutralization of TGF-β or IL-10 would block the effect of anti-CD3 antibody treatment on the development of uveitis. It revealed that IL-10- and TGF-β-dependent Tregs play an indispensable role in the EAU. In addition to anti-CD3 antibody, Hohki et al. found that MR16-1 (anti-mouse IL-6 receptor monoclonal antibody) could effectively relieve EAU symptoms by inhibiting Th17 and promoting Tregs (Hohki et al. 2010). However, in the meantime, results indicated that anti-TNF mAb did not exhibit protective effect of EAU. It is unclear whether different TNFR signals that affect Teffs and Tregs contribute to anti-TNF mAb inability in treating EAU (Yang et al. 2018, 2019b; Jacob et al. 2009). On the contrary, in clinic, the treatment of uveitis with ADA has been widely recognized, which has been proved by many clinical trials (Jaffe et al. 2016). The contradiction between basic experiments and clinical reality also suggests the limitation of EAU model.

All of the above drugs are systemic drugs, which have a certain impact on the systemic immune system. In clinic, local administration is also an ubiquitously essential treatment due to the fewer side effects. He et al. investigated the therapeutic effect of topical administration of SOCS1-KIR (JAK2/STAT1 inhibitor) on EAU. SOCS1-KIR was able to alleviate the clinical manifestations of EAU, reduce the damage of intraocular tissues, and inhibit the development of inflammatory cells in the retina. Specifically, SOCS1-KIR promoted expansion of IL-10-producing Tregs and inhibited the expression of Th1 and Th17 cells and related cytokines and chemokines, such as IFN-γ, IL-17A, TNF-α, IL-1β, IL-6, CCL20, and CXCL9. Of note, SOCS1-KIR curbed EAU manifestations by preventing retinal cells from undergoing apoptosis. Accordingly, SOCS1-KIR did not cause damage to the peripheral immune system as a result of local administration (He et al. 2016). Kasper et al. observed the similar results in the topical administration of EV/mPEGhexPLA (Kasper et al. 2018). Differently, the Tregs in the lymph nodes were increased, but the frequency in the spleen did not change (Kasper et al. 2018).

In general, the application of various drugs to treat EAU has made some progresses. All kinds of immunomodulators, monoclonal antibodies, and inhibitors play encouraging roles in EAU. Basically, on the one hand, they can inhibit the secretion of inflammatory cells (TH1, TH17) and cytokines (INF-γ, TNF-α, IL-6, IL-17, and so on) in the eyes, lymphatics, and spleen. On the other hand, they can exhibit anti-inflammatory function by promoting the differentiation of Tregs and the secretion of related cytokines (IL-10, TGF-β, and so on). Many drugs can protect retinal cells from inflammatory cell infiltration and apoptosis, thus maintaining the integrity of blood–retinal barrier. In particular, the emergence of some local administration drugs, which could reduce systemic side effects, resulting in the delightful expectations.

11.7 Tregs Therapy in EAU Model

The application of Tregs in the treatment of autoimmune diseases has been in full swing (Dominguez-Villar and Hafler 2018). Tregs therapy in the uveitis has long been in progress. As early as 2002, Namba et al. have applied induced Tregs to relieve the incidence and severity of EAU (Namba et al. 2002). Specifically, they utilized α-melanocyte-stimulating hormone (α-MSH) and TGF-β to produce antigen-specific Tregs. Tregs were adoptively transferred into EAU mice, resulting in a less severity of EAU symptoms. Further results showed that Tregs could inhibit the production of IL-4 and IFN-γ and promoted the expression of TGF-β (Namba et al. 2002). Moreover, Asnagli et al. validated the therapeutic potential of collagen II-specific Tregs (Col-Tregs) as a targeted approach for the treatment of EAU.

Mice Col-Tregs expressed CD25, Foxp3, and low surface expression of CD127. The inhibition of Th1 and Th17 by Col-Tregs was proved by in vitro functional assays. Intravenous administration of Col-Tregs could effectively ameliorate the severity of EAU model (Asnagli et al. 2015). We also demonstrated that Tregs could alleviate the inflammatory response of EAU (Su et al. 2019). In vitro, Tregs curbed the proliferation of T cells and related cytokine, such as IL-17 and IFN-γ (Luo et al. 2019; Chen et al. 2017a). We revealed that cAMP could partially contribute to the inhibitory function of Tregs. Intracellular cAMP regulated CD39 expression and CD39-dependent adenosine production in Tregs (Gu et al. 2017). cAMP directly participates in Treg-derived adenosine production by a CD39 signaling-independent extracellular cAMP–adenosine pathway. In particular, elevating the expression of cAMP in Tregs before transfer enhanced the immune suppressive capability in EAU model (Su et al. 2019). All of the above-mentioned Tregs therapies are systematic applications, and there is also study on local application of Tregs therapy for EAU. Gregoire et al. demonstrated in situ injection of preactivated polyclonal Tregs could curb the inflammation of EAU. Interestingly, Tregs injection could not inhibit ocular Teffs and decreased infiltrating immune cells or inflammatory cytokines. However, local administration of Tregs could improve the levels of IL-10, which reduced the production of ROS. Mechanistically, administration of Tregs could drastically repress the production of ROS, which had been proved to be closely related to the development and severity of EAU (Gregoire et al. 2016).

There are also some Tregs-related cell therapies that have shown impressive therapeutic effects. To demonstrate, systemic administration of mesenchymal stem cells (MSCs) could increase the frequency of Tregs by inducing the expression of TGF-β and then inducing TGF-β-dependent Foxp3 (Zhang et al. 2011, 2017; Chen et al. 2013; Zhao et al. 2020; Su et al. 2015). Imai et al. validated the MSCs and suppressed the TH1 and TH17 cells and related inflammatory cytokines (Tasso et al. 2012). Li et al. vindicated the pleasant therapeutic effect of human amniotic epithelial cells (hAECs) in EAU via curbing TH17 cells and promoting Tregs (Li et al. 2018). Thus far, it is optimistic to see that more and more therapeutic mechanisms have been explored. For instances, microRNAs (Shi et al. 2019) and gut microbiome (Nakamura et al. 2016) could also play roles in treating EAU through Tregs.

11.8 Conclusion

At present, the changes of the number and function of Tregs in patients with uveitis have been reported in many studies. When studies focus on a particular type of uveitis, most results showed that Tregs are reduced in frequency and inhibitory function. Especially for patients with active stage or patients with refractory uveitis, this phenomenon is even more obvious. When the subjects were expanded to patients with NIU, one found that the frequency variation of Tregs was intricate and had not been able to obtain consistent and convincing evidence. This may be due to the patient's phenotype, disease activity, molecular markers of Tregs, and many other factors. As such, the TH17/Tregs ratio seems to be a good choice to estimate the disease status and treatment effect on patients. After systematic drug treatment, the TH17/Tregs ratio in patients would decrease, which indicates that the balance between TH17 and Tregs is shifting to Tregs.

Through in-depth exploration of EAU model, the importance of Tregs in the pathogenesis of uveitis has been basically established. The number and function of Tregs in the intraocular and peripheral immune systems have been explored. The number of antigen-specific Tregs would increase gradually and remain at a high level, which contributes to the development, resolution, and remission of EAU. The functional test confirms that ocular and peripheral Tregs exhibit strong immunosuppressive ability. The increase of Tregs number and function in EAU is of great significance to prevent recurrence, which accounts for the monophasic diseases.

Tregs-related therapy in the EAU is attracted a lot of research attention. Immunomodulatory drug, monoclonal antibody, and inhibitor alleviate the symptoms of EAU by regulating the balance of TH17 and Tregs. Specifically, drug promotes the differentiation of Tregs, the expression of TGF-β and Foxp3, and the secretion of IL-10 and TGF-β. On the other hand, drug curbs the generation of TH1, TH17, and related cytokines (IFN-γ, IL-6, IL-17). Tregs therapy is also the focus of current research. Researchers induced Tregs in vitro and injected it into EAU mice, which also achieved ideal therapeutic effect. For another example, MSC is used to regulate the inflammatory response in EAU mice and promote the anti-inflammatory response through Tregs. Notably, topical administration of drugs or Tregs displays inspiring results, which may significantly reduce systemic side effects.

Nevertheless, previous studies mainly focused on the peripheral blood of patients and the balance between Th17 and Tregs in EAU. The mechanism of specific regulation of Tregs in uveitis is still uncertain. In the future, it is necessary to strengthen the functional study of Tregs subtypes in uveitis. Finally, because the effects of the existing treatments in uveitis are not adequately described, obtaining a more accurate functional mechanism of Tregs in uveitis will certainly lead to new approaches to the treatment for patients with uveitis.

Funding This study was supported by the National Key R&D Program of China (2017YFA0105800), the Natural Science Foundation of China (81670897, 81600731 and U1601226), and Guangdong Natural Science Funds for Distinguished Young Scholar (2016A030306006).

Conflict

The authors declare no conflict of interest.

References

Albayrak O, Oray M, Can F, Uludag Kirimli G, Gul A, Tugal-Tutkun I, Onal S (2019) Effect of interferon alfa-2a treatment on adaptive and innate immune systems in patients with Behcet disease uveitis. Invest Ophthalmol Vis Sci 60(1):52–63. https://doi.org/10.1167/iovs.18-25548

Allan SE, Crome SQ, Crellin NK, Passerini L, Steiner TS, Bacchetta R, Roncarolo MG, Levings MK (2007) Activation-induced FOXP3 in human T effector cells does not suppress proliferation or cytokine production. Int Immunol 19(4):345–354. https://doi.org/10.1093/intimm/dxm014

Asnagli H, Jacquin M, Belmonte N, Gertner-Dardenne J, Hubert MF, Sales A, Fall PB, Ginet C, Marchetti I, Menard F, Lara G, Bobak N, Foussat A (2015) Inhibition of noninfectious uveitis using intravenous administration of collagen II-specific type 1 regulatory T cells. Invest Ophthalmol Vis Sci 56(11):6456–6466. https://doi.org/10.1167/iovs.15-16883

Becker MD, Adamus G, Davey MP, Rosenbaum JT (2000) The role of T cells in autoimmune uveitis. Ocul Immunol Inflamm 8(2):93–100

Bennett CL, Christie J, Ramsdell F, Brunkow ME, Ferguson PJ, Whitesell L, Kelly TE, Saulsbury FT, Chance PF, Ochs HD (2001) The immune dysregulation, polyendocrinopathy, enteropathy, X-linked syndrome (IPEX) is caused by mutations of FOXP3. Nat Genet 27(1):20–21. https://doi.org/10.1038/83713

Bertrand PJ, Jamilloux Y, Ecochard R, Richard-Colmant-G, Gerfaud-Valentin M, Guillaud M, Denis P, Kodjikian L, Seve P (2019) Uveitis: Autoimmunity... and beyond. Autoimmun Rev 18(9):102351. https://doi.org/10.1016/j.autrev.2019.102351

Bing SJ, Shemesh I, Chong WP, Horai R, Jittayasothorn Y, Silver PB, Sredni B, Caspi RR (2019) AS101 ameliorates experimental autoimmune uveitis by regulating Th1 and Th17 responses and inducing Treg cells. J Autoimmun 100:52–61. https://doi.org/10.1016/j.jaut.2019.02.006

Brunkow ME, Jeffery EW, Hjerrild KA, Paeper B, Clark LB, Yasayko SA, Wilkinson JE, Galas D, Ziegler SF, Ramsdell F (2001) Disruption of a new forkhead/winged-helix protein, scurfin, results in the fatal lymphoproliferative disorder of the scurfy mouse. Nat Genet 27(1):68–73. https://doi.org/10.1038/83784

Calleja S, Cordero-Coma M, Rodriguez E, Llorente M, Franco M, Ruiz de Morales JG (2012) Adalimumab specifically induces CD3(+) CD4(+) CD25(high) Foxp3(+) CD127(-) T-regulatory cells and decreases vascular endothelial growth factor plasma levels in refractory immuno-mediated uveitis: a non-randomized pilot intervention study. Eye (Lond) 26(3):468–477. https://doi.org/10.1038/eye.2011.320

Chen L, Yang P, Zhou H, He H, Ren X, Chi W, Wang L, Kijlstra A (2008) Diminished frequency and function of CD4+CD25high regulatory T cells associated with active uveitis in Vogt-Koyanagi-Harada syndrome. Invest Ophthalmol Vis Sci 49(8):3475–3482. https://doi.org/10.1167/iovs.08-1793

Chen M, Guo Z, Ju W, Ryffel B, He X, Zheng SG (2012) The development and function of follicular helper T cells in immune responses. Cell Mol Immunol 9(5):375–379. https://doi.org/10.1038/cmi.2012.18

Chen M, Su W, Lin X, Guo Z, Wang J, Zhang Q, Brand D, Ryffel B, Huang J, Liu Z, He X, Le AD, Zheng SG (2013) Adoptive transfer of human gingiva-derived mesenchymal stem cells ameliorates collagen-induced arthritis via suppression of Th1 and Th17 cells and enhancement of regulatory T cell differentiation. Arthritis Rheum 65(5):1181–1193. https://doi.org/10.1002/art.37894

Chen M, Lin X, Liu Y, Li Q, Deng Y, Liu Z, Brand D, Guo Z, He X, Ryffel B, Zheng SG (2014) The function of BAFF on T helper cells in autoimmunity. Cytokine Growth Factor Rev 25(3):301–305. https://doi.org/10.1016/j.cytogfr.2013.12.011

Chen W, Xu Z, Zheng Y, Wang J, Qian W, Olsen N, Brand D, Lin J, Zheng SG (2017a) A protocol to develop T helper and Treg cells in vivo. Cell Mol Immunol 14(12):1013–1016. https://doi.org/10.1038/cmi.2017.116

Chen X, Su W, Wan T, Yu J, Zhu W, Tang F, Liu G, Olsen N, Liang D, Zheng SG (2017b) Sodium butyrate regulates Th17/Treg cell balance to ameliorate uveitis via the Nrf2/HO-1 pathway. Biochem Pharmacol 142:111–119. https://doi.org/10.1016/j.bcp.2017.06.136

Collison LW, Workman CJ, Kuo TT, Boyd K, Wang Y, Vignali KM, Cross R, Sehy D, Blumberg RS, Vignali DA (2007) The inhibitory cytokine IL-35 contributes to regulatory T-cell function. Nature 450(7169):566–569. https://doi.org/10.1038/nature06306

Commodaro AG, Peron JP, Lopes CT, Arslanian C, Belfort R Jr, Rizzo LV, Bueno V (2010) Evaluation of experimental autoimmune uveitis in mice treated with FTY720. Invest Ophthalmol Vis Sci 51(5):2568–2574. https://doi.org/10.1167/iovs.09-4769

Davidson TS, DiPaolo RJ, Andersson J, Shevach EM (2007) Cutting Edge: IL-2 is essential for TGF-beta-mediated induction of Foxp3+ T regulatory cells. J Immunol 178(7):4022–4026. https://doi.org/10.4049/jimmunol.178.7.4022

Dominguez-Villar M, Hafler DA (2018) Regulatory T cells in autoimmune disease. Nat Immunol 19(7):665–673. https://doi.org/10.1038/s41590-018-0120-4

Dong C (2008) TH17 cells in development: an updated view of their molecular identity and genetic programming. Nat Rev Immunol 8(5):337–348. https://doi.org/10.1038/nri2295

Du Z, Wang Q, Huang X, Yi S, Mei S, Yuan G, Su G, Cao Q, Zhou C, Wang Y, Kijlstra A, Yang P (2020) Effect of berberine on spleen transcriptome and gut microbiota composition in experimental autoimmune uveitis. Int Immunopharmacol 81:106270. https://doi.org/10.1016/j.intimp.2020.106270

Durrani OM, Tehrani NN, Marr JE, Moradi P, Stavrou P, Murray PI (2004) Degree, duration, and causes of visual loss in uveitis. Br J Ophthalmol 88(9):1159–1162. https://doi.org/10.1136/bjo.2003.037226

Fang S, Meng X, Zhang Z, Wang Y, Liu Y, You C, Yan H (2016) Vorinostat modulates the imbalance of T cell subsets, suppresses macrophage activity, and ameliorates experimental autoimmune uveoretinitis.

Neuromolecular Med 18(1):134–145. https://doi.org/10.1007/s12017-016-8383-0
Fontenot JD, Gavin MA, Rudensky AY (2003) Foxp3 programs the development and function of CD4+CD25+ regulatory T cells. Nat Immunol 4(4):330–336. https://doi.org/10.1038/ni904
Gilbert RM, Zhang X, Sampson RD, Ehrenstein MR, Nguyen DX, Chaudhry M, Mein C, Mahmud N, Galatowicz G, Tomkins-Netzer O, Calder VL, Lightman S (2018) Clinical remission of sight-threatening non-infectious uveitis is characterized by an upregulation of peripheral T-regulatory cell polarized towards T-bet and TIGIT. Front Immunol 9:907. https://doi.org/10.3389/fimmu.2018.00907
Gregoire S, Terrada C, Martin GH, Fourcade G, Baeyens A, Marodon G, Fisson S, Billiard F, Lucas B, Tadayoni R, Behar-Cohen F, Levacher B, Galy A, LeHoang P, Klatzmann D, Bodaghi B, Salomon BL (2016) Treatment of uveitis by in situ administration of ex vivo-activated polyclonal regulatory T cells. J Immunol 196(5):2109–2118. https://doi.org/10.4049/jimmunol.1501723
Gu J, Lu L, Chen M, Xu L, Lan Q, Li Q, Liu Z, Chen G, Wang P, Wang X, Brand D, Olsen N, Zheng SG (2014) TGF-beta-induced CD4+Foxp3+ T cells attenuate acute graft-versus-host disease by suppressing expansion and killing of effector CD8+ cells. J Immunol 193(7):3388–3397. https://doi.org/10.4049/jimmunol.1400207
Gu J, Ni X, Pan X, Lu H, Lu Y, Zhao J, Guo Zheng S, Hippen KL, Wang X, Lu L (2017) Human CD39(hi) regulatory T cells present stronger stability and function under inflammatory conditions. Cell Mol Immunol 14(6):521–528. https://doi.org/10.1038/cmi.2016.30
Guedes MCE, Arroz MJ, Martins C, Angelo-Dias M, Proenca RD, Borrego LM (2020) Regulatory T cells and IL-17A levels in noninfectious uveitis. Graefes Arch Clin Exp Ophthalmol 258(6):1269–1278. https://doi.org/10.1007/s00417-020-04649-0
He C, Yu CR, Mattapallil MJ, Sun L, Larkin Iii J, Egwuagu CE (2016) SOCS1 mimetic peptide suppresses chronic intraocular inflammatory disease (uveitis). Mediators Inflamm 2016:2939370. https://doi.org/10.1155/2016/2939370
Hill JA, Feuerer M, Tash K, Haxhinasto S, Perez J, Melamed R, Mathis D, Benoist C (2007) Foxp3 transcription-factor-dependent and -independent regulation of the regulatory T cell transcriptional signature. Immunity 27(5):786–800. https://doi.org/10.1016/j.immuni.2007.09.010
Himmel ME, MacDonald KG, Garcia RV, Steiner TS, Levings MK (2013) Helios+ and Helios- cells coexist within the natural FOXP3+ T regulatory cell subset in humans. J Immunol 190(5):2001–2008. https://doi.org/10.4049/jimmunol.1201379
Hohki S, Ohguro N, Haruta H, Nakai K, Terabe F, Serada S, Fujimoto M, Nomura S, Kawahata H, Kishimoto T, Naka T (2010) Blockade of interleukin-6 signaling suppresses experimental autoimmune uveoretinitis by the inhibition of inflammatory Th17 responses. Exp Eye Res 91(2):162–170. https://doi.org/10.1016/j.exer.2010.04.009
Hori S, Nomura T, Sakaguchi S (2003) Control of regulatory T cell development by the transcription factor Foxp3. Science 299(5609):1057–1061. https://doi.org/10.1126/science.1079490
Horie S, Sugita S, Futagami Y, Yamada Y, Mochizuki M (2010) Human retinal pigment epithelium-induced CD4+CD25+ regulatory T cells suppress activation of intraocular effector T cells. Clin Immunol 136(1):83–95. https://doi.org/10.1016/j.clim.2010.03.001
Horwitz DA, Zheng SG, Gray JD (2008) Natural and TGF-beta-induced Foxp3(+)CD4(+) CD25(+) regulatory T cells are not mirror images of each other. Trends Immunol 29(9):429–435. https://doi.org/10.1016/j.it.2008.06.005
Huang W, Solouki S, Koylass N, Zheng SG, August A (2017) ITK signalling via the Ras/IRF4 pathway regulates the development and function of Tr1 cells. Nat Commun 8:15871. https://doi.org/10.1038/ncomms15871
Huang W, Solouki S, Carter C, Zheng SG, August A (2018a) Beyond type 1 regulatory T cells: co-expression of LAG3 and CD49b in IL-10-producing T cell lineages. Front Immunol 9:2625. https://doi.org/10.3389/fimmu.2018.02625
Huang Z, Liu Y, Qi G, Brand D, Zheng SG (2018b) Role of vitamin A in the immune system. J Clin Med 7(9):258. https://doi.org/10.3390/jcm7090258
Itoh M, Takahashi T, Sakaguchi N, Kuniyasu Y, Shimizu J, Otsuka F, Sakaguchi S (1999) Thymus and autoimmunity: production of CD25+CD4+ naturally anergic and suppressive T cells as a key function of the thymus in maintaining immunologic self-tolerance. J Immunol 162(9):5317–5326
Jacob N, Yang H, Pricop L, Liu Y, Gao X, Zheng SG, Wang J, Gao HX, Putterman C, Koss MN, Stohl W, Jacob CO (2009) Accelerated pathological and clinical nephritis in systemic lupus erythematosus-prone New Zealand mixed 2328 mice doubly deficient in TNF receptor 1 and TNF receptor 2 via a Th17-associated pathway. J Immunol 182(4):2532–2541. https://doi.org/10.4049/jimmunol.0802948
Jaffe GJ, Dick AD, Brezin AP, Nguyen QD, Thorne JE, Kestelyn P, Barisani-Asenbauer T, Franco P, Heiligenhaus A, Scales D, Chu DS, Camez A, Kwatra NV, Song AP, Kron M, Tari S, Suhler EB (2016) Adalimumab in patients with active noninfectious uveitis. N Engl J Med 375(10):932–943. https://doi.org/10.1056/NEJMoa1509852
Jia X, Hu M, Wang C, Wang C, Zhang F, Han Q, Zhao R, Huang Q, Xu H, Yuan H, Ren H (2011) Coordinated gene expression of Th17- and Treg-associated molecules correlated with resolution of the monophasic experimental autoimmune uveitis. Mol Vis 17:1493–1507

Kanamori M, Nakatsukasa H, Okada M, Lu Q, Yoshimura A (2016) Induced regulatory T cells: their development, stability, and applications. Trends Immunol 37 (11):803–811. https://doi.org/10.1016/j.it.2016.08.012
Kasper M, Gabriel D, Moller M, Bauer D, Wildschutz L, Courthion H, Bohm MRR, Busch M, Loser K, Thanos S, Gurny R, Heiligenhaus A (2018) Novel everolimus-loaded nanocarriers for topical treatment of murine experimental autoimmune uveoretinitis (EAU). Exp Eye Res 168:49–56. https://doi.org/10.1016/j.exer.2018.01.003
Kawazoe Y, Sugita S, Keino H, Yamada Y, Imai A, Horie S, Mochizuki M (2012) Retinoic acid from retinal pigment epithelium induces T regulatory cells. Exp Eye Res 94(1):32–40. https://doi.org/10.1016/j.exer.2011.11.002
Ke Y, Jiang G, Sun D, Kaplan HJ, Shao H (2008) Ocular regulatory T cells distinguish monophasic from recurrent autoimmune uveitis. Invest Ophthalmol Vis Sci 49 (9):3999–4007. https://doi.org/10.1167/iovs.07-1468
Ke Y, Jiang G, Sun D, Kaplan HJ, Shao H (2011) Anti-CD3 antibody ameliorates experimental autoimmune uveitis by inducing both IL-10 and TGF-beta dependent regulatory T cells. Clin Immunol 138(3):311–320. https://doi.org/10.1016/j.clim.2010.12.016
Keino H, Watanabe T, Sato Y, Okada AA (2011) Oral administration of retinoic acid receptor-alpha/beta-specific ligand Am80 suppresses experimental autoimmune uveoretinitis. Invest Ophthalmol Vis Sci 52 (3):1548–1556. https://doi.org/10.1167/iovs.10-5963
Klein L, Robey EA, Hsieh CS (2019) Central CD4(+) T cell tolerance: deletion versus regulatory T cell differentiation. Nat Rev Immunol 19(1):7–18. https://doi.org/10.1038/s41577-018-0083-6
Kong N, Lan Q, Chen M, Wang J, Shi W, Horwitz DA, Quesniaux V, Ryffel B, Liu Z, Brand D, Zou H, Zheng SG (2012a) Antigen-specific transforming growth factor beta-induced Treg cells, but not natural Treg cells, ameliorate autoimmune arthritis in mice by shifting the Th17/Treg cell balance from Th17 predominance to Treg cell predominance. Arthritis Rheum 64 (8):2548–2558. https://doi.org/10.1002/art.34513
Kong N, Lan Q, Su W, Chen M, Wang J, Yang Z, Park R, Dagliyan G, Conti PS, Brand D, Liu Z, Stohl W, Zou H, Zheng SG (2012b) Induced T regulatory cells suppress osteoclastogenesis and bone erosion in collagen-induced arthritis better than natural T regulatory cells. Ann Rheum Dis 71(9):1567–1572. https://doi.org/10.1136/annrheumdis-2011-201052
Koreth J, Matsuoka K, Kim HT, McDonough SM, Bindra B, Alyea EP III, Armand P, Cutler C, Ho VT, Treister NS, Bienfang DC, Prasad S, Tzachanis D, Joyce RM, Avigan DE, Antin JH, Ritz J, Soiffer RJ (2011) Interleukin-2 and regulatory T cells in graft-versus-host disease. N Engl J Med 365 (22):2055–2066. https://doi.org/10.1056/NEJMoa1108188
Lan Q, Zhou X, Fan H, Chen M, Wang J, Ryffel B, Brand D, Ramalingam R, Kiela PR, Horwitz DA, Liu Z, Zheng SG (2012) Polyclonal CD4+Foxp3+ Treg cells induce TGFbeta-dependent tolerogenic dendritic cells that suppress the murine lupus-like syndrome. J Mol Cell Biol 4(6):409–419. https://doi.org/10.1093/jmcb/mjs040
Lee RW, Nicholson LB, Sen HN, Chan CC, Wei L, Nussenblatt RB, Dick AD (2014) Autoimmune and autoinflammatory mechanisms in uveitis. Semin Immunopathol 36(5):581–594. https://doi.org/10.1007/s00281-014-0433-9
Lerman MA, Larkin J 3rd, Cozzo C, Jordan MS, Caton AJ (2004) CD4+ CD25+ regulatory T cell repertoire formation in response to varying expression of a neo-self-antigen. J Immunol 173(1):236–244. https://doi.org/10.4049/jimmunol.173.1.236
Li Z, Lin F, Zhuo C, Deng G, Chen Z, Yin S, Gao Z, Piccioni M, Tsun A, Cai S, Zheng SG, Zhang Y, Li B (2014) PIM1 kinase phosphorylates the human transcription factor FOXP3 at serine 422 to negatively regulate its activity under inflammation. J Biol Chem 289(39):26872–26881. https://doi.org/10.1074/jbc.M114.586651
Li J, Qiu C, Zhang Z, Yuan W, Ge Z, Tan B, Yang P, Liu J, Zhu X, Qiu C, Lai D, Guo L, Yu L (2018) Subretinal transplantation of human amniotic epithelial cells in the treatment of autoimmune uveitis in rats. Cell Transplant 27(10):1504–1514. https://doi.org/10.1177/0963689718796196
Liang L, Wang H, Peng XY, Zhao M (2011) The changes of Th lymphocyte subsets in patients with Behcet disease. Zhonghua Yan Ke Za Zhi 47(5):393–397
Liang L, Peng XY, Wang H (2019) Th lymphocyte subsets in patients with Vogt-Koyanagi-Harada disease. Int J Ophthalmol 12(2):207–211. https://doi.org/10.18240/ijo.2019.02.04
Lin W, Haribhai D, Relland LM, Truong N, Carlson MR, Williams CB, Chatila TA (2007) Regulatory T cell development in the absence of functional Foxp3. Nat Immunol 8(4):359–368. https://doi.org/10.1038/ni1445
Liu W, Putnam AL, Xu-Yu Z, Szot GL, Lee MR, Zhu S, Gottlieb PA, Kapranov P, Gingeras TR, Fazekas de St Groth B, Clayberger C, Soper DM, Ziegler SF, Bluestone JA (2006) CD127 expression inversely correlates with FoxP3 and suppressive function of human CD4+ T reg cells. J Exp Med 203(7):1701–1711. https://doi.org/10.1084/jem.20060772
Liu Y, Lan Q, Lu L, Chen M, Xia Z, Ma J, Wang J, Fan H, Shen Y, Ryffel B, Brand D, Quismorio F, Liu Z, Horwitz DA, Xu A, Zheng SG (2014) Phenotypic and functional characteristic of a newly identified CD8+ Foxp3- CD103+ regulatory T cells. J Mol Cell Biol 6(1):81–92. https://doi.org/10.1093/jmcb/mjt026
Liu Y, Jarjour W, Olsen N, Zheng SG (2020) Traitor or warrior-Treg cells sneaking into the lesions of psoriatic arthritis. Clin Immunol 215:108425. https://doi.org/10.1016/j.clim.2020.108425
Lu L, Wang J, Zhang F, Chai Y, Brand D, Wang X, Horwitz DA, Shi W, Zheng SG (2010a) Role of

SMAD and non-SMAD signals in the development of Th17 and regulatory T cells. J Immunol 184 (8):4295–4306. https://doi.org/10.4049/jimmunol.0903418

Lu L, Zhou X, Wang J, Zheng SG, Horwitz DA (2010b) Characterization of protective human CD4CD25 FOXP3 regulatory T cells generated with IL-2, TGF-beta and retinoic acid. PLoS One 5(12):e15150. https://doi.org/10.1371/journal.pone.0015150

Lu L, Ma J, Li Z, Lan Q, Chen M, Liu Y, Xia Z, Wang J, Han Y, Shi W, Quesniaux V, Ryffel B, Brand D, Li B, Liu Z, Zheng SG (2011) All-trans retinoic acid promotes TGF-beta-induced Tregs via histone modification but not DNA demethylation on Foxp3 gene locus. PLoS One 6(9):e24590. https://doi.org/10.1371/journal.pone.0024590

Lu L, Lan Q, Li Z, Zhou X, Gu J, Li Q, Wang J, Chen M, Liu Y, Shen Y, Brand DD, Ryffel B, Horwitz DA, Quismorio FP, Liu Z, Li B, Olsen NJ, Zheng SG (2014) Critical role of all-trans retinoic acid in stabilizing human natural regulatory T cells under inflammatory conditions. Proc Natl Acad Sci U S A 111(33):E3432–E3440. https://doi.org/10.1073/pnas.1408780111

Luo Y, Xue Y, Wang J, Dang J, Fang Q, Huang G, Olsen N, Zheng SG (2019) Negligible effect of sodium chloride on the development and function of TGF-beta-induced CD4(+) Foxp3(+) regulatory T cells. Cell Rep 26(7):1869–1879.e1863. https://doi.org/10.1016/j.celrep.2019.01.066

Ma J, Liu Y, Li Y, Gu J, Liu J, Tang J, Wang J, Ryffel B, Shen Y, Brand D, Liu Z, Zheng SG (2014) Differential role of all-trans retinoic acid in promoting the development of CD4+ and CD8+ regulatory T cells. J Leukoc Biol 95(2):275–283. https://doi.org/10.1189/jlb.0513297

Malek TR, Yu A, Vincek V, Scibelli P, Kong L (2002) CD4 regulatory T cells prevent lethal autoimmunity in IL-2Rbeta-deficient mice. Implications for the nonredundant function of IL-2. Immunity 17(2):167–178. https://doi.org/10.1016/s1074-7613(02)00367-9

Matta B, Bora PS, Neuhouser AJ, Bora NS (2019) Inhibitory role of transforming growth factor beta2 in experimental autoimmune anterior uveitis. Graefes Arch Clin Exp Ophthalmol 257(5):953–960. https://doi.org/10.1007/s00417-019-04255-9

Mei S, Huang Y, Li N, Xu Z, Xu J, Dai Q, Wu J, Kijlstra A, Yang P, Hou S (2020) Aminooxy-acetic acid inhibits experimental autoimmune uveitis by modulating the balance between effector and regulatory lymphocyte subsets. Curr Mol Med. https://doi.org/10.2174/1566524020666200211112219

Meng X, Fang S, Zhang Z, Wang Y, You C, Zhang J, Yan H (2017) Preventive effect of chrysin on experimental autoimmune uveitis triggered by injection of human IRBP peptide 1-20 in mice. Cell Mol Immunol 14 (8):702–711. https://doi.org/10.1038/cmi.2015.107

Miyara M, Yoshioka Y, Kitoh A, Shima T, Wing K, Niwa A, Parizot C, Taflin C, Heike T, Valeyre D, Mathian A, Nakahata T, Yamaguchi T, Nomura T, Ono M, Amoura Z, Gorochov G, Sakaguchi S (2009) Functional delineation and differentiation dynamics of human CD4+ T cells expressing the FoxP3 transcription factor. Immunity 30(6):899–911. https://doi.org/10.1016/j.immuni.2009.03.019

Molins B, Mesquida M, Lee RW, Llorenc V, Pelegrin L, Adan A (2015) Regulatory T cell levels and cytokine production in active non-infectious uveitis: in-vitro effects of pharmacological treatment. Clin Exp Immunol 179(3):529–538. https://doi.org/10.1111/cei.12479

Mucida D, Park Y, Kim G, Turovskaya O, Scott I, Kronenberg M, Cheroutre H (2007) Reciprocal TH17 and regulatory T cell differentiation mediated by retinoic acid. Science 317(5835):256–260. https://doi.org/10.1126/science.1145697

Mullen AC, High FA, Hutchins AS, Lee HW, Villarino AV, Livingston DM, Kung AL, Cereb N, Yao TP, Yang SY, Reiner SL (2001) Role of T-bet in commitment of TH1 cells before IL-12-dependent selection. Science 292(5523):1907–1910. https://doi.org/10.1126/science.1059835

Nakamura YK, Metea C, Karstens L, Asquith M, Gruner H, Moscibrocki C, Lee I, Brislawn CJ, Jansson JK, Rosenbaum JT, Lin P (2016) Gut microbial alterations associated with protection from autoimmune uveitis. Invest Ophthalmol Vis Sci 57 (8):3747–3758. https://doi.org/10.1167/iovs.16-19733

Namba K, Kitaichi N, Nishida T, Taylor AW (2002) Induction of regulatory T cells by the immunomodulating cytokines alpha-melanocyte-stimulating hormone and transforming growth factor-beta2. J Leukoc Biol 72(5):946–952

Nanke Y, Kotake S, Goto M, Ujihara H, Matsubara M, Kamatani N (2008) Decreased percentages of regulatory T cells in peripheral blood of patients with Behcet's disease before ocular attack: a possible predictive marker of ocular attack. Mod Rheumatol 18 (4):354–358. https://doi.org/10.1007/s10165-008-0064-x

Ohkura N, Kitagawa Y, Sakaguchi S (2013) Development and maintenance of regulatory T cells. Immunity 38 (3):414–423. https://doi.org/10.1016/j.immuni.2013.03.002

Ouyang W, Beckett O, Ma Q, Li MO (2010) Transforming growth factor-beta signaling curbs thymic negative selection promoting regulatory T cell development. Immunity 32(5):642–653. https://doi.org/10.1016/j.immuni.2010.04.012

Pacholczyk R, Ignatowicz H, Kraj P, Ignatowicz L (2006) Origin and T cell receptor diversity of Foxp3+CD4+CD25+ T cells. Immunity 25(2):249–259. https://doi.org/10.1016/j.immuni.2006.05.016

Papotto PH, Marengo EB, Sardinha LR, Carvalho KI, de Carvalho AE, Castillo-Mendez S, Jank CC, Vanhove B, Goldberg AC, Rizzo LV (2017) Novel CD28 antagonist mPEG PV1-Fab′ mitigates experimental autoimmune uveitis by suppressing CD4+ T

lymphocyte activation and IFN-gamma production. PLoS One 12(3):e0171822. https://doi.org/10.1371/journal.pone.0171822
Park HB, Paik DJ, Jang E, Hong S, Youn J (2004) Acquisition of anergic and suppressive activities in transforming growth factor-beta-costimulated CD4+CD25- T cells. Int Immunol 16(8):1203–1213. https://doi.org/10.1093/intimm/dxh123
Rao NA (2013) Uveitis in developing countries. Indian J Ophthalmol 61(6):253–254. https://doi.org/10.4103/0301-4738.114090
Rathinam SR, Gonzales JA, Thundikandy R, Kanakath A, Murugan SB, Vedhanayaki R, Lim LL, Suhler EB, Al-Dhibi HA, Doan T, Keenan JD, Rao MM, Ebert CD, Nguyen HH, Kim E, Porco TC, Acharya NR, Group FR (2019) Effect of corticosteroid-sparing treatment with mycophenolate mofetil vs methotrexate on inflammation in patients with uveitis: a randomized clinical trial. JAMA 322(10):936–945. https://doi.org/10.1001/jama.2019.12618
Reynolds BC, Turner DG, McPherson RC, Prendergast CT, Phelps RG, Turner NA, O'Connor RA, Anderton SM (2014) Exposure to inflammatory cytokines selectively limits GM-CSF production by induced T regulatory cells. Eur J Immunol 44(11):3342–3352. https://doi.org/10.1002/eji.201444687
Ruggieri S, Frassanito MA, Dammacco R, Guerriero S (2012) Treg lymphocytes in autoimmune uveitis. Ocul Immunol Inflamm 20(4):255–261. https://doi.org/10.3109/09273948.2012.681830
Rytkonen SH, Kulmala P, Autio-Harmainen H, Arikoski P, Enden K, Kataja J, Karttunen T, Nuutinen M, Jahnukainen T (2018) FOXP3(+) T cells are present in kidney biopsy samples in children with tubulointerstitial nephritis and uveitis syndrome. Pediatr Nephrol 33(2):287–293. https://doi.org/10.1007/s00467-017-3796-z
Saadoun D, Rosenzwajg M, Joly F, Six A, Carrat F, Thibault V, Sene D, Cacoub P, Klatzmann D (2011) Regulatory T-cell responses to low-dose interleukin-2 in HCV-induced vasculitis. N Engl J Med 365(22):2067–2077. https://doi.org/10.1056/NEJMoa1105143
Sakaguchi S, Sakaguchi N, Asano M, Itoh M, Toda M (1995) Immunologic self-tolerance maintained by activated T cells expressing IL-2 receptor alpha-chains (CD25). Breakdown of a single mechanism of self-tolerance causes various autoimmune diseases. J Immunol 155(3):1151–1164
Sampson JF, Hasegawa E, Mulki L, Suryawanshi A, Jiang S, Chen WS, Rabinovich GA, Connor KM, Panjwani N (2015) Galectin-8 ameliorates murine autoimmune ocular pathology and promotes a regulatory T cell response. PLoS One 10(6):e0130772. https://doi.org/10.1371/journal.pone.0130772
Sharabi A, Tsokos MG, Ding Y, Malek TR, Klatzmann D, Tsokos GC (2018) Regulatory T cells in the treatment of disease. Nat Rev Drug Discov 17(11):823–844. https://doi.org/10.1038/nrd.2018.148
Sharma RK, Gupta A, Kamal S, Bansal R, Singh N, Sharma K, Virk S, Sachdeva N (2018) Role of regulatory T cells in tubercular uveitis. Ocul Immunol Inflamm 26(1):27–36. https://doi.org/10.1080/09273948.2016.1196711
Shi L, Guo H, Li Z, Wang Y, Wang Y, Cui Y (2019) Adenovirus-mediated down-regulation of miR-21-5p alleviates experimental autoimmune uveoretinitis in mice. Int Immunopharmacol 74:105698. https://doi.org/10.1016/j.intimp.2019.105698
Silver PB, Horai R, Chen J, Jittayasothorn Y, Chan CC, Villasmil R, Kesen MR, Caspi RR (2015) Retina-specific T regulatory cells bring about resolution and maintain remission of autoimmune uveitis. J Immunol 194(7):3011–3019. https://doi.org/10.4049/jimmunol.1402650
Su W, Fan H, Chen M, Wang J, Brand D, He X, Quesniaux V, Ryffel B, Zhu L, Liang D, Zheng SG (2012) Induced CD4+ forkhead box protein-positive T cells inhibit mast cell function and established contact hypersensitivity through TGF-beta1. J Allergy Clin Immunol 130(2):444–452.e447. https://doi.org/10.1016/j.jaci.2012.05.011
Su W, Wan Q, Huang J, Han L, Chen X, Chen G, Olsen N, Zheng SG, Liang D (2015) Culture medium from TNF-alpha-stimulated mesenchymal stem cells attenuates allergic conjunctivitis through multiple antiallergic mechanisms. J Allergy Clin Immunol 136(2):423–432.e428. https://doi.org/10.1016/j.jaci.2014.12.1926
Su W, Chen X, Zhu W, Yu J, Li W, Li Y, Li Z, Olsen N, Liang D, Zheng SG (2019) The cAMP-adenosine feedback loop maintains the suppressive function of regulatory T cells. J Immunol 203(6):1436–1446. https://doi.org/10.4049/jimmunol.1801306
Sugita S, Yamada Y, Kaneko S, Horie S, Mochizuki M (2011) Induction of regulatory T cells by infliximab in Behcet's disease. Invest Ophthalmol Vis Sci 52(1):476–484. https://doi.org/10.1167/iovs.10-5916
Sun M, Yang P, Du L, Zhou H, Ren X, Kijlstra A (2010) Contribution of CD4+CD25+ T cells to the regression phase of experimental autoimmune uveoretinitis. Invest Ophthalmol Vis Sci 51(1):383–389. https://doi.org/10.1167/iovs.09-3514
Suzuki J, Yoshimura T, Simeonova M, Takeuchi K, Murakami Y, Morizane Y, Miller JW, Sobrin L, Vavvas DG (2012) Aminoimidazole carboxamide ribonucleotide ameliorates experimental autoimmune uveitis. Invest Ophthalmol Vis Sci 53(7):4158–4169. https://doi.org/10.1167/iovs.11-9323
Tasso R, Ilengo C, Quarto R, Cancedda R, Caspi RR, Pennesi G (2012) Mesenchymal stem cells induce functionally active T-regulatory lymphocytes in a paracrine fashion and ameliorate experimental autoimmune uveitis. Invest Ophthalmol Vis Sci 53(2):786–793. https://doi.org/10.1167/iovs.11-8211
Thorne JE, Suhler E, Skup M, Tari S, Macaulay D, Chao J, Ganguli A (2016) Prevalence of noninfectious uveitis in the United States: a claims-based analysis. JAMA

Ophthalmol 134(11):1237–1245. https://doi.org/10.1001/jamaophthalmol.2016.3229

Thornton AM, Korty PE, Tran DQ, Wohlfert EA, Murray PE, Belkaid Y, Shevach EM (2010) Expression of Helios, an Ikaros transcription factor family member, differentiates thymic-derived from peripherally induced Foxp3+ T regulatory cells. J Immunol 184(7):3433–3441. https://doi.org/10.4049/jimmunol.0904028

Vang KB, Yang J, Mahmud SA, Burchill MA, Vegoe AL, Farrar MA (2008) IL-2, -7, and -15, but not thymic stromal lymphopoeitin, redundantly govern CD4+Foxp3+ regulatory T cell development. J Immunol 181(5):3285–3290. https://doi.org/10.4049/jimmunol.181.5.3285

Wahren-Herlenius M, Dorner T (2013) Immunopathogenic mechanisms of systemic autoimmune disease. Lancet 382(9894):819–831. https://doi.org/10.1016/S0140-6736(13)60954-X

Walscheid K, Weinhage T, Foell D, Heinz C, Kasper M, Heiligenhaus A (2016) Phenotypic changes of peripheral blood mononuclear cells upon corticosteroid treatment in idiopathic intermediate uveitis. Clin Immunol. https://doi.org/10.1016/j.clim.2016.10.013

Walscheid K, Weinhage T, Foell D, Heinz C, Kasper M, Heiligenhaus A (2019) Effect of adalimumab on peripheral blood mononuclear cells in non-infectious uveitis. Ocul Immunol Inflamm 27(2):330–337. https://doi.org/10.1080/09273948.2017.1374415

Weiss JM, Bilate AM, Gobert M, Ding Y, Curotto de Lafaille MA, Parkhurst CN, Xiong H, Dolpady J, Frey AB, Ruocco MG, Yang Y, Floess S, Huehn J, Oh S, Li MO, Niec RE, Rudensky AY, Dustin ML, Littman DR, Lafaille JJ (2012) Neuropilin 1 is expressed on thymus-derived natural regulatory T cells, but not mucosa-generated induced Foxp3+ T reg cells. J Exp Med 209(10):1723–1742, S1721. https://doi.org/10.1084/jem.20120914

Wildner G, Diedrichs-Mohring M (2019) Resolution of uveitis. Semin Immunopathol 41(6):727–736. https://doi.org/10.1007/s00281-019-00758-z

Willerford DM, Chen J, Ferry JA, Davidson L, Ma A, Alt FW (1995) Interleukin-2 receptor alpha chain regulates the size and content of the peripheral lymphoid compartment. Immunity 3(4):521–530. https://doi.org/10.1016/1074-7613(95)90180-9

Xu W, Lan Q, Chen M, Chen H, Zhu N, Zhou X, Wang J, Fan H, Yan CS, Kuang JL, Warburton D, Togbe D, Ryffel B, Zheng SG, Shi W (2012) Adoptive transfer of induced-Treg cells effectively attenuates murine airway allergic inflammation. PLoS One 7(7):e40314. https://doi.org/10.1371/journal.pone.0040314

Yadav M, Louvet C, Davini D, Gardner JM, Martinez-Llordella M, Bailey-Bucktrout S, Anthony BA, Sverdrup FM, Head R, Kuster DJ, Ruminski P, Weiss D, Von Schack D, Bluestone JA (2012) Neuropilin-1 distinguishes natural and inducible regulatory T cells among regulatory T cell subsets in vivo. J Exp Med 209(10):1713–1722, S1711–S1719. https://doi.org/10.1084/jem.20120822

Yang J, Liu L, Yang Y, Kong N, Jiang X, Sun J, Xie R (2017) Adoptive cell therapy of induced regulatory T cells expanded by tolerogenic dendritic cells on murine autoimmune arthritis. J Immunol Res 2017:7573154. https://doi.org/10.1155/2017/7573154

Yang S, Wang J, Brand DD, Zheng SG (2018) Role of TNF-TNF receptor 2 signal in regulatory T cells and its therapeutic implications. Front Immunol 9:784. https://doi.org/10.3389/fimmu.2018.00784

Yang M, Liu Y, Mo B, Xue Y, Ye C, Jiang Y, Bi X, Liu M, Wu Y, Wang J, Olsen N, Pan Y, Zheng SG (2019a) Helios but not CD226, TIGIT and Foxp3 is a potential marker for CD4(+) Treg cells in patients with rheumatoid arthritis. Cell Physiol Biochem 52(5):1178–1192. https://doi.org/10.33594/000000080

Yang S, Xie C, Chen Y, Wang J, Chen X, Lu Z, June RR, Zheng SG (2019b) Differential roles of TNFalpha-TNFR1 and TNFalpha-TNFR2 in the differentiation and function of CD4(+)Foxp3(+) induced Treg cells in vitro and in vivo periphery in autoimmune diseases. Cell Death Dis 10(1):27. https://doi.org/10.1038/s41419-018-1266-6

Ye C, Brand D, Zheng SG (2018) Targeting IL-2: an unexpected effect in treating immunological diseases. Signal Transduct Target Ther 3:2. https://doi.org/10.1038/s41392-017-0002-5

Yeh S, Li Z, Forooghian F, Hwang FS, Cunningham MA, Pantanelli S, Lew JC, Wroblewski KK, Vitale S, Nussenblatt RB (2009) CD4+Foxp3+ T-regulatory cells in noninfectious uveitis. Arch Ophthalmol 127(4):407–413. https://doi.org/10.1001/archophthalmol.2009.32

Yuan LF, Li GD, Ren XJ, Nian H, Li XR, Zhang XM (2015) Rapamycin ameliorates experimental autoimmune uveoretinitis by inhibiting Th1/Th2/Th17 cells and upregulating CD4+CD25+ Foxp3 regulatory T cells. Int J Ophthalmol 8(4):659–664. https://doi.org/10.3980/j.issn.2222-3959.2015.04.03

Zhang X, Ren X, Li G, Jiao C, Zhang L, Zhao S, Wang J, Han ZC, Li X (2011) Mesenchymal stem cells ameliorate experimental autoimmune uveoretinitis by comprehensive modulation of systemic autoimmunity. Invest Ophthalmol Vis Sci 52(6):3143–3152. https://doi.org/10.1167/iovs.10-6334

Zhang L, Wan F, Song J, Tang K, Zheng F, Guo J, Guo D, Bi H (2016) Imbalance between Th17 CELLS AND REGULATORY T cells during monophasic experimental autoimmune uveitis. Inflammation 39(1):113–122. https://doi.org/10.1007/s10753-015-0229-7

Zhang W, Zhou L, Dang J, Zhang X, Wang J, Chen Y, Liang J, Li D, Ma J, Yuan J, Chen W, Zadeh HH, Olsen N, Zheng SG (2017) Human gingiva-derived mesenchymal stem cells ameliorate streptozoticin-induced T1DM in mice via suppression of T effector cells and up-regulating Treg subsets. Sci Rep 7(1):15249. https://doi.org/10.1038/s41598-017-14979-5

Zhang X, Ouyang X, Xu Z, Chen J, Huang Q, Liu Y, Xu T, Wang J, Olsen N, Xu A, Zheng SG (2019) CD8

+CD103+ iTregs inhibit chronic graft-versus-host disease with lupus nephritis by the increased expression of CD39. Mol Ther 27(11):1963–1973. https://doi.org/10.1016/j.ymthe.2019.07.014

Zhang X, Olsen N, Zheng SG (2020) The progress and prospect of regulatory T cells in autoimmune diseases. J Autoimmun 111:102461. https://doi.org/10.1016/j.jaut.2020.102461

Zhao J, Chen J, Huang F, Wang J, Su W, Zhou J, Qi Q, Cao F, Sun B, Liu Z, Bellanti JA, Zheng SG (2020) Erratum: human gingiva tissue-derived MSC ameliorates immune-mediated bone marrow failure of aplastic anemia via suppression of Th1 and Th17 cells and enhancement of CD4+Foxp3+ regulatory T cells differentiation. Am J Transl Res 12(3):1167

Zheng SG, Gray JD, Ohtsuka K, Yamagiwa S, Horwitz DA (2002) Generation ex vivo of TGF-beta-producing regulatory T cells from CD4+CD25- precursors. J Immunol 169(8):4183–4189. https://doi.org/10.4049/jimmunol.169.8.4183

Zheng SG, Wang JH, Gray JD, Soucier H, Horwitz DA (2004a) Natural and induced CD4+CD25+ cells educate CD4+CD25- cells to develop suppressive activity: the role of IL-2, TGF-beta, and IL-10. J Immunol 172 (9):5213–5221. https://doi.org/10.4049/jimmunol.172.9.5213

Zheng SG, Wang JH, Koss MN, Quismorio F Jr, Gray JD, Horwitz DA (2004b) CD4+ and CD8+ regulatory T cells generated ex vivo with IL-2 and TGF-beta suppress a stimulatory graft-versus-host disease with a lupus-like syndrome. J Immunol 172(3):1531–1539. https://doi.org/10.4049/jimmunol.172.3.1531

Zheng SG, Wang JH, Stohl W, Kim KS, Gray JD, Horwitz DA (2006) TGF-beta requires CTLA-4 early after T cell activation to induce FoxP3 and generate adaptive CD4+CD25+ regulatory cells. J Immunol 176 (6):3321–3329. https://doi.org/10.4049/jimmunol.176.6.3321

Zheng SG, Wang J, Wang P, Gray JD, Horwitz DA (2007) IL-2 is essential for TGF-beta to convert naive CD4+CD25- cells to CD25+Foxp3+ regulatory T cells and for expansion of these cells. J Immunol 178 (4):2018–2027. https://doi.org/10.4049/jimmunol.178.4.2018

Zheng SG, Wang J, Horwitz DA (2008) Cutting edge: Foxp3+CD4+CD25+ regulatory T cells induced by IL-2 and TGF-beta are resistant to Th17 conversion by IL-6. J Immunol 180(11):7112–7116. https://doi.org/10.4049/jimmunol.180.11.7112

Zheng Y, Josefowicz S, Chaudhry A, Peng XP, Forbush K, Rudensky AY (2010) Role of conserved non-coding DNA elements in the Foxp3 gene in regulatory T-cell fate. Nature 463(7282):808–812. https://doi.org/10.1038/nature08750

Zhong H, Liu Y, Xu Z, Liang P, Yang H, Zhang X, Zhao J, Chen J, Fu S, Tang Y, Lv J, Wang J, Olsen N, Xu A, Zheng SG (2018) TGF-beta-induced CD8(+)CD103(+) regulatory T cells show potent therapeutic effect on chronic graft-versus-host disease lupus by suppressing B cells. Front Immunol 9:35. https://doi.org/10.3389/fimmu.2018.00035

Zhou X, Kong N, Wang J, Fan H, Zou H, Horwitz D, Brand D, Liu Z, Zheng SG (2010) Cutting edge: all-trans retinoic acid sustains the stability and function of natural regulatory T cells in an inflammatory milieu. J Immunol 185(5):2675–2679. https://doi.org/10.4049/jimmunol.1000598

Zhu W, Chen X, Yu J, Xiao Y, Li Y, Wan S, Su W, Liang D (2018) Baicalin modulates the Treg/Teff balance to alleviate uveitis by activating the aryl hydrocarbon receptor. Biochem Pharmacol 154:18–27. https://doi.org/10.1016/j.bcp.2018.04.006

Zhuang Z, Wang Y, Zhu G, Gu Y, Mao L, Hong M, Li Y, Zheng M (2017) Imbalance of Th17/Treg cells in pathogenesis of patients with human leukocyte antigen B27 associated acute anterior uveitis. Sci Rep 7:40414. https://doi.org/10.1038/srep40414

Zou W, Wu Z, Xiang X, Sun S, Zhang J (2014) The expression and significance of T helper cell subsets and regulatory T cells CD(4)(+) CD(2)(5)(+) in peripheral blood of patients with human leukocyte antigen B27-positive acute anterior uveitis. Graefes Arch Clin Exp Ophthalmol 252(4):665–672. https://doi.org/10.1007/s00417-014-2567-9

Understanding and Targeting Human Cancer Regulatory T Cells to Improve Therapy

12

H. Ryan Kolb, Nicholas Borcherding, and Weizhou Zhang

Abstract

Regulatory T cells (Tregs) are critical in maintaining immune homeostasis under various pathophysiological conditions. A growing body of evidence demonstrates that Tregs play an important role in cancer progression and that they do so by suppressing cancer-directed immune responses. Tregs have been targeted for destruction by exploiting antibodies against and small-molecule inhibitors of several molecules that are highly expressed in Tregs—including immune checkpoint molecules, chemokine receptors, and metabolites. To date, these strategies have had only limited antitumor efficacy, yet they have also created significant risk of autoimmunity because most of them do not differentiate Tregs in tumors from those in normal tissues. Currently, immune checkpoint inhibitor (ICI)-based cancer immunotherapies have revolutionized cancer treatment, but the resistance to ICI is common and the elevation of Tregs is one of the most important mechanisms. Therapeutic strategies that can selectively eliminate Tregs in the tumor (*i.e.* therapies that do not run the risk of causing autoimmunity by affecting normal tissue), are urgently needed for the development of cancer immunotherapies. This chapter discusses specific properties of human Tregs under the context of cancer and the various ways to target Treg for cancer immunotherapy.

Keywords

Regulatory T cells · Human cancer · Immunotherapy

12.1 Introduction

Regulatory T cells (Tregs) are a subset of immunosuppressive $CD4^+$ T cells that are critical for peripheral immunity, immune homeostasis, and self-tolerance. They play an important role in many conditions and diseases by preventing autoimmunity and overstimulation of the immune system in response to foreign pathogens, promoting resolution of inflammation, and suppressing anti-tumor immunity (Lin et al. 2018; Sakaguchi et al. 2010; Togashi et al. 2019). Indeed, research over the past 20 years has shown that tumors often have an increased density of Tregs, and they help promote the development of the immunosuppressive tumor microenvironment (TME), leading to the evasion of immune system by tumor cells and hence the consequent cancer

H. R. Kolb · W. Zhang (✉)
Department of Pathology, Immunology and Laboratory Medicine, University of Florida, Gainesville, FL, USA
e-mail: zhangw@ufl.edu

N. Borcherding
Department of Pathology and Immunology, Washington University, St. Louis, MO, USA

S.-G. Zheng (ed.), *T Regulatory Cells in Human Health and Diseases*, Advances in Experimental Medicine and Biology 1278, https://doi.org/10.1007/978-981-15-6407-9_12

progression (Chaudhary and Elkord 2016; Togashi et al. 2019).

The existence of suppressive T cells has been dated back as far as to the 1970s; however, the study of these cells was limited by the lack of markers for the identity of these suppressive T cells (Gershon and Kondo 1970; Sakaguchi 2011). In the mid-1990s, a population of $CD4^+$ $CD25^+$ thymic T cells—later referred to as Tregs—was identified to play a role in suppressing autoimmunity (Asano et al. 1996; Sakaguchi et al. 1995). Forkhead Box P3 (FOXP3) was found to be the main transcription factor that drives Treg identity. In 2001, mutations in the mouse *Foxp3* gene were shown to be the cause of lethal autoimmunity and inflammation observed in the scurfy mice (Brunkow et al. 2001). In the same year, mutations in the human *FOXP3* gene were shown to be the cause of immune dysregulation, polyendocrinopathy, and enteropathy X-linked (IPEX) syndrome characterized by autoimmunity in several endocrine organs (Bennett et al. 2001; Wildin et al. 2001). These conditions share a similar phenotype that was observed in mice with depletion of $CD4^+$ $CD25^+$ T cells, which led to the discovery of *FOXP3* expression in $CD4^+CD25^+$ T cells, and that forced expression of FOXP3 in conventional $CD4^+CD25^-$ T cells converts them to functionally suppressive T cells with the phenotypic expression of several characteristic Treg genes such as CD25, cytotoxic T-lymphocyte-associated protein 4 (CTLA4) and glucocorticoid-induced tumor necrosis factor receptor (GITR) (Fontenot et al. 2003; Hori et al. 2003; Khattri et al. 2003). Stable expression of FOXP3, which is achieved by demethylation of an evolutionarily conserved motif in the *FOXP3* gene (referred to as the Treg-specific demethylation region, TSDR), is required to thymic Treg stability (Floess et al. 2007; Ge et al. 2019; Huehn et al. 2009). In mice, FOXP3 is almost exclusively expressed in Tregs; however, in humans both Tregs and conventional $CD4^+$ T cells express FOXP3 following T-cell receptor (TCR) stimulation. As a result, the $CD4^+$ $FOXP3^+$ T-cell population may also contain some activated conventional T cells (Morgan et al. 2005; Roncador et al. 2005; Stockis et al. 2019). Demethylation of the *FOXP3* TSDR is the most distinguishing feature of human Tregs (Stockis et al. 2019). For the identification and isolation of functional human Tregs, neither demethylation of the TSDR nor FOXP3 expression is suitable; a combination of surface markers is required (Yang et al. 2019). In human, Tregs are generally identified by—though not perfect—the expression of CD4 and CD25 and low to no expression of the α-chain of the interleukin-7 receptor (IL-7R; CD127) (Liu et al. 2006; Romano et al. 2017).

12.2 Tregs in Cancer

The role of Tregs in suppressing anti-tumor immunity was first shown by Onizuka et al. and Shimizu et al. wherein they demonstrated that depletion of $CD25^+$ T cells in mice resulted in increased tumor rejection and reduced tumor growth. Similarly, adoptive transfer of Treg-depleted ($CD25^+$ depletion) splenocytes had the same effect (Onizuka et al. 1999; Shimizu et al. 1999). The role of Tregs in human cancers has been studied extensively and been reviewed numerous times (Chaudhary and Elkord 2016; Togashi et al. 2019). Tregs have been shown to be increased in the peripheral blood and lymph nodes of cancer patients and to accumulate in many solid tumors where they account for 10–50% of the tumor-infiltrating $CD4^+$ cells (Badoual et al. 2006; Hiraoka et al. 2006; Ichihara et al. 2003; Ling et al. 2007; Schaefer et al. 2005). The role of Tregs in suppressing anti-tumor immunity in humans is supported by several studies. Ladoire et al. showed that the pathologic complete response (PCR) to neoadjuvant chemotherapy in breast cancer patients was correlated with decreased Tregs and increased $CD8^+$ T cells. Depletion of Tregs using a previously FDA-approved CD25-blocking antibody improved the response to an experimental cancer vaccine in metastatic breast cancer patients (Rech et al. 2012). Transient depletion of Tregs via an IL-2-diphtheria toxin conjugate reduced metastatic lesions in melanoma patients (Rasku et al. 2008). The elevated abundance of FOXP3$^+$ Tregs is generally associated with a poor prognosis in most non-mucosal-derived solid tumors (Chaudhary and Elkord 2016; Shang et al. 2015). This association between tumor-infiltrating Treg abundance and prognosis is particularly true when using the ratio between Tregs and conventional T cells, where a higher ratio is significantly correlated with a worse prognosis in breast cancer, lung cancer, melanoma, pancreatic cancer, and ovarian cancer (Curiel et al. 2004; Jiang et al. 2014; Leffers et al. 2009; Sayour et al. 2015; Tang et al. 2014; Tao et al. 2012; Yang et al. 2006). While Tregs tend to be higher in the peripheral blood in cancer patients, this is not always associated with the abundance of tumor-infiltrating Tregs (Adeegbe and Nishikawa 2013; Togashi et al. 2019). In contrast, a higher number of FOXP3$^+$ Tregs can be associated with good prognosis as well, such as in gastric and colorectal cancers (Haas et al. 2009; Salama et al. 2009). This may be due to the role of Tregs in suppressing tumor initiating and promoting inflammation in the colon associated with changes in the gut microbiome (Ladoire et al. 2011). Alternatively, recent studies have shown that colorectal tumors have high infiltration of FOXP3$^+$ non-Tregs that are inflammatory and associated with a good prognosis (Saito et al. 2016). Due to the difficulty of distinguishing Tregs in human with just FOXP3 expression (Morgan et al. 2005; Roncador et al. 2005), immunohistochemistry staining for FOXP3 may not be a viable method for determining the prognostic value of Treg infiltration in colorectal cancer. In support of this, infiltration of actual suppressive Tregs (defined by high expression of FOXP3 and negative for CD45RA) is associated with poor prognosis in colorectal cancer (Saito et al. 2016).

12.2.1 Cellular Source of Tumor-Infiltrating Tregs

Tregs can develop within the thymus by positive selection (thymic Tregs or tTregs) or arise from peripheral conventional CD4$^+$ FOXP3$^-$ T cells following prolonged T-cell receptor (TCR) stimulation in the presence of certain cytokines (pTregs, also referred to as induced Tregs (iTregs)) (Lee et al. 2011; Zheng et al. 2002, 2004). tTregs develop when the TCR of CD4 and CD8 double-positive cells in the thymus have a high-affinity interaction with self-antigens leading to the upregulation of CD25 as well as other Treg-associated receptors such as GITR (Burchill et al. 2008; Lio and Hsieh 2008). A second step for the development of stable CD25$^+$ FOXP3$^+$ Tregs involves IL-2 and STAT5 signaling, leading to stable FOXP3 expression (Burchill et al. 2007; Lio and Hsieh 2008). This development process results in a unique TCR repertoire relative to those from conventional CD4$^+$ T cells (Hsieh et al. 2006; Park et al. 2020; Wong et al. 2007). Zheng SG group first reported that iTregs arise from conventional CD4$^+$ FOXP3$^-$ T cells after prolonged TCR stimulation under certain cytokine conditions, such as in the presence of TGF-β and IL-2 (Davidson et al. 2007; Zheng et al. 2002, 2007). Thus, iTregs can share the TCR repertoire with peripheral conventional CD4$^+$ T cells. The stable expression of FOXP3 and thus the development of long-lived Tregs requires demethylation of the TSDR, which only happens in tTregs (Floess et al. 2007; Ge et al. 2019; Huehn et al. 2009).

Tumor-infiltrating Tregs can arise from several different sources, conversion of tumor-infiltrating CD4$^+$ FOXP3$^-$ T cells, recruitment of tTregs, and expansion of tissue-resident Tregs (Stockis et al. 2019). Studies have shown that several types of leukemias and lymphomas could induce the differentiation of conventional CD4$^+$CD25$^-$ T cells into Tregs (Deng 2018). For instance, malignant B cells from follicular lymphoma and non-Hodgkin's lymphoma could induce the expression of FOXP3 in CD4$^+$CD25$^-$ T cells (Ai et al. 2009; Mittal et al. 2008). Whether conversion of conventional CD4 T cells into Tregs occurs in human solid tumors is still debatable. In mice, adoptive transfer of conventional CD4$^+$CD25$^-$ T cells into tumor-bearing mice leads to the conversion of some of these cells to FOXP3$^+$ Tregs (Valzasina et al. 2006); however, whether this occurs in human and whether the conversion of conventional $CD4^+CD25^-$ T cells to Tregs is a major source of tumor-infiltrating Tregs are unknown. Many tumor cells or other cells with TME can express TGF-β, so it is conceivable that the TME could induce the conversion of conventional $CD4^+CD25^-$ T cells to Tregs. $CD4^+CD25^-$ T cells and Tregs do not seem to share the same TCR repertoire in human and mouse tumors (Ahmadzadeh et al. 2019; Plitas et al. 2016), suggesting that tumor-infiltrating Tregs may not arise from $CD4^+CD25^-$ T cells. Ahmadzadeh et al. found that the clonality of tumor-infiltrating Tregs from melanoma, gastric, and ovarian cancers had little overlap with tumor-infiltrating or peripheral blood conventional $CD4^+CD25^-$ T cells, but tumor-infiltrating Tregs did share clones with their peripheral blood counterparts (Ahmadzadeh et al. 2019). Most importantly, the TCRs from tumor-infiltrating Tregs showed specificity to tumor antigens and could be expanded in an antigen-specific manner (Ahmadzadeh et al. 2019). This would suggest that tumor-infiltrating Tregs may arise from both the recruitment and clonal expansion of peripheral or tissue resident tTregs; or a second explanation is that tumor-infiltrating Tregs are able to extravasate from tumors and enter circulation. The expansion of tissue resident Tregs in tumors is supported by a study that revealed the tumor-infiltrating Tregs had a similar gene-expression pattern as normal tissue Tregs (Plitas et al. 2016). It should be noted that there is plasticity between specific subsets of $CD4^+$ T cells and Tregs, in particular between Th17 cells and Tregs (Wan et al. 2020). It was recently reported that Th17 cells could be converted into suppressive IL-17^+ $FOXP3^+$ and IL-17^- $FOXP3^+$ Tregs in the TME, indicating that the conversion of Th17 cells into Tregs could be an additional source of tumor-infiltrating Tregs (Downs-Canner et al. 2017).

12.2.2 Chemokine Receptors in Tumor-Infiltrating Tregs

The identification of chemokines and their receptors that potentially mediate the recruitment and retention of Tregs into the TME is an area of active research (Stockis et al. 2019). Tumor-infiltrating Tregs express a panel of chemokine receptors such as CC chemokine receptor 4 (CCR4) (ligands CCL22/CCL17), CCR5 (ligand CCL5), CCR6 (ligand CCL20), CCR8 (ligand CCL1), and CCR10 (ligand CCL28). Many studies have attempted to use these chemokine receptors to explain the recruitment of effector Tregs to the TME; however, these Treg chemokine receptors may have a more pronounced role of retaining Tregs within TME since all ligands are highly expressed within TME as well. CCR4—working through its ligands CCL22 or CCL17—is the best studied chemokine signaling in Treg recruitment into the TME. Several studies of ovarian, prostate, breast, gastric, and bladder cancers have shown that tumor-infiltrating Tregs and Tregs from malignant ascites express CCR4 and that the ligand CCL22, which is highly expressed in tumors by tumor cells or macrophages, can act as a chemoattractant for Tregs (Curiel et al. 2004; Gobert et al. 2009; Maeda et al. 2019; Miller et al. 2006; Mizukami et al. 2008). Recent studies have shown that secretion of CCL5 by tumors or cancer-associated fibroblasts can recruit the Tregs though its receptor CCR5 in mouse models of pancreatic adenocarcinoma, squamous cell carcinoma, colorectal cancer, and breast cancer (Tan et al. 2011; Wang et al. 2017; Ward et al. 2015) and that CCL5 could recruit Tregs to metastatic sites in the lung (Halvorsen et al. 2016). CCR6, a known chemokine receptor shared by memory Th1, Th2, Th17, and Tregs, was able to recruit Tregs into the TME via macrophage-produced CCL20 (Chen et al. 2013; Lee et al. 2017; Liu et al. 2011; Zhang et al. 2015). CCL28 can be induced via tumor-associated hypoxia within the TME and plays a role in the recruitment of Tregs though its receptor CCR10 (Facciabene et al. 2011). CCR8 was recently identified to be exclusively elevated in human tumor-infiltrating Tregs in breast cancer and several other cancer types (De Simone et al. 2016; Plitas et al. 2016). CCL1, expressed by Tregs, provides an autocrine signaling to upregulate its own receptor CCR8 on Tregs and STAT3-dependent upregulation of Foxp3,

CD39, IL-10, and granzyme B (Barsheshet et al. 2017) and is a major chemotaxis factor for Tregs in human breast cancer (Kuehnemuth et al. 2018). CCR8 can be targeted by monoclonal antibodies that have shown to reduce tumor-infiltrating Tregs (Villarreal et al. 2018).

12.2.3 Mechanisms of Action (Summarized in Fig. 12.1)

Tregs suppress effector T cells (Teff cells) via many different actions, either in a contact-dependent or -independent fashion. Many co-stimulatory (OX-40, GITR, 4-1BB, etc.) or co-inhibitory molecules (CTLA-4, PD-1, TIGIT, LAG3, TIM-3, etc.) are constitutively expressed on tumor-infiltrating Tregs. These co-inhibitory receptor–ligand pairs either promote the expansion of Tregs or suppress effector cells directly in a contact-dependent manner. Most studies support the contact-dependent mechanism for both human and mouse Tregs when using in vitro suppression assay (Dieckmann et al. 2001; Jonuleit et al. 2001; Piccirillo and Shevach 2001; Takahashi et al. 1998; Thornton and Shevach 1998). Tregs can also secrete peptides (TGF-β, IL-10, IL-35) or metabolize ATP to adenosine via CD39 and CD73, which provides an immunosuppressive microenvironment (Su et al. 2019). Many in vivo studies strongly support the role of cytokines or metabolites in Teff cell suppression (Asseman et al. 1999; Belkaid et al. 2002; Collison et al. 2007, 2009; Kingsley et al. 2002; Lan et al. 2012; Li et al. 2007; Powrie et al. 1996). We summarize the mechanisms of action for Tregs in Fig. 12.1. Most of these mechanisms were well-established in animal models with strong genetic evidence, whereas not all mechanisms are validated in human Tregs particularly relevant to human cancers. Here we briefly discuss the various mechanisms of action for Tregs and elaborate further in the next section for those related to human cancers.

The best-studied mechanism is via the co-inhibitory molecule CTLA-4. CTLA-4 is a high-affinity inhibitory receptor for the co-stimulatory molecules CD80 and CD86 expressed on antigen-presenting cells (APCs) that otherwise bind to CD28 on Teff cells to induce a co-stimulatory signal for T-cell activation, in conjunction with the primary activating signal from MHC-antigen complexes binding to the TCR on Teff cells (Ge et al. 2019; Togashi

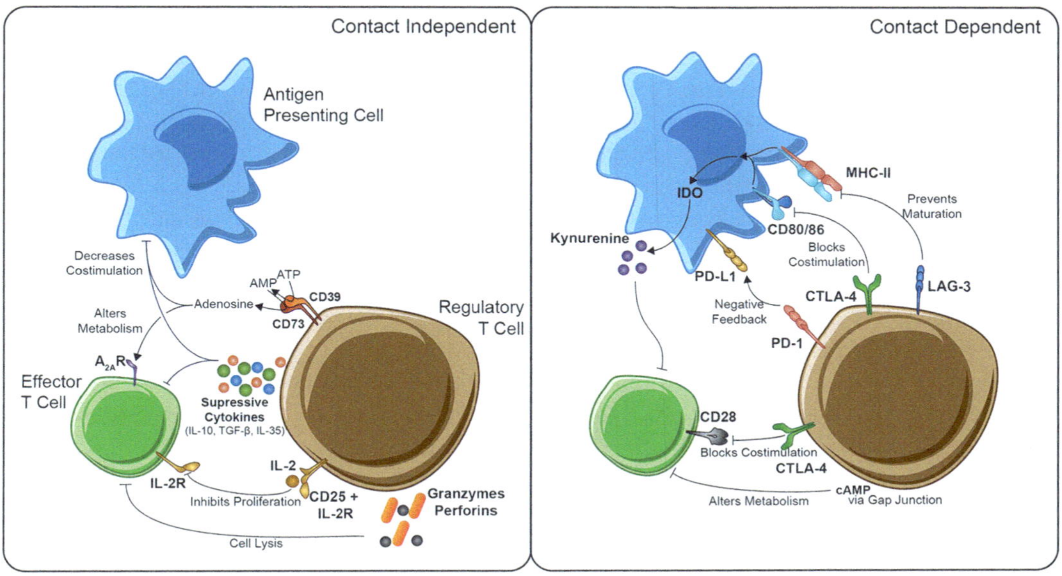

Fig. 12.1 The immunosuppressive mechanisms by Tregs

et al. 2019). Actually, CYLA-4-B7.1 signal also drives Treg development (Zheng et al. 2006). Tumor-infiltrating Tregs also express many other co-inhibitory molecules including T-cell immunoreceptor with Ig and ITIM domains (TIGIT) (Kurtulus et al. 2015), Tim-3 (HAVCR2) (Das et al. 2017; Gao et al. 2012; Sakuishi et al. 2013a), LAG-3 (CD223) (Camisaschi et al. 2010), and PD-1 (Kamada et al. 2019a; Lowther et al. 2016). TIGIT competes for binding of CD155 with CD226, preventing CD226-mediated co-stimulation of Teff cells and can also induce the expression of the suppressive cytokine IL-10 in dendritic cells (DCs) (Levin et al. 2011; Yu et al. 2009). LAG-3 binds to MHCII on APCs with a higher affinity than CD4, thus preventing antigen-specific stimulation of CD4 T cells (Huard et al. 1994; Sasidharan Nair and Elkord 2018). LAG-3 can also induce the secretion of indoleamine 2,3-dioxygenase (IDO) from DCs which can impair the function of Teff cells by producing kynuremine (Ge et al. 2019; Munn and Mellor 2013). Studies have also shown that Tregs express a large amount of cyclic adenosine monophosphate (cAMP) which they can directly transfer to Teff cells via gap junctions leading to downregulation of IL-2 and decreased proliferation (Klein and Bopp 2016). These mechanisms of suppression by Tregs require contact between the Tregs and Teff cells or APCs, and until recently, it was not known how Tregs come into the proximity of Teff cells to mediate suppression. Patterson et al. found that Tregs secrete the chemokines CCL3 and CCL4 which can actively promote the migration of Teff cells to close proximity with the Tregs to mediate suppression (Patterson et al. 2016). It is unknown whether tumor-infiltrating Tregs use the same mechanism or not, but our unpublished results indicate a common mechanism because tumor-infiltrating Tregs express several Teff chemokines—including CCL3, CCL4, and CXCL10—at much higher levels than those expressed by splenic Tregs (unpublished data).

Several contact-independent mechanisms of suppression have also been identified. Tregs highly and constitutively express CD25, which is a high-affinity receptor for IL-2. IL-2 is primarily produced by conventional T cells and is a critical cytokine for the proliferation of T and B cells. The high expression of CD25 on Tregs acts to sequester IL-2 from conventional T cells preventing their proliferation (Ge et al. 2019; Yau et al. 2012). The role of sequestration of IL-2 by CD25 in Treg suppression is supported by in vitro studies, showing that an excess of IL-2 can overcome Treg-mediated suppression of conventional T-cell proliferation (Takahashi et al. 1998; Yamaguchi et al. 2012). Tregs can also secrete several immunosuppressive cytokines including IL-10, TGF-β, and IL-35 (Chaudhary and Elkord 2016; Ge et al. 2019). Tumor-infiltrating Tregs from several human cancers including colorectal cancer, hepatocellular carcinoma, and pancreatic cancer can suppress the activity of autologous T cells by secreting TGF-β and IL-10 (Amedei et al. 2013; Kakita et al. 2012; Scurr et al. 2014; Yi et al. 2013). While IL-10 can inhibit DC activation, it can activate Teff cells under certain conditions and may not be a major mechanism of Treg-mediated suppression (Ge et al. 2019; Ouyang and O'Garra 2019). Tregs can also express the ectonucleotidases CD39 and CD73 which combine to convert extracellular adenosine triphosphate (ATP) into adenosine (Allard et al. 2020). Adenosine can bind to the adenosine receptors A_{2A} and A_{2B}, leading to increased intracellular cAMP which downregulates IL-2 in effector T cells (Blay et al. 1997; Klein and Bopp 2016; Ohta et al. 2006). Co-expression of CD39 and CD73 on human Tregs is rare, though studies have shown that CD39 is highly expressed on tumor-infiltrating Tregs in several human cancers and that $CD39^+$ cells can interact with $CD73^+$ cells or exosomes in the TME to produce adenosine (Jie et al. 2013; Schuler et al. 2014; Sundström et al. 2016).

12.3 Targeting Tregs for Cancer Immunotherapy

12.3.1 Immune Checkpoints as Therapeutic Targets for Tregs

12.3.1.1 Cytotoxic T-Lymphocyte-Associated Protein 4 (CTLA-4)

Discovery of CTLA-4 and its mechanisms of action. The Golstein group initially cloned CTLA-4 from mouse-activated $CD8^+$ T cells and called cytotoxic T-lymphocyte-associated protein 4 (CTLA-4) (Brunet et al. 1987) that was later confirmed to be present within human genome (Dariavach et al. 1988). CTLA-4 was later defined as a negative regulator of T-cell activation (Krummel and Allison 1995)—later named as immune checkpoint—that completes with CD28 to bind to CD80/CD86, with 20-fold higher binding affinity than CD28 (Linsley et al. 1991). CTLA-4 knockout mice exhibited an alteration in the T-cell development in the thymus and resulted in highly proliferative and active T cells in periphery (Tivol et al. 1995). James Allison's group established the suppressive role of CTLA-4 in cancer immunosurveillance in 1996 and found that anti-CTLA-4 antibody induced a strong antitumor immunity (Leach et al. 1996), the primary reason for which James Allison won the Nobel Prize in Physiology or Medicine in 2018. The connection between CTLA-4 and Tregs was established by the Sakaguchi group in 2008, where Treg-specific deletion of *CTLA4* gene—driven by a constitutively expressed *FoxP3-IRES-Cre*—led to a similar phenotype as germline deletion of CTLA-4. These data established that the immunosuppressive function of CTLA-4 is mainly through its expression within tTregs (Wing et al. 2008). In contrast, peripheral Treg-specific deletion of *CTLA4* gene in adult mice—driven by a tamoxifen-inducible *Foxp3-eGFP/Cre-ERT2*—resulted in the expansion of both conventional $CD4^+$ T cells and peripheral Tregs. Transcriptomic analysis further confirmed that *CTLA4* deletion led to a compensatory overexpression of immunosuppressive molecules including LAG3, PD-1, IL-10, etc., which are essential to maintain the suppressive phenotype of CTLA-4^- peripheral Tregs (Paterson et al. 2015).

At the molecular level, CTLA-4 is believed to be important for counteracting the co-stimulatory signal of CD28 to CD80/CD86 on antigen-presenting cells, either by direct suppression of antigen-presenting cells via CD80/CD86-mediated signaling transduction (Onishi et al. 2008; Wing et al. 2008) or by removing surface CD80/CD86 via trans-endocytosis (Qureshi et al. 2011). At the cellular level, CTLA-4 expression on the Teff cells, either conventional $CD4^+$ T cells or $CD8^+$ T cells, is important to limit the priming stage for T-cell activation and proliferation within secondary lymphoid tissues. A similar mechanism of CTLA-4 on Tregs is conceived to be within the secondary lymphoid tissues where CTLA-4 on Tregs inhibits APC function via CD80/C86 binding (Onishi et al. 2008; Wing et al. 2008). Nevertheless, CTLA-4 is not the only—sometimes not even the major—mechanism for the suppressive function of Tregs since polyclonal T-cell activation using anti-CD3 and anti-CD28 co-activation can be potently suppressed by Tregs, but such in vitro system is not involved in APC and the CD80/CD86 proteins.

CTLA-4 in cancer immunotherapy. The James Allison group established the potential of antagonizing CTLA-4 to activate antitumor immunity in 1996 (Leach et al. 1996). The FDA approved ipilimumab, a fully human anti-CTLA-4 antibody, to treat metastatic melanoma in 2011. This is a milestone of immune checkpoint inhibitors. As another immune checkpoint, i.e., the PD-1/PD-L1, has gained more success than targeting CTLA-4, lessons can still be learned from the mechanism of action for ipilimumab. We have pointed out above that CTLA-4 plays an immunosuppressive role in tTregs which have been proven the major Treg populations in various cancer types; hence, it is not surprising to find that the efficacy of ipilimumab is positively correlated with the reduced Treg abundance in tumor microenvironment. Animal studies further support that anti-CTLA-4 works on both Teff activation and Treg depletion for its maximal efficacy (Bulliard et al. 2013; Selby et al. 2013; Simpson et al. 2013). One caveat of ipilimumab therapy is the high rate of treatment-related adverse events, and many patients receiving ipilimumab experienced level 3 or 4 immune-related adverse events. In the EORTC 18071 trial, five patients died of ipilimumab treatment-related colitis, myocarditis, or multiorgan failure associated with Guillain–Barre syndrome. Ipilimumab is a human IgG1 antibody with predicted deleting activity, it is likely that these severe immune-related adverse events are the cause of ipilimumab-mediated Treg depletion via FC-gamma receptors (Arce Vargas et al. 2018). Tremelimumab, a human IgG2 isotype without deleting activity, can also bind to FC-gamma receptors and deplete Tregs (Arce Vargas et al. 2018). Even though a recent study argued against the role of ipilimumab on human cancers, the sampling time (many weeks after the last ipilimumab treatment) may miss the point of Treg depletion from these clinical samples (Sharma et al. 2019). The Sakaguchi group re-engineered the Fc-portion of ipilimumab to enhance its binding affinity to human FC-gamma receptor IIIa, which leads to antibody-mediated cytotoxicity (ADCC)-mediated killing of Tregs as well as exhausted CD8$^+$ T cells (Ha et al. 2019). The distinct difference of CTLA-4 expression on Tregs versus Teff cells is that Tregs constitutively express CTLA-4 on the surface, whereas Teff cells only express CTLA-4 on the surface upon activation and to a much lower level than that on Tregs. This feature gives a window for Treg depletion first, followed by CD8 T-cell activation by other means such as vaccination or anti-PD-1 therapy that will be mentioned later (Ha et al. 2019).

12.3.1.2 Lymphocyte-Activation Gene 3 (LAG-3, CD223)

LAG-3 is a surface protein that is expressed by activated CD4$^+$ and CD8$^+$ T cells, as well as by Tregs. LAG-3 could be the third most promising immune checkpoint in cancer immunotherapy after CTLA-4 and PD-1/PD-L1. As an immune checkpoint, LAG-3 binds to MHCII on antigen-presenting cells (Liang et al. 2008) to block the TCR and CD4-co-receptor-mediated signals for T-cell activation at the priming phase of the tumor-immune cycle. In addition, cancer cells can produce another ligand, namely fibrinogen-like protein 1 (FGL1), as the major immune-inhibitory ligand to bind with LAG-3 independent of MHCII (Wang et al. 2019) and inhibit T-cell activation at the effector phase (Topalian et al. 2016). Elevated expression of LAG-3 in tumor-infiltrating lymphocytes is significantly associated with disease progression of many human cancers (Chen and Chen 2014; Gandhi et al. 2006; Hemon et al. 2011; Matsuzaki et al. 2010; Shapiro et al. 2017). Many inhibitory molecules/antibodies against LAG-3 have been developed and showed some clinical benefit either alone or in combination with other immune checkpoint inhibitors (Table 12.1). In combination with anti-PD-1 therapy, anti-LAG-3 facilitates the eradication of established tumors that are resistant to either single antibody treatment by inducing an active anti-cancer immune response (Matsuzaki et al. 2010; Woo et al. 2012). The expression of LAG-3 on Tregs is induced upon the activation of Teff cells. Genetic deletion of *LAG3* or the treatment with anti-LAG-3 antibody inhibits the proliferative and suppressive capacities of Tregs, supporting that LAG-3 is important for Treg-mediated immune suppression under physiological conditions (Huang et al. 2004). The role of LAG-3 in Tregs, however, can be reversed when Tregs are placed under chronic inflammation such as in autoimmune diabetes (Zhang et al. 2017). Treg-specific deletion of LAG-3 in non-obese diabetic mice (NOD, autoimmune type 1 diabetic model) resulted in Treg expansion in the islets but not peripheral

Table 12.1 Summary of clinical trials related to tumor-infiltrating Tregs

NCT	Cancer	Compound	Target	Additional agents	Phase	Status
NCT02946671	Solid tumors	Mogamulizumab	CCR4	Nivolumab (PD-1)	I	Completed
NCT01626664	Adult T-cell leukemia-lymphoma	Mogamulizumab	CCR4	None	II	Completed
NCT00888927	Peripheral T-cell lymphoma	Mogamulizumab	CCR4	None	I/II	Completed
NCT00355472	Relapsed adult T-cell leukemia-lymphoma and peripheral T-cell lymphoma	Mogamulizumab	CCR4	None	I	Completed
NCT01728805	Cutaneous T-cell lymphoma	Mogamulizumab	CCR4	None	III	Active, not recruiting
NCT04185220	Adult T-cell leukemia-lymphoma and cutaneous T-cell lymphoma	Mogamulizumab	CCR4	Recombinant IL-15	I	Recruiting
NCT04256018	Cutaneous T-cell lymphoma	Mogamulizumab	CCR4	Low-dose total skin electron beam	II	Not yet recruiting
NCT01611142	Peripheral T-cell lymphoma	Mogamulizumab	CCR4	None	II	Completed
NCT02476123	Advanced solid tumors	Mogamulizumab	CCR4	Nivolumab (PD-1)	I	Completed
NCT00920790	CCR4+ Adult T-cell leukemia-lymphoma	Mogamulizumab	CCR4	None	II	Completed
NCT02301130	Advanced solid tumors	Mogamulizumab	CCR4	Durvalumab (PD-1) and tremelimumab (CTLA-4)	I	Completed
NCT04128072	Cutaneous T-cell lymphoma	Mogamulizumab	CCR4	Low-dose total skin electron beam	II	Not yet recruiting
NCT02281409	Advanced and metastatic solid tumors	Mogamulizumab	CCR4	None	I/II	Completed
NCT03309878	Relapsed or refractory diffuse large B-cell lymphoma	Mogamulizumab	CCR4	Pembrolizumab (PD-1)	I/II	Recruiting
NCT02444793	Advanced solid tumors	Mogamulizumab	CCR4	Utomilumab (4-1BB)	I	Terminated
NCT02358473	Non–small cell lung cancer	Mogamulizumab	CCR4	Docetaxel	I	Completed
NCT02867007	Locally advanced or metastatic solid tumors	Mogamulizumab	CCR4	KHK2455 (IDO)	I	Active, not recruiting
NCT03767582	Pancreatic adenocarcinoma	BMS-813160	CCR2/ CCR5	GVAX	I/II	Recruiting
NCT03274804	Microsatellite stable metastatic colorectal cancer	Maraviroc	CCR5	Pembrolizumab (PD-1)	I	Active, not recruiting
NCT03631407	Microsatellite stable metastatic colorectal cancer	Vicriviroc	CCR5	Pembrolizumab (PD-1)	II	Active, not recruiting
NCT01736813	Metastatic colorectal cancer	Maraviroc	CCR5	None	I	Completed
NCT03838367	Metastatic triple-negative breast cancer	Leronlimab	CCR5	Carboplatin	I/II	Recruiting
NCT00128622	CEA-expressing malignancies	Denileukin diftitox	CD25	Tumor vaccine	I	Completed
NCT00847106	Advanced melanoma	Daclizumab	CD25	DC-based anti-tumor vaccine	I/II	Completed
NCT00082914	Metastatic melanoma and kidney cancer	Denileukin diftitox	CD25	None	II	Completed
NCT00278369	Metastatic renal cancer	Denileukin diftitox	CD25	None	I	Completed
NCT00425672	Breast cancer	Denileukin diftitox	CD25	None	I/II	Completed

(continued)

Table 12.1 (continued)

NCT	Cancer	Compound	Target	Additional agents	Phase	Status
NCT00726037	Metastatic pancreatic cancer	Denileukin diftitox	CD25	None	II	Withdrawn
NCT03621982	Select advanced solid tumors	ADCT-301	CD25-ADC	None	I	Recruiting
NCT03884556	Lymphomas and solid tumors	TTX-030	CD39	Pembrolizumab (PD-1), docetaxel, gemcitabine, paclitaxel	I	Recruiting
NCT02503774	Advanced solid tumors	Oleclumab	CD73	Durvalumab (PD-1)	I	Active, not recruiting
NCT04262388	Pancreatic adenocarcinoma, small cell lung cancer, and head and neck cancer	Oleclumab	CD73	Durvalumab (PD-1)	II	Not yet recruiting
NCT04262375	Non–small cell lung cancer and renal clear cell carcinoma	Oleclumab	CD73	Durvalumab (PD-1)	II	Not yet recruiting
NCT04148937	Advanced solid tumors	LY3475070	CD73	Pembrolizumab (PD-1)	I	Recruiting
NCT03454451	Advanced malignancies	CPI-006	CD73	Ciforadenant (A2A receptor) and pembrolizumab (PD-1)	I	Recruiting
NCT03616886	Metastatic triple-negative breast cancer	Oleclumab	CD73	Paclitaxel, carboplatin, and durvalumab (PD-1)	I/II	Recruiting
NCT03875573	Luminal B breast cancer	Oleclumab	CD73	Radiotherapy and durvalumab (PD-1)	II	Recruiting
NCT03835949	Advanced or metastatic cancers	TJ004309	CD73	Atezolizumab (PD-L1)	I	Recruiting
NCT03267589	Relapsed ovarian cancer	MEDI9447	CD73	Durvalumab (PD-1)	II	Recruiting
NCT03549000	Non–small cell lung cancer, triple-negative breast cancer, pancreatic adenocarcinoma, ovarian cancer, renal clear cell carcinoma, metastatic castration-resistant prostate cancer, microsatellite stable colorectal cancer	NZV930	CD73	Spartalizumab (PD-1) and NIR178 (A2A receptor)	I	Recruiting
NCT04104672	Pancreatic adenocarcinoma	AB680	CD73	Zimberelimab (PD-1), nab-paclitaxel, and gemcitabine	I	Recruiting
NCT02754141	Advanced solid tumors	BMS-986179	CD73	Nivolumab (PD-1) and rHuPH20	I/II	Recruiting
NCT03954704	Advanced solid tumors	GS-1423	CD73-TGFB	None	I	Recruiting
NCT02740270	Advanced solid tumors and lymphomas	GWN323	GITR	Spartalizumab (PD-1)	I	Active, not recruiting
NCT02697591	Advanced or metastatic solid tumors	INCAGN01876	GITR	None	I/II	Active, not recruiting

(continued)

Table 12.1 (continued)

NCT	Cancer	Compound	Target	Additional agents	Phase	Status
NCT03277352	Advanced or metastatic malignancies	INCAGN01876	GITR	Epacadostat (IDO1) and pembrolizumab (PD-1)	I/II	Active, not recruiting
NCT03126110	Advanced or metastatic malignancies	INCAGN01876	GITR	Ipilimumab (CTLA-4) and nivolumab (PD-1)	I/II	Active, not recruiting
NCT01239134	Malignant melanoma	TRX518	GITR	None	I	Completed
NCT02583165	Advanced tumors	MEDI1873	GITR	None	I	Completed
NCT04335039	Glioblastoma	INCAGN01876	GITR	INCAGN01876 (PD-1), SRS	II	Not yet recruiting
NCT04021043	Advanced lung, chest, and liver cancers	BMS-986156	GITR	Ipilimumab (CTLA-4), nivolumab (PD-1), SRS	I/II	Recruiting
NCT03799003	Advanced solid tumors	ASP195	GITR	Pembrolizumab (PD-1)	I	Recruiting
NCT03295942	Locally advanced or metastatic tumors	OMP-336B11	GITR	None	I	Terminated
NCT01216436	Metastatic melanoma	GITR-L-transfected DC	GITR	Anti-CTLA-4-transfected DC	I	Terminated
NCT02553499	Advanced solid tumors	MK-1248	GITR	Pembrolizumab (PD-1)	I	Terminated
NCT03489369	Advanced solid tumors and lymphomas	Sym022	Lag-3	None	I	Active, not recruiting
NCT02460224	Advanced malignancies	LAG525	Lag-3	Spartalizumab (PD-1)	I/II	Active, not recruiting
NCT02614833	Metastatic breast cancer	IMP321	Lag-3	Paclitaxel	II	Active, not recruiting
NCT02060188	Colorectal cancer	BMS-986016	Lag-3	Nivolumab (PD-1)	II	Active, not recruiting
NCT00351949	Metastatic renal cancer	IMP321	Lag-3	None	I	Completed
NCT00349934	Metastatic breast cancer	IMP321	Lag-3	None	I	Completed
NCT03252938	Advanced solid tumors	IMP321	Lag-3	Avelumab (PD-1)	I	Recruiting
NCT03250832	Advanced solid tumors	TSR-033	Lag-3	Anti-PD-1	I	Recruiting
NCT03005782	Advanced malignancies	REGN3767	Lag-3	REGN2810 (PD-1)	I	Recruiting
NCT01968109	Non–small cell lung cancer, gastric cancer, hepatocellular carcinoma, renal cell carcinoma	Relatlimab	Lag-3	Nivolumab (PD-1)	I/II	Recruiting
NCT02817633	Advanced solid tumors	SR-033	Lag-3	TSR-022 (Tim-3)	I	Recruiting
NCT03311412	Advanced solid tumors and lymphomas	Sym022	Lag-3	Sym021 (PD-1)	I	Recruiting
NCT02658981	Recurrent GBM	Urelumab	Lag-3	Nivolumab (PD-1)	I	Recruiting
NCT03607890	Advanced mismatch repair deficient cancers	Relatlimab	Lag-3	Nivolumab (PD-1)	II	Recruiting
NCT03538028	Advanced malignancies	INCAGN02385	Lag-3	None	I	Recruiting
NCT00732082	Pancreatic adenocarcinoma	IMP321	Lag-3	Gemcitabine	I	Terminated
NCT03849469	Select solid tumors	XmAb22841	Lag-3-CTLA-4	Pembrolizumab (PD-1)	I	Recruiting
NCT04082364	HER2+ gastric/GEJ cancer	MGD013	Lag-3-PD-1	Margetuximab (HER2)	II/III	Recruiting

(continued)

Table 12.1 (continued)

NCT	Cancer	Compound	Target	Additional agents	Phase	Status
NCT02274155	Advanced head and neck cancers	MEDI6469	OX40	None	I	Active, not recruiting
NCT02559024	Metastatic colorectal cancer	MEDI6469	OX40	None	I	Active, not recruiting
NCT02315066	Locally advanced or metastatic tumors	PF-04518600	OX40	PF-05082566 (4-1BB)	I	Active, not recruiting
NCT02528357	Advanced solid tumors	GSK3174998	OX40	Pembrolizumab (PD-1)	I	Active, not recruiting
NCT01862900	Metastatic breast cancer	MEDI6469	OX40	SBRT	I	Completed
NCT01303705	Metastatic prostate cancer	Anti-OX40	OX40	Radiation and cyclophosphamide	I	Completed
NCT01644968	Advanced cancers	Anti-OX40	OX40	None	I	Completed
NCT02410512	Locally advanced or metastatic solid tumors	MOXR0916	OX40	Atezolizumab (PD-L1)	I	Completed
NCT02221960	Select advanced solid tumors	MEDI6383	OX40	Durvalumab (PD-L1)	I	Completed
NCT02318394	Selected advanced solid tumors	MEDI0562	OX40	None	I	Completed
NCT02705482	Advanced solid tumors	MEDI0562	OX40	Tremelimumab (CTLA-4) and durvalumab (PD-1)	I	Completed
NCT03241173	Advanced or metastatic malignancies	INCAGN01949	OX40	Nivolumab (PD-1) and Ipilimumab (CTLA-4)	I/II	Completed
NCT04215978	Advanced solid tumors	BGB-A445	OX40	Tislelizumab (PD-1)	I	Not yet recruiting
NCT03092856	Metastatic kidney cancer	PF-04518600	OX40	Axitinib	II	Recruiting
NCT03831295	Advanced or metastatic solid tumors	BMS 986178	OX40	SD-101 (TLR9)	I	Recruiting
NCT03971409	Triple-negative breast cancer	PF-04518600	OX40	Avelumab (PD-1), binimetinib (MEK), and utomilumab (4-1BB)	II	Recruiting
NCT03410901	B-cell non-Hodgkin lymphoma	BMS 986178	OX40	SD-101 (TLR9) and radiation therapy	I	Recruiting
NCT04198766	Locally advanced or metastatic solid tumors	INBRX-106	OX40	Pembrolizumab (PD-1)	I	Recruiting
NCT03336606	Head and neck squamous cell carcinoma	MEDI6469	OX40	None	I	Recruiting
NCT03267589	Relapsed ovarian cancer	MEDI0562	OX40	Durvalumab (PD-1) and tremelimumab (CTLA-4)	II	Recruiting
NCT02554812	Locally advanced or metastatic solid tumors	PF-04518600	OX40	Avelumab (PD-1)	II	Recruiting
NCT03636503	Follicular lymphoma	PF-04518600	OX40	Rituximab (CD20), utomilumab (4-1BB), and avelumab (PD-1)	I	Recruiting

(continued)

Table 12.1 (continued)

NCT	Cancer	Compound	Target	Additional agents	Phase	Status
NCT03447314	Advanced solid tumors	GSK3174998	OX40	GSK1795091 (TLR4)	I	Recruiting
NCT02923349	Advanced solid tumors	INCAGN01949	OX40	None	I/II	Recruiting
NCT03758001	Advanced solid tumors	IBI101	OX40	Sintilimab (PD-1)	I	Recruiting
NCT03217747	Advanced malignancies	PF-04518600	OX40	Utomilumab (4-1BB), avelumab (PD-1), and radiation	I/II	Recruiting
NCT03390296	Acute myeloid leukemia	PF-04518600	OX40	Avelumab (PD-1) and azacytidine	I/II	Recruiting
NCT02205333	Aggressive B-cell lymphoma	MEDI6469	OX40	Durvalumab (PD-L1), rituximab (CD20), and tremelimumab (CTLA-4)	I/II	Terminated
NCT01689870	Metastatic melanoma	Anti-OX40	OX40	Ipilimumab (CTLA-4)	I/II	Withdrawn
NCT01416844	Metastatic melanoma	Anti-OX40	OX40	None	II	Withdrawn
NCT03782467	Advanced solid tumors	ATOR-1015	OX40-CTLA-4	None	I	Recruiting
NCT04116710	Advanced solid tumors	HS-130	OX40L-Ag fusion	HS-110	I	Recruiting
NCT03323398	Advanced malignancies	mRNA-2416	OX40L mRNA	Durvalumab (PD-1)	I/II	Recruiting
NCT03739931	Advanced malignancies	mRNA-2416	OX40L mRNA	Durvalumab (PD-L1)	I	Recruiting
NCT03894618	Advanced solid tumors and lymphomas	SL-279252	PD1-Fc-OX40L	None	I	Recruiting
NCT04140500	Advanced solid tumors	RO7247669	PD1-LAG3	None	I	Recruiting
NCT03563716	Non–small cell lung cancer	MTIG7192A	TIGIT	Atezolizumab (PD-L1)	II	Active, not recruiting
NCT04294810	Non–small cell lung cancer	Tiragolumab	Tigit	Atezolizumab (PD-L1)	III	Not yet recruiting
NCT04047862	Advanced solid tumors	BGB-A1217	TIGIT	Tislelizumab (PD-1)	I	Recruiting
NCT04256421	Small cell lung cancer	Tiragolumab	TIGIT	Atezolizumab (PD-L1), etoposide, carboplatin	III	Recruiting
NCT04262856	Non–small cell lung cancer	Zimberelimab	Tigit	Zimberelimab (PD-1) and AB928 (A2b receptor)	II	Recruiting
NCT03628677	Advanced malignancies	AB154	TIGIT	Zimberelimab (PD-1)	I	Recruiting
NCT03119428	Advanced solid tumors	OMP-313M32	TIGIT	Nivolumab (PD-1)	I	Terminated
NCT00986518	Metastatic colorectal cancer	Treg-depleted autologous cell transplant		None	I/II	Completed

tissues, ultimately reducing the autoimmune diabetes (Zhang et al. 2017). Cancer Tregs are considered to have near maximal suppressive activity (Delgoffe et al. 2013) with a population expressing high levels of LAG-3. Since most clinical trials related to anti-LAG-3 antibodies are earlier in clinical trials, there is insufficient information as of how anti-LAG-3 antibodies influence cancer Tregs. LAG-3$^+$ Tregs are significantly enriched in blood from melanoma and colon cancer patients and exhibit an effector/memory phenotype, along with the production of immunosuppressive cytokines TGF-β and IL-10 (Camisaschi et al. 2010). In colorectal cancer, LAG-3 and TIM-3 are co-expressed in more than 50% of cancer Tregs, along with other immunosuppressive molecules such as TGF-β, IL-10, and CTLA-4 (Ma et al. 2018). In addition to classic Foxp3-positive Tregs, co-expression of CD49b and LAG-3 identifies human regulatory type 1 (Tr-1) T cells (Gagliani et al. 2013) that are highly suppressive. It is anticipated that LAG-3-targeting interventions may result in cancer-specific Teff activation as well as Treg inhibition.

12.3.1.3 T-Cell Immunoreceptor with Ig and ITIM Domains (TIGIT)

TIGIT was first identified as a coinhibitory molecule expressed on Teff cells that gained attention by suppressing autoimmune responses (Joller et al. 2011; Levin et al. 2011). TIGIT binds to co-stimulatory ligand CD155 on DCs, which leads to the reduced production of IL-12, but induces IL-10 production (Yu et al. 2009). TIGIT was later found to be expressed on human Tregs with superior immune suppression toward Th1 and Th17 helper cells but not Th2 cells (Joller et al. 2014). TIGIT marks highly dysfunctional CD8 T cells in tumors as well as a highly immune suppressive subpopulation of TI-Tregs, but genetic evidence supports that TIGIT expression on Tregs is dominant in suppressing antitumor immunity (Kurtulus et al. 2015). The fact that TIGIT knockout mice are normal in development and do not develop autoimmune diseases, in addition to the highly immunosuppressive nature of TIGIT$^+$ Tregs, makes TIGIT a great candidate for Treg-based cancer immunotherapy. Several anti-human TIGIT antibodies have been developed and entered early clinical trials, most of which have negligible effect; however, the interest remains from the pharmaceutical industry likely due to its synergistic effects with anti-PD-1/PD-L1 blockade (Table 12.1).

12.3.1.4 T-Cell Immunoglobulin and Mucin-Domain Containing-3 (TIM-3)

TIM-3 is an immunoglobulin and mucin domain family and is originally identified on CD4 and CD8 T cells (Monney et al. 2002) with immune modulatory function. TIM-3 is later found to be expressed by Tregs and innate immune cells including dendritic cells, natural killer cells, monocytes, macrophages, and mast cells (Wolf et al. 2020). Four ligands have been identified—including galectin-9 (Gal-9) (Jayaraman et al. 2010), high-mobility group protein B1 (HMGB1) (Chiba et al. 2012), Ceacam-1 (Huang et al. 2015), and phosphatidylserine (DeKruyff et al. 2010)—that mediate different immune-modulatory function of CD4$^+$ or CD8$^+$ T cells. TIM-3 is expressed on tumor-infiltrating Tregs of many cancer types, with studies showing that TIM-3$^+$ Tregs are more immunosuppressive than their TIM-3$^-$ counterparts and are co-expressing other immune checkpoints such as TIGIT, CTLA-4, and PD-1 (Gao et al. 2012; Kurtulus et al. 2015; Liu et al. 2018; Ma et al. 2018; Sakuishi et al. 2013b). The genetic evidence of TIM-3 in the role of Tregs is lacking, and there is no clinical evidence that TIM-3 inhibition has direct impact on Treg function.

12.3.1.5 Programmed Cell Death-1 (PD-1)

PD-1 is another immune checkpoint protein that was initially identified on active CD8$^+$ and CD4$^+$ T cells. Ligation of PD-L1, mainly expressed by cancer cells or myeloid cells, with PD-1 leads to T-cell exhaustion and dysfunction. There are many outstanding reviews related to the PD-1/PD-L1 axis in the field of cancer immunotherapy (Chamoto et al. 2020; Iwai et al. 2017; Sanmamed and Chen 2018; Zou et al. 2016). Briefly, anti-PD-1/PD-L1 antibodies mainly disrupt the PD-L1 ligation, which reverses the exhaustion phenotype of Teff cells—a process referred to as rejuvenation. As rejuvenation becomes the primary explanation for the mechanism of T-cell activation under anti-PD-1/PD-L1 therapies, a very recent paper provides a secondary opinion showing that anti-PD-1 antibodies (pembrolizumab and cemiplimab) were able to deplete tumor-infiltrating $CD8^+$ T-cell clones and replace them with novel $CD8^+$ T-cell clones against tumor neoantigens (Yost et al. 2019). In relation to Tregs, PD-1 was initially identified as an intracellular protein in resting Tregs and, upon TCR stimulation, moved to the surface of active Tregs (Raimondi et al. 2006). The role of PD-1/PD-L1 axis in the induced Tregs has been reviewed recently (Gianchecchi and Fierabracci 2018) and will not be covered here due to the irrelevance in most solid cancers. Tumor-infiltrating Tregs consist of a significant $PD\text{-}1^+$ population. Interestingly, limited literature points to a role of PD-1 in Treg suppression, including (1) in malignant gliomas where PD-1 marks dysfunctional Tregs with IFN-γ expression (Lowther et al. 2016); (2) The Nishikawa group identified $PD\text{-}1^+$ Tregs in gastric cancer that were amplified by anti-PD-1 antibody treatment, leading to the hyperprogression of cancers upon anti-PD-1 therapy (Kamada et al. 2019b); (3) similar Treg accumulation also occurs in the hyperprogressive adult T-cell leukemia/lymphoma when treated with anti-PD-1 therapy (Rauch et al. 2019). These data are consistent with animal models where Treg-specific deletion of *PDCD1* (gene encoding PD-1) led to the expansion of Tregs that are more suppressive to Teff cells (Kamada et al. 2019b). The PD-1 and PD-L1 axis may not be a very good therapeutic target for Treg-based immunotherapy.

12.3.2 Co-stimulatory Receptors as Therapeutic Targets for Tregs

Another field of interest in cancer immunotherapy is the agonistic activation of co-stimulatory receptors such as GITR, OX-40, and 4-1BB. Interestingly, TI-Tregs from human cancers preferentially express these co-stimulatory receptors at much higher levels relative to Tregs from peripheral blood. Several studies have shown that agonistic activation of these receptors results in the expansion of $CD8^+$ T cells while at the same time eliminating/inhibiting TI-Tregs (Arce Vargas et al. 2017; Bulliard et al. 2013, 2014). As these co-stimulatory agonists mainly activate $CD8^+$ T cells (Table 12.1), and the effects on human TI-Tregs are largely missing, we will not cover these receptors in detail.

12.3.3 Chemokine Receptors as Potential Therapeutic Targets for Tregs

We have briefly discussed the roles of chemokine receptors in the recruitment/retention of tumor-infiltrating Tregs. Here we choose three candidates, CCR4, CCR5, and CCR8, for further discussion.

12.3.3.1 CCR4

CCR4 is identified as the major chemokine receptor for Th2 and Tregs (Yoshie and Matsushima 2015), two $CD4^+$ T-cell subtypes that have tumor-promoting functions. The Zou group first established the function of CCR4—working through CCL22 produced by the TME—in the recruitment of Tregs to ovarian cancers (Curiel et al. 2004), which is further confirmed within several other cancer types (Curiel et al. 2004; Gobert et al. 2009; Maeda et al. 2019; Miller et al. 2006; Mizukami et al. 2008). The major drive to develop anti-CCR4 agents is the elevated expression of CCR4 in mature T-cell neoplasms including adult T-cell leukemia/lymphoma (ATL), cutaneous T-cell lymphomas (CTCLs), and peripheral T-cell lymphomas (PTCLs) (Ogura et al. 2014; Ohshima et al. 2004; Shimauchi et al. 2005; Yoshie et al. 2002). Mogamulizumab is a humanized anti-CCR4 antibody that was approved by the Japanese Pharmaceuticals and Medical Devices Agency (PMDA) to treat $CCR4^+$ ATL in 2012 (Ishida et al. 2012). Mogamulizumab is also effective in treating other T-cell neoplasms with CCR4 expression including relapsed CTCLs and PTCLs (Ogura et al. 2014). FDA approved its usage for the treatment of relapsed or refractory mycosis fungoides and Sézary disease in 2018 (Kasamon et al. 2019). It should be noted that the benefit of mogamulizumab comes hand-in-hand with some severe skin-relevant adverse effects including some fatal cases, largely attributed to its on-target elimination of skin-resident Tregs (Honda et al. 2015; Ishida et al. 2013; Maemoto et al. 2019). The mogamulizumab-mediated Treg depletion, however, can be repurposed to treat cancers and the Sakaguchi group confirmed that targeting CCR4 by anti-CCR4 monoclonal antibody selectively depletes effector-type Tregs and evokes the immune response to cancer (Sugiyama et al. 2013). Many following studies confirmed the Treg-depleting effect is through antibody-mediated cytotoxicity (ADCC) (Chang et al. 2016; Kurose et al. 2015; Maeda et al. 2019; Ni et al. 2015; Ogura et al. 2014; Remer et al. 2014; Winsett et al. 2017). The first reported Phase I cancer trial using mogamulizumab showed promising Treg depletion and limited toxicity, with additional on-target depletion of Th2 and Th17 cells (Kurose et al. 2015). Another Phase I study was recently reported and showed that mogamulizumab, in combination of nivolumab, provides a relative safety profile—with manageable level 3 or 4 treatment-related adverse events in 29% patients, showing evidence of anti-tumor activity and on-target Treg depletion (Doi et al. 2019). There are several other on-going early trials assessing the toxicity and anti-tumor activity in solid cancers (Table 12.1). A recent report puts some doubt on the recovery of Tregs after mogamulizumab treatment in a patient with severe graft-versus-host disease (GVHD), where the elimination of residual mogamulizumab by plasma exchange did not result in prompt recovery of donor Tregs (Sugiura et al. 2019). This situation, if it also turns out to be true, may be the primary reason for mogamulizumab-treatment related adverse events in the skin (Honda et al. 2015; Ishida et al. 2013; Maemoto et al. 2019) or in the long run may lead to chronic autoimmune diseases as seen in cancer patients treated with ICIs (Michot et al. 2016).

12.3.3.2 CCR5

CCR5 is expressed within and mediates the functions of several immune cell types, including T cells, macrophages, eosinophils, myeloid-derived suppressor cells (MDSC), and dendritic cells (Jiao et al. 2019). Cancer cells can have elevated CCR5 expression that provides them the proliferative, migratory, and/or invasive properties (Jiao et al. 2018; Nishikawa et al. 2019; Singh et al. 2018; Tang et al. 2016; Yang et al. 2017; You et al. 2018; Zhang et al. 2018). The initial burst of developing CCR5 inhibitors—either small molecules or antibodies—was due to the definition of CCR5 as a receptor for human immunodeficiency virus (HIV) with mutations that can resist HIV infection (Dean et al. 1996; Samson et al. 1996). Many CCR5 inhibitors are re-purposed for clinical studies in cancer patients (Table 12.1), though all these trials are not initially designed to target Tregs. CCR5 is expressed by tumor-infiltrating Tregs in several cancer types (Schlecker et al. 2012; Tan et al. 2009). We have shown that CCL5—a ligand for CCR5—from the TME can recruit Tregs to tumors (Tan et al. 2011). Preclinical studies using CCR5 inhibitor TAK-779 disrupts CCR5-dependent recruitment of Tregs (Tan et al. 2009). These results—along with strong genetic evidence that CCR5 deletion reduces tumorigenesis—indicate a potential therapeutic effect of targeting CCR5 on certain cancer patients. The first reported Phase I trial using maraviroc in colon cancer liver metastasis showed some therapeutic effects such as decreased proliferative index and elevated immune response to metastatic tumors (Halama et al. 2016). The result did not include an analysis of Tregs. While other trials are on-going, it is expected that there will be more clinical data to explore the impact of targeting CCR5 on Treg depletion in human cancers. The expression pattern of CCR5, however, dictates a nonspecific nature, a potential caveat leading to complicated clinical outcomes.

12.3.3.3 CCR8

CCR8 was initially identified as a human monocyte and thymus chemokine receptors (Tiffany et al. 1997). Similar to CCR4, CCR8 is also selectively expressed in Th2 cells (Zingoni et al. 1998) and recently found to be elevated in tumor-infiltrating Tregs in many human cancer types (Plitas et al. 2016), tissue-resident memory T cells in human skin (McCully et al. 2018), dendritic cells during allergic immune response (Sokol et al. 2018), as well as granulocytes (Blanco-Perez et al. 2019). $CCR8^+$ Tregs were later identified as a major regulator of autoimmune onset in the experimental autoimmune encephalomyelitis (EAE), a mouse model used for the study of multiple sclerosis (Barsheshet et al. 2017). Among Tregs, CCR8 expression is very specific to tumor-infiltrating Tregs relative to peripheral blood and normal tissue counterparts (De Simone et al. 2016; Plitas et al. 2016). This distinct expression of CCR8 in tumor-infiltrating Tregs is very intriguing as targeting CCR8—via antibody-mediated ADCC—will result in specific deletion of tumor-infiltrating Tregs while sparing normal tissue Tregs as shown in colon cancer (Villarreal et al. 2018). CCR8-targeted therapy holds great promise in Treg-based cancer immunotherapy; however, the clinical benefit for targeting CCR8 is yet-to-be established.

12.3.4 Other Targets for Tumor-Infiltrating Tregs

We have listed several other potential targets for tumor-infiltrating Tregs (Fig. 12.2), including anti-CD25 antibodies that block IL-2 sequestration, anti-TGF-β antibody that prevents the downstream immunosuppressive effect, and potential apoptosis inducers that target the highly proliferative but vulnerable tumor-infiltrating T cells.

12.4 Perspectives

It has been known for decades that tumor-infiltrating Tregs are outstanding suppressors for the antitumor immune responses. It is perceivable that tTregs—undergoing positive selection after

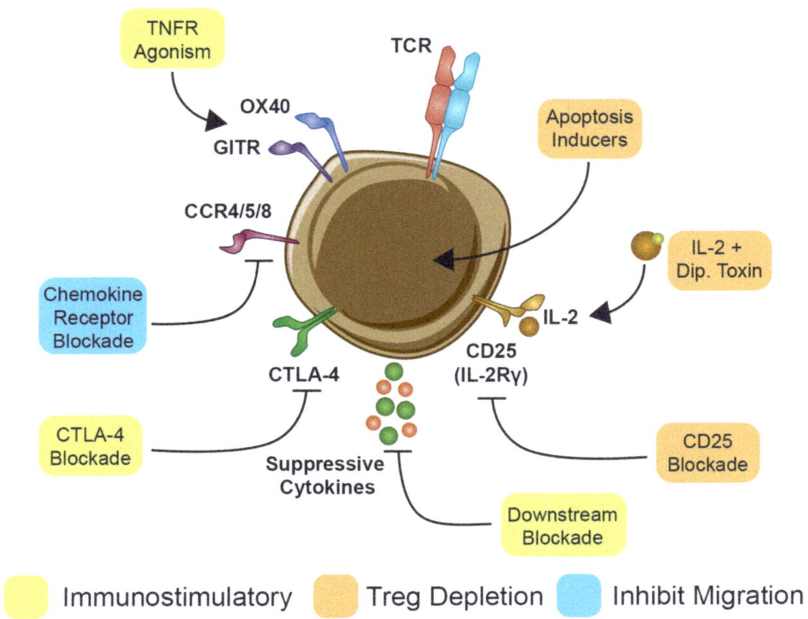

Fig. 12.2 Potential targets and therapeutics of tumor-infiltrating Tregs

encountering MHCII with self-antigens—are the major populations of tumor-infiltrating Tregs as most tumor antigens are self-antigens with a minor fraction of neoantigens. Now that we know the critical function of tumor-infiltrating Tregs in immune suppression, targeting or depleting tumor-infiltrating Tregs represents a viable approach to release anti-tumor immunity. The obstacles are (1) the efficacy of depletion/inhibition of tumor-infiltrating Tregs; (2) the specificity—including to normal Tregs and other immune cell populations; and (3) the identification of target cancer patient populations.

The first obstacle is relatively easy to conquer. For example, many outstanding publications have shown efficient Treg deletion using anti-CTLA-4 antibody, anti-CCR4 antibody, and others under the context of cancer immunotherapy in human clinical trials and mouse models of human cancer as discussed above.

The second obstacle has been a long-standing issue because all the current drug targets for Tregs are not specific to tumor-infiltrating or tumor-activated—those that are not located within tumors but are activated by tumor antigens—Tregs. This is the major reason that these Treg-targeting drugs often lead to treatment-related adverse events in cancer patients and that in many cases, these adverse events cause major problems to patients. The Sakaguchi group has tried to divide the $CD4^{+}FOXP3^{+}$ cells in the peripheral blood of healthy individuals into three distinct populations—based on the differential expression of CD45RA, FOXP3, or CD25—these including naïve/resting Tregs ($CD45RA^{+}FOXP3^{low}CD25^{low}$), effector Tregs which have high suppressive activity ($CD45RA^{-}FOXP3^{hi}CD25^{hi}$), and non-Tregs with no suppressive activity ($CD45RA^{-}FOXP3^{low}CD25^{low}$) (Miyara et al. 2009; Tanaka and Sakaguchi 2019). Numerous studies have shown that tumor-infiltrating Tregs tend to be effector Tregs which express immunosuppressive molecules, such as CTLA-4, LAG-3, and TIGIT, at higher levels than peripheral blood Tregs. As discussed previously, therapies targeting these molecules may have some selectivity in targeting tumor-infiltrating Tregs (Tanaka and Sakaguchi 2019). The problem still remains because effector Tregs exist at a significant amount within normal peripheral bloods and other tissues; hence, targeting these Tregs are inevitably leading to immune-related adverse responses. This leads to the ultimate demand for the identification of markers specific for tumor-infiltrating Tregs. Several other studies have looked at transcriptional differences between tumor infiltrating Tregs and peripheral blood Tregs or normal tissue Tregs (De Simone et al. 2016; Magnuson et al. 2018; Plitas et al. 2016). Zheng et al. used single-cell RNA sequencing of tumor-infiltrating, peripheral blood, and normal tissue lymphocytes from hepatocellular carcinoma patients to characterize tumor-infiltrating Tregs and found several differentially regulated genes; some that were identified by other studies including *CTLA4*, GITR (*TNFRSF18*), *TIGIT*, *LAYN*, 4-1BB (*TNFRSF9*), OX40 (*TNFRSF4*), and *CCR8*, as well as those that have not been identified before including *STAT3* and *RGS1* (Zheng et al. 2017). Apart from the expression of CCR8 in tumor-infiltrating Tregs and its potential prognostic value in breast cancer, Plitas et al. also found that a subset of tumor-infiltrating Tregs in breast, lung, and melanoma patients exclusively expressed *CD177*, a cell surface protein previously only studied in neutrophils (Plitas et al. 2016). Recently we identified an important function of epithelial-cell expression of CD177 in tumor suppression via attenuating β-catenin (Kluz et al. 2020) and that the expression of CD177 in tumor-infiltrating Tregs is critical in mediating the immunosuppressive function of Tregs in human cancers (Borcherding et al. 2018). These types of studies may be used to identify novel suppressive proteins and signaling pathways that are uniquely upregulated in tumor-infiltrating Tregs. Another potential avenue is to exploit the unconventional targets of tumor-infiltrating Tregs that are previously considered to be undruggable. For example, nuclear receptor 4A family (NR4A) has been shown to play a critical role in maintaining the abundance of tumor-infiltrating Tregs (Hibino et al. 2018). Our analysis indicates that NR4A family genes are among the top of the differentially expressed genes in tumor-infiltrating Tregs in human cancers, along with a list of genes whose protein products are present intracellularly (Borcherding et al. 2018; Vishwakarma et al. 2019). These nuclear receptors and non-kinase intracellular proteins are traditionally considered as undruggable targets; however, the advance in bioengineering and medicinal chemistry makes them possible to be targeted such as using the *pro*teolysis-*ta*rgeting *c*himera (PROTAC) technology as we have recently published to better target BCL-X_L in cancer therapy (Khan et al. 2019).

The third obstacle is to choose cancer types and patient populations within certain cancer types. Like all currently available cancer immunotherapies, targeting Tregs will not benefit all cancer patients and sometimes may do more harm than good due to the broad immunosuppressive activity of Tregs including cancer-promoting immune cells. The primary consideration should be given for the abundance of immunosuppressive populations of Tregs and the immune landscaping within tumors. The threshold of Treg abundance should be the prerequisite for patients receiving anti-Treg therapy. The immune landscaping is able to dictate the potential influence of Treg depletion on shaping antitumor immunity. For example, some tumors exhibit dependence on other immunosuppressive cells such as MDSCs or on both Tregs and MDSCs, where depleting Tregs will be insufficient to invoke antitumor immunity.

The future Treg-based cancer immunotherapy should be able to compensate for ICI therapies since cancer types that benefit the most from ICI therapies show only an average of 25% response rate. As the primary ICIs targets are exhausted T cells, Treg-based therapy may benefit some patient populations as the frontline choice where ICIs are predicted to fail. Moreover, Tregs have been accused of the culprit for certain hyperprogressive cancers after nivolumab therapy (Kamada et al. 2019b), suggesting that Treg-based therapy could also benefit these hyperprogressive patients after ICI therapy.

References

Adeegbe DO, Nishikawa H (2013) Natural and induced T regulatory cells in cancer. Front Immunol 4:190

Ahmadzadeh M, Pasetto A, Jia L, Deniger DC, Stevanović S, Robbins PF, Rosenberg SA (2019) Tumor-infiltrating human CD4. Sci Immunol 4: eaao4310

Ai WZ, Hou JZ, Zeiser R, Czerwinski D, Negrin RS, Levy R (2009) Follicular lymphoma B cells induce the conversion of conventional CD4+ T cells to T-regulatory cells. Int J Cancer 124:239–244

Allard D, Allard B, Stagg J (2020) On the mechanism of anti-CD39 immune checkpoint therapy. J Immunother Cancer 8:e000186

Amedei A, Niccolai E, Benagiano M, Della Bella C, Cianchi F, Bechi P, Taddei A, Bencini L, Farsi M, Cappello P et al (2013) Ex vivo analysis of pancreatic cancer-infiltrating T lymphocytes reveals that ENO-specific Tregs accumulate in tumor tissue and inhibit Th1/Th17 effector cell functions. Cancer Immunol Immunother 62:1249–1260

Arce Vargas F, Furness AJS, Solomon I, Joshi K, Mekkaoui L, Lesko MH, Miranda Rota E, Dahan R, Georgiou A, Sledzinska A et al (2017) Fc-optimized anti-CD25 depletes tumor-infiltrating regulatory T cells and synergizes with PD-1 blockade to eradicate established tumors. Immunity 46:577–586

Arce Vargas F, Furness AJS, Litchfield K, Joshi K, Rosenthal R, Ghorani E, Solomon I, Lesko MH, Ruef N, Roddie C et al (2018) Fc effector function contributes to the activity of human anti-CTLA-4 Antibodies. Cancer cell 33:649–663.e644

Asano M, Toda M, Sakaguchi N, Sakaguchi S (1996) Autoimmune disease as a consequence of developmental abnormality of a T cell subpopulation. J Exp Med 184:387–396

Asseman C, Mauze S, Leach MW, Coffman RL, Powrie F (1999) An essential role for interleukin 10 in the function of regulatory T cells that inhibit intestinal inflammation. J Exp Med 190:995–1004

Badoual C, Hans S, Rodriguez J, Peyrard S, Klein C, Agueznay NIH, Mosseri V, Laccourreye O, Bruneval P, Fridman WH et al (2006) Prognostic value of tumor-infiltrating CD4+ T-cell subpopulations in head and neck cancers. Clin Cancer Res 12:465–472

Barsheshet Y, Wildbaum G, Levy E, Vitenshtein A, Akinseye C, Griggs J, Lira SA, Karin N (2017) CCR8(+)FOXp3(+) Treg cells as master drivers of immune regulation. Proc Natl Acad Sci U S A 114:6086–6091

Belkaid Y, Piccirillo CA, Mendez S, Shevach EM, Sacks DL (2002) CD4+CD25+ regulatory T cells control Leishmania major persistence and immunity. Nature 420:502–507

Bennett CL, Christie J, Ramsdell F, Brunkow ME, Ferguson PJ, Whitesell L, Kelly TE, Saulsbury FT, Chance PF, Ochs HD (2001) The immune dysregulation, polyendocrinopathy, enteropathy,

X-linked syndrome (IPEX) is caused by mutations of FOXP3. Nat Genet 27:20–21

Blanco-Perez F, Kato Y, Gonzalez-Menendez I, Laino J, Ohbayashi M, Burggraf M, Krause M, Kirberg J, Iwakura Y, Martella M et al (2019) CCR8 leads to eosinophil migration and regulates neutrophil migration in murine allergic enteritis. Sci Rep 9:9608

Blay J, White TD, Hoskin DW (1997) The extracellular fluid of solid carcinomas contains immunosuppressive concentrations of adenosine. Cancer Res 57:2602–2605

Borcherding N, Ahmed K, Voigt AP, Vishwakarma A, Kolb R, Kluz P, Pandey G, Gibson-Corley KN, Klesney-Tait J, Zhu Y et al (2018) Transcriptional heterogeneity in cancer-associated regulatory T cells is predictive of survival. BioRxiv, Cold Spring Harbor Laboratory

Brunet JF, Denizot F, Luciani MF, Roux-Dosseto M, Suzan M, Mattei MG, Golstein P (1987) A new member of the immunoglobulin superfamily--CTLA-4. Nature 328:267–270

Brunkow ME, Jeffery EW, Hjerrild KA, Paeper B, Clark LB, Yasayko SA, Wilkinson JE, Galas D, Ziegler SF, Ramsdell F (2001) Disruption of a new forkhead/winged-helix protein, scurfin, results in the fatal lymphoproliferative disorder of the scurfy mouse. Nat Genet 27:68–73

Bulliard Y, Jolicoeur R, Windman M, Rue SM, Ettenberg S, Knee DA, Wilson NS, Dranoff G, Brogdon JL (2013) Activating Fc gamma receptors contribute to the antitumor activities of immunoregulatory receptor-targeting antibodies. J Exp Med 210:1685–1693

Bulliard Y, Jolicoeur R, Zhang J, Dranoff G, Wilson NS, Brogdon JL (2014) OX40 engagement depletes intratumoral Tregs via activating FcgammaRs, leading to antitumor efficacy. Immunol Cell Biol 92:475–480

Burchill MA, Yang J, Vogtenhuber C, Blazar BR, Farrar MA (2007) IL-2 receptor beta-dependent STAT5 activation is required for the development of Foxp3+ regulatory T cells. J Immunol 178:280–290

Burchill MA, Yang J, Vang KB, Moon JJ, Chu HH, Lio CW, Vegoe AL, Hsieh CS, Jenkins MK, Farrar MA (2008) Linked T cell receptor and cytokine signaling govern the development of the regulatory T cell repertoire. Immunity 28:112–121

Camisaschi C, Casati C, Rini F, Perego M, De Filippo A, Triebel F, Parmiani G, Belli F, Rivoltini L, Castelli C (2010) LAG-3 expression defines a subset of CD4(+) CD25(high)Foxp3(+) regulatory T cells that are expanded at tumor sites. J Immunol 184:6545–6551

Chamoto K, Hatae R, Honjo T (2020) Current issues and perspectives in PD-1 blockade cancer immunotherapy. Int J Clin Oncol 25:790–800

Chang DK, Peterson E, Sun J, Goudie C, Drapkin RI, Liu JF, Matulonis U, Zhu Q, Marasco WA (2016) Anti-CCR4 monoclonal antibody enhances antitumor immunity by modulating tumor-infiltrating Tregs in an ovarian cancer xenograft humanized mouse model. Oncoimmunology 5:e1090075

Chaudhary B, Elkord E (2016) Regulatory T cells in the tumor microenvironment and cancer progression: role and therapeutic targeting. Vaccines (Basel) 4:28

Chen J, Chen Z (2014) The effect of immune microenvironment on the progression and prognosis of colorectal cancer. Med Oncol 31:82

Chen B, Zhang D, Zhou J, Li Q, Zhou L, Li SM, Zhu L, Chou KY, Zhou L, Tao L, Lu LM (2013) High CCR6/CCR7 expression and Foxp3+ Treg cell number are positively related to the progression of laryngeal squamous cell carcinoma. Oncol Rep 30:1380–1390

Chiba S, Baghdadi M, Akiba H, Yoshiyama H, Kinoshita I, Dosaka-Akita H, Fujioka Y, Ohba Y, Gorman JV, Colgan JD et al (2012) Tumor-infiltrating DCs suppress nucleic acid-mediated innate immune responses through interactions between the receptor TIM-3 and the alarmin HMGB1. Nat Immunol 13:832–842

Collison LW, Workman CJ, Kuo TT, Boyd K, Wang Y, Vignali KM, Cross R, Sehy D, Blumberg RS, Vignali DA (2007) The inhibitory cytokine IL-35 contributes to regulatory T-cell function. Nature 450:566–569

Collison LW, Pillai MR, Chaturvedi V, Vignali DA (2009) Regulatory T cell suppression is potentiated by target T cells in a cell contact, IL-35- and IL-10-dependent manner. J Immunol 182:6121–6128

Curiel TJ, Coukos G, Zou L, Alvarez X, Cheng P, Mottram P, Evdemon-Hogan M, Conejo-Garcia JR, Zhang L, Burow M et al (2004) Specific recruitment of regulatory T cells in ovarian carcinoma fosters immune privilege and predicts reduced survival. Nat Med 10:942–949

Dariavach P, Mattei MG, Golstein P, Lefranc MP (1988) Human Ig superfamily CTLA-4 gene: chromosomal localization and identity of protein sequence between murine and human CTLA-4 cytoplasmic domains. Eur J Immunol 18:1901–1905

Das M, Zhu C, Kuchroo VK (2017) Tim-3 and its role in regulating anti-tumor immunity. Immunol Rev 276:97–111

Davidson TS, DiPaolo RJ, Andersson J, Shevach EM (2007) Cutting edge: IL-2 is essential for TGF-beta-mediated induction of Foxp3+ T regulatory cells. J Immunol 178:4022–4026

De Simone M, Arrigoni A, Rossetti G, Gruarin P, Ranzani V, Politano C, Bonnal RJP, Provasi E, Sarnicola ML, Panzeri I et al (2016) Transcriptional landscape of human tissue lymphocytes unveils uniqueness of tumor-infiltrating T regulatory cells. Immunity 45:1135–1147

Dean M, Carrington M, Winkler C, Huttley GA, Smith MW, Allikmets R, Goedert JJ, Buchbinder SP, Vittinghoff E, Gomperts E et al (1996) Genetic restriction of HIV-1 infection and progression to AIDS by a deletion allele of the CKR5 structural gene. Hemophilia growth and development study, Multicenter AIDS Cohort Study, Multicenter Hemophilia Cohort

Study, San Francisco City Cohort, ALIVE Study. Science 273:1856–1862

DeKruyff RH, Bu X, Ballesteros A, Santiago C, Chim YL, Lee HH, Karisola P, Pichavant M, Kaplan GG, Umetsu DT et al (2010) T cell/transmembrane, Ig, and mucin-3 allelic variants differentially recognize phosphatidylserine and mediate phagocytosis of apoptotic cells. J Immunol 184:1918–1930

Delgoffe GM, Woo SR, Turnis ME, Gravano DM, Guy C, Overacre AE, Bettini ML, Vogel P, Finkelstein D, Bonnevier J et al (2013) Stability and function of regulatory T cells is maintained by a neuropilin-1-semaphorin-4a axis. Nature 501:252–256

Deng G (2018) Tumor-infiltrating regulatory T cells: origins and features. Am J Clin Exp Immunol 7:81–87

Dieckmann D, Plottner H, Berchtold S, Berger T, Schuler G (2001) Ex vivo isolation and characterization of CD4(+)CD25(+) T cells with regulatory properties from human blood. J Exp Med 193:1303–1310

Doi T, Muro K, Ishii H, Kato T, Tsushima T, Takenoyama M, Oizumi S, Gemmoto K, Suna H, Enokitani K et al (2019) A phase I study of the anti-CC chemokine receptor 4 antibody, mogamulizumab, in combination with nivolumab in patients with advanced or metastatic solid tumors. Clin Cancer Res 25:6614–6622

Downs-Canner S, Berkey S, Delgoffe GM, Edwards RP, Curiel T, Odunsi K, Bartlett DL, Obermajer N (2017) Suppressive IL-17A. Nat Commun 8:14649

Facciabene A, Peng X, Hagemann IS, Balint K, Barchetti A, Wang LP, Gimotty PA, Gilks CB, Lal P, Zhang L, Coukos G (2011) Tumour hypoxia promotes tolerance and angiogenesis via CCL28 and T(reg) cells. Nature 475:226–230

Floess S, Freyer J, Siewert C, Baron U, Olek S, Polansky J, Schlawe K, Chang HD, Bopp T, Schmitt E et al (2007) Epigenetic control of the foxp3 locus in regulatory T cells. PLoS Biol 5:e38

Fontenot JD, Gavin MA, Rudensky AY (2003) Foxp3 programs the development and function of CD4 +CD25+ regulatory T cells. Nat Immunol 4:330–336

Gagliani N, Magnani CF, Huber S, Gianolini ME, Pala M, Licona-Limon P, Guo B, Herbert DR, Bulfone A, Trentini F et al (2013) Coexpression of CD49b and LAG-3 identifies human and mouse T regulatory type 1 cells. Nat Med 19:739–746

Gandhi MK, Lambley E, Duraiswamy J, Dua U, Smith C, Elliott S, Gill D, Marlton P, Seymour J, Khanna R (2006) Expression of LAG-3 by tumor-infiltrating lymphocytes is coincident with the suppression of latent membrane antigen-specific CD8+ T-cell function in Hodgkin lymphoma patients. Blood 108:2280–2289

Gao X, Zhu Y, Li G, Huang H, Zhang G, Wang F, Sun J, Yang Q, Zhang X, Lu B (2012) TIM-3 expression characterizes regulatory T cells in tumor tissues and is associated with lung cancer progression. PLoS One 7:e30676

Ge X, Zhao Y, Chen C, Wang J, Sun L (2019) Cancer immunotherapies targeting tumor-associated regulatory T cells. Onco Targets Ther 12:11033–11044

Gershon RK, Kondo K (1970) Cell interactions in the induction of tolerance: the role of thymic lymphocytes. Immunology 18:723–737

Gianchecchi E, Fierabracci A (2018) Inhibitory receptors and pathways of lymphocytes: the role of PD-1 in Treg development and their involvement in autoimmunity onset and cancer progression. Front Immunol 9:2374

Gobert M, Treilleux I, Bendriss-Vermare N, Bachelot T, Goddard-Leon S, Arfi V, Biota C, Doffin AC, Durand I, Olive D et al (2009) Regulatory T cells recruited through CCL22/CCR4 are selectively activated in lymphoid infiltrates surrounding primary breast tumors and lead to an adverse clinical outcome. Cancer Res 69:2000–2009

Ha D, Tanaka A, Kibayashi T, Tanemura A, Sugiyama D, Wing JB, Lim EL, Teng KWW, Adeegbe D, Newell EW et al (2019) Differential control of human Treg and effector T cells in tumor immunity by Fc-engineered anti-CTLA-4 antibody. Proc Natl Acad Sci U S A 116:609–618

Haas M, Dimmler A, Hohenberger W, Grabenbauer GG, Niedobitek G, Distel LV (2009) Stromal regulatory T-cells are associated with a favourable prognosis in gastric cancer of the cardia. BMC Gastroenterol 9:65

Halama N, Zoernig I, Berthel A, Kahlert C, Klupp F, Suarez-Carmona M, Suetterlin T, Brand K, Krauss J, Lasitschka F et al (2016) Tumoral immune cell exploitation in colorectal cancer metastases can be targeted effectively by anti-CCR5 therapy in cancer patients. Cancer cell 29:587–601

Halvorsen EC, Hamilton MJ, Young A, Wadsworth BJ, LePard NE, Lee HN, Firmino N, Collier JL, Bennewith KL (2016) Maraviroc decreases CCL8-mediated migration of CCR5(+) regulatory T cells and reduces metastatic tumor growth in the lungs. Oncoimmunology 5:e1150398

Hemon P, Jean-Louis F, Ramgolam K, Brignone C, Viguier M, Bachelez H, Triebel F, Charron D, Aoudjit F, Al-Daccak R, Michel L (2011) MHC class II engagement by its ligand LAG-3 (CD223) contributes to melanoma resistance to apoptosis. J Immunol 186:5173–5183

Hibino S, Chikuma S, Kondo T, Ito M, Nakatsukasa H, Omata-Mise S, Yoshimura A (2018) Inhibition of Nr4a receptors enhances antitumor immunity by breaking Treg-mediated immune tolerance. Cancer Res 78:3027–3040

Hiraoka N, Onozato K, Kosuge T, Hirohashi S (2006) Prevalence of FOXP3+ regulatory T cells increases during the progression of pancreatic ductal adenocarcinoma and its premalignant lesions. Clin Cancer Res 12:5423–5434

Honda T, Hishizawa M, Kataoka TR, Ohmori K, Takaori-Kondo A, Miyachi Y, Kabashima K (2015) Stevens-Johnson syndrome associated with mogamulizumab-induced deficiency of regulatory T cells in an adult

T-cell leukaemia patient. Acta Derm Venereol 95:606–607

Hori S, Nomura T, Sakaguchi S (2003) Control of regulatory T cell development by the transcription factor Foxp3. Science 299:1057–1061

Hsieh CS, Zheng Y, Liang Y, Fontenot JD, Rudensky AY (2006) An intersection between the self-reactive regulatory and nonregulatory T cell receptor repertoires. Nat Immunol 7:401–410

Huang CT, Workman CJ, Flies D, Pan X, Marson AL, Zhou G, Hipkiss EL, Ravi S, Kowalski J, Levitsky HI et al (2004) Role of LAG-3 in regulatory T cells. Immunity 21:503–513

Huang YH, Zhu C, Kondo Y, Anderson AC, Gandhi A, Russell A, Dougan SK, Petersen BS, Melum E, Pertel T et al (2015) CEACAM1 regulates TIM-3-mediated tolerance and exhaustion. Nature 517:386–390

Huard B, Tournier M, Hercend T, Triebel F, Faure F (1994) Lymphocyte-activation gene 3/major histocompatibility complex class II interaction modulates the antigenic response of CD4+ T lymphocytes. Eur J Immunol 24:3216–3221

Huehn J, Polansky JK, Hamann A (2009) Epigenetic control of FOXP3 expression: the key to a stable regulatory T-cell lineage? Nat Rev Immunol 9:83–89

Ichihara F, Kono K, Takahashi A, Kawaida H, Sugai H, Fujii H (2003) Increased populations of regulatory T cells in peripheral blood and tumor-infiltrating lymphocytes in patients with gastric and esophageal cancers. Clin Cancer Res 9:4404–4408

Ishida T, Joh T, Uike N, Yamamoto K, Utsunomiya A, Yoshida S, Saburi Y, Miyamoto T, Takemoto S, Suzushima H et al (2012) Defucosylated anti-CCR4 monoclonal antibody (KW-0761) for relapsed adult T-cell leukemia-lymphoma: a multicenter phase II study. J Clin Oncol 30:837–842

Ishida T, Ito A, Sato F, Kusumoto S, Iida S, Inagaki H, Morita A, Akinaga S, Ueda R (2013) Stevens-Johnson syndrome associated with mogamulizumab treatment of adult T-cell leukemia/lymphoma. Cancer Sci 104:647–650

Iwai Y, Hamanishi J, Chamoto K, Honjo T (2017) Cancer immunotherapies targeting the PD-1 signaling pathway. J Biomed Sci 24:26

Jayaraman P, Sada-Ovalle I, Beladi S, Anderson AC, Dardalhon V, Hotta C, Kuchroo VK, Behar SM (2010) Tim3 binding to galectin-9 stimulates antimicrobial immunity. J Exp Med 207:2343–2354

Jiang Y, Du Z, Yang F, Di Y, Li J, Zhou Z, Pillarisetty VG, Fu D (2014) FOXP3+ lymphocyte density in pancreatic cancer correlates with lymph node metastasis. PLoS One 9:e106741

Jiao X, Velasco-Velazquez MA, Wang M, Li Z, Rui H, Peck AR, Korkola JE, Chen X, Xu S, DuHadaway JB et al (2018) CCR5 governs DNA damage repair and breast cancer stem cell expansion. Cancer Res 78:1657–1671

Jiao X, Nawab O, Patel T, Kossenkov AV, Halama N, Jaeger D, Pestell RG (2019) Recent advances targeting CCR5 for cancer and its role in immuno-oncology. Cancer Res 79:4801–4807

Jie HB, Gildener-Leapman N, Li J, Srivastava RM, Gibson SP, Whiteside TL, Ferris RL (2013) Intratumoral regulatory T cells upregulate immunosuppressive molecules in head and neck cancer patients. Br J Cancer 109:2629–2635

Joller N, Hafler JP, Brynedal B, Kassam N, Spoerl S, Levin SD, Sharpe AH, Kuchroo VK (2011) Cutting edge: TIGIT has T cell-intrinsic inhibitory functions. J Immunol 186:1338–1342

Joller N, Lozano E, Burkett PR, Patel B, Xiao S, Zhu C, Xia J, Tan TG, Sefik E, Yajnik V et al (2014) Treg cells expressing the coinhibitory molecule TIGIT selectively inhibit proinflammatory Th1 and Th17 cell responses. Immunity 40:569–581

Jonuleit H, Schmitt E, Stassen M, Tuettenberg A, Knop J, Enk AH (2001) Identification and functional characterization of human CD4(+)CD25(+) T cells with regulatory properties isolated from peripheral blood. J Exp Med 193:1285–1294

Kakita N, Kanto T, Itose I, Kuroda S, Inoue M, Matsubara T, Higashitani K, Miyazaki M, Sakakibara M, Hiramatsu N et al (2012) Comparative analyses of regulatory T cell subsets in patients with hepatocellular carcinoma: a crucial role of CD25(-) FOXP3(-) T cells. Int J Cancer 131:2573–2583

Kamada T, Togashi Y, Tay C, Ha D, Sasaki A, Nakamura Y, Sato E, Fukuoka S, Tada Y, Tanaka A et al (2019a) PD-1. Proc Natl Acad Sci U S A 116:9999–10008

Kamada T, Togashi Y, Tay C, Ha D, Sasaki A, Nakamura Y, Sato E, Fukuoka S, Tada Y, Tanaka A et al (2019b) PD-1(+) regulatory T cells amplified by PD-1 blockade promote hyperprogression of cancer. Proc Natl Acad Sci U S A 116:9999–10008

Kasamon YL, Chen H, de Claro RA, Nie L, Ye J, Blumenthal GM, Farrell AT, Pazdur R (2019) FDA approval summary: Mogamulizumab-kpkc for mycosis fungoides and Sezary syndrome. Clin Cancer Res 25:7275–7280

Khan S, Zhang X, Lv D, Zhang Q, He Y, Zhang P, Liu X, Thummuri D, Yuan Y, Wiegand JS et al (2019) A selective BCL-XL PROTAC degrader achieves safe and potent antitumor activity. Nat Med 25:1938–1947

Khattri R, Cox T, Yasayko SA, Ramsdell F (2003) An essential role for Scurfin in CD4+CD25+ T regulatory cells. Nat Immunol 4:337–342

Kingsley CI, Karim M, Bushell AR, Wood KJ (2002) CD25+CD4+ regulatory T cells prevent graft rejection: CTLA-4- and IL-10-dependent immunoregulation of alloresponses. J Immunol 168:1080–1086

Klein M, Bopp T (2016) Cyclic AMP represents a crucial component of Treg cell-mediated immune regulation. Front Immunol 7:315

Kluz PN, Kolb R, Xie Q, Borcherding N, Liu Q, Luo Y, Kim MC, Wang L, Zhang Y, Li W et al (2020) Cancer cell-intrinsic function of CD177 in attenuating beta-catenin signaling. Oncogene 39:2877–2889

Krummel MF, Allison JP (1995) CD28 and CTLA-4 have opposing effects on the response of T cells to stimulation. J Exp Med 182:459–465
Kuehnemuth B, Piseddu I, Wiedemann GM, Lauseker M, Kuhn C, Hofmann S, Schmoeckel E, Endres S, Mayr D, Jeschke U, Anz D (2018) CCL1 is a major regulatory T cell attracting factor in human breast cancer. BMC Cancer 18:1278
Kurose K, Ohue Y, Wada H, Iida S, Ishida T, Kojima T, Doi T, Suzuki S, Isobe M, Funakoshi T et al (2015) Phase Ia study of FoxP3+ CD4 Treg depletion by infusion of a humanized anti-CCR4 antibody, KW-0761, in cancer patients. Clin Cancer Res 21:4327–4336
Kurtulus S, Sakuishi K, Ngiow SF, Joller N, Tan DJ, Teng MW, Smyth MJ, Kuchroo VK, Anderson AC (2015) TIGIT predominantly regulates the immune response via regulatory T cells. J Clin Invest 125:4053–4062
Ladoire S, Martin F, Ghiringhelli F (2011) Prognostic role of FOXP3+ regulatory T cells infiltrating human carcinomas: the paradox of colorectal cancer. Cancer Immunol Immunother 60:909–918
Lan Q, Zhou X, Fan H, Chen M, Wang J, Ryffel B, Brand D, Ramalingam R, Kiela PR, Horwitz DA et al (2012) Polyclonal CD4+Foxp3+ Treg cells induce TGFbeta-dependent tolerogenic dendritic cells that suppress the murine lupus-like syndrome. J Mol Cell Biol 4:409–419
Leach DR, Krummel MF, Allison JP (1996) Enhancement of antitumor immunity by CTLA-4 blockade. Science 271:1734–1736
Lee HM, Bautista JL, Hsieh CS (2011) Thymic and peripheral differentiation of regulatory T cells. Adv Immunol 112:25–71
Lee JJ, Kao KC, Chiu YL, Jung CJ, Liu CJ, Cheng SJ, Chang YL, Ko JY, Chia JS (2017) Enrichment of human CCR6(+) regulatory T cells with superior suppressive activity in oral cancer. J Immunol 199:467–476
Leffers N, Gooden MJ, de Jong RA, Hoogeboom BN, ten Hoor KA, Hollema H, Boezen HM, van der Zee AG, Daemen T, Nijman HW (2009) Prognostic significance of tumor-infiltrating T-lymphocytes in primary and metastatic lesions of advanced stage ovarian cancer. Cancer Immunol Immunother 58:449–459
Levin SD, Taft DW, Brandt CS, Bucher C, Howard ED, Chadwick EM, Johnston J, Hammond A, Bontadelli K, Ardourel D et al (2011) Vstm3 is a member of the CD28 family and an important modulator of T-cell function. Eur J Immunol 41:902–915
Li MO, Wan YY, Flavell RA (2007) T cell-produced transforming growth factor-beta1 controls T cell tolerance and regulates Th1- and Th17-cell differentiation. Immunity 26:579–591
Liang B, Workman C, Lee J, Chew C, Dale BM, Colonna L, Flores M, Li N, Schweighoffer E, Greenberg S et al (2008) Regulatory T cells inhibit dendritic cells by lymphocyte activation gene-3 engagement of MHC class II. J Immunol 180:5916–5926
Lin S, Wu H, Wang C, Xiao Z, Xu F (2018) Regulatory T cells and acute lung injury: cytokines, uncontrolled inflammation, and therapeutic implications. Front Immunol 9:1545
Ling KL, Pratap SE, Bates GJ, Singh B, Mortensen NJ, George BD, Warren BF, Piris J, Roncador G, Fox SB et al (2007) Increased frequency of regulatory T cells in peripheral blood and tumour infiltrating lymphocytes in colorectal cancer patients. Cancer Immun 7:7
Linsley PS, Brady W, Urnes M, Grosmaire LS, Damle NK, Ledbetter JA (1991) CTLA-4 is a second receptor for the B cell activation antigen B7. J Exp Med 174:561–569
Lio CW, Hsieh CS (2008) A two-step process for thymic regulatory T cell development. Immunity 28:100–111
Liu W, Putnam AL, Xu-yu Z, Szot GL, Lee MR, Zhu S, Gottlieb PA, Kapranov P, Gingeras TR, de St. Groth BF et al (2006) CD127 expression inversely correlates with FoxP3 and suppressive function of human CD4 (+) T reg cells. J Exp Med 203:1701–1711
Liu J, Zhang N, Li Q, Zhang W, Ke F, Leng Q, Wang H, Chen J, Wang H (2011) Tumor-associated macrophages recruit CCR6+ regulatory T cells and promote the development of colorectal cancer via enhancing CCL20 production in mice. PLoS One 6: e19495
Liu Z, McMichael EL, Shayan G, Li J, Chen K, Srivastava R, Kane LP, Lu B, Ferris RL (2018) Novel effector phenotype of Tim-3(+) regulatory T cells leads to enhanced suppressive function in head and neck cancer patients. Clin Cancer Res 24:4529–4538
Lowther DE, Goods BA, Lucca LE, Lerner BA, Raddassi K, van Dijk D, Hernandez AL, Duan X, Gunel M, Coric V et al (2016) PD-1 marks dysfunctional regulatory T cells in malignant gliomas. JCI Insight 1:e85935
Ma Q, Liu J, Wu G, Teng M, Wang S, Cui M, Li Y (2018) Co-expression of LAG3 and TIM3 identifies a potent Treg population that suppresses macrophage functions in colorectal cancer patients. Clin Exp Pharmacol Physiol 45:1002–1009
Maeda S, Murakami K, Inoue A, Yonezawa T, Matsuki N (2019) CCR4 blockade depletes regulatory T cells and prolongs survival in a canine model of bladder cancer. Cancer Immunol Res 7:1175–1187
Maemoto H, Ariga T, Kusada T, Heianna J, Manabe Y, Miyakawa A, Nakachi S, Morishima S, Iraha S, Ganaha F et al (2019) Radiation-induced dermatitis after administration of mogamulizumab for adult T-cell leukaemia/lymphoma: a multi-institutional retrospective study. Jpn J Clin Oncol 49:153–159
Magnuson AM, Kiner E, Ergun A, Park JS, Asinovski N, Ortiz-Lopez A, Kilcoyne A, Paoluzzi-Tomada E, Weissleder R, Mathis D, Benoist C (2018) Identification and validation of a tumor-infiltrating Treg transcriptional signature conserved across species and

tumor types. Proc Natl Acad Sci U S A 115:E10672–E10681

Matsuzaki J, Gnjatic S, Mhawech-Fauceglia P, Beck A, Miller A, Tsuji T, Eppolito C, Qian F, Lele S, Shrikant P et al (2010) Tumor-infiltrating NY-ESO-1-specific CD8+ T cells are negatively regulated by LAG-3 and PD-1 in human ovarian cancer. Proc Natl Acad Sci U S A 107:7875–7880

McCully ML, Ladell K, Andrews R, Jones RE, Miners KL, Roger L, Baird DM, Cameron MJ, Jessop ZM, Whitaker IS et al (2018) CCR8 expression defines tissue-resident memory T cells in human skin. J Immunol 200:1639–1650

Michot JM, Bigenwald C, Champiat S, Collins M, Carbonnel F, Postel-Vinay S, Berdelou A, Varga A, Bahleda R, Hollebecque A et al (2016) Immune-related adverse events with immune checkpoint blockade: a comprehensive review. Eur J Cancer 54:139–148

Miller AM, Lundberg K, Ozenci V, Banham AH, Hellstrom M, Egevad L, Pisa P (2006) CD4 +CD25high T cells are enriched in the tumor and peripheral blood of prostate cancer patients. J Immunol 177:7398–7405

Mittal S, Marshall NA, Duncan L, Culligan DJ, Barker RN, Vickers MA (2008) Local and systemic induction of CD4+CD25+ regulatory T-cell population by non-Hodgkin lymphoma. Blood 111:5359–5370

Miyara M, Yoshioka Y, Kitoh A, Shima T, Wing K, Niwa A, Parizot C, Taflin C, Heike T, Valeyre D et al (2009) Functional delineation and differentiation dynamics of human CD4+ T cells expressing the FoxP3 transcription factor. Immunity 30:899–911

Mizukami Y, Kono K, Kawaguchi Y, Akaike H, Kamimura K, Sugai H, Fujii H (2008) CCL17 and CCL22 chemokines within tumor microenvironment are related to accumulation of Foxp3+ regulatory T cells in gastric cancer. Int J Cancer 122:2286–2293

Monney L, Sabatos CA, Gaglia JL, Ryu A, Waldner H, Chernova T, Manning S, Greenfield EA, Coyle AJ, Sobel RA et al (2002) Th1-specific cell surface protein Tim-3 regulates macrophage activation and severity of an autoimmune disease. Nature 415:536–541

Morgan ME, van Bilsen JH, Bakker AM, Heemskerk B, Schilham MW, Hartgers FC, Elferink BG, van der Zanden L, de Vries RR, Huizinga TW et al (2005) Expression of FOXP3 mRNA is not confined to CD4 +CD25+ T regulatory cells in humans. Hum Immunol 66:13–20

Munn DH, Mellor AL (2013) Indoleamine 2,3 dioxygenase and metabolic control of immune responses. Trends Immunol 34:137–143

Ni X, Jorgensen JL, Goswami M, Challagundla P, Decker WK, Kim YH, Duvic MA (2015) Reduction of regulatory T cells by Mogamulizumab, a defucosylated anti-CC chemokine receptor 4 antibody, in patients with aggressive/refractory mycosis fungoides and Sezary syndrome. Clin Cancer Res 21:274–285

Nishikawa G, Kawada K, Nakagawa J, Toda K, Ogawa R, Inamoto S, Mizuno R, Itatani Y, Sakai Y (2019) Bone marrow-derived mesenchymal stem cells promote colorectal cancer progression via CCR5. Cell Death Dis 10:264

Ogura M, Ishida T, Hatake K, Taniwaki M, Ando K, Tobinai K, Fujimoto K, Yamamoto K, Miyamoto T, Uike N et al (2014) Multicenter phase II study of mogamulizumab (KW-0761), a defucosylated anti-cc chemokine receptor 4 antibody, in patients with relapsed peripheral T-cell lymphoma and cutaneous T-cell lymphoma. J Clin Oncol 32:1157–1163

Ohshima K, Karube K, Kawano R, Tsuchiya T, Suefuji H, Yamaguchi T, Suzumiya J, Kikuchii M (2004) Classification of distinct subtypes of peripheral T-cell lymphoma unspecified, identified by chemokine and chemokine receptor expression: Analysis of prognosis. Int J Oncol 25:605–613

Ohta A, Gorelik E, Prasad SJ, Ronchese F, Lukashev D, Wong MK, Huang X, Caldwell S, Liu K, Smith P et al (2006) A2A adenosine receptor protects tumors from antitumor T cells. Proc Natl Acad Sci U S A 103:13132–13137

Onishi Y, Fehervari Z, Yamaguchi T, Sakaguchi S (2008) Foxp3+ natural regulatory T cells preferentially form aggregates on dendritic cells in vitro and actively inhibit their maturation. Proc Natl Acad Sci U S A 105:10113–10118

Onizuka S, Tawara I, Shimizu J, Sakaguchi S, Fujita T, Nakayama E (1999) Tumor rejection by in vivo administration of anti-CD25 (interleukin-2 receptor alpha) monoclonal antibody. Cancer Res 59:3128–3133

Ouyang W, O'Garra A (2019) IL-10 family cytokines IL-10 and IL-22: from basic science to clinical translation. Immunity 50:871–891

Park JE, Botting RA, Dominguez Conde C, Popescu DM, Lavaert M, Kunz DJ, Goh I, Stephenson E, Ragazzini R, Tuck E et al (2020) A cell atlas of human thymic development defines T cell repertoire formation. Science 367:eaay3224

Paterson AM, Lovitch SB, Sage PT, Juneja VR, Lee Y, Trombley JD, Arancibia-Carcamo CV, Sobel RA, Rudensky AY, Kuchroo VK et al (2015) Deletion of CTLA-4 on regulatory T cells during adulthood leads to resistance to autoimmunity. J Exp Med 212:1603–1621

Patterson SJ, Pesenacker AM, Wang AY, Gillies J, Mojibian M, Morishita K, Tan R, Kieffer TJ, Verchere CB, Panagiotopoulos C, Levings MK (2016) T regulatory cell chemokine production mediates pathogenic T cell attraction and suppression. J Clin Invest 126:1039–1051

Piccirillo CA, Shevach EM (2001) Cutting edge: control of CD8+ T cell activation by CD4+CD25+ immunoregulatory cells. J Immunol 167:1137–1140

Plitas G, Konopacki C, Wu K, Bos PD, Morrow M, Putintseva EV, Chudakov DM, Rudensky AY (2016) Regulatory T cells exhibit distinct features in human breast cancer. Immunity 45:1122–1134

Powrie F, Carlino J, Leach MW, Mauze S, Coffman RL (1996) A critical role for transforming growth factor-beta but not interleukin 4 in the suppression of T helper type 1-mediated colitis by CD45RB(low) CD4+ T cells. J Exp Med 183:2669–2674

Qureshi OS, Zheng Y, Nakamura K, Attridge K, Manzotti C, Schmidt EM, Baker J, Jeffery LE, Kaur S, Briggs Z et al (2011) Trans-endocytosis of CD80 and CD86: a molecular basis for the cell-extrinsic function of CTLA-4. Science 332:600–603

Raimondi G, Shufesky WJ, Tokita D, Morelli AE, Thomson AW (2006) Regulated compartmentalization of programmed cell death-1 discriminates CD4+CD25 + resting regulatory T cells from activated T cells. J Immunol 176:2808–2816

Rasku MA, Clem AL, Telang S, Taft B, Gettings K, Gragg H, Cramer D, Lear SC, McMasters KM, Miller DM, Chesney J (2008) Transient T cell depletion causes regression of melanoma metastases. J Transl Med 6:12

Rauch DA, Conlon KC, Janakiram M, Brammer JE, Harding JC, Ye BH, Zang X, Ren X, Olson S, Cheng X et al (2019) Rapid progression of adult T-cell leukemia/lymphoma as tumor-infiltrating Tregs after PD-1 blockade. Blood 134:1406–1414

Rech AJ, Mick R, Martin S, Recio A, Aqui NA, Powell DJ, Colligon TA, Trosko JA, Leinbach LI, Pletcher CH et al (2012) CD25 blockade depletes and selectively reprograms regulatory T cells in concert with immunotherapy in cancer patients. Sci Transl Med 4:134ra162

Remer M, Al-Shamkhani A, Glennie M, Johnson P (2014) Mogamulizumab and the treatment of CCR4-positive T-cell lymphomas. Immunotherapy 6:1187–1206

Romano M, Tung SL, Smyth LA, Lombardi G (2017) Treg therapy in transplantation: a general overview. Transpl Int 30:745–753

Roncador G, Brown PJ, Maestre L, Hue S, Martínez-Torrecuadrada JL, Ling KL, Pratap S, Toms C, Fox BC, Cerundolo V et al (2005) Analysis of FOXP3 protein expression in human CD4+CD25+ regulatory T cells at the single-cell level. Eur J Immunol 35:1681–1691

Saito T, Nishikawa H, Wada H, Nagano Y, Sugiyama D, Atarashi K, Maeda Y, Hamaguchi M, Ohkura N, Sato E et al (2016) Two FOXP3(+)CD4(+) T cell subpopulations distinctly control the prognosis of colorectal cancers. Nat Med 22:679–684

Sakaguchi S (2011) Regulatory T cells: history and perspective. Methods Mol Biol 707:3–17

Sakaguchi S, Sakaguchi N, Asano M, Itoh M, Toda M (1995) Immunologic self-tolerance maintained by activated T cells expressing IL-2 receptor alpha-chains (CD25). Breakdown of a single mechanism of self-tolerance causes various autoimmune diseases. J Immunol 155:1151–1164

Sakaguchi S, Miyara M, Costantino CM, Hafler DA (2010) FOXP3+ regulatory T cells in the human immune system. Nat Rev Immunol 10:490–500

Sakuishi K, Ngiow SF, Sullivan JM, Teng MW, Kuchroo VK, Smyth MJ, Anderson AC (2013a) TIM3. Oncoimmunology 2:e23849

Sakuishi K, Ngiow SF, Sullivan JM, Teng MW, Kuchroo VK, Smyth MJ, Anderson AC (2013b) TIM3(+) FOXP3(+) regulatory T cells are tissue-specific promoters of T-cell dysfunction in cancer. Oncoimmunology 2:e23849

Salama P, Phillips M, Grieu F, Morris M, Zeps N, Joseph D, Platell C, Iacopetta B (2009) Tumor-infiltrating FOXP3+ T regulatory cells show strong prognostic significance in colorectal cancer. J Clin Oncol 27:186–192

Samson M, Libert F, Doranz BJ, Rucker J, Liesnard C, Farber CM, Saragosti S, Lapoumeroulie C, Cognaux J, Forceille C et al (1996) Resistance to HIV-1 infection in caucasian individuals bearing mutant alleles of the CCR-5 chemokine receptor gene. Nature 382:722–725

Sanmamed MF, Chen L (2018) A paradigm shift in cancer immunotherapy: from enhancement to normalization. Cell 175:313–326

Sasidharan Nair V, Elkord E (2018) Immune checkpoint inhibitors in cancer therapy: a focus on T-regulatory cells. Immunol Cell Biol 96:21–33

Sayour EJ, McLendon P, McLendon R, De Leon G, Reynolds R, Kresak J, Sampson JH, Mitchell DA (2015) Increased proportion of FoxP3+ regulatory T cells in tumor infiltrating lymphocytes is associated with tumor recurrence and reduced survival in patients with glioblastoma. Cancer Immunol Immunother 64:419–427

Schaefer C, Kim GG, Albers A, Hoermann K, Myers EN, Whiteside TL (2005) Characteristics of CD4+CD25+ regulatory T cells in the peripheral circulation of patients with head and neck cancer. Br J Cancer 92:913–920

Schlecker E, Stojanovic A, Eisen C, Quack C, Falk CS, Umansky V, Cerwenka A (2012) Tumor-infiltrating monocytic myeloid-derived suppressor cells mediate CCR5-dependent recruitment of regulatory T cells favoring tumor growth. J Immunol 189:5602–5611

Schuler PJ, Saze Z, Hong CS, Muller L, Gillespie DG, Cheng D, Harasymczuk M, Mandapathil M, Lang S, Jackson EK, Whiteside TL (2014) Human CD4+ CD39+ regulatory T cells produce adenosine upon co-expression of surface CD73 or contact with CD73 + exosomes or CD73+ cells. Clin Exp Immunol 177:531–543

Scurr M, Ladell K, Besneux M, Christian A, Hockey T, Smart K, Bridgeman H, Hargest R, Phillips S, Davies M et al (2014) Highly prevalent colorectal cancer-infiltrating LAP^{+} Foxp3^{-} T cells exhibit more potent immunosuppressive activity than Foxp3^{+} regulatory T cells. Mucosal Immunol 7:428–439

Selby MJ, Engelhardt JJ, Quigley M, Henning KA, Chen T, Srinivasan M, Korman AJ (2013) Anti-CTLA-4 antibodies of IgG2a isotype enhance antitumor activity through reduction of intratumoral regulatory T cells. Cancer Immunol Res 1:32–42

Shang B, Liu Y, Jiang SJ, Liu Y (2015) Prognostic value of tumor-infiltrating FoxP3+ regulatory T cells in cancers: a systematic review and meta-analysis. Sci Rep 5:15179

Shapiro M, Herishanu Y, Katz BZ, Dezorella N, Sun C, Kay S, Polliack A, Avivi I, Wiestner A, Perry C (2017) Lymphocyte activation gene 3: a novel therapeutic target in chronic lymphocytic leukemia. Haematologica 102:874–882

Sharma A, Subudhi SK, Blando J, Scutti J, Vence L, Wargo J, Allison JP, Ribas A, Sharma P (2019) Anti-CTLA-4 immunotherapy does not deplete FOXP3(+) regulatory T cells (Tregs) in human cancers. Clin Cancer Res 25:1233–1238

Shimauchi T, Hirokawa Y, Tokura Y (2005) Purpuric adult T-cell leukaemia/lymphoma: expansion of unusual CD4/CD8 double-negative malignant T cells expressing CCR4 but bearing the cytotoxic molecule granzyme B. Br J Dermatol 152:350–352

Shimizu J, Yamazaki S, Sakaguchi S (1999) Induction of tumor immunity by removing CD25+CD4+ T cells: a common basis between tumor immunity and autoimmunity. J Immunol 163:5211–5218

Simpson TR, Li F, Montalvo-Ortiz W, Sepulveda MA, Bergerhoff K, Arce F, Roddie C, Henry JY, Yagita H, Wolchok JD et al (2013) Fc-dependent depletion of tumor-infiltrating regulatory T cells co-defines the efficacy of anti-CTLA-4 therapy against melanoma. J Exp Med 210:1695–1710

Singh SK, Mishra MK, Eltoum IA, Bae S, Lillard JW Jr, Singh R (2018) CCR5/CCL5 axis interaction promotes migratory and invasiveness of pancreatic cancer cells. Sci Rep 8:1323

Sokol CL, Camire RB, Jones MC, Luster AD (2018) The chemokine receptor CCR8 promotes the migration of dendritic cells into the lymph node parenchyma to initiate the allergic immune response. Immunity 49:449–463.e446

Stockis J, Roychoudhuri R, Halim TYF (2019) Regulation of regulatory T cells in cancer. Immunology 157:219–231

Su W, Chen X, Zhu W, Yu J, Li W, Li Y, Li Z, Olsen N, Liang D, Zheng SG (2019) The cAMP-adenosine feedback loop maintains the suppressive function of regulatory T cells. J Immunol 203:1436–1446

Sugiura H, Matsuoka KI, Sando Y, Meguri Y, Ikegawa S, Nakamura M, Iwamoto M, Yoshioka T, Asano T, Kondo E et al (2019) Plasma exchange eliminates residual mogamulizumab but does not warrant prompt recovery of peripheral Treg levels. Transfus Apher Sci 58:472–474

Sugiyama D, Nishikawa H, Maeda Y, Nishioka M, Tanemura A, Katayama I, Ezoe S, Kanakura Y, Sato E, Fukumori Y et al (2013) Anti-CCR4 mAb selectively depletes effector-type FoxP3+CD4+ regulatory T cells, evoking antitumor immune responses in humans. Proc Natl Acad Sci U S A 110:17945–17950

Sundström P, Stenstad H, Langenes V, Ahlmanner F, Theander L, Ndah TG, Fredin K, Börjesson L, Gustavsson B, Bastid J, Quiding-Järbrink M (2016) Regulatory T cells from colon cancer patients inhibit effector T-cell migration through an adenosine-dependent mechanism. Cancer Immunol Res 4:183–193

Takahashi T, Kuniyasu Y, Toda M, Sakaguchi N, Itoh M, Iwata M, Shimizu J, Sakaguchi S (1998) Immunologic self-tolerance maintained by CD25+CD4+ naturally anergic and suppressive T cells: induction of autoimmune disease by breaking their anergic/suppressive state. Int Immunol 10:1969–1980

Tan MC, Goedegebuure PS, Belt BA, Flaherty B, Sankpal N, Gillanders WE, Eberlein TJ, Hsieh CS, Linehan DC (2009) Disruption of CCR5-dependent homing of regulatory T cells inhibits tumor growth in a murine model of pancreatic cancer. J Immunol 182:1746–1755

Tan W, Zhang W, Strasner A, Grivennikov S, Cheng JQ, Hoffman RM, Karin M (2011) Tumour-infiltrating regulatory T cells stimulate mammary cancer metastasis through RANKL-RANK signalling. Nature 470:548–553

Tanaka A, Sakaguchi S (2019) Targeting Treg cells in cancer immunotherapy. Eur J Immunol 49:1140–1146

Tang Y, Xu X, Guo S, Zhang C, Tian Y, Ni B, Lu B, Wang H (2014) An increased abundance of tumor-infiltrating regulatory T cells is correlated with the progression and prognosis of pancreatic ductal adenocarcinoma. PLoS One 9:e91551

Tang S, Xiang T, Huang S, Zhou J, Wang Z, Xie R, Long H, Zhu B (2016) Ovarian cancer stem-like cells differentiate into endothelial cells and participate in tumor angiogenesis through autocrine CCL5 signaling. Cancer Lett 376:137–147

Tao H, Mimura Y, Aoe K, Kobayashi S, Yamamoto H, Matsuda E, Okabe K, Matsumoto T, Sugi K, Ueoka H (2012) Prognostic potential of FOXP3 expression in non-small cell lung cancer cells combined with tumor-infiltrating regulatory T cells. Lung Cancer 75:95–101

Thornton AM, Shevach EM (1998) CD4+CD25+ immunoregulatory T cells suppress polyclonal T cell activation in vitro by inhibiting interleukin 2 production. J Exp Med 188:287–296

Tiffany HL, Lautens LL, Gao JL, Pease J, Locati M, Combadiere C, Modi W, Bonner TI, Murphy PM (1997) Identification of CCR8: a human monocyte and thymus receptor for the CC chemokine I-309. J Exp Med 186:165–170

Tivol EA, Borriello F, Schweitzer AN, Lynch WP, Bluestone JA, Sharpe AH (1995) Loss of CTLA-4 leads to massive lymphoproliferation and fatal multiorgan tissue destruction, revealing a critical negative regulatory role of CTLA-4. Immunity 3:541–547

Togashi Y, Shitara K, Nishikawa H (2019) Regulatory T cells in cancer immunosuppression - implications for anticancer therapy. Nat Rev Clin Oncol 16:356–371

Topalian SL, Taube JM, Anders RA, Pardoll DM (2016) Mechanism-driven biomarkers to guide immune checkpoint blockade in cancer therapy. Nat Rev Cancer 16:275–287

Valzasina B, Piconese S, Guiducci C, Colombo MP (2006) Tumor-induced expansion of regulatory T cells by conversion of CD4+CD25- lymphocytes is thymus and proliferation independent. Cancer Res 66:4488–4495

Villarreal DO, L'Huillier A, Armington S, Mottershead C, Filippova EV, Coder BD, Petit RG, Princiotta MF (2018) Targeting CCR8 induces protective antitumor immunity and enhances vaccine-induced responses in colon cancer. Cancer Res 78:5340–5348

Vishwakarma A, Bocherding N, Chimenti MS, Vishwakarma P, Nepple K, Salem A, Jenkins RW, Zhang W, Zakharia Y (2019) Mapping the immune landscape of clear cell renal cell carcinoma by single-cell RNA-seq. Cold Spring Harbor Laboratory

Wan Z, Zhou Z, Liu Y, Lai Y, Luo Y, Peng X, Zou W (2020) Regulatory T cells and T helper 17 cells in viral infection. Scand J Immunol 91:e12873

Wang X, Lang M, Zhao T, Feng X, Zheng C, Huang C, Hao J, Dong J, Luo L, Li X et al (2017) Cancer-FOXP3 directly activated CCL5 to recruit FOXP3. Oncogene 36:3048–3058

Wang J, Sanmamed MF, Datar I, Su TT, Ji L, Sun J, Chen L, Chen Y, Zhu G, Yin W et al (2019) Fibrinogen-like protein 1 is a major immune inhibitory ligand of LAG-3. Cell 176:334–347.e312

Ward ST, Li KK, Hepburn E, Weston CJ, Curbishley SM, Reynolds GM, Hejmadi RK, Bicknell R, Eksteen B, Ismail T et al (2015) The effects of CCR5 inhibition on regulatory T-cell recruitment to colorectal cancer. Br J Cancer 112:319–328

Wildin RS, Ramsdell F, Peake J, Faravelli F, Casanova JL, Buist N, Levy-Lahad E, Mazzella M, Goulet O, Perroni L et al (2001) X-linked neonatal diabetes mellitus, enteropathy and endocrinopathy syndrome is the human equivalent of mouse scurfy. Nat Genet 27:18–20

Wing K, Onishi Y, Prieto-Martin P, Yamaguchi T, Miyara M, Fehervari Z, Nomura T, Sakaguchi S (2008) CTLA-4 control over Foxp3+ regulatory T cell function. Science 322:271–275

Winsett FT, Lewis DJ, Duvic M (2017) Mogamulizumab for the treatment of relapsed or refractory adult T-cell leukemia-lymphoma. Exp Rev Hematol 10:757–760

Wolf Y, Anderson AC, Kuchroo VK (2020) TIM3 comes of age as an inhibitory receptor. Nat Rev Immunol 20:173–185

Wong J, Obst R, Correia-Neves M, Losyev G, Mathis D, Benoist C (2007) Adaptation of TCR repertoires to self-peptides in regulatory and nonregulatory CD4+ T cells. J Immunol 178:7032–7041

Woo SR, Turnis ME, Goldberg MV, Bankoti J, Selby M, Nirschl CJ, Bettini ML, Gravano DM, Vogel P, Liu CL et al (2012) Immune inhibitory molecules LAG-3 and PD-1 synergistically regulate T-cell function to promote tumoral immune escape. Cancer Res 72:917–927

Yamaguchi T, Yanagisawa K, Sugiyama R, Hosono Y, Shimada Y, Arima C, Kato S, Tomida S, Suzuki M, Osada H, Takahashi T (2012) NKX2-1/TITF1/TTF-1-Induced ROR1 is required to sustain EGFR survival signaling in lung adenocarcinoma. Cancer Cell 21:348–361

Yang ZZ, Novak AJ, Stenson MJ, Witzig TE, Ansell SM (2006) Intratumoral CD4+CD25+ regulatory T-cell-mediated suppression of infiltrating CD4+ T cells in B-cell non-Hodgkin lymphoma. Blood 107:3639–3646

Yang T, Chen M, Yang X, Zhang X, Zhang Z, Sun Y, Xu B, Hua J, He Z, Song Z (2017) Down-regulation of KLF5 in cancer-associated fibroblasts inhibit gastric cancer cells progression by CCL5/CCR5 axis. Cancer Biol Ther 18:806–815

Yang M, Liu Y, Mo B, Xue Y, Ye C, Jiang Y, Bi X, Liu M, Wu Y, Wang J et al (2019) Helios but not CD226, TIGIT and Foxp3 is a potential marker for CD4(+) Treg cells in patients with rheumatoid arthritis. Cell Physiol Biochem 52:1178–1192

Yau JW, Rogers SL, Kawasaki R, Lamoureux EL, Kowalski JW, Bek T, Chen SJ, Dekker JM, Fletcher A, Grauslund J et al (2012) Global prevalence and major risk factors of diabetic retinopathy. Diabetes Care 35:556–564

Yi Y, He HW, Wang JX, Cai XY, Li YW, Zhou J, Cheng YF, Jin JJ, Fan J, Qiu SJ (2013) The functional impairment of HCC-infiltrating γδ T cells, partially mediated by regulatory T cells in a TGFβ- and IL-10-dependent manner. J Hepatol 58:977–983

Yoshie O, Matsushima K (2015) CCR4 and its ligands: from bench to bedside. Int Immunol 27:11–20

Yoshie O, Fujisawa R, Nakayama T, Harasawa H, Tago H, Izawa D, Hieshima K, Tatsumi Y, Matsushima K, Hasegawa H et al (2002) Frequent expression of CCR4 in adult T-cell leukemia and human T-cell leukemia virus type 1-transformed T cells. Blood 99:1505–1511

Yost KE, Satpathy AT, Wells DK, Qi Y, Wang C, Kageyama R, McNamara KL, Granja JM, Sarin KY, Brown RA et al (2019) Clonal replacement of tumor-specific T cells following PD-1 blockade. Nat Med 25:1251–1259

You Y, Li Y, Li M, Lei M, Wu M, Qu Y, Yuan Y, Chen T, Jiang H (2018) Ovarian cancer stem cells promote tumour immune privilege and invasion via CCL5 and regulatory T cells. Clin Exp Immunol 191:60–73

Yu X, Harden K, Gonzalez LC, Francesco M, Chiang E, Irving B, Tom I, Ivelja S, Refino CJ, Clark H et al (2009) The surface protein TIGIT suppresses T cell activation by promoting the generation of mature immunoregulatory dendritic cells. Nat Immunol 10:48–57

Zhang CY, Qi Y, Li XN, Yang Y, Liu DL, Zhao J, Zhu DY, Wu K, Zhou XD, Zhao S (2015) The role of CCL20/CCR6 axis in recruiting Treg cells to tumor sites of NSCLC patients. Biomed Pharmacother 69:242–248

Zhang Q, Chikina M, Szymczak-Workman AL, Horne W, Kolls JK, Vignali KM, Normolle D, Bettini M, Workman CJ, Vignali DAA (2017) LAG3 limits regulatory

T cell proliferation and function in autoimmune diabetes. Sci Immunol 2:eaah4569
Zhang W, Xu J, Fang H, Tang L, Chen W, Sun Q, Zhang Q, Yang F, Sun Z, Cao L et al (2018) Endothelial cells promote triple-negative breast cancer cell metastasis via PAI-1 and CCL5 signaling. FASEB J 32:276–288
Zheng SG, Gray JD, Ohtsuka K, Yamagiwa S, Horwitz DA (2002) Generation ex vivo of TGF-beta-producing regulatory T cells from CD4+CD25- precursors. J Immunol 169:4183–4189
Zheng SG, Wang JH, Gray JD, Soucier H, Horwitz DA (2004) Natural and induced CD4+CD25+ cells educate CD4+CD25- cells to develop suppressive activity: the role of IL-2, TGF-beta, and IL-10. J Immunol 172:5213–5221
Zheng SG, Wang JH, Stohl W, Kim KS, Gray JD, Horwitz DA (2006) TGF-beta requires CTLA-4 early after T cell activation to induce FoxP3 and generate adaptive CD4+CD25+ regulatory cells. J Immunol 176:3321–3329
Zheng SG, Wang J, Wang P, Gray JD, Horwitz DA (2007) IL-2 is essential for TGF-beta to convert naive CD4+CD25- cells to CD25+Foxp3+ regulatory T cells and for expansion of these cells. J Immunol 178:2018–2027
Zheng C, Zheng L, Yoo JK, Guo H, Zhang Y, Guo X, Kang B, Hu R, Huang JY, Zhang Q et al (2017) Landscape of infiltrating T cells in liver cancer revealed by single-cell sequencing. Cell 169:1342–1356.e1316
Zingoni A, Soto H, Hedrick JA, Stoppacciaro A, Storlazzi CT, Sinigaglia F, D'Ambrosio D, O'Garra A, Robinson D, Rocchi M et al (1998) The chemokine receptor CCR8 is preferentially expressed in Th2 but not Th1 cells. J Immunol 161:547–551
Zou W, Wolchok JD, Chen L (2016) PD-L1 (B7-H1) and PD-1 pathway blockade for cancer therapy: mechanisms, response biomarkers, and combinations. Sci Transl Med 8:328rv324

TNF–TNFR2 Signal Plays a Decisive Role in the Activation of CD4$^+$Foxp3$^+$ Regulatory T Cells: Implications in the Treatment of Autoimmune Diseases and Cancer

13

Md Sahidul Islam, Yang Yang, and Xin Chen

Abstract

The puzzling biphasic or dual roles of tumor necrosis factor α (TNF) in the inflammatory and immune responses are likely to be mediated by distinct signaling pathways transduced by one of its two receptors, e.g., TNF receptor type I (TNFR1) and TNF receptor type II (TNFR2). Unlike TNFR1 that is ubiquitously expressed on almost all types of cells, the expression of TNFR2 is rather restricted to certain types of cells, such as T lymphocytes. There is now compelling evidence that TNFR2 is preferentially expressed by CD4$^+$Foxp3$^+$ regulatory T cells (Tregs), and TNFR2 plays a decisive role in the activation, expansion, in vivo function, and phenotypical stability of Tregs. In this chapter, the current understanding of the molecular basis and signaling pathway of TNF–TNFRs signal is introduced. Latest studies that have further supported and substantiated the pivotal role of TNF–TNFR2 interaction in Tregs biology and its molecular basis are discussed. The research progress regarding TNFR2-targeting treatment for autoimmune diseases and cancer is analyzed. Future study should focus on the further understanding of molecular mechanism underlying Treg-stimulatory effect of TNFR2 signal, as well as on the translation of research findings into therapeutic benefits of human patients with autoimmune diseases, allergy, allograft rejection, and cancer.

Keywords

TNF · CD4$^+$Foxp3$^+$ regulatory T cells · TNFR2 · Autoimmune diseases · Cancer

13.1 Introduction

CD4$^+$Foxp3$^+$ regulatory T cells (Tregs) are immunosuppressive cells that play an indispensable role in the maintenance of immune homeostasis (Sakaguchi et al. 2008; Togashi and Nishikawa 2017). Tregs accumulate in the tumor environment and promote tumor immune evasion by inhibiting anti-tumor immune responses (Zou 2006). Therefore, modulation of Tregs activity has important therapeutic implications. For example, upregulation of Tregs activity is helpful in attenuation of autoimmune inflammatory responses, while downregulation of Treg activity can be beneficial in cancer immunotherapy (Sakaguchi et al. 2010; Wing et al. 2019; Kong et al. 2012; Lu et al. 2014). Biological pathways crucial for the modulation of Tregs function can

Md Sahidul Islam and Yang Yang contributed equally with all other contributors.

M. S. Islam · Y. Yang · X. Chen (✉)
State Key Laboratory of Quality Research in Chinese Medicine, Institute of Chinese Medical Sciences, University of Macau, Macau SAR, China
e-mail: xchen@um.edu.mo

S.-G. Zheng (ed.), *T Regulatory Cells in Human Health and Diseases*, Advances in Experimental Medicine and Biology 1278, https://doi.org/10.1007/978-981-15-6407-9_13

be leveraged to control Tregs activity for therapeutic purposes.

In 2007s, we (Xin Chen and Joost J Oppenheim) for the first time reported that TNF stimulates the activation and expansion of Tregs, and this effect of TNF is mediated by TNFR2 (Chen et al. 2007). Furthermore, we for the first time found and reported that TNFR2-expressing Tregs have the maximally suppressive function, while TNFR2^{-} Tregs have minimal or no suppressive function at all (Chen et al. 2008, 2010a, b). Further, TNF preferentially upregulated TNFR2 expression on Tregs (Hamano et al. 2011) and TNF–TNFR2 interaction was required for the in vivo function as well as phenotypic stability of Tregs (Chen et al. 2013). Our findings have been supported and substantiated by our recent studies (Chen et al. 2015, 2016; Nie et al. 2018; Hamano et al. 2015; Yan et al. 2015; Chen and Oppenheim 2016, 2017; Zaragoza et al. 2016; Chen and Plebanski 2019) and by the majority of researchers in the field (Chen and Plebanski 2019; Fischer et al. 2019a; Tseng et al. 2019; Jacob et al. 2009). Now there is compelling evidence that the signal of TNFR2 plays a decisive role in the activation, expansion, and phenotypical stability of Tregs in response to the stimulation of TNF (reviewed by us in (Chen and Oppenheim 2010, 2011a, b, c)). Recent studies in the field have further advanced the understanding of the molecular basis and signaling events underlying the Treg-stimulatory effect of TNF–TNFR2 interaction. For instance, it was found that TNFR2 maintained the expression of Foxp3 in Tregs by limiting DNA methylation at Foxp3 promoters (Tseng et al. 2019). On the other hand, the role of TNFR2-expressing Tregs in the pathogenesis of autoimmune diseases and cancer, as well as the value of TNFR2-targeting therapy in the treatment of major human diseases, was intensively studied. Translational studies have been focused on the development of biologics and small molecules that can act on TNFR1 and TNFR2 as a potential therapeutic agent for the treatment of autoimmune disease, graft-versus-host disease (GVHD), and cancer (Zou et al. 2018). In this chapter, the current understanding of the molecular basis and signaling pathway of TNF–TNFRs signal is systemically introduced. Latest studies that have further substantiated the pivotal role of TNF–TNFR2 interaction in Tregs biology and its molecular basis have been discussed as well. The research progress regarding TNFR2-targeting treatment for autoimmune diseases and cancer is further analyzed.

13.2 Biology of Tumor Necrosis Factor and Its Receptors

13.2.1 Tumor Necrosis Factor

TNF is a pleiotropic cytokine that can regulate many cellular and biological processes such as activation and function of immune cells, cell differentiation, proliferation, apoptosis, and energy metabolism (Nie et al. 2018; Aggarwal 1991; Bach et al. 2013; Battegay et al. 1995; Chen et al. 2009; Drabarek et al. 2012; Sade-Feldman et al. 2013; Sethi and Hotamisligil 1999; Wang et al. 2008; Vaughan et al. 2013; Yamashita and Passegué 2019). Two distinct forms of TNF have been identified: soluble TNF (sTNF) and transmembrane TNF (mTNF) (Horiuchi et al. 2010; Kriegler et al. 1988). Transmembrane TNF is a 26 kDa cell surface type II polypeptide that possesses 230 amino acids (aa) in the sequence. It consists of cytoplasmic domains (30 aa), transmembrane domains (26 aa), and extracellular domains (177 aa). In sequence, the first 76 amino acids represent a highly conserved hydrophobic sequence that anchors the precursor polypeptide in the membrane (Horiuchi et al. 2010). mTNF is predominately expressed as stable β-jellyroll like homotrimer on activated macrophages and lymphocytes, but also can be expressed by a variety of other immune cells (Horiuchi et al. 2010). Each subunit consists of two packed β-sheets of five antiparallel β-strands with three additional β-strands in the N-terminal part (Puimège et al. 2014). The homotrimer looks like a triangular cone or bell in which the three subunits are arranged edge to face. The receptor binding sites of TNF are in the lower half of the triangular cone, in the groove between two subunits (Idriss and Naismith 2000). The precursor mTNF molecules undergo proteolytic cleavage to yield a 17-kDa soluble TNF (sTNF) molecules. sTNF (157aa) is cleaved by the proteolytic action of TNF-α-converting enzyme (TACE) (aka. ADAM metallopeptidase domain 17, ADAM17) between residues Ala76 and Val77 (Wang et al. 1985). Moreover, the further proteolytical processing of mTNF by signal peptide peptidase-like protease (SPPL2A or SPPL2B) leads to the release of TNF intracellular domains (ICD1 and ICD2) in the cytosol. This event can also lead to the release of TNF C-domain 1 and C-domain 2 secreted into the extracellular space (Friedmann et al. 2006). In dendritic cells, these intracellular domains induce IL-12 production (Friedmann et al. 2006). On the surface, the mTNF is phosphorylated on serine residues and dephosphorylated when it binds with soluble tumor necrosis factor receptor 1 (TNFR1) (Watts et al. 1999). Even though the mouse and human forms of TNF share 79% amino acid sequence similarity, only the mice form of TNF is N-glycosylated (Aggarwal 1991). Studies suggested that in biological and metabolic responses of the cell, mTNF may mediate paracrine and autocrine signals, leaving sTNF to mediate endocrine effects (Cawthorn and Sethi 2008).

13.2.2 Tumor Necrosis Factor Receptors

To exert pleiotropic effects of TNF on various cell types, both forms of TNF can bind to its two types of receptors that belong to the TNF receptor superfamily namely TNF receptor type I (TNFR1 aka. TNFRSF1A, CD120a, p55/p60) or TNF receptor type II (TNFR2 aka. TNFRSF1B, CD120b, p75/p80) (Grell et al. 1995, 1998). In general, the binding of TNF–TNFR1 mediates cytotoxic, pro-inflammatory, and pro-apoptotic effects, whereas TNF–TNFR2 is involved in the activation and proliferation of lymphocytes (Idriss and Naismith 2000; Locksley et al. 2001). TNFR1 is ubiquitously expressed by all types of cells. In contrast, TNFR2 is only expressed by subpopulations of cells in the lymphoid system, endothelial cells, and neurons in normal mammals (Vanamee and Faustman 2017). On the cell surface, TNFR1 and TNFR2 are expressed as single-spanning type I transmembrane proteins, but these can also be soluble forms resulting from alternative splicing or shedding (Gregory et al. 2012; Wajant and Siegmund 2019). Of note, structurally both these receptor molecules contain four cysteine-rich domains (CRDs) termed CRD1 through CRD4. Functionally, to form the TNFR self-complex on the cell surface, the CRD1 (aka. Preligand Assembly Domain/PLAD) domain is critical while CRD2 and CRD3 are TNF-binding domains. The function of the CRD4 remains unclear (Mukai et al. 2010). A study found that CRD1 and CRD2 were topologically and structurally similar in both receptors as a representative of the CRDs of conventional members of the TNFR superfamily (Mukai et al. 2010). However, they also noted that the A2 module of CRD3 from different TNFR1 and TNFR2 is observed in a certain type of TNFR superfamily members, such as the CD30, CD40, and 4-1BB proteins (Mukai et al. 2010). Because of the diversity of their modules, the essential binding location for TNF in both receptors varies. Interestingly, the location of Arg77 in the CRD2 of TNFR1was compensated by Arg113 of the TNFR2 CRD3. Besides, the loop structure of the ligand-binding region in TNFR1 consisted of five residues (Arg77 to Gly81) while in TNFR2 it is composed of only three residues (Ser79 to Asp81) (Mukai et al. 2010). However, the major difference in TNFR1 and TNFR2 structure is the presence of the death domain (DD). TNFR1 contains a DD in its cytoplasmic part while TNFR2 has no DD (Wajant and Siegmund 2019; Tartaglia et al. 1993). In contrast, TNFR2 harbors TNF receptor-associated factor (TRAF)-inducing domains (Xie 2013). The DD of TNFR1 is a conserved type of protein–protein interaction domain that enables TNFR1 to cytotoxic signaling pathways as well as signaling pathways that activate transcription factors of the nuclear factor of kappa B (NF-κB) family or kinases of the MAP kinase (MAPK) family (Wajant and Scheurich 2011).

13.2.3 TNF–TNFRs Signal Transduction

It is well established that TNFR1 has an almost 20-fold stronger affinity for sTNF compared to TNFR2. sTNF can transduce the TNF–TNFR1 signal even in the picomolar range. In contrast, mTNF stimulates TNFR1 and TNFR2 equally well (Grell et al. 1995, 1998). This high affinity of TNFR1 for sTNF is mainly caused by the stabilization of the ligand/receptor complex (Grell et al. 1998). Interestingly, the activation of TNFR2 is more likely to mTNF dependence due to its high demand for ligand-mediated crosslinking to mediate signaling cluster formation (Fischer et al. 2017; Krippner-Heidenreich et al. 2002).

The TNF–TNFR1 forms two different signaling complexes depending on the cellular context: (1) the first complex (Complex I) to control the expression of antiapoptotic proteins and (2) the second complex (Complex II or DISC) to trigger cell death processes (Micheau and Tschopp 2003). In Complex I interaction, TNF interacts with TNFR1 through the adaptor protein TNFR type 1-associated death domain protein (TRADD), with receptor-interacting protein-1 (RIP1) and with TNF receptor-associated factor-2 and factor-3 (TRAF2 and TRAF3), cellular inhibitor of apoptosis 1 and 2 (cIAP1 and cIAP2). This complex transduces signal via (1) activation of the transcriptional factor AP-1 through the activation of MAP3Ks and JNK and (2) binding with LUBAC complex for the activation of the transcriptional factor NF-κB through the IκBα degradation (Cabal-Hierro and Lazo 2012; Haas et al. 2009; Hsu et al. 1996; Mahoney et al. 2008). To induce apoptosis and necroptosis, upon the internalization of the receptor, the dissociation of TRADD from complex and the recruitment of FADD (Fas-associated death domain protein) and procaspase-8 happen followed by the assembly of TRAF2/5, and cIAP1/2 (Cabal-Hierro and Lazo 2012). Moreover, deubiquitination of RIP1 by cylindromatosis (CYLD) lysine-63 deubiquitinase recruited RIP3 kinase in this complex (Hitomi et al. 2008). Under conditions in which caspases activation is inhibited, RIP1 and RIP3 could be phosphorylated and a new signaling complex named necroptosome is formed, which leads to cell death through a caspase-independent process known as necroptosis (Cho et al. 2009; Vandenabeele et al. 2010). In this complex, the inactivation of RIP1 and RIP3 leads to the activation of caspase 8 and thus to the triggering of the apoptotic process as well (Micheau and Tschopp 2003).

As noted above, TNF acts as a double sword cytokine in the cell due to the induction of a signal through its ligands-receptors binding variabilities. TNFR1 and TNFR2 have distinct intracellular signaling pathways for TNF-mediating cellular fate, although there are some overlap and crosstalk (Faustman and Davis 2010). Like TNFR1, TNFR2 has a similar extracellular component though it has no cytoplasmic death domain. It is well established that sTNF induces no or weak signaling for TNFR2. In general, binding of mTNF with the intracellular domains of TNFR2 recruits existing cytoplasmic TRAF2 which, in turn, binds TRAF1, TRAF3, cIAP1, and cIAP2 to form the complexes (Rothe et al. 1995, 1994). To transduce the cell survival signal, these proteins activate several other signaling proteins, which liberate NF-κB from its inhibitor protein IκBα in the cytoplasm and translocate to the nucleus (Gehr et al. 1992). TNF–TNFR2 signaling not only activates this NF-κB pathway, but also this interaction can activate the reciprocal PI3K/Akt pathway (Fischer et al. 2011, 2014). This pathway maintains survival and enhances proliferation along with facilitation of the formation of TNFR2–Etk–VEGFR2 complex which participates in cell adhesion, migration, survival, and proliferation (Zhang et al. 2003; Zhou et al. 2008).

13.3 The Decisive Role of TNF–TNFR2 Interaction in the Activation of Tregs

Tregs are a specialized subset of $CD4^+$ T cells which play a central role in the regulation of immune response and for the maintenance of immune homeostasis. Impaired function or reduced number of Tregs leads to the development of autoimmune diseases, while an increased number of highly suppressive Tregs helps immune evasion of the tumor (Sakaguchi et al. 2008; Panduro et al. 2016). Targeting of Tregs has great potential in the treatment of major human diseases (Taams et al. 2006).

Treg cells can be differentiated in the thymus (tTreg) or induced extrathymically (peripheral, pTreg) or ex vivo in the presence of TGF-β and IL-2. All of them express the transcription factor Foxp3 (forkhead box P3) (Fontenot et al. 2003; Gavin et al. 2007; Sakaguchi et al. 1995; Zheng et al. 2002, 2004, 2007). Foxp3 acts as a master regulator of the suppressive phenotype in the development and function of Tregs along with the stabilization of the Tregs lineage (Fontenot et al. 2003; Gavin et al. 2007; Bennett et al. 2001; Hori et al. 2003). Differentiation of Treg cells depend mostly on the interplay of the T-cell receptor (TCR) with self-antigens. Furthermore, in addition to TCR signaling, CD28, GITR, OX40, CTLA-4, and TNFR2 make important contributions to the differentiation of Tregs cells (Mahmud et al. 2014; Tai et al. 2005; Zheng et al. 2006). Tregs suppressive function needs direct cell-cell contact but is also related to immunosuppressive cytokines (Schmidt et al. 2012; Horwitz et al. 2003). Over the past decades, the suppressor's mechanism of Tregs has been intensively studied, and different mechanisms have been reported, including the immunosuppressive cytokines such as IL-10, IL-35, and TGF-β, co-inhibitory receptor (CTLA-4), metabolic disruption (production of adenosine), cytotoxicity (granzyme and perforin), IL-2 deprivation (Schmidt et al. 2012), hydroxyprostaglandin dehydrogenase (HPGD) (Schmidleithner et al. 2019), and others. However, these findings are likely to just represent part of the suppressive capacity of Tregs, and so far with no consensus on one universal mechanism (Schmidt et al. 2012; Shevach 2018). Further research is needed toward a thorough understanding and identification of the molecular basis and definitive suppressor mechanism of Tregs.

With the success of anti-TNF therapy against autoimmune diseases (Monaco et al. 2015), the role of TNF on Treg cells has been studied intensively (Chen and Oppenheim 2010, 2011a, b, c). There is now compelling evidence that TNF through TNFR2 plays an indispensable role in the activation, expansion, phenotypical stability, and in vivo function of human and mouse Tregs (Chen et al. 2007, 2008, 2010a, 2013; Hamano et al. 2011; Zaragoza et al. 2016; Chen and Oppenheim 2011b; Okubo et al. 2013). TNFR2 signal is important not only for the generation and functional activity of Tregs under steady state but also for the proliferative expansion, inhibitory activity, and phenotypical stability of these cells in the inflammation. We have previously shown that Treg cells deficient in TNFR2 were unable to efficiently reconstitute in the lymphopenic Rag 1 KO mice. More importantly, unlike WT Tregs, Tregs deficient in TNFR2 failed to inhibit colitis in Rag 1 KO recipient mice that were co-transferred with naïve WT CD4 T cells. This was associated with the loss of Foxp3 expression and the expression of pro-inflammatory IL-17A in TNFR2-deficient Treg cells in Rag 1 KO mice (Chen et al. 2013). Our discovery was confirmed by a recent study in a mouse model of arthritis (Tseng et al. 2019). Furthermore, this study further revealed the molecular mechanism: TNFR2 signaling promotes the sustainable expression of Foxp3 in the inflammatory environment by restricting DNA methylation at the Foxp3 promoter (Tseng et al. 2019). Urbano and colleagues found that TNF–TNFR2 interaction suppresses the kinome activity of antigen-activated Tregs (CD4$^+$Foxp3$^+$CD45RO$^+$RA$^-$) and controls IL-17 expression of the human Tregs. They also revealed that the TCR, JAK, MAPK, and PKC signaling pathways are associated with this suppression of kinase activity (PCM et al. 2019). In contrast, neutralizing anti-TNF antibody increased IL-17A expression by effector Tregs (CD4$^+$Foxp3$^+$CD45RO$^+$RA$^-$) (PCM et al. 2019). The protective role for TNF administration via TNFR2 in systemic lupus erythematosus (SLE)-prone (New Zealand Black × New Zealand White) F(1) mice has also been established (Jacob et al. 2009). Therefore, TNF–

TNFR2 signaling is important not only for the phenotypical stability of Tregs in mice but also for human Tregs. Recently, Yang et al. reported that TNF–TNFR2 interaction also played an important role in the differentiation, proliferation, and function of induced Tregs (iTregs) in both in vitro and the inflammatory in vivo settings (Yang et al. 2019, 2018). In contrast, this study found that TNFR1 promoted the differentiation of proinflammatory Th1 and Th17 cells (Yang et al. 2019). Thus, this study provides further support to the idea that TNFR1 antagonists or TNFR2 antagonists may be useful in the treatment of autoimmune diseases (Zou et al. 2018).

TNF provided by pathogenic T cells appears to be its major source in the activation and expansion of Tregs in autoimmune diseases (Grinberg-Bleyer et al. 2010). Among in vitro differentiated Th1, Th2, and Th17 cells, the Th17 cells produced the highest levels of TNF (Zhou et al. 2014). Furthermore, Th17 cells potently stimulate the proliferative expansion of Tregs in both in vitro and in vivo settings, through TNF–TNFR2 interaction. Intriguingly, Th17 cells were required for the maintenance of Foxp3 expression by Tregs, while Tregs were needed for the production of IL-17A and TNF from Th17 cells (Zhou et al. 2014). Thus, proinflammatory Th17 cells and immunosuppressive Tregs reciprocally stimulate each other. Since cellular therapies with Tregs and Th17 cells are under investigation, this bidirectional crosstalk and mutual stimulatory effect of these two subsets of Th cells should be considered (Chen and Oppenheim 2014).

The molecular basis and signaling pathway of TNF–TNFR2 in Tregs activation remain to be fully understood. TNFR2 signaling was recently shown to maintain demethylation at *Foxp3* promoter and TNFR2 agonist antibody was proved to favor Tregs homogeneous expansion in vitro (Tseng et al. 2019). A recent study showed that the epigenetic regulation played a role in an autocrine TNF–TNFR2 feedback loop that can promote the stability of a highly suppressive phenotype for Tregs (TIGIT^{hi}, FOXP3^{hi}, Helios^{hi}, and EZH2^{hi}) as shown by the in vitro experiment (Urbano et al. 2018). In addition, since the compound SB203580 had both in vitro and in vivo effects in the inhibition of TNF-induced expansion of Tregs, thus p38-MAPK signaling pathways may represent a major component of the signaling pathway of TNFR2 in mediating the Treg-stimulatory effect of TNF (He et al. 2018). Recently, it was reported that the TNFR superfamily (TNFRSF, including TNFR2, 4-1BB, GITR, and DR3, but not OX40) shared common signal transduction pathways to increase proliferation and survival of Tregs and canonical NF-κB played a major role (Lubrano di Ricco et al. 2020). Our study found a key role of IKKα in maintaining homeostasis of Tregs pool as well as Tregs expansion in the inflammatory environment, suggesting that the activation of noncanonical NF-κB attributes to Treg-boosting effect of TNF/TNFR2 stimulation (Chen et al. 2015). These studies shed some light on the molecular mechanisms and signaling pathways required for TNFR2 in the activation of Tregs (Chen and Plebanski 2019); however, further investigation is needed for a more thorough understanding.

13.4 Activation of Tregs Through TNFR2 in the Treatment of Inflammatory Diseases

To date, five anti-TNF drugs, namely infliximab, adalimumab, certolizumab pegol, golimumab, and etanercept, have been approved in the treatment of human autoimmune disorders such as rheumatoid arthritis (RA), psoriasis, inflammatory bowel disease (IBD), and ankylosing spondylitis (AS) (Monaco et al. 2015). As a mainstay treatment, anti-TNF therapy has revolutionized the clinical management of inflammatory diseases (Papamichael et al. 2019; Li et al. 2017; Willrich et al. 2015). Although effective in most patients, a considerable fraction of patients do not respond to the initial treatment or loss responsiveness over time (Roda et al. 2016). Furthermore, anti-TNF therapy can also result in some severe adverse effects including opportunistic infection, reactivation or development of de novo autoimmune diseases, lymphoma, and even malignancies (Brown et al. 2002; Shakoor et al. 2002; Sicotte and Voskuhl 2001; Slifman et al. 2003). Our findings regarding the Treg-boosting effect of TNF–TNFR2 have provided a new framework to analyze the non-responsiveness and side effects of anti-TNF therapy (Chen and Oppenheim 2011a). The role of TNF and its receptors in the regulation of Tregs function in different clinical contexts was recently summarized and reviewed (Chen and Plebanski 2019; Yang et al. 2018). Theoretically, the blockade of the TNF–TNFR1 pathway without interruption of the TNF–TNFR2 pathway might overcome, at least partially, the non-responsiveness or adverse effects of anti-TNF biologics used in the treatment of inflammatory diseases.

Tregs can be used as a cellular therapy to prevent or treat diverse clinical conditions in transplantation, allergy, infectious disease, GvHD, neurodegenerative disease, and autoimmunity (McMurchy et al. 2011; Su et al. 2012; Gu et al. 2014; Xu et al. 2016). However, the clinical application of Treg cell-based therapy in autoimmunity is in dilemma due to the difficulty in obtaining and safely applying sufficient quantities of functional Tregs, whether generated ex vivo or stimulated in vivo. Previous strategies to expand the Treg cells population include the stimulation with IL-2, anti-CD3, anti-CD28, TL_1A-Ig, etc. (Faustman and Davis 2010; Faustman 2018). Tregs are known to express high levels of TNFR2, and TNF–TNFR2 signaling axis stimulates the activation and expansion of Tregs and increases their Foxp3 expression (Chen et al. 2007, 2008, 2010a; Annunziato et al. 2002). Therefore, TNFR2 agonist may be therapeutically harnessed to induce the proliferation and suppressive function of Tregs. Okubo and colleagues reported that "TNFR2 agonists" (potent agonistic TNFR2 monoclonal antibodies (mAb)) had the capacity to stimulate the expansion of Foxp3$^+$ Tregs present in cultures of CD4 cells, accompanied by the upregulation of CHUK (conserved helix-loop-helix ubiquitous kinase), TRAF2, TRAF3, BIRC3 (cIAP2), and Foxp3 mRNA expression (Okubo et al. 2013).

Type 1 diabetes (T1D) is an autoimmune disease in which Tregs activation is defected through an increase in the resting Tregs (rTregs; CD4$^+$Foxp3$^+$CD45RO$^-$RA$^+$) and a decrease in the activated Tregs (aTregs; CD4$^+$Foxp3$^+$CD45RO$^+$RA$^-$) (Okubo et al. 2016; Faustman et al. 1989). When rTregs are converted to aTregs, they express higher levels of TNFR2 (Chen et al. 2008, 2010a, b, 2013; Okubo et al. 2013). Thus, TNFR2 agonist has the potential to combat T1D (Faustman 2018; Okubo et al. 2016; Ban et al. 2008). More recently, the effect of "TNFR2 agonists" on the activation and expansion of Tregs isolated from patients with type 1 diabetes (T1D) was studied. In comparison with control, TNFR2 agonist dose-dependently expanded rTregs to potent aTregs (Okubo et al. 2016). This suggests that the in vivo treatment with TNFR2 agonists had the capacity to correct the defects of Tregs.

Activation and expansion of Tregs through the TNF–TNFR2 signaling axis could decrease the clinical GvHD score and GvHD-related mortality (Cohen and Wood 2018). It is also possible that TNFR2 agonists can induce TNFR2$^+$ Tregs since one study has demonstrated that TNFR2$^+$ Tregs had a superior capacity to inhibit immune response in related TNFR2$^-$ Treg cells (Horwitz et al. 2013). Chopra and colleagues constructed a nonameric TNFR2-specific variant of mouse TNF (STAR2), which had both in vitro and in vivo activity to stimulate the proliferation of Tregs via TNFR2-dependent and IL-2-independent manner. In a mouse model of allogenic hematopoietic stem cell transplantation, treatment of irradiated recipient mice with STAR2 markedly prolonged the survival and reduced GvHD severity (Chopra et al. 2016). Importantly, STAR2 had no harmful effect on the ability of donor T cells present in the transplant to mediate graft versus tumor effect or to eliminate pathogens (Chopra et al. 2016).

The neuroprotective role of TNFR2 signaling has been reported in several studies (Fischer et al. 2014; Maier et al. 2013; Patel et al. 2012). Fischer and colleagues developed a number of TNFR2 agonistic and TNFR1 antagonistic biologics, such as TNC-scTNFR$_2$, EHD2-scTNFR$_2$, and ATROSAB, and examined their protective effect on autoimmune neurological disorder such as multiple sclerosis (MS) and neuropathic pain (Fischer et al. 2011, 2017, 2019a, b; Williams et al. 2018). They found that TNC-scTNFR2 could selectively mimic mTNF and stimulate TNFR2 activation, consequently protect the death of neuronal cells from oxidative stress (Fischer et al. 2011). EHD2-scTNF$_{R2}$ (TNFR2 agonist) treatment improved memory and cell viability, and a reduction in the loss of cholinergic fibers and inflammation in a mouse model of NMDA-induced acute neurodegeneration (Dong et al. 2016). In a more recent study, they found that the activation of TNFR2 using EHD2-scTNFR2 in mice with chronic neuropathic pain promoted long-lasting pain recovery via Treg-dependent responses (Fischer et al. 2019a). Moreover, systemic EHD2-scTNFR2 treatment promoted Tregs expansion and stimulated the proliferation and differentiation of oligodendrocyte progenitor cells in experimental autoimmune encephalomyelitis (EAE), a mouse model of MS (Fischer et al. 2019b). It was shown recently that TROS, a nanobody-based selective inhibitor of TNFR1, was able to inhibit mouse experimental autoimmune encephalomyelitis and this effect is attributable to the diversion of TNF to interact with TNFR2 (Steeland et al. 2017). Thus, a considerable number of studies have shown the efficacy of TNFR2 agonistic agents in the treatment of several autoimmune and inflammatory diseases. Hopefully, this TNFR2-targeting treatment would be more effective and safer than globally blocking TNF activity, and further investigation is warranted.

13.5 Elimination of Tregs Activity Through Blockade of TNFR2 Signal in Cancer Immunotherapy

Tregs play a major role in the establishment and maintenance of immunosuppressive tumor microenvironments (Wing et al. 2019; Togashi et al. 2019). A large number of Tregs accumulate in the tumor microenvironment, and they potently dampen naturally occurring and therapeutically induced anti-tumor immune responses (Nishikawa and Sakaguchi 2010). Therefore, tumor-infiltrating Tregs represent a major cellular mechanism of tumor immune evasion. In addition, myeloid-derived suppressor cells (MDSCs) and mesenchymal stem cells (MSCs) operate collaboratively with Tregs in promotion of tumor immune evasion. Thus, the elimination of tumor-infiltrating Tregs, MDSCs, and MSCs or suppression of their function have become a strategy to enhance anti-tumor immune responses (Munn et al. 2018; Liu et al. 2016). We (Xin Chen and Joost J Oppenheim) have reported that tumor-infiltrating Tregs express high levels of TNFR2 and are potent immunosuppressive cells in the tumor microenvironment (Chen et al. 2008). These TNFR2-expressing Tregs are abundantly present in different tumor cell types including colon cancer, multiple myeloma, renal cell carcinoma, Hodgkin's lymphoma and cutaneous non-Hodgkin's lymphoma, and ovarian cancer (Uhlén et al. 2005; Hamilton et al. 2011; Rauert et al. 2011; Wang and Al-Lamki 2013; Govindaraj et al. 2013; Nakayama et al. 2014; Ungewickell et al. 2015; Sheng et al. 2018). A recent study on 788 commercially available human cancer cell lines from diverse cancer tissues examined the expression of TNFR2 in the tumor microenvironment. They showed that TNFR2 was overexpressed on both human tumor cells and human tumor-infiltrating Tregs, but negligibly expressed on beneficial Teffs. The expression pattern was widespread and variable, the greatest expression was on hematopoietic and lymphoid cell lines, as well as on lung and breast tumor lines (Yang et al. 2020). Moreover, it has been reported that the immunosuppressive function of MDSCs and MSCs are dependent on the TNF–TNFR2 signaling axis (Zhao et al. 2012; Kelly et al. 2010). MSCs treated with TNF in vitro also enhanced the suppressive function (Su et al. 2015). Consequently, therapeutically targeting TNFR2 with antagonistic agents represents a strategy to eliminate or suppress the functional activities of these immunosuppressive cells in tumor environment (Vanamee and Faustman 2017; Yang et al. 2020; He et al. 2019; Al-Hatamleh et al. 2019a, b).

Faustman and her colleagues developed TNFR2 antagonistic antibodies which can inhibit Tregs proliferation, reduced sTNFR2 secretion from normal cells, and enabled T effector cell (Teff) expansion (Torrey et al. 2017). The antibodies also inhibit the activation of NF-κB pathways and gene expression associated with TNFR2 signaling in Tregs. Furthermore, the TNFR2 antibodies killed Tregs isolated from metastatic sites (ascites) in ovarian cancer patients more potently than it killed Tregs from healthy donor samples. Additionally, antagonistic antibodies also killed OVCAR3 ovarian cancer cells (Torrey et al. 2017). In a recent in vitro study by this team, they examined TNFR2 antagonists on freshly isolated lymphocytes from patients (Stage IVA) with Sézary syndrome (cutaneous T-cell lymphoma) and healthy controls. TNFR2 antagonistic antibody killed TNFR2$^+$ Sézary syndrome (SS) tumor cells and this gives a free rein to a rapid expansion of Teff cells to correct Treg/Teff ratios like normal healthy controls (Torrey et al. 2019). In another study, they have investigated TNFR2 antibody antagonist for in vitro killing function using SW480 colon cancer cell line (low TNFR2 expression), MOTN-1 leukemia cell line, and JEKO-1 lymphoma cell line (high TNFR2 expression). TNFR2 antagonistic antibody induced the death of SW480 cell line after 7-day treatment and induced the death of MOTN-1 and JEKO-1 within 48–72 h. In the tumor environment, the antagonist antibody selectively induced the death of TNFR2-expressing tumor cells and Tregs but sparing Teffs (and promoted the proliferation of Teff cells) (Yang et al. 2020). We showed that the administration of TNFR2-neutralizing antibody and a toll-like receptor 9 agonists (CpG oligodeoxynucleotide) has the potential to improve the therapeutic effect against breast cancer and colon cancer cells in mice. These combination therapy markedly decreased the number of TNFR2$^+$ Treg cells and increased the number of interferon-γ-positive (IFN-γ$^+$) CD8$^+$ cytotoxic T lymphocytes infiltrating the tumor, resulting in a long-term tumor-free survival in the mouse cohort (Nie et al. 2018). The Faustman's group optimized the TNFR2 antagonist antibody by shuffling the isoform and stabilizing the hinge that resulted in the improvement of activity and safety (Yang et al. 2020).

It has been shown that depletion of TNFR2$^+$ Tregs was associated with the antitumor effect of therapeutics. For example, tumor eradication after cyclophosphamide depended on concurrent depletion of TNFR2$^+$ Tregs in a mouse tumor model (van der Most et al. 2009). Treating acute myeloid leukemia patients with azacytidine and panobinostat effectively eliminated TNFR2$^+$ Treg cells in peripheral blood and bone marrow (Govindaraj et al. 2014a). These beneficial clinical responses came from more active Teff as determined from increased production of interferon-γ and IL-2. In another study on acute myeloid leukemia patients, they found that combination treatment with azacytidine and lenalidomide decreased TNFR2 expression and activity in Treg cells, and this action improved the clinical outcomes (Govindaraj et al. 2014b).

Taken together, these preclinical and clinical data support the idea that TNFR2-expressing Tregs are tumor-associated suppressors, and they represent the major cellular mechanism in tumor immune evasion. Elimination of Tregs activity by blockade of TNFR2 may markedly enhance the efficacy of cancer immunotherapy. TNFR2-targeting treatment may also directly act on tumor cells. These dual beneficial effects of TNFR2 antagonism in human patients should be further studied.

13.6 Conclusion

Current experimental evidence further supports and substantiates the decisive role of TNF–TNFR2 interactions in the activation, expansion, in vivo function, and phenotypic stability of Tregs. Both protective and detrimental effects of the TNF–TNFR2 signaling in Tregs, by maintaining immune homeostasis and dampening autoimmune or inflammatory responses or by promoting tumor immune evasion, have been intensively studied. The molecular basis and signaling pathway of TNFR2 required for Tregs activation were also investigated. However, the exact mechanism, as well as the key signaling events remains to be further clarified, and such study may lead to the identification of a novel molecular target for the discovery of drugs with the capacity to modulate Tregs activity. The therapeutic potential of TNFR2-targeting pharmacological agents in the treatment of inflammatory diseases and cancer have been revealed. Further study should focus on the translation of research findings into the therapeutic benefits of human patients. Hopefully, in the near future, TNFR2-targeting therapy can be proved to be useful in the treatment of major human diseases such as autoimmune disorders, allergy, allograft rejection, GVHD, and cancer.

Acknowledgments This project was funded by The Science and Technology Development Fund, Macau (File no. 201/2017/A3 and 0056/2019/AFJ) and The University of Macau (File no. MYRG2016-00023-ICMS-QRCM, MYRG2017-00120-ICMS and MYRG2019-00169-ICMS).

References

Aggarwal BB (1991) Structure of tumor necrosis factor and its receptor. Biotherapy 3(2):113–120. https://doi.org/10.1007/BF02172083

Al-Hatamleh MAI, E A R ENS, Boer JC, Ferji K, Six J-L, Chen X et al (2019a) Synergistic effects of nanomedicine targeting TNFR2 and DNA demethylation inhibitor-an opportunity for cancer treatment. Cell 9(1):33. https://doi.org/10.3390/cells9010033

Al-Hatamleh MAI, Ahmad S, Boer JC, Lim J, Chen X, Plebanski M et al (2019b) A perspective review on the role of nanomedicine in the modulation of TNF-TNFR2 axis in breast cancer immunotherapy. J Oncol 2019:6313242. https://doi.org/10.1155/2019/6313242

Annunziato F, Cosmi L, Liotta F, Lazzeri E, Manetti R, Vanini V et al (2002) Phenotype, localization, and mechanism of suppression of CD4+CD25+ human thymocytes. J Exp Med 196(3):379–387. https://doi.org/10.1084/jem.20020110

Bach E, Nielsen RR, Vendelbo MH, Møller AB, Jessen N, Buhl M et al (2013) Direct effects of TNF-α on local fuel metabolism and cytokine levels in the placebo-controlled, bilaterally infused human leg: increased insulin sensitivity, increased net protein breakdown, and increased IL-6 release. Diabetes 62(12):4023–4029. https://doi.org/10.2337/db13-0138

Ban L, Zhang J, Wang L, Kuhtreiber W, Burger D, Faustman DL (2008) Selective death of autoreactive T cells in human diabetes by TNF or TNF receptor 2 agonism. Proc Natl Acad Sci U S A 105(36):13644–13649. https://doi.org/10.1073/pnas.0803429105

Battegay EJ, Raines EW, Colbert T, Ross R (1995) TNF-alpha stimulation of fibroblast proliferation. Dependence on platelet-derived growth factor (PDGF) secretion and alteration of PDGF receptor expression. J Immunol 154(11):6040–6047

Bennett CL, Christie J, Ramsdell F, Brunkow ME, Ferguson PJ, Whitesell L et al (2001) The immune dysregulation, polyendocrinopathy, enteropathy, X-linked syndrome (IPEX) is caused by mutations of FOXP3. Nat Genet 27(1):20–21. https://doi.org/10.1038/83713

Brown SL, Greene MH, Gershon SK, Edwards ET, Braun MM (2002) Tumor necrosis factor antagonist therapy and lymphoma development: twenty-six cases reported to the Food and Drug Administration. Arthritis Rheum 46(12):3151–3158. https://doi.org/10.1002/art.10679

Cabal-Hierro L, Lazo PS (2012) Signal transduction by tumor necrosis factor receptors. Cell Signal 24(6):1297–1305. https://doi.org/10.1016/j.cellsig.2012.02.006

Cawthorn WP, Sethi JK (2008) TNF-alpha and adipocyte biology. FEBS Lett 582(1):117–131. https://doi.org/10.1016/j.febslet.2007.11.051

Chen X, Oppenheim JJ (2010) TNF-alpha: an activator of CD4+FoxP3+TNFR2+ regulatory T cells. Curr Dir Autoimmun 11:119–134. https://doi.org/10.1159/000289201

Chen X, Oppenheim JJ (2011a) Contrasting effects of TNF and anti-TNF on the activation of effector T cells and regulatory T cells in autoimmunity. FEBS Lett 585(23):3611–3618. https://doi.org/10.1016/j.febslet.2011.04.025

Chen X, Oppenheim JJ (2011b) The phenotypic and functional consequences of tumour necrosis factor receptor type 2 expression on CD4(+) FoxP3(+) regulatory T cells. Immunology 133(4):426–433. https://doi.org/10.1111/j.1365-2567.2011.03460.x

Chen X, Oppenheim JJ (2011c) Resolving the identity myth: key markers of functional CD4+FoxP3+ regulatory T cells. Int Immunopharmacol 11(10):1489–1496. https://doi.org/10.1016/j.intimp.2011.05.018

Chen X, Oppenheim JJ (2014) Th17 cells and Tregs: unlikely allies. J Leukoc Biol 95(5):723–731. https://doi.org/10.1189/jlb.1213633

Chen X, Oppenheim JJ (2016) Therapy: paradoxical effects of targeting TNF signalling in the treatment of autoimmunity. Nat Rev Rheumatol 12(11):625–626. https://doi.org/10.1038/nrrheum.2016.145

Chen X, Oppenheim JJ (2017) Targeting TNFR2, an immune checkpoint stimulator and oncoprotein, is a promising treatment for cancer. Sci Signal 10(462):1–3. https://doi.org/10.1126/scisignal.aal2328

Chen X, Plebanski M (2019) The role of TNF-TNFR2 signal in immunosuppressive cells and its therapeutic

implications. Front Immunol 10:2126. https://doi.org/10.3389/fimmu.2019.02126
Chen X, Bäumel M, Männel DN, Howard OZ, Oppenheim JJ (2007) Interaction of TNF with TNF receptor type 2 promotes expansion and function of mouse CD4+ CD25+ T regulatory cells. J Immunol 179(1):154–161. https://doi.org/10.4049/jimmunol.179.1.154
Chen X, Subleski JJ, Kopf H, Howard OMZ, Männel DN, Oppenheim JJ (2008) Cutting edge: expression of TNFR2 defines a unique subset of mouse CD4 +CD25+ T regulatory cells with maximal suppressive functions: applicability to tumor infiltrating T regulatory cells. J Immunol 180(10):6467–6471. https://doi.org/10.4049/jimmunol.180.10.6467
Chen X, Xun K, Chen L, Wang Y (2009) TNF-α, a potent lipid metabolism regulator. Cell Biochem Funct 27 (7):407–416. https://doi.org/10.1002/cbf.1596
Chen X, Subleski JJ, Hamano R, Howard OMZ, Wiltrout RH, Oppenheim JJ (2010a) Co-expression of TNFR2 and CD25 identifies more of the functional CD4 +FOXP3+ regulatory T cells in human peripheral blood. Eur J Immunol 40(4):1099–1106. https://doi.org/10.1002/eji.200940022
Chen X, Hamano R, Subleski JJ, Hurwitz AA, Howard OM, Oppenheim JJ (2010b) Expression of costimulatory TNFR2 induces resistance of CD4 +FoxP3- conventional T cells to suppression by CD4 +FoxP3+ regulatory T cells. J Immunol 185 (1):174–182. https://doi.org/10.4049/jimmunol.0903548
Chen X, Wu X, Zhou Q, Howard OM, Netea MG, Oppenheim JJ (2013) TNFR2 is critical for the stabilization of the CD4+Foxp3+ regulatory T. cell phenotype in the inflammatory environment. J Immunol 190 (3):1076–1084. https://doi.org/10.4049/jimmunol.1202659
Chen X, Willette-Brown J, Wu X, Hu Y, Howard OM, Oppenheim JJ (2015) IKKα is required for the homeostasis of regulatory T cells and for the expansion of both regulatory and effector CD4 T cells. FASEB J 29 (2):443–454. https://doi.org/10.1096/fj.14-259564.
Chen X, Nie Y, Xiao H, Bian Z, Scarzello AJ, Song NY et al (2016) TNFR2 expression by CD4 effector T cells is required to induce full-fledged experimental colitis. Sci Rep 6:32834. https://doi.org/10.1038/srep32834
Cho Y, Challa S, Moquin D, Genga R, Ray TD, Guildford M et al (2009) Phosphorylation-driven assembly of the RIP1-RIP3 complex regulates programmed necrosis and virus-induced inflammation. Cell 137 (6):1112–1123. https://doi.org/10.1016/j.cell.2009.05.037
Chopra M, Biehl M, Steinfatt T, Brandl A, Kums J, Amich J et al (2016) Exogenous TNFR2 activation protects from acute GvHD via host T reg cell expansion. J Exp Med 213(9):1881–1900. https://doi.org/10.1084/jem.20151563
Cohen JL, Wood KJ (2018) TNFR2: the new Treg switch? OncoImmunology 7(1):e1373236. https://doi.org/10.1080/2162402X.2017.1373236
Dong Y, Fischer R, Naude PJ, Maier O, Nyakas C, Duffey M et al (2016) Essential protective role of tumor necrosis factor receptor 2 in neurodegeneration. Proc Natl Acad Sci U S A 113(43):12304–12309. https://doi.org/10.1073/pnas.1605195113
Drabarek B, Dymkowska D, Szczepanowska J, Zablocki K (2012) TNFα affects energy metabolism and stimulates biogenesis of mitochondria in EA.hy926 endothelial cells. Int J Biochem Cell Biol 44 (9):1390–1397. https://doi.org/10.1016/j.biocel.2012.05.022
Faustman DL (2018) TNF, TNF inducers, and TNFR2 agonists: a new path to type 1 diabetes treatment. Diabetes Metab Res Rev 34(1):e2941. https://doi.org/10.1002/dmrr.2941
Faustman D, Davis M (2010) TNF receptor 2 pathway: drug target for autoimmune diseases. Nat Rev Drug Discov 9(6):482–493. https://doi.org/10.1038/nrd3030
Faustman D, Eisenbarth G, Daley J, Breitmeyer J (1989) Abnormal T-lymphocyte subsets in type I diabetes. Diabetes 38(11):1462–1468. https://doi.org/10.2337/diab.38.11.1462
Fischer R, Maier O, Siegemund M, Wajant H, Scheurich P, Pfizenmaier K (2011) A TNF receptor 2 selective agonist rescues human neurons from oxidative stress-induced cell death. PLoS One 6(11):e27621. https://doi.org/10.1371/journal.pone.0027621
Fischer R, Wajant H, Kontermann R, Pfizenmaier K, Maier O (2014) Astrocyte-specific activation of TNFR2 promotes oligodendrocyte maturation by secretion of leukemia inhibitory factor. Glia 62 (2):272–283. https://doi.org/10.1002/glia.22605
Fischer R, Marsal J, Guttà C, Eisler SA, Peters N, Bethea JR et al (2017) Novel strategies to mimic transmembrane tumor necrosis factor-dependent activation of tumor necrosis factor receptor 2. Sci Rep 7(1):6607. https://doi.org/10.1038/s41598-017-06993-4
Fischer R, Sendetski M, Del Rivero T, Martinez GF, Bracchi-Ricard V, Swanson KA et al (2019a) TNFR2 promotes Treg-mediated recovery from neuropathic pain across sexes. Proc Natl Acad Sci U S A 116 (34):17045–17050. https://doi.org/10.1073/pnas.1902091116
Fischer R, Padutsch T, Bracchi-Ricard V, Murphy KL, Martinez GF, Delguercio N et al (2019b) Exogenous activation of tumor necrosis factor receptor 2 promotes recovery from sensory and motor disease in a model of multiple sclerosis. Brain Behav Immun 81:247–259. https://doi.org/10.1016/j.bbi.2019.06.021
Fontenot JD, Gavin MA, Rudensky AY (2003) Foxp3 programs the development and function of CD4 +CD25+ regulatory T cells. Nat Immunol 4 (4):330–336. https://doi.org/10.1038/ni904
Friedmann E, Hauben E, Maylandt K, Schleeger S, Vreugde S, Lichtenthaler SF et al (2006) SPPL2a and SPPL2b promote intramembrane proteolysis of TNFα in activated dendritic cells to trigger IL-12 production. Nat Cell Biol 8(8):843–848. https://doi.org/10.1038/ncb1440

Gavin MA, Rasmussen JP, Fontenot JD, Vasta V, Manganiello VC, Beavo JA et al (2007) Foxp3-dependent programme of regulatory T-cell differentiation. Nature 445(7129):771–775. https://doi.org/10.1038/nature05543

Gehr G, Gentz R, Brockhaus M, Loetscher H, Lesslauer W (1992) Both tumor necrosis factor receptor types mediate proliferative signals in human mononuclear cell activation. J Immunol 149(3):911–917

Govindaraj C, Scalzo-Inguanti K, Madondo M, Hallo J, Flanagan K, Quinn M et al (2013) Impaired Th1 immunity in ovarian cancer patients is mediated by TNFR2+ Tregs within the tumor microenvironment. Clin Immunol 149(1):97–110. https://doi.org/10.1016/j.clim.2013.07.003

Govindaraj C, Tan P, Walker P, Wei A, Spencer A, Plebanski M (2014a) Reducing TNF receptor 2+ regulatory T cells via the combined action of azacitidine and the HDAC inhibitor, panobinostat for clinical benefit in acute myeloid leukemia patients. Clin Cancer Res 20(3):724–735. https://doi.org/10.1158/1078-0432.CCR-13-1576

Govindaraj C, Madondo M, Kong YY, Tan P, Wei A, Plebanski M (2014b) Lenalidomide-based maintenance therapy reduces TNF receptor 2 on CD4 T cells and enhances immune effector function in acute myeloid leukemia patients. Am J Hematol 89(8):795–802. https://doi.org/10.1002/ajh.23746

Gregory AP, Dendrou CA, Attfield KE, Haghikia A, Xifara DK, Butter F et al (2012) TNF receptor 1 genetic risk mirrors outcome of anti-TNF therapy in multiple sclerosis. Nature 488(7412):508–511. https://doi.org/10.1038/nature11307

Grell M, Douni E, Wajant H, Löhden M, Clauss M, Maxeiner B et al (1995) The transmembrane form of tumor necrosis factor is the prime activating ligand of the 80 kDa tumor necrosis factor receptor. Cell 83(5):793–802. https://doi.org/10.1016/0092-8674(95)90192-2

Grell M, Wajant H, Zimmermann G, Scheurich P (1998) The type 1 receptor (CD120a) is the high-affinity receptor for soluble tumor necrosis factor. Proc Natl Acad Sci U S A 95(2):570–575. https://doi.org/10.1073/pnas.95.2.570

Grinberg-Bleyer Y, Saadoun D, Baeyens A, Billiard F, Goldstein JD, Gregoire S et al (2010) Pathogenic T cells have a paradoxical protective effect in murine autoimmune diabetes by boosting Tregs. J Clin Invest 120(12):4558–4568. https://doi.org/10.1172/jci42945

Gu J, Lu L, Chen M, Xu L, Lan Q, Li Q et al (2014) TGF-beta-induced CD4+Foxp3+ T cells attenuate acute graft-versus-host disease by suppressing expansion and killing of effector CD8+ cells. J Immunol 193(7):3388–3397. https://doi.org/10.4049/jimmunol.1400207

Haas TL, Emmerich CH, Gerlach B, Schmukle AC, Cordier SM, Rieser E et al (2009) Recruitment of the linear ubiquitin chain assembly complex stabilizes the TNF-R1 signaling complex and is required for TNF-mediated gene induction. Mol Cell 36(5):831–844. https://doi.org/10.1016/j.molcel.2009.10.013

Hamano R, Huang J, Yoshimura T, Oppenheim JJ, Chen X (2011) TNF optimally activatives regulatory T cells by inducing TNF receptor superfamily members TNFR2, 4-1BB and OX40. Eur J Immunol 41(7):2010–2020. https://doi.org/10.1002/eji.201041205

Hamano R, Wu X, Wang Y, Oppenheim JJ, Chen X (2015) Characterization of MT-2 cells as a human regulatory T cell-like cell line. Cell Mol Immunol 12(6):780–782. https://doi.org/10.1038/cmi.2014.123

Hamilton KE, Simmons JG, Ding S, Van Landeghem L, Lund PK (2011) Cytokine induction of tumor necrosis factor receptor 2 is mediated by STAT3 in colon cancer cells. Mol Cancer Res 9(12):1718–1731. https://doi.org/10.1158/1541-7786

He T, Liu S, Chen S, Ye J, Wu X, Bian Z et al (2018) The p38 MAPK inhibitor SB203580 abrogates tumor necrosis factor-induced proliferative expansion of mouse CD4(+)Foxp3(+) regulatory T cells. Front Immunol 9:1556. https://doi.org/10.3389/fimmu.2018.01556

He J, Li R, Chen Y, Hu Y, Chen X (2019) TNFR2-expressing CD4+Foxp3+ regulatory T cells in cancer immunology and immunotherapy. In: Teplow DB (ed) Progress in molecular biology and translational science. Academic Press, Cambridge, pp 101–117

Hitomi J, Christofferson DE, Ng A, Yao J, Degterev A, Xavier RJ et al (2008) Identification of a molecular signaling network that regulates a cellular necrotic cell death pathway. Cell 135(7):1311–1323. https://doi.org/10.1016/j.cell.2008.10.044

Hori S, Nomura T, Sakaguchi S (2003) Control of regulatory T cell development by the transcription factor Foxp3. Science 299(5609):1057–1061. https://doi.org/10.1126/science.1079490

Horiuchi T, Mitoma H, Harashima S-i, Tsukamoto H, Shimoda T (2010) Transmembrane TNF-alpha: structure, function and interaction with anti-TNF agents. Rheumatology (Oxford) 49(7):1215–1228. https://doi.org/10.1093/rheumatology/keq031

Horwitz DA, Zheng SG, Gray JD (2003) The role of the combination of IL-2 and TGF-beta or IL-10 in the generation and function of CD4+ CD25+ and CD8+ regulatory T cell subsets. J Leukoc Biol 74(4):471–478. https://doi.org/10.1189/jlb.0503228

Horwitz DA, Pan S, Ou JN, Wang J, Chen M, Gray JD et al (2013) Therapeutic polyclonal human CD8+ CD25+ Fox3+ TNFR2+ PD-L1+ regulatory cells induced ex-vivo. Clin Immunol 149(3):450–463. https://doi.org/10.1016/j.clim.2013.08.007.

Hsu H, Huang J, Shu HB, Baichwal V, Goeddel DV (1996) TNF-dependent recruitment of the protein kinase RIP to the TNF receptor-1 signaling complex. Immunity 4(4):387–396. https://doi.org/10.1016/S1074-7613(00)80252-6

Idriss HT, Naismith JH (2000) TNFα and the TNF receptor superfamily: structure-function relationship(s).

Microsc Res Tech 50(3):184–195. https://doi.org/10.1002/1097-0029(20000801)50:3<184::AID-JEMT2>3.0.CO;2-H

Jacob N, Yang H, Pricop L, Liu Y, Gao X, Zheng SG et al (2009) Accelerated pathological and clinical nephritis in systemic lupus erythematosus-prone New Zealand mixed 2328 mice doubly deficient in TNF receptor 1 and TNF receptor 2 via a Th17-associated pathway. J Immunol 182(4):2532–2541. https://doi.org/10.4049/jimmunol.0802948

Kelly ML, Wang M, Crisostomo PR, Abarbanell AM, Herrmann JL, Weil BR et al (2010) TNF receptor 2, not TNF receptor 1 enhances mesenchymal stem cell-mediated cardiac protection following acute ischemia. Shock 33(6):602. https://doi.org/10.1097/SHK.0b013e3181cc0913

Kong N, Lan Q, Su W, Chen M, Wang J, Yang Z et al (2012) Induced T regulatory cells suppress osteoclastogenesis and bone erosion in collagen-induced arthritis better than natural T regulatory cells. Ann Rheum Dis 71(9):1567–1572. https://doi.org/10.1136/annrheumdis-2011-201052

Kriegler M, Perez C, DeFay K, Albert I, Lu SD (1988) A novel form of TNF/cachectin is a cell surface cytotoxic transmembrane protein: ramifications for the complex physiology of TNF. Cell 53(1):45–53. https://doi.org/10.1016/0092-8674(88)90486-2

Krippner-Heidenreich A, Tubing F, Bryde S, Willi S, Zimmermann G, Scheurich P (2002) Control of receptor-induced signaling complex formation by the kinetics of ligand/receptor interaction. J Biol Chem 277(46):44155–44163. https://doi.org/10.1074/jbc.M207399200

Li P, Zheng Y, Chen X (2017) Drugs for autoimmune inflammatory diseases: from small molecule compounds to anti-TNF biologics. Front Pharmacol 8:460. https://doi.org/10.3389/fphar.2017.00460

Liu C, Workman CJ, Vignali DA (2016) Targeting regulatory T cells in tumors. FEBS J 283(14):2731–2748. https://doi.org/10.1111/febs.13656

Locksley RM, Killeen N, Lenardo MJ (2001) The TNF and TNF receptor superfamilies: integrating mammalian biology. Cell 104(4):487–501. https://doi.org/10.1016/S0092-8674(01)00237-9.

Lu L, Lan Q, Li Z, Zhou X, Gu J, Li Q et al (2014) Critical role of all-trans retinoic acid in stabilizing human natural regulatory T cells under inflammatory conditions. Proc Natl Acad Sci U S A 111(33):E3432–E3440. https://doi.org/10.1073/pnas.1408780111

Lubrano di Ricco M, Ronin E, Collares D, Divoux J, Gregoire S, Wajant H et al (2020) Tumor necrosis factor receptor family costimulation increases regulatory T-cell activation and function via NF-kappaB. Eur J Immunol. https://doi.org/10.1002/eji.201948393

Mahmud SA, Manlove LS, Schmitz HM, Xing Y, Wang Y, Owen DL et al (2014) Costimulation via the tumor-necrosis factor receptor superfamily couples TCR signal strength to the thymic differentiation of regulatory T cells. Nat Immunol 15(5):473–481. https://doi.org/10.1038/ni.2849

Mahoney DJ, Cheung HH, Lejmi Mrad R, Plenchette S, Simard C, Enwere E et al (2008) Both cIAP1 and cIAP2 regulate TNFα-mediated NF-κB activation. Proc Natl Acad Sci U S A 105(33):11778–11783. https://doi.org/10.1073/pnas.0711122105

Maier O, Fischer R, Agresti C, Pfizenmaier K (2013) TNF receptor 2 protects oligodendrocyte progenitor cells against oxidative stress. Biochem Biophys Res Commun 440(2):336–341. https://doi.org/10.1016/j.bbrc.2013.09.083

McMurchy AN, Bushell A, Levings MK, Wood KJ (2011) Moving to tolerance: clinical application of T regulatory cells. Semin Immunol 23(4):304–313. https://doi.org/10.1016/j.smim.2011.04.001

Micheau O, Tschopp J (2003) Induction of TNF receptor I-mediated apoptosis via two sequential signaling complexes. Cell 114(2):181–190. https://doi.org/10.1016/S0092-8674(03)00521-X

Monaco C, Nanchahal J, Taylor P, Feldmann M (2015) Anti-TNF therapy: past, present and future. Int Immunol 27(1):55–62. https://doi.org/10.1093/intimm/dxu102

van der Most RG, Currie AJ, Mahendran S, Prosser A, Darabi A, Robinson BW et al (2009) Tumor eradication after cyclophosphamide depends on concurrent depletion of regulatory T cells: a role for cycling TNFR2-expressing effector-suppressor T cells in limiting effective chemotherapy. Cancer Immunol Immunother 58(8):1219–1228. https://doi.org/10.1007/s00262-008-0628-9

Mukai Y, Nakamura T, Yoshikawa M, Yoshioka Y, Tsunoda S-i, Nakagawa S et al (2010) Solution of the structure of the TNF-TNFR2 complex. Sci Signal 3(148):ra83-ra. https://doi.org/10.1126/scisignal.2000954

Munn DH, Sharma MD, Johnson TS (2018) Treg destabilization and reprogramming: implications for cancer immunotherapy. Cancer Res 78(18):5191–5199. https://doi.org/10.1158/0008-5472

Nakayama S, Yokote T, Tsuji M, Akioka T, Miyoshi T, Hirata Y et al (2014) Expression of tumour necrosis factor-α and its receptors in Hodgkin lymphoma. Br J Haematol 167(4):574–577. https://doi.org/10.1111/bjh.13015

Nie Y, He J, Shirota H, Trivett AL, Yang D, Klinman DM et al (2018) Blockade of TNFR2 signaling enhances the immunotherapeutic effect of CpG ODN in a mouse model of colon cancer. Sci Signal 11(511):eaan0790. https://doi.org/10.1126/scisignal.aan0790

Nishikawa H, Sakaguchi S (2010) Regulatory T cells in tumor immunity. Int J Cancer 127(4):759–767. https://doi.org/10.1002/ijc.25429.

Okubo Y, Mera T, Wang L, Faustman DL (2013) Homogeneous expansion of human T-regulatory cells via tumor necrosis factor receptor 2. Sci Rep 3:3153. https://doi.org/10.1038/srep03153

Okubo Y, Torrey H, Butterworth J, Zheng H, Faustman DL (2016) Treg activation defect in type 1 diabetes: correction with TNFR2 agonism. Clin Transl Immunol 5(1):e56. https://doi.org/10.1038/cti.2015.43

Panduro M, Benoist C, Mathis D (2016) Tissue Tregs. Annu Rev Immunol 34:609–633. https://doi.org/10.1146/annurev-immunol-032712-095948

Papamichael K, Vogelzang EH, Lambert J, Wolbink G, Cheifetz AS (2019) Therapeutic drug monitoring with biologic agents in immune mediated inflammatory diseases. Expert Rev Clin Immunol 15(8):837–848. https://doi.org/10.1080/1744666X.2019.1630273

Patel JR, Williams JL, Muccigrosso MM, Liu L, Sun T, Rubin JB et al (2012) Astrocyte TNFR2 is required for CXCL12-mediated regulation of oligodendrocyte progenitor proliferation and differentiation within the adult CNS. Acta Neuropathol 124(6):847–860. https://doi.org/10.1007/s00401-012-1034-0

PCM U, Xuehui H et al (2019) TNFα-signaling modulates the kinase activity of human effector Treg and regulates IL-17A expression. Front Immunol 10:3047. https://doi.org/10.3389/fimmu.2019.03047

Puimège L, Libert C, Van Hauwermeiren F (2014) Regulation and dysregulation of tumor necrosis factor receptor-1. Cytokine Growth Factor Rev 25 (3):285–300. https://doi.org/10.1016/j.cytogfr.2014.03.004

Rauert H, Stühmer T, Bargou R, Wajant H, Siegmund D (2011) TNFR1 and TNFR2 regulate the extrinsic apoptotic pathway in myeloma cells by multiple mechanisms. Cell Death Dis 2(8):e194-e. https://doi.org/10.1038/cddis.2011.78

Roda G, Jharap B, Neeraj N, Colombel JF (2016) Loss of response to anti-TNFs: definition, epidemiology, and management. Clin Transl Gastroenterol 7:e135. https://doi.org/10.1038/ctg.2015.63

Rothe M, Wong SC, Henzel WJ, Goeddel DV (1994) A novel family of putative signal transducers associated with the cytoplasmic domain of the 75 kDa tumor necrosis factor receptor. Cell 78(4):681–692. https://doi.org/10.1016/0092-8674(94)90532-0

Rothe M, Sarma V, Dixit V, Goeddel D (1995) TRAF2-mediated activation of NF-kappa B by TNF receptor 2 and CD40. Science 269(5229):1424–1427. https://doi.org/10.1126/science.7544915

Sade-Feldman M, Kanterman J, Ish-Shalom E, Elnekave M, Horwitz E, Baniyash M (2013) Tumor necrosis factor-alpha blocks differentiation and enhances suppressive activity of immature myeloid cells during chronic inflammation. Immunity 38 (3):541–554. https://doi.org/10.1016/j.immuni.2013.02.007

Sakaguchi S, Sakaguchi N, Asano M, Itoh M, Toda M (1995) Immunologic self-tolerance maintained by activated T cells expressing IL-2 receptor alpha-chains (CD25). Breakdown of a single mechanism of self-tolerance causes various autoimmune diseases. J Immunol 155(3):1151–1164

Sakaguchi S, Yamaguchi T, Nomura T, Ono M (2008) Regulatory T cells and immune tolerance. Cell 133 (5):775–787. https://doi.org/10.1016/j.cell.2008.05.009

Sakaguchi S, Miyara M, Costantino CM, Hafler DA (2010) FOXP3+ regulatory T cells in the human immune system. Nat Rev Immunol 10(7):490–500. https://doi.org/10.1038/nri2785

Schmidleithner L, Thabet Y, Schönfeld E, Köhne M, Sommer D, Abdullah Z et al (2019) Enzymatic activity of HPGD in Treg cells suppresses tconv cells to maintain adipose tissue homeostasis and prevent metabolic dysfunction. Immunity 50(5):1232–48.e14. https://doi.org/10.1016/j.immuni.2019.03.014

Schmidt A, Oberle N, Krammer PH (2012) Molecular mechanisms of treg-mediated T cell suppression. Front Immunol 3:51. https://doi.org/10.3389/fimmu.2012.00051

Sethi JK, Hotamisligil GS (eds) (1999) The role of TNFα in adipocyte metabolism. Seminars in cell & developmental biology, vol 10. Elsevier, Amsterdam, p 19

Shakoor N, Michalska M, Harris CA, Block JA (2002) Drug-induced systemic lupus erythematosus associated with etanercept therapy. Lancet 359 (9306):579–580. https://doi.org/10.1016/S0140-6736(02)07714-0

Sheng Y, Li F, Qin Z (2018) TNF receptor 2 makes tumor necrosis factor a friend of tumors. Front Immunol 9:1170. https://doi.org/10.3389/fimmu.2018.01170

Shevach EM (2018) Foxp3(+) T regulatory cells: still many unanswered questions-A perspective after 20 years of study. Front Immunol 9:1048. https://doi.org/10.3389/fimmu.2018.01048

Sicotte NL, Voskuhl RR (2001) Onset of multiple sclerosis associated with anti-TNF therapy. Neurology 57 (10):1885–1888. https://doi.org/10.1212/wnl.57.10.1885

Slifman NR, Gershon SK, Lee JH, Edwards ET, Braun MM (2003) Listeria monocytogenes infection as a complication of treatment with tumor necrosis factor alpha-neutralizing agents. Arthritis Rheum 48 (2):319–324. https://doi.org/10.1002/art.10758

Steeland S, Van Ryckeghem S, Van Imschoot G, De Rycke R, Toussaint W, Vanhoutte L et al (2017) TNFR1 inhibition with a Nanobody protects against EAE development in mice. Sci Rep 7(1):13646. https://doi.org/10.1038/s41598-017-13984-y

Su W, Fan H, Chen M, Wang J, Brand D, He X et al (2012) Induced CD4+ forkhead box protein-positive T cells inhibit mast cell function and established contact hypersensitivity through TGF-beta1. J Allergy Clin Immunol 130(2):444–52.e7. https://doi.org/10.1016/j.jaci.2012.05.011

Su W, Wan Q, Huang J, Han L, Chen X, Chen G et al (2015) Culture medium from TNF-alpha-stimulated mesenchymal stem cells attenuates allergic conjunctivitis through multiple antiallergic mechanisms. J Allergy Clin Immunol 136(2):423–32.e8. https://doi.org/10.1016/j.jaci.2014.12.1926

Taams LS, Palmer DB, Akbar AN, Robinson DS, Brown Z, Hawrylowicz CM (2006) Regulatory T cells in human disease and their potential for therapeutic manipulation. Immunology 118(1):1–9. https://doi.org/10.1111/j.1365-2567.2006.02348.x

Tai X, Cowan M, Feigenbaum L, Singer A (2005) CD28 costimulation of developing thymocytes induces Foxp3 expression and regulatory T cell differentiation independently of interleukin 2. Nat Immunol 6 (2):152–162. https://doi.org/10.1038/ni1160

Tartaglia LA, Ayres TM, Wong GH, Goeddel DV (1993) A novel domain within the 55 kd TNF receptor signals cell death. Cell 74(5):845–853. https://doi.org/10.1016/0092-8674(93)90464-2

Togashi Y, Nishikawa H (2017) Regulatory T cells: molecular and cellular basis for Immunoregulation. Curr Top Microbiol Immunol 410:3–27. https://doi.org/10.1007/82_2017_58

Togashi Y, Shitara K, Nishikawa H (2019) Regulatory T cells in cancer immunosuppression—implications for anticancer therapy. Nat Rev Clin Oncol 16 (6):356–371. https://doi.org/10.1038/s41571-019-0175-7

Torrey H, Butterworth J, Mera T, Okubo Y, Wang L, Baum D et al (2017) Targeting TNFR2 with antagonistic antibodies inhibits proliferation of ovarian cancer cells and tumor-associated Tregs. Sci Signal 10(462): eaaf8608. https://doi.org/10.1126/scisignal.aaf8608

Torrey H, Khodadoust M, Tran L, Baum D, Defusco A, Kim YH et al (2019) Targeted killing of TNFR2-expressing tumor cells and Tregs by TNFR2 antagonistic antibodies in advanced Sézary syndrome. Leukemia 33(5):1206–1218. https://doi.org/10.1038/s41375-018-0292-9

Tseng WY, Huang YS, Clanchy F, McNamee K, Perocheau D, Ogbechi J et al (2019) TNF receptor 2 signaling prevents DNA methylation at the Foxp3 promoter and prevents pathogenic conversion of regulatory T cells. Proc Natl Acad Sci U S A 116 (43):21666–21672. https://doi.org/10.1073/pnas.1909687116

Uhlén M, Björling E, Agaton C, Szigyarto CA-K, Amini B, Andersen E et al (2005) A human protein atlas for normal and cancer tissues based on antibody proteomics. Mol Cell Proteomics 4(12):1920–1932. https://doi.org/10.1074/mcp.M500279-MCP200

Ungewickell A, Bhaduri A, Rios E, Reuter J, Lee CS, Mah A et al (2015) Genomic analysis of mycosis fungoides and Sezary syndrome identifies recurrent alterations in TNFR2. Nat Genet 47(9):1056–1060. https://doi.org/10.1038/ng.3370

Urbano PCM, Koenen HJPM, Joosten I, He X (2018) An autocrine TNFα–tumor necrosis factor receptor 2 loop promotes epigenetic effects inducing human Treg stability in vitro. Front Immunol 9:573. https://doi.org/10.3389/fimmu.2018.00573

Vanamee ÉS, Faustman DL (2017) TNFR2: A novel target for cancer immunotherapy. Trends Mol Med 23 (11):1037–1046. https://doi.org/10.1016/j.molmed.2017.09.007

Vandenabeele P, Galluzzi L, Vanden Berghe T, Kroemer G (2010) Molecular mechanisms of necroptosis: an ordered cellular explosion. Nat Rev Mol Cell Biol 11 (10):700–714. https://doi.org/10.1038/nrm2970

Vaughan RA, Garcia-Smith R, Trujillo KA, Bisoffi M (2013) Tumor necrosis factor alpha increases aerobic glycolysis and reduces oxidative metabolism in prostate epithelial cells. Prostate 73(14):1538–1546. https://doi.org/10.1002/pros.22703

Wajant H, Scheurich P (2011) TNFR1-induced activation of the classical NF-kappaB pathway. FEBS J 278 (6):862–876. https://doi.org/10.1111/j.1742-4658.2011.08015.x

Wajant H, Siegmund D (2019) TNFR1 and TNFR2 in the control of the life and death balance of macrophages. Front Cell Dev Biol 7:91. https://doi.org/10.3389/fcell.2019.00091.

Wang J, Al-Lamki RS (2013) Tumor necrosis factor receptor 2: its contribution to acute cellular rejection and clear cell renal carcinoma. Biomed Res Int 2013:1. https://doi.org/10.1155/2013/821310

Wang A, Creasey A, Ladner M, Lin L, Strickler J, Van Arsdell J et al (1985) Molecular cloning of the complementary DNA for human tumor necrosis factor. Science 228(4696):149–154. https://doi.org/10.1126/science.3856324

Wang L, Du F, Wang X (2008) TNF-α induces two distinct Caspase-8 activation pathways. Cell 133 (4):693–703. https://doi.org/10.1016/j.cell.2008.03.036

Watts AD, Hunt NH, Wanigasekara Y, Bloomfield G, Wallach D, Roufogalis BD et al (1999) A casein kinase I motif present in the cytoplasmic domain of members of the tumour necrosis factor ligand family is implicated in 'reverse signalling'. EMBO J 18 (8):2119–2126. https://doi.org/10.1093/emboj/18.8.2119

Williams SK, Fairless R, Maier O, Liermann PC, Pichi K, Fischer R et al (2018) Anti-TNFR1 targeting in humanized mice ameliorates disease in a model of multiple sclerosis. Sci Rep 8(1):13628. https://doi.org/10.1038/s41598-018-31957-7

Willrich MA, Murray DL, Snyder MR (2015) Tumor necrosis factor inhibitors: clinical utility in autoimmune diseases. Transl Res 165(2):270–282. https://doi.org/10.1016/j.trsl.2014.09.006

Wing JB, Tanaka A, Sakaguchi S (2019) Human FOXP3 (+) regulatory T cell heterogeneity and function in autoimmunity and cancer. Immunity 50(2):302–316. https://doi.org/10.1016/j.immuni.2019.01.020

Xie P (2013) TRAF molecules in cell signaling and in human diseases. J Mol Signal 8(1):7. https://doi.org/10.1186/1750-2187-8-7

Xu A, Liu Y, Chen W, Wang J, Xue Y, Huang F et al (2016) TGF-beta-induced regulatory T cells directly suppress B cell responses through a noncytotoxic mechanism. J Immunol 196(9):3631–3641. https://doi.org/10.4049/jimmunol.1501740

Yamashita M, Passegué E (2019) TNF-α coordinates hematopoietic stem cell survival and myeloid regeneration. Cell Stem Cell 25(3):357–72.e7. https://doi.org/10.1016/j.stem.2019.05.019

Yan F, Du R, Wei F, Zhao H, Yu J, Wang C et al (2015) Expression of TNFR2 by regulatory T cells in peripheral blood is correlated with clinical pathology of lung cancer patients. Cancer Immunol Immunother 64 (11):1475–1485. https://doi.org/10.1007/s00262-015-1751-z

Yang S, Wang J, Brand DD, Zheng SG (2018) Role of TNF–TNF receptor 2 signal in regulatory T cells and its therapeutic implications. Front Immunol 9:784. https://doi.org/10.3389/fimmu.2018.00784

Yang S, Xie C, Chen Y, Wang J, Chen X, Lu Z et al (2019) Differential roles of TNFα-TNFR1 and TNFα-TNFR2 in the differentiation and function of CD4+Foxp3+ induced Treg cells in vitro and in vivo periphery in autoimmune diseases. Cell Death Dis 10(1):27. https://doi.org/10.1038/s41419-018-1266-6

Yang M, Tran L, Torrey H, Song Y, Perkins H, Case K et al (2020) Optimizing TNFR2 antagonism for immunotherapy with tumor microenvironment specificity. J Leukoc Biol 107:1–10. https://doi.org/10.1002/jlb.5ab0320-415rrrrr.

Zaragoza B, Chen X, Oppenheim JJ, Baeyens A, Gregoire S, Chader D et al (2016) Suppressive activity of human regulatory T cells is maintained in the presence of TNF. Nat Med 22(1):16–17. https://doi.org/10.1038/nm.4019

Zhang R, Xu Y, Ekman N, Wu Z, Wu J, Alitalo K et al (2003) Etk/Bmx transactivates vascular endothelial growth factor 2 and recruits phosphatidylinositol 3-kinase to mediate the tumor necrosis factor-induced angiogenic pathway. J Biol Chem 278 (51):51267–51276. https://doi.org/10.1074/jbc.M310678200

Zhao X, Rong L, Zhao X, Li X, Liu X, Deng J et al (2012) TNF signaling drives myeloid-derived suppressor cell accumulation. J Clin Invest 122(11):4094–4104. https://doi.org/10.1172/JCI64115.

Zheng SG, Gray JD, Ohtsuka K, Yamagiwa S, Horwitz DA (2002) Generation ex vivo of TGF-beta-producing regulatory T cells from CD4+CD25- precursors. J Immunol 169(8):4183–4189. https://doi.org/10.4049/jimmunol.169.8.4183

Zheng SG, Wang JH, Gray JD, Soucier H, Horwitz DA (2004) Natural and induced CD4+CD25+ cells educate CD4+CD25− cells to develop suppressive activity: the role of IL-2, TGF-β, and IL-10. J Immunol 172 (9):5213–5221. https://doi.org/10.4049/jimmunol.172.9.5213

Zheng SG, Wang JH, Stohl W, Kim KS, Gray JD, Horwitz DA (2006) TGF-beta requires CTLA-4 early after T cell activation to induce FoxP3 and generate adaptive CD4+CD25+ regulatory cells. J Immunol 176 (6):3321–3329. https://doi.org/10.4049/jimmunol.176.6.3321

Zheng SG, Wang J, Wang P, Gray JD, Horwitz DA (2007) IL-2 is essential for TGF-beta to convert naive CD4 +CD25- cells to CD25+Foxp3+ regulatory T cells and for expansion of these cells. J Immunol 178 (4):2018–2027. https://doi.org/10.4049/jimmunol.178.4.2018

Zhou Z, Gengaro P, Wang W, Wang XQ, Li C, Faubel S et al (2008) Role of NF-kappaB and PI 3-kinase/Akt in TNF-alpha-induced cytotoxicity in microvascular endothelial cells. Am J Physiol Renal Physiol 295(4): F932–F941. https://doi.org/10.1152/ajprenal.00066.2008

Zhou Q, Hu Y, Howard OMZ, Oppenheim JJ, Chen X (2014) In vitro generated Th17 cells support the expansion and phenotypic stability of CD4(+)Foxp3(+) regulatory T cells in vivo. Cytokine 65(1):56–64. https://doi.org/10.1016/j.cyto.2013.09.008.

Zou W (2006) Regulatory T cells, tumour immunity and immunotherapy. Nat Rev Immunol 6(4):295–307. https://doi.org/10.1038/nri1806

Zou H, Li R, Hu H, Hu Y, Chen X (2018) Modulation of regulatory T cell activity by TNF receptor type II-targeting pharmacological agents. Front Immunol 9:594. https://doi.org/10.3389/fimmu.2018.00594

The Pursuit of Regulatory T Cells in the Induction of Transplant Tolerance 14

Preston R. Arnold and Xian C. Li

Abstract

Organ transplantation is a preferred treatment option for patients with end-stage organ failure. However, transplant induces a robust rejection response that necessitates life-long immunosuppression, which often leads to a plethora of comorbidities. Thus, the goal of transplantation is to achieve a state of tolerance wherein the host permanently accepts the transplanted organ while maintaining normal immune responses to other antigens. Regulatory T cells (Tregs) play an important role in realizing this goal and are being explored in both animal models and human clinical trials. In this chapter, we discuss the key principles of transplant rejection and Treg biology, as well as the status of human clinical trials utilizing Tregs as cellular therapy. We discuss how the current immunosuppressive drugs are utilized in transplantation in favoring an increased Treg to T effector cell ratio, different approaches in generation of therapeutic Tregs, and various facets in Treg trial designs in the clinic. Such clinical trials provided many opportunities to leverage our understanding of Tregs in transplantation. They also demonstrated Tregs as a safe cellular therapy for human use, but the efficacy of this treatment has yet to be fully realized.

Keywords

Transplantation · Tolerance · Treg · Clinical trials · Immunosuppression

14.1 Introduction

Solid organ transplantation remains a preferred choice of treatment for patients with end-stage organ failure and its use in the clinic continues to increase over time (Watson and Dark 2012). In 2019, nearly 40,000 solid organ transplants were performed in the United States (United Network for Organ Sharing 2020). Among them, kidney transplants accounted for over half of the solid organs transplanted, while liver, heart, and lung transplants accounted for the majority of the remaining transplants. Despite the ever-increasing demand for organ transplantation in the clinic, our ability to perform them is limited by the severe shortage of organ donors. In fact, the patients waitlisted for organ transplants in the United States (~120,000 patients) far exceed the

P. R. Arnold
Immunobiology and Transplant Science Center and Department of Surgery, Houston Methodist Hospital, Texas Medical Center, Houston, TX, USA

X. C. Li (✉)
Immunobiology and Transplant Science Center and Department of Surgery, Houston Methodist Hospital, Texas Medical Center, Houston, TX, USA

Department of Surgery, Weill Cornell Medical College of Cornell University, New York, NY, USA
e-mail: xcli@houstonmethodist.org

S.-G. Zheng (ed.), *T Regulatory Cells in Human Health and Diseases*, Advances in Experimental Medicine and Biology 1278, https://doi.org/10.1007/978-981-15-6407-9_14

number of donor organs available. As a consequence, a substantial number of patients die while waiting for a life-saving transplant. Furthermore, for those with an organ transplant, graft rejection mounted by the patient's immune system remains a major threat to long-term engraftment, and because of that, patients must take immunosuppression drugs to broadly suppress the immune system for life to prevent graft rejection. Unfortunately, despite the much-improved graft survival in the era of modern immunosuppressive regimens, chronic graft rejection continues to occur, especially in the long-term immunosuppressed patients. Additionally, lifetime maintenance of immunosuppressive treatment markedly increases the risks for opportunistic infections and malignancies in transplant patients. Thus, one of the critical objectives in the field of organ transplantation is to induce a state of immunological tolerance toward the transplanted donor organs, where the transplant patients would permanently accept the donor organs without taking lifelong immunosuppressive drugs, while allowing immune responses to other antigens including pathogens to proceed in a normal fashion. Clearly, achieving this goal in the clinic would markedly improve the long-term outcome of organ transplants.

In this chapter, we focus on key features of transplant rejection, how the process is initiated, and the complexity of cell types involved in rejection, emphasizing the regulatory mechanisms that potentially suppress rejection and promote transplant tolerance. There are multiple cell types identified so far that exhibit regulatory properties. They are likely involved in the induction and maintenance of immune tolerance via different mechanisms, and the longevity (or stability) of tolerance may demand the collective efforts of diverse regulatory cell types. However, Foxp3+ T regulatory cells (Tregs) are among the best-studied cells and play a dominant role in immune tolerance. They are therefore the central focus of our discussion here. We specifically highlight the ongoing efforts in utilizing Tregs in clinical trials in transplant patients, discussing progresses and challenges, as well as unresolved questions that warrant further investigations. This discussion is centered on solid organ transplantation, but we believe that the principle of Tregs in transplant tolerance should also be relevant to hematopoietic cell transplantation.

14.2 Transplant Rejection and the Unique Roles of Tregs

14.2.1 Features of Transplant Rejection

Transplant rejection presents in one of several ways, each representing unique immunological responses (Nankivell and Alexander 2010). Hyperacute rejections are driven by preexisting donor-specific antibodies (DSA) in the host and lead to a robust rejection of the transplanted organs within hours of transplantation. This type of rejection has been almost completely prevented in the clinic due to effective serological testing for DSAs before transplantation. Acute rejection occurs over the course of a few months and presents with progressing signs of allograft failure. It is generally defined as either T-cell mediated or antibody mediated, based on the presence of lymphocyte infiltrate or antibody/complement staining when the organ is examined histologically, although mixed T-cell/antibody rejections are common. Antibodies in this setting are developed after organ transplantation and primarily directed against the graft vasculature. Importantly, T cells have been implicated in the pathogenesis of acute rejection, and both CD4+ and CD8+ subsets are often involved. T cells must be activated first by donor antigens before becoming effector cells, a process in which Tregs may be able to regulate. Prevention of acute rejection is mediated primarily through the institution of immunosuppressive drugs, which requires lifelong maintenance immunosuppression. Despite these efforts, acute rejection episodes are not uncommon and are a major source of morbidity among transplant recipients. Finally, chronic rejection develops years or decades after organ transplantation; it is an insidious process manifesting fibrosis within the graft and damage to the graft vasculature, leading to gradual decreasing of graft function and ultimately loss of the allograft. Although also immunologically driven, the mechanisms of chronic rejection are poorly understood, and because of that, chronic rejection is untreatable and constitutes the primary cause of graft loss in the long term.

Rejection is initiated upon recognition of donor major histocompatibility complex (MHC) by host T cells. MHC antigens, which consists of class I and class II molecules, are encoded in humans by the human leukocyte antigen (HLA) complex. MHC class I molecules are expressed in nearly all nucleated cells and present antigens to $CD8^+$ cytotoxic T cells. MHC class II molecules are expressed primarily in professional antigen-presenting cells (APCs) such as dendritic cells, macrophages, and B cells and present antigens to $CD4^+$ T cells, including Tregs. The HLA gene complex is highly polymorphic consisting of over 10 loci, with thousands of alleles discovered across human populations. Thus, organ transplants inevitably present novel MHC antigens to the host immune system, against which T cells are robustly activated. Efforts to mitigate this through the identification of closely matched donor/recipient pairs has led to decreased frequency of graft rejection.

The pathways of antigen presentation are key to understanding transplant rejection from the perspective of regulatory or effector cells. There are two primary pathways of antigen presentation: indirect and direct (Auchincloss and Sultan 1996). Indirect alloantigen presentation entails the traditional presentation of processed alloantigens by host APCs on host MHC class I and II molecules to host T cells, a pathway that is similar to protective T-cell responses. Direct alloantigen presentation occurs as host T cells directly recognize the intact donor MHC molecules present in donor APCs. This licenses T cells to recognize the donor MHC molecules expressed by the graft tissue. While professional APCs express high levels of MHC class II, epithelial cells and endothelial cells are also capable of expressing MHC class II, thus acting as APCs in allografts and allowing $CD4^+$ T cells including Tregs to directly recognize the graft (Quiroga et al. 2006). Because a finite number of donor APCs with no way to replenish themselves are transferred in solid organ transplants, the opportunity for direct presentation is relatively short-lived. Accordingly, direct presentation has been considered a predominant contributor to acute rejection while the indirect pathway may contribute more strongly to chronic or prolonged responses (Siu et al. 2018; Joffre et al. 2008). A modified version of the direct pathway, termed semi-direct presentation, has been described wherein host APCs acquire donor MHC class I/II molecules via endosomes and exosomes (Brown et al. 2011; Marino et al. 2016). This pathway may allow sustained direct pathway activation through epitope linkage as host dendritic cells expressing donor MHC class I in addition to donor MHC class II presented on host MHC class II (indirect pathway) can activate cytotoxic T cells through the direct pathway, while receiving help from $CD4^+$ T cells through the direct or indirect pathway (Siu et al. 2018). It also provides important insights into potential mechanisms through with regulatory T cells can control the transplant response.

The contribution of each of these pathways to different stages of transplant rejection, particularly in human subjects, remains poorly understood. However, it appears that the direct pathway is able to more powerfully induce T-cell activation (Rulifson et al. 2002). Deep sequencing of T-cell repertoires has indicated that greater disparity between self and allogenic HLA alleles increases the strength of the T-cell response; however, even highly similar HLA alleles are able to induce a specific alloreactive response (Arrieta-Bolanos et al. 2018). The potential implications of these pathways may also be tissue specific, as many tissues maintain their own specialized APCs (such as Kupffer cells and Langerhans cells) and may differ in their ability to express MHC class II molecules. Nevertheless, consideration of these unique means of antigen presentation is crucial when considering how Tregs might regulate the rejection response.

14.2.2 Tregs in Transplant Survival

Tregs represent a heterogeneous group of cells that contain various subsets specialized to specific immune responses. Much research continues to expand our understanding of these complex cells. Tregs are defined primarily by expression of the transcription factor Foxp3, as well as the surface markers $CD4^{+}CD25^{+}CD127^{low}$. Functionally, they are potent suppressor cells and are considered here in transplant studies. The importance of Tregs in reducing or preventing graft rejection in transplantation was first identified in animal studies wherein animals that had developed tolerance demonstrated increased Tregs in both the lymphoid organs and the tolerated allograft (Graca et al. 2002; Lee et al. 2005; Xia et al. 2008). Evaluation of human transplant recipients has similarly suggested that increased Tregs in the peripheral blood and graft parenchyma contribute to hypo-responsiveness to donor antigens and prolonged graft survival (Hoerning et al. 2012; Bestard et al. 2007). The contribution of Tregs residing within or outside the allograft is not fully understood. Limited animal models have suggested that early recruitment of Tregs to the allograft is important to limit the infiltration of T effector cells (Golshayan et al. 2009; Shao et al. 2019). Tregs subsequently accumulate in the lymph nodes where they further suppress the immune response by suppressing activation of T effector cells and APCs (Shao et al. 2019; Zhang et al. 2009). Tregs inhibit immune responses through multiple mechanisms, some of which are particularly pertinent to transplantation. These include the expression of anti-inflammatory cytokines such as IL-10, IL-35, and TGF-β; expression of CD39 and CD73 ectoenzymes to convert extracellular ATP, which is highly inflammatory, into the immunosuppressive adenosine; outcompeting T effector cells for IL-2; expression of CTLA4; and directly killing APCs and activated T effector cells through perforin and granzyme-dependent pathway. However, much remains to be learned regarding the specific mechanisms by which Tregs inhibit the transplant response in vivo. Mouse studies using adoptively transferred Tregs in skin, heart, kidney, or corneal transplant models have demonstrated the effects of Tregs both at the graft and in the draining lymph nodes (Golshayan et al. 2009; Shao et al. 2019; Zheng et al. 2006; Liao et al. 2017). At the graft, they limit T effector and APC infiltration, potentially through the secretion of TGF-β and IL-10 within the grafts. In the draining lymph nodes, Tregs work to limit T effector cell activation, proliferation, and differentiation by preventing the maturation of APCs, leading to a reduced MHC class II expression. This leads to a decrease in Th1 IFN-γ^{+} cells which promote the cellular response critical for transplant rejection. Within the draining lymph nodes, secretion of anti-inflammatory cytokines and inhibition of APCs may lead the induction of Tregs which recognize additional antigens not recognized by the initial Tregs and perpetuate tolerance. Nevertheless, the Treg response in transplantation is not uniquely different from that in other settings, and many of the traditional mechanisms appear to be at play promoting in promoting tolerance to organ transplants.

A higher percentage of Tregs taken from human blood were able to recognize donor alloantigens through the direct pathway than through the indirect pathway (Veerapathran et al. 2011). Tregs utilizing the direct pathway are sufficient to induce tolerance in some animal models (Lee et al. 2014). This should not suggest, however, that the indirect pathway is of minimal importance to graft acceptance. To the contrary, many studies have demonstrated the importance of the indirect pathway in inducing and maintaining tolerance, with some studies suggesting superiority of the indirect pathway, particularly in long-term graft acceptance (Wise et al. 1998; Tsang et al. 2008). Considering the presence of Tregs in both the graft and secondary lymphoid tissues, both indirect and direct antigen presentations likely play important roles in establishing the transplant response. This was demonstrated in at least one mouse model where directly stimulated Tregs were sufficient to prevent acute rejection while indirectly stimulated

Tregs were additionally necessary to prevent chronic rejection (Joffre et al. 2008). Ultimately, it is the balance between Tregs and T effector cells that is critical to graft acceptance, as models of allograft rejection (Lee et al. 2014; Wang et al. 2008; Fan et al. 2010). The pool size or Tregs and T effector cells in transplant survival highlights the importance of reducing T effector cells, particularly during the acute response, and justifies the use of immunosuppressive regimens, as T effector cells are superior in numbers in graft rejection (Li et al. 2001). Chronic rejection, on the other hand, is poorly understood and contributes significantly to late graft loss. Tregs may play a role in preventing this graft loss, as various models have been able to induce long-term tolerance through Treg therapies (Joffre et al. 2008; Xia et al. 2008; Lee et al. 2014; Tsang et al. 2008). In addition to their immunomodulatory roles, Tregs are known to play roles in tissue repair and regeneration and are being explored as treatments to ischemic injuries and myocardial infarctions (Hong and Kim 2018). This ability presents the attractive hypothesis that Tregs could contribute to preventing the vascular remodeling and fibrosis in chronic rejection. However, additional studies will be needed to assess this possibility.

A critical consideration in transplantation is the difference between thymus-derived Tregs (tTregs) and peripherally converted Tregs (pTregs). It might be expected that pTregs are the primary contributors to transplant tolerance, as they are derived from donor reactive T effector cells. tTregs are matured in the thymus and selected primarily by self-antigens, though some may exhibit cross-reactivity to alloantigens. However, pTregs developed from naïve T cells after alloantigen encounter follow similar kinetics as activation of T effector cells and may not be timely in effectively suppressing an allograft-specific immune response. Additionally, pTregs require an anti-inflammatory environment to express FoxP3, which is most potently induced by TGF-β and IL-2 signaling (Zheng et al. 2002, 2004, 2007; Davidson et al. 2007). The post-transplant environment, by contrast, is a markedly pro-inflammatory as the allograft is subjected to ischemic reperfusion injury in addition to surgical traumas (Mengel et al. 2011). These conditions are not conducive to pTreg development or maintenance. Of particular concern is that IL-6 in conjunction with TGF-β drives pTregs toward a pathogenic Th17 cells which promote graft rejection (Wang et al. 2008; Hanidziar and Koulmanda 2010). In contrast, tTregs have been shown in some studies to predominate over pTreg responses in graft survival (Fan et al. 2010). This is potentially due to the direct recognition of donor MHCs and highlights the potential benefits of close HLA matching between donors and recipients (Mohr Gregoriussen and Bohr 2017). tTregs show greater stability in the context of inflammatory milieu and may therefore be especially critical to establishing tolerance in the early stages of the transplant response (Barbi et al. 2014). After that, pTregs may begin to play a more predominant role in maintaining tolerance. The observation of "infectious tolerance" in animal models appears to support this notion (Wise et al. 1998; Waldmann et al. 2006).

Despite these diverse immunoregulatory mechanisms, the default course of transplantation is rejection, which renders the induction of transplant survival a challenging task. The initial inflammation in the graft induced by ischemia injury and surgical trauma in conjunction with robust T effector cell responses are too strong for any Tregs to contain without effective interventions in the effector arm of rejection. However, efforts to increase the total number and the suppressive activities of Tregs, in conjunction with therapeutics to limit inflammation and effector T cells, present a viable strategy for inducing and maintaining transplant tolerance. These efforts, which are the focus of various clinical trials, shift the Treg and T effector balance in favor of tolerance aimed at avoiding lifelong immunosuppression in transplant patients.

14.3 Clinical Trials Involving Tregs as Cellular Therapeutics

14.3.1 Production of Tregs Ex Vivo

One of the initial challenges in translating Treg therapies into clinical trials was the generation of clinical grade Tregs. The initial report of Treg isolation in accordance with good manufacturing protocol involved a magnetic enrichment system to select a population of $CD4^{+}CD25^{+}$ T cells from standard leukapheresis products (Hoffmann et al. 2006). However, early efforts to isolate Tregs for clinical use were hampered by both contamination with conventional T effector cells and an inability to extract sufficient numbers of Tregs from peripheral blood. Subsequent protocols sought to address this by including additional markers such as $CD127^{low}$ and by expanding the Tregs ex vivo using anti-CD3/CD28 beads (Putnam et al. 2009). Expansion not only increases cell number by potentially million-fold, but when completed in the presence of modifying reagents such as the mTOR inhibitor rapamycin or *all-trans* retinoic acid (atRA), it allows selective expansion of Tregs, thereby dramatically reducing the number of contaminating T effector cells (Putnam et al. 2009; Hippen et al. 2011; Lu et al. 2014). In addition, ex vivo expansion provides the opportunity to groom a population of Tregs that are selective for alloantigens. Alloantigen-specific Tregs are significantly more potent in reducing the allogenic immune response than are polyclonal Tregs (Xia et al. 2008; Veerapathran et al. 2011; Sagoo 2011). During Treg induction ex vivo, alloantigens can be presented in vitro via the direct or indirect pathway, which may have implications for Treg use and the timing of their administration in clinical trials, particularly as our understanding of how each of these pathways contribute to various stages of the allogenic immune response expands (Siu et al. 2018; Joffre et al. 2008; Veerapathran et al. 2011). Despite these progresses, ex vivo manipulations of Tregs continue to face many technical and regulatory challenges that will need to be addressed in future studies, especially those related to stability and viability of such expanded Tregs following adoptive transfer.

14.3.2 The Impact of Immunosuppression Drugs on Treg Functions

Several classes of therapeutics and administration strategies are frequently utilized in transplantation. Because immunosuppressive drugs are an indispensable facet of transplant medicine, understanding how these pharmacological agents may impact Treg function is critical to understanding the role of endogenous or therapeutically administered Tregs. The commonly prescribed immunosuppressive drugs and their mechanisms of action are particularly relevant to Treg therapies. The combination of immunotherapeutic agents is generally selected to act in two unique phases. The initial phase, or induction therapy, is designed to deplete or severely diminish the T-cell pool. This is given before or at the same time as transplantation when inflammation is at peak levels. After this initial induction therapy, transplant patients are treated with three classes of drugs (calcineurin inhibitors, anti-proliferatives, and steroids), which will be continued lifelong to prevent rejection.

Induction therapy agents include rabbit anti-thymocyte globin (rATG) and anti-CD25 antibodies. rATG broadly depletes T cells, including subsets of Tregs. However, this agent tends to result in a more favorable Treg to T effector ratio weeks following rATG therapy, partly due to far more efficient depletion of T effector cells (Tang et al. 2012; Gurkan et al. 2010; Krystufkova et al. 2012). Anti-CD25 antibodies antagonize IL-2 signaling in an effort to target activated T cells. Anti-CD25 antibodies also reduce Treg populations (Krystufkova et al. 2012; Bouvy et al. 2014), but studies have observed a persistence of FoxP3 expression after anti-CD25 treatment, suggesting that Tregs may survive this treatment by downregulating CD25 (Krystufkova et al. 2012; Vondran et al. 2010). The impact of these agents on Tregs has important implications for the timing of exogenous Treg administration. Considering that the goal of exogenous Treg therapy is to favorably shift the Treg to Teff ratio, administration of Tregs in conjunction with depleting therapies could be counter-productive or antagonistic. This may be especially true of anti-CD25 therapy because the highly expressed CD25 is a common feature of Tregs. Additionally, induction therapies are administered at a time of excessive inflammation in the graft. Such conditions may not be favorable for exogenously expanded Tregs which may be susceptible to reversion to an effector phenotype in the presence of inflammatory signals. For these reasons, many clinical trials have chosen to administer Tregs after induction therapy. However, at least one animal model has demonstrated that initial induction therapy synergizes with Treg therapy infused at a later timepoint (Xia et al. 2008).

Maintenance therapies include calcineurin inhibitors (CNI), mTOR inhibitors, mycophenolate, and corticosteroids. CNIs work by inhibiting the phosphatase calcineurin, which is necessary for the activation of the transcription factor NFAT. This has several important implications for Tregs as NFAT normally binds to the promoters of FoxP3 and CD25 to promote their expression. Accordingly, treatment with CNIs has been shown to decrease Treg functions as well as stability in transplant patients (Akimova et al. 2012; Satake et al. 2014). mTOR inhibitors, on the other hand, are well documented to promote FoxP3 expression (Huang et al. 2020). Treatment with mTOR inhibitors promotes Treg levels as compared to CNIs in transplant patients (Satake et al. 2014; Ruggenenti et al. 2007; Shan et al. 2014). One animal study demonstrated that mTOR inhibitors combined with Tregs was much better than CNI with Tregs in prolonging graft survival (Ma et al. 2009). However, the increase in Tregs has not always correlated to improved graft function in human clinical trials, perhaps because measurements in these trials have been limited to peripheral blood Tregs rather than intragraft levels (Shan et al. 2014). Notably, while mTOR inhibitors promote FoxP3 expression, they may have deleterious effects on Treg expansion or suppressive activities due to the complex interplay of metabolic and Treg functions (Shi and Chi 2019; Furukawa et al. 2016). In animal models, the in vivo effects of mTOR inhibitors on Tregs seem to be dependent on dosage and timing (Shan et al. 2014). Thus, while generally considered favorable for FoxP3 expression, much remains to be learned about the effects of mTOR inhibitors on the overall Treg activities in vivo.

Corticosteroids are known to have many anti-inflammatory properties, and their effects on Tregs are complex and likely context dependent (Cari et al. 2019). Corticosteroids appear to promote FoxP3 expression through the transcription factor glucocorticoid-induced leucine zipper (GILZ), which leads to increased levels of anti-inflammatory cytokines such as IL-10 and TGF-β (Ugor 2018). One study observed increased Tregs in patients treated with a corticosteroid bolus, although the effect appeared to be short lived (Seissler et al. 2012). Other studies have suggested that corticosteroids increase Tregs in allografts indirectly through myeloid-derived suppressor cells (Nakao et al. 2018). Mycophenolate inhibits de novo purine synthesis, which is especially detrimental to lymphocytes as they rely exclusively on this pathway for expansion (Allison and Eugui 2005). In vitro studies have suggested that mycophenolate may promote a favorable Treg/Th17 ratio, and transplant patients on mycophenolate therapy have accordingly demonstrated decreased IL-17 in peripheral blood (Abadja et al. 2011, 2009). Various studies have demonstrated that mycophenolate favorably alters the Treg/T effector ratio in transplant patients relative to other maintenance therapies (Demirkiran et al. 2009; Fourtounas et al. 2010; Zeng et al. 2019).

Overall, many of the immunosuppressive therapies utilized during transplantation favorably alter the Treg/Teff ratio, most likely due to preferential inhibition of T effector cells over Tregs, rather than an actual promotion of Tregs. Additionally, while many studies have considered FoxP3 as markers for Tregs, fewer studies have considered the effect of these drugs on the suppressive capacity of Tregs (Arroyo Hornero et al.

2017). Studies disentangling the effects of immunosuppressive drugs on FoxP3 expression from immunosuppressive capacity are especially difficult to conduct and interpret in human patients, where patients are subjected to a cocktail of drugs and the Treg capacity is interpreted through surrogate measures such as graft rejection rates and measures of graft functions. In vivo studies thus far have also been primarily limited to interpreting the effects of these drugs on endogenous Tregs. Their impact could be markedly different when exogenous Tregs are introduced into transplant patients. With clinical trials addressing exogenously expanded Tregs for transplantation are still primarily in phase I to determine safety, later trials experimenting with unique drug cocktails in conjunction with Treg therapies will be informative. Additionally, the timing of when to introduce Treg therapies (during induction or maintenance) will be a critical determinant to the success of Treg therapies.

14.3.3 Clinical Trials Utilizing Tregs in Transplantation

Early studies in rodents as well as in non-human primates demonstrated that exogenously expanded Tregs could be safely re-introduced to the host and that the Tregs were stable and functional in vivo (Riley et al. 2009; Bashuda et al. 2005; Ma et al. 2011). However, the exogenously transferred Tregs rapidly diminished within the first few weeks, after which a more stable population persisted (Zhang et al. 2015). It was found that cryopreserved Tregs could be thawed, expanded, and maintained their phenotype and the in vivo kinetics comparably to freshly expanded Tregs (Guo et al. 2015). However, caution was also warranted as at least one model demonstrated decreased graft function and increased effector/memory function following Treg infusion (Ezzelarab et al. 2016). This apparently contradictory result is likely due to factors related to the timing of infusion, expansion protocols, dosing, and combination with other therapeutic agents. In addition, different transplant models and immunosuppressive therapies utilized may also contribute to differences in Treg outcomes. For example, early Treg infusion following transplantation in conjunction with rATG may have contributed to perturbation of the expanded Tregs. Nevertheless, these preclinical studies laid the groundwork for human trials on how Tregs can be best utilized in transplant patients. Indeed, various clinical trials have been conducted utilizing exogenously expanded Tregs. As of this writing, three phase I trials have been completed and published: one in liver and two in kidney transplants (Chandran et al. 2017; Todo et al. 2016; Mathew et al. 2018). It is important to note that as phase I trials, these studies primarily aim to study the safety of Treg therapies in humans rather than the efficacy of such therapies. Accordingly, a relatively small number of patients are evaluated, making it difficult to draw meaningful conclusions regarding the efficacy of these treatments. Notwithstanding, the initial results are promising as no patients showed significant adverse events in response to Treg infusions. Each of these trials utilized unique protocols to produce Tregs that are discussed here.

The liver clinical trial, conducted by Todo et al. (2016), was conducted in ten patients who received Treg therapy between 2010 and 2012 and was followed through 2015. In this trial, lymphocytes were collected from the donors, irradiated, and then cultured with recipient lymphocytes in the presence of CD80/CD86 blockade. Due to impurities in the cell collection process, this process likely resulted in both direct and indirect stimulation of recipient lymphocytes with donor alloantigens. The cell culture process enriched the Treg population by an average of approximately fourfold, resulting in an infusion product containing approximately 15% Tregs. Presumably, the remaining T cells were rendered anergic by the costimulatory blockade, although this was not analyzed. Criteria of the infusion product included >80% cell viability, $\sim 1 \times 10^6$ Tregs/kg body weight per patient, and standard immunosuppression protocol consisting of steroids, mycophenolate, and 1 month of CNI following transplantation. Patients also received a single dose of cyclophosphamide given 5 days posttransplant. The Treg-enriched cell product was infused on day 13 posttransplant. No adverse events were observed in response to cell infusion. Large variations were seen between patients in the total number of Tregs 3 months after transplant, although the percent of $FoxP3^+$ cells tended to increase relative to pre-transplant levels. Weaning off CNIs was begun in this study at 6 months posttransplantation. Over the 12-month weaning period, seven of the ten patients were successfully weaned off of CNIs. The three patients who were not successfully weaned all received liver transplants due to autoimmune diseases. This suggests an interesting association between Treg therapies and underlying primary disease in this cohort. Of the seven weaned patients, five demonstrated non-responsiveness toward donor alloantigens by mixed lymphocyte reaction, while the remaining two demonstrated hypo-responsiveness. Importantly, the patients retained greater reactivity toward third-party antigens than they did against donor antigens, demonstrating specificity of the immune tolerance. There are several limitations to this study. As a phase I clinical trial with only ten patients, no controls (patients without Treg infusion) were included, thus it is difficult to draw any definitive conclusions about the efficacy of the treatment. Additionally, patients were splenectomized during transplant surgery. This splenomectomy, combined with the impurity of the Treg cell-infusion product, clouds any attributions to Tregs. However, the study does provide a framework in which Treg infusions are part of safe therapies leading to donor-specific hypo-responsiveness.

The TASKp trial involved polyclonally expanded Tregs in three kidney transplant patients (Chandran et al. 2017). Tregs were isolated from peripheral blood by cell sorting as $CD4^+CD25^+CD127^{lo}$ cells and were then expanded in vitro in the presence of IL-2 and deuterated glucose to allow for subsequent cell tracking. The expanded Treg products were approximately 95% FoxP3+ and over 99% viable and were infused 14 days after expansion with a dose of approximately 3.2×10^8 Tregs per patient. All patients were on immunosuppressive therapy consisting of CNI, mycophenolate, and corticosteroid before and after Treg infusion. No adverse events were observed as a result of cell infusion, and patient survival and graft survival were 100% at 1 year post-infusion. Deuterated Tregs were detectable for approximately 3 months following injection. Importantly, deuterium signals were not detected in other T-cell populations, indicating stability of the injected Tregs. This study provided an additional framework demonstrating the safety of Treg infusion into transplant patients. It is unique in that Tregs were collected later in the course of transplantation. These Tregs had already been exposed to maintenance therapy in vivo, but were still able to be expanded, although their immunosuppressive capacity was not determined. Such trials may be especially applicable to efforts in preventing chronic rejection, although the efficacy of this treatments in such conditions is unknown.

The TRACT trial represents an important clinical trial in human Treg therapy for kidney transplant patients (Mathew et al. 2018). In this trial, peripheral blood monoclonal cells (PBMCs) were taken from nine patients approximately 1 month prior to transplantation and cryopreserved. Induction therapy was initiated at the time of transplant with alemtuzumab (a monoclonal antibody to CD52 which deletes T and B cells) and corticosteroid. Maintenance therapy of CNI and mycophenolate was also administered at the time of transplantation, with the CNI switched to an mTOR inhibitor at 30 days posttransplant. Tregs were enriched by deleting $CD8^+$ and $CD19^+$ cells followed by $CD25^+$ cell enhancement. The cells were then polyclonally expanded for 21 days using IL-2 and TGF-β, including the presence of an mTOR inhibitor for the first 9 days. This resulted in a highly purified Treg product with marked demethylation of the FoxP3 locus, indicating Treg stability. The expanded Tregs also maintained clonal diversity and increased in immunosuppressive capacity throughout the expansion process, as measured by mixed lymphocyte reactions. Importantly, the Tregs demonstrated the ability to induce non-expanded recipient T cells into Tregs during mixed lymphocyte reactions. Three tiered doses (with three patients in each tier) of 0.5×10^9, 1.0×10^9, or 5.0×10^9 Tregs were infused into patients at day 60 posttransplantation. At 90 days posttransplant, Treg levels were comparable to or exceeded pretransplant levels, in contrast to total $CD4^+$ and $CD8^+$ levels which remained depressed from induction therapy. Treg levels remained elevated 1 year posttransplant. This increase was attributed to the Treg infusion, as a historical control group that received the same transplant and immunosuppression regimen without Treg infusion did not show any increase in Tregs over time. Patients also maintained immunological response to various pathogenic antigens (such as CMV), which is a concern in transplant patients. In all doses, there was 100% patient and graft survival at 2 years without evidence of opportunistic infection or malignancy, demonstrating Treg infusion over varied doses to be safe.

Clearly, these studies vary dramatically in their Treg purification and expansion methods ranging from highly pure polyclonal Treg products to alloantigen-specific Treg-enhanced cellular products while utilizing varied cytokines and pharmacological agents to promote Treg expansion. Additionally, Treg infusion therapies were given nearly congruous with transplant (13 days posttransplant) to more than 6 months after surgery and with wide varieties in dosing and in immunosuppression regimens. Despite the tremendous heterogeneity between them, all studies found Treg infusion therapies to be safe in humans with no severe adverse reactions. Tregs appear to be stable, do not convert to effector cells, albeit in a limited number of patients. These are all encouraging signs that have allowed the advancement of additional phase I and phase II clinical trials.

14.4 Ongoing Clinical Trials

Several new phase I and phase II clinical trials for therapeutic use of Tregs in kidney and liver transplant are currently being conducted. Although results are still forthcoming, we will briefly discuss a few of the predominant trials, their hypotheses, and their protocols.

The aforementioned TASKp trial has moved forward to a combined phase I/II trial, which involves patients who develop kidney graft inflammation at their 6 month posttransplant biopsy (Flavio Vincenti 2019). The advantages of this include the limited treatment options currently available for such patients and ability to quickly assess changes in inflammation (and therefore treatment efficacy) following Treg infusion (Tang and Vincenti 2017). Although originally conceived and planned as a trial to compare polyclonally expanded Tregs, donor alloantigen-specific Tregs, and standard of care, the most recent update of the trial dropped the donor alloantigen-specific Treg arm. Thirty patients will be recruited, of which 15 will receive standard of care and the remaining 15 will receive an infusion of approximately 5.50×10^8 polyclonally expanded Tregs. Similar to the TASKp trial, patients will be maintained on CNI and mycophenolate maintenance therapy before and after Treg infusion, but in this study the arm receiving polyclonal Tregs will be converted from the CNI to an mTOR inhibitor after Treg infusion. Outcomes measured will include graft rejection/inflammation as well as infection rates. This trial is large enough to provide some of the earliest data on Treg efficacy in human transplant patients.

The ONE study represents a multicenter clinical trial focused on the use of different regulatory cell therapies in living donor kidney transplant recipients (Edward et al. 2018). Four centers are utilizing Tregs, each generated in a unique way. The King's College London (KCL) group produced polyclonal Tregs by magnetically sorting $CD4^+CD25^+$ cells and then expanding them in the presence of IL-2 and an mTOR inhibitor. Cells were cryopreserved until infusion. The Charite-Berlin (CB) group similarly produced polyclonal nTreg cells which were expanded in vitro and then infused fresh into patients. The University of California San Francisco (UCSF) group produced donor-specific Tregs which were directly stimulated by inactivated donor B cells. In the Massachusetts General Hospital (MGH) group, recipient PBMCs are cocultured with irradiated donor PBMCs in the presence of costimulatory blockade (belatacept, CTLA4-Ig) before expansion and selection of $CD4^{+}CD25^{+}$ Tregs. This produces a cell product of approximately 90% Tregs with alloantigen specificity. The initial goal was to recruit 6–12 patients into each treatment group, compared with 61 patients receiving standard of care therapy. Dosing and timing of Treg infusions varied between the centers, but all centers infused their respective cell products within 2 weeks of transplantation. All patients, including those receiving standard of care, were given an identical maintenance regimen that included continual mycophenolate and CNI therapy combined with an initial corticosteroid dose for 3 months. Standard of care patients also underwent anti-CD25 induction therapy. Patients who received Treg cell therapy were optionally tapered to CNI monotherapy starting at 9 months posttransplant.

Results for the ONE trial are not yet published in full, but preliminary results are now reported on CORDIS. Preliminary results for patients receiving polyclonal Treg expansions have been very favorable. Over one third of patients were on CNI monotherapy without rejection events. Additionally, patients who received Treg cell therapy reported lower incidence of infections, likely because these patients were able to be maintained with decreased immunosuppression. Preliminary results of acute rejection episodes were comparable between standard of care and cell therapy groups. The KCL group showed a special promise as none of the 12 patients receiving Treg cell therapy experienced a rejection event. Generally speaking, the polyclonal Treg therapies were more successful in conducting the trials, as each was able to treat 10+ patients. The UCSF trial treated two patients, but one showed signs of acute rejection shortly after infusion, and the trial was then suspended. The MGH trials was able to infuse three patients without adverse events but suspended the trial to improve the manufacturing protocol for Tregs, and the improved protocols will be pursued in another clinical trial.

Several other clinical trials are attempting to extend these initial results. The KCL has taken their initial results into a phase IIb, termed the TWO trial, which will utilize a similar Treg expansion protocol but will infuse the cells at 6 months posttransplant, at which time patients will be maintained on CNI monotherapy (ISRCTN 2020). Patients will be followed at least 1 year following Treg infusion with acute rejection being the primary endpoint. This trial will give additional critical information on the efficacy of Treg therapies. Additional clinical trials are underway to explore Treg therapies in liver transplantation as well. The ThRIL clinical trial, also conducted by KCL, will infuse one of two doses of polyclonal Tregs into liver transplant patients (Lombardi 2019). Patients will be treated with rATG, mTOR inhibitor, and CNI and will be assessed for acute rejection and therapeutic toxicities. The MGH branch of the LITTMUS trial aims to infuse $CD4^{+}CD25^{+}CD127^{lo}$ donor-specific Tregs stimulated in the presence of costimulatory blockade into liver transplant patients at a single dose of $>2.5 \times 10^{6}$ Tregs (Markmann 2019). This study aims to determine the effectiveness of Tregs in helping wean liver transplant patients off immunosuppressive therapies. The ARTEMIS trial similarly looks to infuse donor-specific Tregs in liver transplant patients to reduce or completely wean off immunosuppressive drugs (Sandy Feng and Tang 2020). Unfortunately, several other promising therapies, including branches of LITTMUS trial and the dELTA trial (related to the ARTIMES trial), have been discontinued due to difficulties in manufacturing the Treg product.

14.5 Conclusions

Tregs are critical to preventing rejection of transplanted organs; they are also indispensable in the induction of transplant tolerance. As a therapy, their potential to induce a state of tolerance reduces or eliminates the need for deleterious immunosuppressive regimens in recipients and makes Tregs a very enticing option. However, many challenges remain in manufacturing clinical grade Tregs. Additionally, the complex relationship between immunosuppressive therapeutics and Treg activities has important implications for utilizing Tregs in conjunction with these agents. From a basic science perspective, a greater understanding of how Tregs are induced and maintained, their mechanisms of action in preventing acute and chronic rejections, as well as how antigen presentation through the direct or indirect pathway affects these processes, is needed. Such understanding will enhance the ability to manufacture Tregs and the dose and timing of their application in clinical trials. Animal models and initial clinical trials indicate that the Treg infusion therapies are well tolerated, although additional monitoring will be necessary to determine the long-term risk of malignancy or other complications such as chronic infections. Preliminary results indicate that Treg therapies do not increase the risk of acute rejection, but many additional clinical trials, including those studying various timing and immunosuppressive regimens, are warranted to truly assess the efficacy of these therapies. Clinical trials evaluating the efficacy of donor alloantigen-specific Tregs are especially needed, as these therapies have shown greater promise in animal studies, but have proven difficult to manufacture and apply in human clinical trials. Trials are also needed to examine the efficacy of Treg therapies in other solid organ transplants such as heart and lung. In spite of the many challenges and unknowns that remain, the power of Tregs and the initial results are promising and justify optimism for Tregs as a promising therapy for solid organ transplantation.

Acknowledgments We acknowledge all lab members at the Immunobiology and Transplant Science Center for helpful discussions during the preparation of this chapter and grant supports from the National Institutes of Health.

References

Abadja F, Videcoq C, Alamartine E, Berthoux F, Mariat C (2009) Differential effect of cyclosporine and mycophenolic acid on the human regulatory T cells and TH-17 cells balance. Transplant Proc 41:3367–3370. https://doi.org/10.1016/j.transproceed.2009.08.031

Abadja F, Atemkeng S, Alamartine E, Berthoux F, Mariat C (2011) Impact of mycophenolic acid and tacrolimus on Th17-related immune response. Transplantation 92:396–403. https://doi.org/10.1097/TP.0b013e3182247b5f

Akimova T et al (2012) Differing effects of rapamycin or calcineurin inhibitor on T-regulatory cells in pediatric liver and kidney transplant recipients. Am J Transplant 12:3449–3461. https://doi.org/10.1111/j.1600-6143.2012.04269.x

Allison AC, Eugui EM (2005) Mechanisms of action of mycophenolate mofetil in preventing acute and chronic allograft rejection. Transplantation 80:S181–S190. https://doi.org/10.1097/01.tp.0000186390.10150.66

Arrieta-Bolanos E et al (2018) Alloreactive T cell receptor diversity against structurally similar or dissimilar HLA-DP antigens assessed by deep sequencing. Front Immunol 9:280. https://doi.org/10.3389/fimmu.2018.00280

Arroyo Hornero R et al (2017) CD45RA distinguishes CD4+CD25+CD127−/low TSDR demethylated regulatory T cell subpopulations with differential stability and susceptibility to tacrolimus-mediated inhibition of suppression. Transplantation 101:302–309. https://doi.org/10.1097/tp.0000000000001278

Auchincloss H, Sultan H (1996) Antigen processing and presentation in transplantation. Curr Opin Immunol 8:681–687. https://doi.org/10.1016/S0952-7915(96)80086-0

Barbi J, Pardoll D, Pan F (2014) Treg functional stability and its responsiveness to the microenvironment. Immunol Rev 259:115–139. https://doi.org/10.1111/imr.12172

Bashuda H et al (2005) Renal allograft rejection is prevented by adoptive transfer of anergic T cells in nonhuman primates. J Clin Invest 115:1896–1902. https://doi.org/10.1172/jci23743

Bestard O et al (2007) Achieving donor-specific hyporesponsiveness is associated with FOXP3+ regulatory T cell recruitment in human renal allograft infiltrates. J Immunol 179:4901–4909. https://doi.org/10.4049/jimmunol.179.7.4901

Bouvy AP et al (2014) The impact of induction therapy on the homeostasis and function of regulatory T cells in kidney transplant patients. Nephrol Dial Transplant 29:1587–1597. https://doi.org/10.1093/ndt/gfu079

Brown K, Sacks SH, Wong W (2011) Coexpression of donor peptide/recipient MHC complex and intact donor MHC: evidence for a link between the direct and indirect pathways. Am J Transplant 11:826–831. https://doi.org/10.1111/j.1600-6143.2011.03437.x

Cari L, De Rosa F, Nocentini G, Riccardi C (2019) Context-dependent effect of glucocorticoids on the proliferation, differentiation, and apoptosis of regulatory T cells: a review of the empirical evidence and clinical applications. Int J Mol Sci 20. https://doi.org/10.3390/ijms20051142

Chandran S et al (2017) Polyclonal regulatory T cell therapy for control of inflammation in kidney transplants. Am J Transplant 17:2945–2954. https://doi.org/10.1111/ajt.14415

Davidson TS, DiPaolo RJ, Andersson J, Shevach EM (2007) Cutting edge: IL-2 is essential for TGF-beta-mediated induction of Foxp3+ T regulatory cells. J Immunol 178:4022–4026. https://doi.org/10.4049/jimmunol.178.7.4022

Demirkiran A et al (2009) Conversion from calcineurin inhibitor to mycophenolate mofetil-based immunosuppression changes the frequency and phenotype of CD4 +FOXP3+ regulatory T cells. Transplantation 87:1062–1068. https://doi.org/10.1097/TP.0b013e31819d2032

Edward K Geissler BJ, Bayer C, Renzulli M (2018) Final report summary - THE ONE STUDY (a unified approach to evaluating cellular immunotherapy in solid organ transplantation). https://cordis.europa.eu/project/id/260687/reporting

Ezzelarab MB et al (2016) Regulatory T cell infusion can enhance memory T cell and alloantibody responses in lymphodepleted nonhuman primate heart allograft recipients. Am J Transplant 16:1999–2015. https://doi.org/10.1111/ajt.13685

Fan Z et al (2010) In vivo tracking of 'color-coded' effector, natural and induced regulatory T cells in the allograft response. Nat Med 16:718–722. https://doi.org/10.1038/nm.2155

Flavio Vincenti SC (2019) Treg therapy in subclinical inflammation in kidney transplantation (TASK). https://clinicaltrials.gov/ct2/show/NCT02711826

Fourtounas C et al (2010) Different immunosuppressive combinations on T-cell regulation in renal transplant recipients. Am J Nephrol 32:1–9. https://doi.org/10.1159/000313940

Furukawa A, Wisel SA, Tang Q (2016) Impact of immune-modulatory drugs on regulatory T cell. Transplantation 100:2288–2300. https://doi.org/10.1097/tp.0000000000001379

Golshayan D et al (2009) Transplantation tolerance induced by regulatory T cells: in vivo mechanisms and sites of action. Int Immunopharmacol 9:683–688. https://doi.org/10.1016/j.intimp.2008.12.006

Graca L, Cobbold SP, Waldmann H (2002) Identification of regulatory T cells in tolerated allografts. J Exp Med 195:1641–1646. https://doi.org/10.1084/jem.20012097

Guo H, Zhang H, Lu L, Ezzelarab MB, Thomson AW (2015) Generation, cryopreservation, function and in vivo persistence of ex vivo expanded cynomolgus monkey regulatory T cells. Cell Immunol 295:19–28. https://doi.org/10.1016/j.cellimm.2015.02.006

Gurkan S et al (2010) Immune reconstitution following rabbit antithymocyte globulin. Am J Transplant 10:2132–2141. https://doi.org/10.1111/j.1600-6143.2010.03210.x

Hanidziar D, Koulmanda M (2010) Inflammation and the balance of Treg and Th17 cells in transplant rejection and tolerance. Curr Opin Organ Transplant 15:411–415. https://doi.org/10.1097/MOT.0b013e32833b7929

Hippen KL et al (2011) Massive ex vivo expansion of human natural regulatory T cells (T(regs)) with minimal loss of in vivo functional activity. Sci Transl Med 3:83ra41. https://doi.org/10.1126/scitranslmed.3001809

Hoerning A et al (2012) Peripherally circulating CD4(+) FOXP3(+) CXCR3(+) T regulatory cells correlate with renal allograft function. Scand J Immunol 76:320–328. https://doi.org/10.1111/j.1365-3083.2012.02732.x

Hoffmann P et al (2006) Isolation of CD4+CD25+ regulatory T cells for clinical trials. Biol Blood Marrow Transplant 12:267–274. https://doi.org/10.1016/j.bbmt.2006.01.005

Hong J, Kim BS (2018) Regulatory T cell-mediated tissue repair. Adv Exp Med Biol 1064:221–233. https://doi.org/10.1007/978-981-13-0445-3_14

Huang H, Long L, Zhou P, Chapman NM, Chi H (2020) mTOR signaling at the crossroads of environmental signals and T-cell fate decisions. Immunol Rev 295:15. https://doi.org/10.1111/imr.12845

ISRCTN (2020) TWO study: cell therapy trial in renal transplantation. http://www.isrctn.com/ISRCTN11038572

Joffre O et al (2008) Prevention of acute and chronic allograft rejection with CD4+CD25+Foxp3+ regulatory T lymphocytes. Nat Med 14:88–92. https://doi.org/10.1038/nm1688

Krystufkova E et al (2012) Regulatory T cells in kidney transplant recipients: the effect of induction immunosuppression therapy. Nephrol Dial Transplant 27:2576–2582. https://doi.org/10.1093/ndt/gfr693

Lee I et al (2005) Recruitment of Foxp3+ T regulatory cells mediating allograft tolerance depends on the CCR4 chemokine receptor. J Exp Med 201:1037–1044. https://doi.org/10.1084/jem.20041709

Lee K, Nguyen V, Lee KM, Kang SM, Tang Q (2014) Attenuation of donor-reactive T cells allows effective control of allograft rejection using regulatory T cell therapy. Am J Transplant 14:27–38. https://doi.org/10.1111/ajt.12509

Li XC, Strom TB, Turka LA, Wells AD (2001) T cell death and transplantation tolerance. Immunity 14:407–416. https://doi.org/10.1016/S1074-7613(01)00121-2

Liao T et al (2017) In vivo attenuation of antibody-mediated acute renal allograft rejection by ex vivo TGF-β-induced CD4+Foxp3+ regulatory T cells. Front Immunol 8. https://doi.org/10.3389/fimmu.2017.01334

Lombardi G (2019) Safety and efficacy study of regulatory T cell therapy in liver transplant patients (ThRIL). https://clinicaltrials.gov/ct2/show/NCT02166177

Lu L et al (2014) Critical role of *all−trans* retinoic acid in stabilizing human natural regulatory T cells under inflammatory conditions. Proc Natl Acad Sci 111: E3432–E3440. https://doi.org/10.1073/pnas.1408780111

Ma A et al (2009) Combined therapy of CD4+CD25+ regulatory T cells with low-dose sirolimus, but not calcineurin inhibitors, preserves suppressive function of regulatory T cells and prolongs allograft survival in mice. Int Immunopharmacol 9:553–563. https://doi.org/10.1016/j.intimp.2009.01.033

Ma A et al (2011) Adoptive transfer of CD4+CD25+ regulatory cells combined with low-dose sirolimus and anti-thymocyte globulin delays acute rejection of renal allografts in Cynomolgus monkeys. Int Immunopharmacol 11:618–629. https://doi.org/10.1016/j.intimp.2010.11.001

Marino J et al (2016) Donor exosomes rather than passenger leukocytes initiate alloreactive T cell responses after transplantation. Sci Immunol 1:aaf8759. https://doi.org/10.1126/sciimmunol.aaf8759

Markmann J (2019) Liver Transplantation With Tregs at MGH (LITTMUS-MGH). https://clinicaltrials.gov/ct2/show/NCT03577431

Mathew JM et al (2018) A phase I clinical trial with ex vivo expanded recipient regulatory T cells in living donor kidney transplants. Sci Rep 8:7428. https://doi.org/10.1038/s41598-018-25574-7

Mengel M et al (2011) The molecular phenotype of 6-week protocol biopsies from human renal allografts: reflections of prior injury but not future course. Am J Transplant 11:708–718. https://doi.org/10.1111/j.1600-6143.2010.03339.x

Mohr Gregoriussen AM, Bohr HG (2017) A novel model on DST-induced transplantation tolerance by the transfer of self-specific donor tTregs to a haplotype-matched organ recipient. Front Immunol 8:9–9. https://doi.org/10.3389/fimmu.2017.00009

Nakao T et al (2018) Dexamethasone prolongs cardiac allograft survival in a murine model through myeloid-derived suppressor cells. Transplant Proc 50:299–304. https://doi.org/10.1016/j.transproceed.2017.11.014

Nankivell BJ, Alexander SI (2010) Rejection of the kidney allograft. N Engl J Med 363:1451–1462. https://doi.org/10.1056/NEJMra0902927

Putnam AL et al (2009) Expansion of human regulatory T-cells from patients with type 1 diabetes. Diabetes 58:652–662. https://doi.org/10.2337/db08-1168

Quiroga I et al (2006) Expression of MHC class I-related Chain B (MICB) molecules on renal transplant biopsies. Transplantation 81:1196–1203. https://doi.org/10.1097/01.tp.0000205788.05322.42

Riley JL, June CH, Blazar BR (2009) Human T regulatory cell therapy: take a billion or so and call me in the morning. Immunity 30:656–665. https://doi.org/10.1016/j.immuni.2009.04.006

Ruggenenti P et al (2007) Sirolimus versus cyclosporine therapy increases circulating regulatory T cells, but does not protect renal transplant patients given alemtuzumab induction from chronic allograft injury. Transplantation 84:956–964. https://doi.org/10.1097/01.tp.0000284808.28353.2c

Rulifson IC, Szot GL, Palmer E, Bluestone JA (2002) Inability to induce tolerance through direct antigen presentation. Am J Transplant 2:510–519. https://doi.org/10.1034/j.1600-6143.2002.20604.x

Sagoo P et al (2011) Human regulatory T cells with alloantigen specificity are more potent inhibitors of alloimmune skin graft damage than polyclonal regulatory T cells. Sci Transl Med 3:83ra42. https://doi.org/10.1126/scitranslmed.3002076

Sandy Feng JB, Tang Q (2020) Donor alloantigen reactive tregs (darTregs) for calcineurin inhibitor (CNI) reduction (ARTEMIS). https://clinicaltrials.gov/ct2/show/NCT02474199

Satake A, Schmidt AM, Nomura S, Kambayashi T (2014) Inhibition of calcineurin abrogates while inhibition of mTOR promotes regulatory T cell expansion and graft-versus-host disease protection by IL-2 in allogeneic bone marrow transplantation. PLoS One 9:e92888. https://doi.org/10.1371/journal.pone.0092888

Seissler N et al (2012) Methylprednisolone treatment increases the proportion of the highly suppressive HLA-DR(+)-Treg-cells in transplanted patients. Transpl Immunol 27:157–161. https://doi.org/10.1016/j.trim.2012.09.003

Shan J et al (2014) The effects of rapamycin on regulatory T cells: its potential time-dependent role in inducing transplant tolerance. Immunol Lett 162:74–86. https://doi.org/10.1016/j.imlet.2014.07.006

Shao C et al (2019) Local delivery of regulatory T cells promotes corneal allograft survival. Transplantation 103:182–190. https://doi.org/10.1097/TP.0000000000002442

Shi H, Chi H (2019) Metabolic control of Treg cell stability, plasticity, and tissue-specific heterogeneity. Front Immunol 10. https://doi.org/10.3389/fimmu.2019.02716

Siu JHY, Surendrakumar V, Richards JA, Pettigrew GJ (2018) T cell allorecognition pathways in solid organ transplantation. Front Immunol 9. https://doi.org/10.3389/fimmu.2018.02548

Tang Q, Vincenti F (2017) Transplant trials with Tregs: perils and promises. J Clin Invest 127:2505–2512. https://doi.org/10.1172/jci90598

Tang Q et al (2012) Altered balance between effector T cells and FOXP3+ HELIOS+ regulatory T cells after thymoglobulin induction in kidney transplant recipients. Transpl Int 25:1257–1267. https://doi.org/10.1111/j.1432-2277.2012.01565.x

Todo S et al (2016) A pilot study of operational tolerance with a regulatory T-cell-based cell therapy in living donor liver transplantation. Hepatology 64:632–643. https://doi.org/10.1002/hep.28459

Tsang JY et al (2008) Conferring indirect allospecificity on CD4+CD25+ Tregs by TCR gene transfer favors transplantation tolerance in mice. J Clin Invest 118:3619–3628. https://doi.org/10.1172/jci33185

Ugor E et al (2018) Glucocorticoid hormone treatment enhances the cytokine production of regulatory T cells by upregulation of Foxp3 expression. Immunobiology 223:422–431. https://doi.org/10.1016/j.imbio.2017.10.010

United Network for Organ Sharing: transplant trends (2020). https://unos.org/data/transplant-trends/
Veerapathran A, Pidala J, Beato F, Yu XZ, Anasetti C (2011) Ex vivo expansion of human Tregs specific for alloantigens presented directly or indirectly. Blood 118:5671–5680. https://doi.org/10.1182/blood-2011-02-337097
Vondran FW et al (2010) Impact of Basiliximab on regulatory T-cells early after kidney transplantation: down-regulation of CD25 by receptor modulation. Transpl Int 23:514–523. https://doi.org/10.1111/j.1432-2277.2009.01013.x
Waldmann H, Adams E, Fairchild P, Cobbold S (2006) Infectious tolerance and the long-term acceptance of transplanted tissue. Immunol Rev 212:301–313. https://doi.org/10.1111/j.0105-2896.2006.00406.x
Wang S et al (2008) Reduction of Foxp3-expressing regulatory T cell infiltrates during the progression of renal allograft rejection in a mouse model. Transpl Immunol 19:93–102. https://doi.org/10.1016/j.trim.2008.03.004
Watson CJ, Dark JH (2012) Organ transplantation: historical perspective and current practice. Br J Anaesth 108 (Suppl 1):i29–i42. https://doi.org/10.1093/bja/aer384
Wise MP, Bemelman F, Cobbold SP, Waldmann H (1998) Cutting edge: linked suppression of skin graft rejection can operate through indirect recognition. J Immunol 161:5813–5816
Xia G, He J, Leventhal JR (2008) Ex vivo-expanded natural CD4+CD25+ regulatory T cells synergize with host T-cell depletion to promote long-term survival of allografts. Am J Transplant 8:298–306. https://doi.org/10.1111/j.1600-6143.2007.02088.x
Zeng Q et al (2019) Effects of tacrolimus (FK506) and mycophenolate mofetil (MMF) on regulatory T cells and co-inhibitory receptors in the peripheral blood of human liver allograft patients. Immunopharmacol Immunotoxicol 41:380–385. https://doi.org/10.1080/08923973.2018.1533026
Zhang N et al (2009) Regulatory T cells sequentially migrate from inflamed tissues to draining lymph nodes to suppress the alloimmune response. *Immunity* 30:458–469. https://doi.org/10.1016/j.immuni.2008.12.022
Zhang H et al (2015) Sequential monitoring and stability of ex vivo-expanded autologous and nonautologous regulatory T cells following infusion in nonhuman primates. Am J Transplant 15:1253–1266. https://doi.org/10.1111/ajt.13113
Zheng SG, Gray JD, Ohtsuka K, Yamagiwa S, Horwitz DA (2002) Generation ex vivo of TGF-beta-producing regulatory T cells from CD4+CD25- precursors. J Immunol 169:4183–4189. https://doi.org/10.4049/jimmunol.169.8.4183
Zheng SG, Wang JH, Gray JD, Soucier H, Horwitz DA (2004) Natural and induced CD4+CD25+ cells educate CD4+CD25− cells to develop suppressive activity: the role of IL-2, TGF-β, and IL-10. J Immunol 172:5213–5221. https://doi.org/10.4049/jimmunol.172.9.5213
Zheng SG et al (2006) Transfer of regulatory T cells generated ex vivo modifies graft rejection through induction of tolerogenic CD4+CD25+ cells in the recipient. Int Immunol 18:279–289. https://doi.org/10.1093/intimm/dxh368
Zheng SG, Wang J, Wang P, Gray JD, Horwitz DA (2007) IL-2 is essential for TGF-beta to convert naive CD4+CD25- cells to CD25+Foxp3+ regulatory T cells and for expansion of these cells. J Immunol 178:2018–2027. https://doi.org/10.4049/jimmunol.178.4.2018

Regulatory T Cells for the Induction of Transplantation Tolerance

15

Weitao Que and Xiao-Kang Li

Abstract

Organ transplantation is the optimal treatment for terminal and irreversible organ failure. Achieving transplantation tolerance has long been the ultimate goal in the field of transplantation. Regulatory T cell (Treg)-based therapy is a promising novel approach for inducing donor organ-specific tolerance. Tregs play critical roles in the maintenance of immune homeostasis and self-tolerance, by promoting transplantation tolerance through a variety of mechanisms on different target cells, including anti-inflammatory cytokine production, induction of apoptosis, disruption of metabolic pathways, and mutual interaction with dendritic cells. The continued success of Treg-based therapy in the clinical setting is critically dependent on preclinical studies that support its translational potential. However, although some initial clinical trials of adoptive Treg therapy have successively demonstrated safety and efficacy for immunosuppressant minimization and transplantation tolerance induction, most Treg-based hematopoietic stem cell and solid organ clinical trials are still in their infancy. These clinical trials have not only focused on safety and efficacy but also included optimization and standardization protocols of good manufacturing practice regarding cell isolation, expansion, dosing, timing, specificity, quality control, concomitant immunosuppressants, and post-administration monitoring. We herein report a brief introduction of Tregs, including their phenotypic and functional characterization, and focus on the clinical translation of Treg-based therapeutic applications in the setting of transplantation.

Keywords

Allograft · Tregs · Tolerance · Transplantation

15.1 Introduction

Organ transplantation has achieved remarkable success over the past half-century and has become the optimal treatment choice for terminal and irreversible organ failure. However, despite this success, the lifelong requirement for continued administration of nonspecific immunosuppressants after transplantation can have major side effects, which include not only severe infections, chronic renal failure, cardiovascular diseases, metabolic syndrome, and malignancy but also the psychological burdens that impair quality of life (Ojo et al. 2003; Watt et al. 2010). Still worse is the fact that the effectiveness

W. Que · X.-K. Li (✉)
Division of Transplantation Immunology, National Research Institute for Child Health and Development, Tokyo, Japan
e-mail: ri-k@ncchd.go.jp

S.-G. Zheng (ed.), *T Regulatory Cells in Human Health and Diseases*, Advances in Experimental Medicine and Biology 1278, https://doi.org/10.1007/978-981-15-6407-9_15

of the immunosuppressive drugs required to maintain the tolerant state is suboptimal at preventing donor-specific antibody production and later-phase chronic organ rejection, which can eventually lead to graft failure (Kwun and Knechtle 2009; Wood and Goto 2012; Valenzuela and Reed 2017). In this regard, the quest for immunosuppressant-free organ transplantation, whereby patients achieve donor organ-specific tolerance without compromising their protective immunity, remains the ultimate goal in the field of transplantation.

In the pursuit of this Holy Grail of transplantation, regulatory immune cell-based therapies are emerging as promising strategies for inducing donor organ-specific tolerance. The regulatory immune cells, including regulatory T cells (Tregs) (Kingsley et al. 2002; Long and Wood 2009; Roncarolo et al. 2014), regulatory macrophages (Broichhausen et al. 2012), regulatory B cells (Kwun and Knechtle 2009; Sicard et al. 2015), myeloid-derived suppressor cells (Brem-Exner et al. 2008; Fleming and Mosser 2011; Broichhausen et al. 2012), and regulatory dendritic cells (Marín et al. 2018; Rosen et al. 2018; Thomson et al. 2019; Ochando et al. 2019), have been extensively examined for their ability to promote transplantation tolerance and have clearly shown benefits of safety and the graft survival (Wood et al. 2012; Papp et al. 2017). Among these cells, Tregs are particularly attractive for their critical roles in maintaining immune homeostasis and self-tolerance. Since the biology of Tregs has been extensively reviewed in several publications elsewhere (Wing and Sakaguchi 2010; Josefowicz et al. 2012; Georgiev et al. 2019) here, we focus on the clinical translation of Treg-based therapeutic applications in the setting of transplantation.

15.2 Phenotype and Heterogeneity

Broadly speaking, Tregs constitute a group of specialized T cells that act to suppress the immune response, thereby maintaining homeostasis and self-tolerance. These include a spectrum of regulatory T cells that include $CD4^{+}CD25^{+}Foxp3^{+}$ classical Tregs, interleukin-10 (IL-10)-producing $CD4^{+}$ T cells (Tr1), tumor growth factor-β (TGF-β)-producing T helper 3 cells (Th3), $CD8^{+}CD122^{+}$ Tregs, $CD8^{+}CD25^{+}Foxp3^{+}$ Tregs, double-negative Tregs (DNTs), natural killer T (NKT) cells, and γδT cells (Shalev et al. 2011). $CD4^{+}CD25^{+}Foxp3^{+}$ classical Tregs, which we will focus on in the present report, are the most well-studied, and clinical trials concerning these cells have advanced the farthest.

Treg cells comprise approximately 5–10% of circulating $CD4^{+}$ T cells and play indispensable roles in the regulation of immune homeostasis and transplantation tolerance (Sakaguchi et al. 2010). In 1995, Sakaguchi et al. were the first to identify the immunologic function of CD4+CD25+ Tregs in the maintenance of self-tolerance (Sakaguchi et al. 1995). Later, the X-linked gene forkhead box P3 (Foxp3) was identified as the master control gene of Tregs, specifically with regard to Treg development and function (Hori et al. 2003; Fontenot et al. 2003). Later, the expression of CD127, α-chain of IL-7R, was shown to be inversely correlated with the Foxp3 expression and Treg suppressive function (Liu et al. 2006; Seddiki et al. 2006). As such, the combination of CD4, CD25, and CD127 is considered to be a stringent marker demarcating a bona fide population of Tregs with an optimal stability and function.

More recently, the demethylation status of the Treg-specific demethylated region (TSDR) was reported to be crucial for regulating gene expression of the Foxp3 locus (Polansky et al. 2008). Notably, thymus-derived Tregs have a fully demethylated TSDR, while peripherally derived Tregs display only a partially demethylated region (Shevach and Thornton 2014). In addition, Miyara et al. described three phenotypically and functionally distinct subpopulations of Tregs in human peripheral blood based upon their CD45RA and Foxp3 expression (Miyara et al. 2009). They divided Tregs into $CD45RA^{+}Foxp3^{low}$ resting Tregs and $CD45RA^{-}Foxp3^{high}$ activated Tregs, both of which demonstrated suppressive in vitro properties, as well as a population of cytokine-secreting nonsuppressive CD45RA$^-$Foxp3^{low} Treg cells.

15.3 Mechanisms of Treg in Underlying Transplantation Tolerance

The mechanisms involving Tregs have been extensively studied and reviewed elsewhere (Sakaguchi et al. 2009; Schmidt et al. 2012). Tregs exert their suppressive function on different target cells through a variety of mechanisms, including the production of anti-inflammatory cytokines, the induction of apoptosis, and disruption of metabolic pathways and mutual interaction with dendritic cells (DCs). Most of these general suppressive mechanisms apply to alloimmune responses. The anti-inflammatory cytokines secreted by Treg cells include IL-10, IL-35, and TGF-β that act to suppress T-cell proliferation and maturation of DCs. Blocking IL-10 and TGF-β abrogates the Treg-mediated tolerance to skin and heart allografts (Bickerstaff et al. 2000; Hara et al. 2001). Tregs can also directly kill target cells via perforin- and granzyme-dependent apoptotic pathways (Grossman et al. 2004; Gondek et al. 2005). Granzyme-deficient mice are unable to establish long-term tolerance in a model of Treg-dependent skin graft tolerance (Gondek et al. 2008). Tregs mediate the disruption of metabolic pathways through the ectoenzymes CD39/CD73-mediated conversion of ADP/ATP to AMP and AMP to adenosine (Antonioli et al. 2013). Tregs from CD39-deficient mice showed a lower suppressive ability than those from wild type mice and failed to induce tolerance of allogeneic skin grafts (Deaglio et al. 2007). Tregs also display their immunosuppressive function by rendering DCs tolerogenic via the induction of indoleamine 2,3-dioxygenase expression in DCs or through transendocytosis of CD80/CD86 costimulatory molecules (Fallarino et al. 2003; Qureshi et al. 2011), following which tolerogenic DCs promote the generation of Tregs (Hasegawa and Matsumoto 2018). Min et al. reported the existence of an inhibitory feedback loop between tolerogenic DCs and Tregs in transplantation tolerance, in which tolerogenic DCs induce Tregs, which in turn induce tolerogenic DCs (Min et al. 2003). Alternatively, Tregs can perform their suppressive activities in a non-antigen-specific manner, namely bystander suppression, which allows them to suppress effector T cells of diverse specificities (Thornton and Shevach 2000). In addition, a Treg-mediated suppressive microenvironment promotes the emergence of Tregs specific for other antigens or other types of regulatory cells (Gershon and Kondo 1971; Jonuleit et al. 2002; Andersson et al. 2008). This phenomenon, known as "infectious tolerance," may contribute to the long-term maintenance of original Treg-induced tolerance.

15.4 The Role Treg Cells in Preclinical Transplantation Models

In 2002, Graca et al. first identified the persistent presence of Tregs in tolerated allografts and demonstrated that Tregs in tolerant skin grafts could prevent nontolerant lymphocytes from rejecting fresh skin allografts, without impediment of rejection of third-party skin (Graca et al. 2002). This study was the first to highlight the critical role of Tregs in transplantation tolerance. Since then, numerous studies have demonstrated the importance of Tregs in transplantation tolerance induction and maintenance in multiple organ transplantation models. Li et al. reported that the frequency of Tregs was increased in both the periphery and in tolerant liver grafts from day 5 to day 100 post-transplantation (Li et al. 2008). Depletion of Tregs using anti-CD25 antibodies resulted in acute liver allograft rejection in tolerant mice. They also demonstrated that CTLA4, TGF-β, IL-4, and apoptosis of graft-infiltrating T cells were involved in Treg-mediated transplantation tolerance. Joffre et al. reported the prevention of acute and chronic allograft rejection with CD4$^+$CD25$^+$Foxp3$^+$ Tregs (Joffre et al. 2008). They showed that in vitro allo-stimulated Tregs were able to induce long-term tolerance to bone marrow and subsequent skin and cardiac allografts. They also showed that Tregs specific only for directly presented alloantigens were able to prevent acute rejection but failed to prevent chronic rejection, whereas Tregs specific for directly and indirectly presented alloantigens were capable of preventing both acute and chronic allograft rejection. Additional evidence supporting the application of Tregs to transplantation tolerance has been made available from studies using humanized mouse models of organ transplantation (Shultz et al. 2007; Issa et al. 2010; Wu et al. 2013). Nadig et al. showed that human Tregs expanded ex vivo could prevent transplant arteriosclerosis, which is a pathologic hallmark of chronic transplant dysfunction affecting the long-term outcome after transplantation (Nadig et al. 2010).

15.5 The Role of Treg Cells in Clinical Transplantation Trials

In clinical liver transplantation, operational tolerance can occur in selected recipients after immunosuppression has completely weaned off. These recipients exhibit a significantly increased frequency of circulating and intrahepatic Tregs compared to non-tolerant recipients or healthy volunteers (Li et al. 2004). In addition, increased levels of circulating and intra-graft Tregs are associated with stable graft function in both kidney and lung transplantation (F et al. 2006; Bestard et al. 2007). In clinical trials performed in 2009, Trzonkowski et al. first demonstrated the effectiveness of adoptive transfer of ex vivo expanded $CD4^{+}CD25^{+}CD127^{low}$ Tregs in the treatment of graft-versus-host disease (GvHD) (Trzonkowski et al. 2009). In 2016, after publishing the results of a clinical trial of Treg-based therapy in solid organ transplantation, Todo et al. showed that adoptive transfer of ex vivo-generated Tregs was safe and effective for immunosuppressant minimization and operational tolerance induction in living donor liver recipients with nonimmunological liver diseases (Todo et al. 2016). Soon after, Chandran et al. reported the treatment of subclinical inflammation in kidney transplant patients with polyclonal autologous Tregs (Chandran et al. 2017). At present, most clinical trials regarding the infusion of Tregs in hematopoietic stem cell transplantation and solid organ transplantation are still ongoing (Table 15.1). These clinical trials have not only focused on safety and efficacy but have also included optimization and standardization protocols of good manufacturing practice regarding cell isolation, expansion, dosing, timing, specificity, quality control, concomitant immunosuppressants, and post-administration monitoring (Fig. 15.1).

15.5.1 Isolation

Currently, the majority of clinical trials harvest Tregs from adult peripheral blood (ABP), which is an easily performed procedure to obtain these cells. However, since Tregs are relatively rare in ABP, and since ABP Tregs comprise highly heterogenous populations of antigen experienced cells, some researchers have used umbilical cord blood (UCB) as an alternative source for the isolation of Tregs, since UCB is rich in Tregs and virtually devoid of antigen-experienced memory T cells (Riley et al. 2009). Most UCB-derived Tregs consist of naïve cells with long telomeres and are easily separable from effector T cells (Teffs) due to the low complexity of UCB T cells. Brunstein et al. isolated $5–7.5 \times 10^{6}$ Tregs per unit of UBC with a mean purity of 66% (Brunstein et al. 2011). UCB Tregs have shown a higher expansion capacity and suppressive function than their PB counterparts (Lin et al. 2014). Clinical trials have demonstrated the safety and efficacy of UCB Tregs in autoimmune diseases and transplantation (Brunstein et al. 2016; Seay et al. 2017). One drawback associated with the use of UCB as a source is that such Tregs are allogeneic cells unless derived from banked autologous UCB.

Another source of Tregs is discarded pediatric thymus (Dijke et al. 2016). Discarded pediatric thymus is an excellent source, with 1 g of thymus able to yield around 500 times more Tregs than 1 ml of BP. These Tregs also show a superior

Table 15.1 Current clinical trials of Treg-based therapy in transplantation

Trial ID	Phase	Sponsor	Organ	Treg type	Doses	Recruitment status
NCT02474199	Phase I/II	National Institute of Allergy and Infectious Diseases	Liver	Donor-specific Tregs	300–500 × 10^6 cells	Completed
NCT03577431	Phase I/II	National Institute of Allergy and Infectious Diseases	Liver	Donor-specific Tregs	2.5–500 × 10^6 cells	Recruiting
NCT02166177	Phase I/II	Guy's and St Thomas' NHS Foundation Trust	Liver	Polyclonal Tregs	0.5–4.5 × 10^6/kg	Completed
NCT02145325	Phase I	Northwestern University	Kidney	Polyclonal Tregs	500–5000 × 10^6 cells	Completed
NCT02711826	Phase I/II	National Institute of Allergy and Infectious Diseases	Kidney	Polyclonal vs. donor-specific Tregs	1 × 10^6 cells/kg	Recruiting
NCT03284242	Not applicable	Roberto Gedaly	Kidney	Unknown	Unknown	Recruiting
NCT02091232	Phase I	Massachusetts General Hospital	Kidney	Donor-specific Tregs	300–900 × 10^6 cells	Active, not recruiting
NCT02244801	Phase I	University of California, San Francisco	Kidney	Donor-specific Tregs	300–900 × 10^6 cells	Completed
NCT02371434	Phase I/II	Petra Reinke	Kidney	Polyclonal Tregs	0.5–3 × 10^6 cells/kg	Completed
NCT02129881	Phase I/II	Guy's and St Thomas' NHS Foundation Trust	Kidney	Polyclonal Tregs	1–10 × 10^6 cells	Completed
NCT03867617	Phase I/II	Thomas Wekerle	Kidney	Unknown	Unknown	Recruiting
NCT01446484	Phase I/II	Pirogov Russian National Research Medical University	Kidney	Polyclonal Tregs	Two doses of 200 × 10^6 cells	Unknown
NCT02088931	Phase I	University of California, San Francisco	Kidney	Polyclonal Tregs	320 × 10^6 cells	Completed
ISRCTN11038572	Phase II	University of Oxford	Kidney	Polyclonal Tregs	5–10 × 10^6/kg	Suspended
NCT03444064	Phase I	University of Alberta	Islet	Polyclonal Tregs	400–1600 × 10^6 cells	Recruiting
NCT02749084	Phase I/II	Mario Arpinati	Stem cell	Alloantigen Tregs	0.17–0.66 × 10^6 cells/kg/month	Recruiting
NCT02385019	Phase I/II	Instituto de Medicina Molecular João Lobo Antunes	Stem cell	Alloantigen Tregs	0.5–3 × 10^6 cells/kg	Recruiting
NCT01903473	Phase II	University of Liege	Stem cell	Alloantigen Tregs	≥0.5 × 10^6 cells/kg	Recruiting
NCT00602693	Phase I	Masonic Cancer Center, University of Minnesota	Stem cell	Alloantigen Tregs	0.1–300 × 10^6 cells/kg	Completed
NCT02526329	Phase I	Everett Meyer	Stem cell	Alloantigen Tregs	Unknown	Suspended
NCT01937468	Phase I	Dana-Farber Cancer Institute	Stem cell	Alloantigen Tregs	Unknown	Active, not recruiting
NCT01660607	Phase I/II	Everett Meyer	Stem cell	Alloantigen Tregs	1–10 × 10^6 cells/kg	Recruiting
NCT01634217	Phase I	Masonic Cancer Center, University of Minnesota	Stem cell	Alloantigen Tregs	3–100 × 10^6 cells/kg	Completed
NCT01795573	Phase I	H. Lee Moffitt Cancer Center and Research Institute	Stem cell	Alloantigen Tregs	Unknown	Active, not recruiting

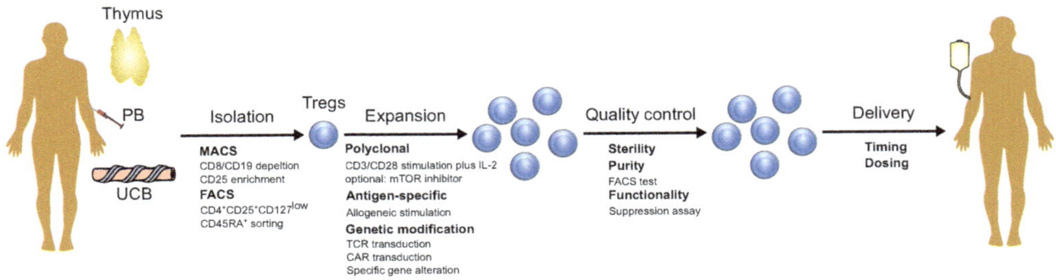

Fig. 15.1 Current workflow of Treg-based therapy. In the first step, peripheral blood (PB), umbilical cord blood (UCB), and thymus are used as sources for Treg isolation. Treg isolation methods include magnetic-activated cell sorting (MACS) or fluorescence-activated cell sorting (FACS) based on surface marker expression profiles. Tregs can be expanded in a polyclonal manner or antigen-specific manner. Polyclonal expansion is performed by stimulation with anti-CD3/CD28 and interleukin-2 (IL-2). Antigen-specific expansion is performed by stimulation with allogeneic cells. Genetic Treg modifications could be used to improve Treg specificity, potency, stability, trafficking, and safety. Quality control is performed prior to in vivo transfer of Treg

suppressive function compared to ABP Tregs. Furthermore, thymus-isolated Tregs express lower levels of class I HLA and include fewer HLA-DR$^+$ cells than BP Tregs. The immunogenic property of thymus isolated Tregs may allow them to be used in "off-the-shelf" tolerogenic therapy (Dijke et al. 2017).

The methods of Treg isolation that comply with Good Manufacturing Practice (GMP) guidelines are usually based on magnetic bead depletion/enrichment protocols (Hoffmann et al. 2006a). The CliniMACS system is the most widely applied instrument for magnetic separation of Tregs. In this method, undesired CD8$^+$/CD19$^+$ cells are depleted, and CD25$^+$ cells are then enriched by positive selection, eventually yielding CD4$^+$CD25$^+$ Tregs. However, isolated CD4$^+$CD25$^+$ Tregs may be contaminated with Teffs expressing low levels of CD25. The CliniMACS system does not allow for Treg isolation based on multiple markers. FACS has proven to be an effective method of further Treg isolation with higher purity based on multiple markers. Researchers have developed a FACS-based protocol for isolating CD4$^+$CD25$^+$CD127^{low} Tregs, which yield a Treg purity ranging 95–100% (Putnam et al. 2009). In addition, Hoffmann et al. suggested the inclusion of the CD45RA marker to isolate CD4$^+$CD25$^+$CD127^{low}CD45RA$^+$ naïve Tregs for adoptive therapy, which displayed a greater expansion potential, Foxp3 stability, and suppressive function than their CD45RA$^-$ antigen-experienced or memory counterparts (Hoffmann et al. 2006b; Jb et al. 2015).

15.5.2 Expansion

Isolated Tregs usually require ex vivo expansion to meet the needs for adoptive Treg therapy. Tregs can be expanded ex vivo in a polyclonal manner with bead-bound or soluble anti-CD3/CD28 monoclonal antibody stimulation in the presence of IL-2 (Hoffmann et al. 2004). Hippen et al. achieved a >50 million-fold expansion of PB Tregs with the addition of artificial antigen-presenting cells expressing the Fc receptor and CD86 (Hippen et al. 2011b). However, despite such impressive results, these protocols have several drawbacks.

First, such protocols are limited to expanding polyclonal Tregs. Second, contaminated Teffs may also proliferate vigorously alongside Tregs, which can cause safety issues, especially for MACS-isolated CD4$^+$CD25$^+$ Tregs. Finally, repeated stimulation may lead to the loss of Foxp3 expression and emergence of pro-inflammatory cytokines (Hoffmann et al. 2009).

Multiple approaches have been proposed to address these defects. Some researchers use allogenic B cells as antigen-presenting cells to generate and expand clinical grade alloantigen-specific Tregs (Putnam et al. 2013; Landwehr-Kenzel et al. 2014). Aside from the FACS-based isolation protocol, another important solution to the issue of Teff contamination is to optimize the culture conditions using mammalian target of rapamycin (mTOR) inhibitors, such as rapamycin, which also improves Treg stability (Battaglia et al. 2005; Strauss et al. 2007; Tresoldi et al. 2011; Scottà et al. 2013). Treg expansion in the presence of rapamycin selectively suppresses non-Treg proliferation while enhancing the Treg expansion, stability, and suppressive function (Battaglia et al. 2005, 2006; Zeiser et al. 2008; Bocian et al. 2010). Therefore, the addition of mTOR inhibitors, such as rapamycin, has become a part of some GMP protocols.

15.5.3 Timing

The optimal time points for Treg infusion in organ transplantation is still being investigated. Most preclinical studies of Treg-based therapy infuse Tregs around the time of transplantation. The reason for the pre-transplant administration of Tregs is to allow Tregs time to reduce the degree of Teff activation. In addition, the early post-transplantation period is accompanied by ischemia-reperfusion injury-induced inflammation, which can recruit Tregs to the graft, facilitating immunoregulation and tissue repair. If induction with depleting agents is applied at the time of the transplant, Tregs should be infused after transplantation to avoid their subsequent depletion by depleting agents. Treg infusion may also synergize with T-cell depletion to promote the long-term survival of allografts (Xia et al. 2008; Ma et al. 2011). In addition, later-phase Treg administration has been tested for immunosuppressant withdrawal and tolerance induction (NCT02474199).

15.5.4 Dosing

Another point of consideration is the optimal dosing of Treg infusion. The ratio of Tregs to Teffs influences the balance between tolerance and rejection. Preclinical studies have shown that a high ratio of 1:2 to 1:1, i.e., 33–50% Tregs, is necessary to prevent transplantation rejection (Riley et al. 2009; Hippen et al. 2011c; Tang and Lee 2012). It is therefore assumed that a dose of 49 to 79 $\times$ 10^9 Tregs will be needed to increase the Treg pool to clinically efficacious numbers for a patient with a body weight of 70 kg (Tang and Lee 2012). Furthermore, evidence from preclinical studies has also suggested that around 10- to 100-fold more polyclonal Tregs are needed to achieve the same efficacy as antigen-specific Tregs (Nishimura et al. 2004; Golshayan et al. 2007; Tarbell et al. 2007). Moreover, in the presence of lymphodepletion, such as anti-thymocyte globulin (ATG) induction, a single fusion of around 3–5 $\times$ 10^9 Tregs can effectively boost the Treg percentage to more than 33%. Given that most of these data were from animal studies, the most effective Treg dose in human transplantation remains unclear. Several dose-escalation studies in humans have been carried out, and no serious adverse events have been observed, suggesting the safety of Treg-based therapies. Further comprehensive dose escalation studies should be performed to assess the benefit–risk balance of Treg dosing in human transplantation.

15.5.5 Quality Control

Although quality control is required prior to the in vivo transfer of Tregs, it is not yet standardized. Purity testing is commonly performed by FACS. The assessment of the epigenetic status of the Foxp3 locus, especially TSDR demethylation, which is closely related to the stability and thus the function of Tregs, is another important quality control approach. Traditional Treg suppression assays, which involve analyzing their capacity to inhibit proliferation or cytokine production of co-cultured Teffs, are time-consuming and associated with substantial bias (Brusko et al. 2007; Venken et al. 2007). Novel rapid functional assays should be developed to ensure the quality of adoptive Tregs.

15.5.6 Genetically Modified Tregs

Genetic engineering is a revolutionary biotechnology in current therapeutic medicine. With regard to Treg-based therapies, genetic Treg modifications can be used to improve the specificity, potency, stability trafficking, and safety of Tregs. As mentioned previously, antigen-specific Tregs are more robust than polyclonal Tregs in their tolerance induction but are difficult to expand due to their scarcity. Genetically modifying T-cell receptors (TCRs) allows for the generation of antigen-specific Tregs by delivering designated TCR-α and -β chain genes. Genetically TCR-modified antigen-specific Tregs have been examined in preclinical models of autoimmune diseases and transplantation (Zhou et al. 2004; Tang et al. 2004; Fujio et al. 2006; Golshayan et al. 2007; Stephens et al. 2009). In addition, genetic TCR modification can be combined with in vitro allostimulation to generate dual-specific Tregs. Tsang et al. generated Tregs with direct and indirect allospecificity, which have increased efficiency to promote graft tolerance (Tsang et al. 2008).

Chimeric antigen receptor (CAR) technology is another promising approach to generate antigen-specific Tregs. CAR Tregs were engineered with artificial receptors that have an antigen recognition extracellular domain and T-cell activation signaling intracellular domains. The antigen recognition domain has the ability to recognize both MHC-restricted and MHC-nonrestricted antigens, which gives CAR Tregs broader applicability than TCR-modified Tregs. CAR-engineered effector T cells have helped achieve great advances in blood cancer treatment (Porter et al. 2011; Brentjens et al. 2013; Maude et al. 2014). Serval preclinical studies have demonstrated their safety and efficacy in promoting transplantation tolerance, although clinical studies are still lagging behind (MacDonald et al. 2016; Boardman et al. 2017; Noyan et al. 2017).

In addition to specificity modifications, genetic engineering can also target selected genes to improve the potency and stability of Tregs. Okada et al. engineered mouse primary T cells by targeting the Foxp3 promoter locus to induce stable Foxp3 expression, even under inflammatory conditions, which is also associated with an enhanced suppressive function (Okada et al. 2017). Furthermore, there are many other valuable features imbued by genetic engineering that have been tested in Teffs but not translated to Tregs, such as the insertion of suicide genes to maintain the safety, overexpression of chemokine receptors to enhance trafficking capacity, and the incorporation of tissue repair molecules (Tey 2014; Raffin et al. 2020).

15.5.7 Concomitant Immunosuppressants

Although Treg-based therapy is being developed with the goal of minimizing or even avoiding the need for immunosuppressants after transplantation, the use of concomitant immunosuppressants is inevitable for preventing acute rejection in the current stage of clinical trials. The influence of immunosuppressants on Tregs differs among agents, which should be taken into consideration when designing Treg-based therapy trials in transplantation.

The effects of variety immunosuppressants used for transplantation on Tregs have been comprehensively reviewed (Furukawa et al. 2016; Camirand and Riella 2017). The introduction of calcineurin inhibitors (CNIs) is considered a milestone in immunosuppression, effectively inhibiting acute rejection and improving the short-term allograft survival. CNIs inhibit the calcium-dependent nuclear translocation of nuclear factor of activated T cells (NFAT) following T-cell receptor engagement, thereby blocking the Teff function and transcription of various cytokines, including IL-2, IFN-γ, and TNF-α (Hermann-Kleiter and Baier 2010; Macian

2005). However, Tregs also require the calcineurin/NFAT signaling pathway for their survival and proliferation, and the detrimental impact of CNIs on Tregs has been observed in both experimental and clinical settings (San Segundo et al. 2007; Calvo-Turrubiartes et al. 2009; Miroux et al. 2012). mTOR inhibitors are a safe alternative to standard CNI-based therapy due to their relative lack of nephrotoxicity. mTOR inhibitors has been shown to preferentially inhibit Teffs while favoring the Treg expansion and function (Thomson et al. 2009). Kidney transplant patients on rapamycin-based therapy showed a significantly higher frequency of circulating Tregs than patients on CNI-based therapy at 1 year post-transplantation (Segundo et al. 2006). Rapamycin can also promote ex vivo Treg generation and expansion (Battaglia et al. 2005; Hippen et al. 2011a). Aside from maintenance immunosuppressants, whether or not induction immunosuppressants should be given in conjunction with Tregs is highly debated, and conflicting results have been reported. For example, induction using depleting agents, such as ATG and alemtuzumab, has been reported to increase Treg populations (Watanabe et al. 2006; Bloom et al. 2008; Gurkan et al. 2010; Shimony et al. 2012). However, induction with basiliximab, which blocks IL-2 receptor, reduces the Treg population (Bluestone et al. 2008; van den Hoogen et al. 2015).

15.5.8 Post-administration Monitoring

As a novel therapy, the safety and efficiency of Treg-based therapy should be considered. To address safety concerns, it is essential to evaluate and reduce the potential risk of infection, malignancy, and cytokine storm complications caused by polyclonal Tregs or bystander suppression. Efficiency concerns can be addressed by clinical outcome measures, such as the graft function and biopsy findings. In addition, researchers have developed various monitoring methods for evaluating the efficiency of Treg-based therapy, such as in vivo cell tracking, the IFN-γ ELISPOT assay, and FACS-based immune phenotype profiling (Ashoor et al. 2013; Bestard et al. 2013; Streitz et al. 2013; Bluestone et al. 2015).

15.6 Conclusion

Tolerance has long been the "Holy Grail" for transplantation. In the past decade, advances in Treg biological studies have supported the translational potential of Treg-based therapy in promoting transplantation tolerance in the clinical setting. Animal studies and early human trials in transplantation have shown very promising results. Currently ongoing clinical trials will give us more data on the safety and efficacy of Treg-based therapy. The further optimization of highly reproducible GMP protocols with regard to cell isolation, expansion, dosing, timing, specificity, quality control, concomitant immunosuppressants, and post-administration monitoring is needed. With the rapid advances in science and technology in this field, our ultimate goal of transplantation tolerance will soon become a reality.

References

Andersson J, Tran DQ, Pesu M et al (2008) CD4+ FoxP3+ regulatory T cells confer infectious tolerance in a TGF-beta-dependent manner. J Exp Med 205:1975–1981. https://doi.org/10.1084/jem.20080308

Antonioli L, Pacher P, Vizi ES, Haskó G (2013) CD39 and CD73 in immunity and inflammation. Trends Mol Med 19:355–367. https://doi.org/10.1016/j.molmed.2013.03.005

Ashoor I, Najafian N, Korin Y et al (2013) Standardization and cross validation of alloreactive IFNγ ELISPOT assays within the clinical trials in organ transplantation consortium. Am J Transplant 13:1871–1879. https://doi.org/10.1111/ajt.12286

Battaglia M, Stabilini A, Roncarolo M-G (2005) Rapamycin selectively expands CD4+CD25+FoxP3+ regulatory T cells. Blood 105:4743–4748. https://doi.org/10.1182/blood-2004-10-3932

Battaglia M, Stabilini A, Migliavacca B et al (2006) Rapamycin promotes expansion of functional CD4+CD25+FOXP3+ regulatory T cells of both healthy subjects and type 1 diabetic patients. J Immunol 177:8338–8347. https://doi.org/10.4049/jimmunol.177.12.8338

Bestard O, Cruzado JM, Mestre M et al (2007) Achieving donor-specific hyporesponsiveness is associated with

FOXP3+ regulatory T cell recruitment in human renal allograft infiltrates. J Immunol 179:4901–4909. https://doi.org/10.4049/jimmunol.179.7.4901

Bestard O, Crespo E, Stein M et al (2013) Cross-validation of IFN-γ Elispot assay for measuring alloreactive memory/effector T cell responses in renal transplant recipients. Am J Transplant 13:1880–1890. https://doi.org/10.1111/ajt.12285

Bickerstaff AA, VanBuskirk AM, Wakely E, Orosz CG (2000) Transforming growth factor-beta and interleukin-10 subvert alloreactive delayed type hypersensitivity in cardiac allograft acceptor mice. Transplantation 69:1517–1520. https://doi.org/10.1097/00007890-200004150-00055

Bloom DD, Chang Z, Fechner JH et al (2008) CD4+CD25+FOXP3+ regulatory T cells increase de novo in kidney transplant patients after immunodepletion with Campath-1H. Am J Transplant 8:793–802. https://doi.org/10.1111/j.1600-6143.2007.02134.x

Bluestone JA, Liu W, Yabu JM et al (2008) The effect of costimulatory and interleukin 2 receptor blockade on regulatory T cells in renal transplantation. Am J Transplant 8:2086–2096. https://doi.org/10.1111/j.1600-6143.2008.02377.x

Bluestone JA, Buckner JH, Fitch M et al (2015) Type 1 diabetes immunotherapy using polyclonal regulatory T cells. Sci Transl Med 7:315ra189. https://doi.org/10.1126/scitranslmed.aad4134

Boardman DA, Philippeos C, Fruhwirth GO et al (2017) Expression of a chimeric antigen receptor specific for donor HLA class I enhances the potency of human regulatory T cells in preventing human skin transplant rejection. Am J Transplant 17:931–943. https://doi.org/10.1111/ajt.14185

Bocian K, Borysowski J, Wierzbicki P et al (2010) Rapamycin, unlike cyclosporine A, enhances suppressive functions of in vitro-induced CD4+CD25+ Tregs. Nephrol Dial Transplant 25:710–717. https://doi.org/10.1093/ndt/gfp586

Brem-Exner BG, Sattler C, Hutchinson JA et al (2008) Macrophages driven to a novel state of activation have anti-inflammatory properties in mice. J Immunol 180:335–349. https://doi.org/10.4049/jimmunol.180.1.335

Brentjens RJ, Davila ML, Riviere I et al (2013) CD19-targeted T cells rapidly induce molecular remissions in adults with chemotherapy-refractory acute lymphoblastic leukemia. Sci Transl Med 5:177ra38. https://doi.org/10.1126/scitranslmed.3005930

Broichhausen C, Riquelme P, Geissler EK, Hutchinson JA (2012) Regulatory macrophages as therapeutic targets and therapeutic agents in solid organ transplantation. Curr Opin Organ Transplant 17:332–342. https://doi.org/10.1097/MOT.0b013e328355a979

Brunstein CG, Miller JS, Cao Q et al (2011) Infusion of ex vivo expanded T regulatory cells in adults transplanted with umbilical cord blood: safety profile and detection kinetics. Blood 117:1061–1070. https://doi.org/10.1182/blood-2010-07-293795

Brunstein CG, Miller JS, McKenna DH et al (2016) Umbilical cord blood-derived T regulatory cells to prevent GVHD: kinetics, toxicity profile, and clinical effect. Blood 127:1044–1051. https://doi.org/10.1182/blood-2015-06-653667

Brusko TM, Hulme MA, Myhr CB et al (2007) Assessing the in vitro suppressive capacity of regulatory T cells. Immunol Investig 36:607–628. https://doi.org/10.1080/08820130701790368

Calvo-Turrubiartes M, Romano-Moreno S, García-Hernández M et al (2009) Quantitative analysis of regulatory T cells in kidney graft recipients: a relationship with calcineurin inhibitor level. Transpl Immunol 21:43–49. https://doi.org/10.1016/j.trim.2009.02.002

Camirand G, Riella LV (2017) Treg-centric view of immunosuppressive drugs in transplantation: a balancing act. Am J Transplant 17:601–610. https://doi.org/10.1111/ajt.14029

Chandran S, Tang Q, Sarwal M et al (2017) Polyclonal regulatory T cell therapy for control of inflammation in kidney transplants. Am J Transplant 17:2945–2954. https://doi.org/10.1111/ajt.14415

Deaglio S, Dwyer KM, Gao W et al (2007) Adenosine generation catalyzed by CD39 and CD73 expressed on regulatory T cells mediates immune suppression. J Exp Med 204:1257–1265. https://doi.org/10.1084/jem.20062512

Dijke IE, Hoeppli RE, Ellis T et al (2016) Discarded human thymus is a novel source of stable and Long-lived therapeutic regulatory T cells. Am J Transplant 16:58–71. https://doi.org/10.1111/ajt.13456

Dijke E, Hoeppli R, Larsen I et al (2017) OR42 expanded thymic regulatory T cells (Tregs) have low class I HLA expression levels and are HLA-DR negative – potential for 'off-the-shelf' cellular therapy? Hum Immunol 78:40. https://doi.org/10.1016/j.humimm.2017.06.048

F M, M M, N S et al (2006) Peripheral CD4+ CD25+ Treg cell expansion in lung transplant recipients is not affected by calcineurin inhibitors. Int Immunopharmacol 6:2002–2010. https://doi.org/10.1016/j.intimp.2006.07.019

Fallarino F, Grohmann U, Hwang KW et al (2003) Modulation of tryptophan catabolism by regulatory T cells. Nat Immunol 4:1206–1212. https://doi.org/10.1038/ni1003

Fleming BD, Mosser DM (2011) Regulatory macrophages: setting the threshold for therapy. Eur J Immunol 41:2498–2502. https://doi.org/10.1002/eji.201141717

Fontenot JD, Gavin MA, Rudensky AY (2003) Foxp3 programs the development and function of CD4+CD25+ regulatory T cells. Nat Immunol 4:330–336. https://doi.org/10.1038/ni904

Fujio K, Okamoto A, Araki Y et al (2006) Gene therapy of arthritis with TCR isolated from the inflamed paw. J Immunol 177:8140–8147. https://doi.org/10.4049/jimmunol.177.11.8140

Furukawa A, Wisel SA, Tang Q (2016) Impact of immune-modulatory drugs on Treg. Transplantation

100:2288–2300. https://doi.org/10.1097/TP.0000000000001379

Georgiev P, Charbonnier L-M, Chatila TA (2019) Regulatory T cells: the many faces of Foxp3. J Clin Immunol 39(7):623–640. https://doi.org/10.1007/s10875-019-00684-7

Gershon RK, Kondo K (1971) Infectious immunological tolerance. Immunology 21:903–914

Golshayan D, Jiang S, Tsang J et al (2007) In vitro–expanded donor alloantigen–specific CD4+CD25+ regulatory T cells promote experimental transplantation tolerance. Blood 109:827–835. https://doi.org/10.1182/blood-2006-05-025460

Gondek DC, Lu L-F, Quezada SA et al (2005) Cutting edge: contact-mediated suppression by CD4+CD25+ regulatory cells involves a Granzyme B-dependent, perforin-independent mechanism. J Immunol 174:1783–1786. https://doi.org/10.4049/jimmunol.174.4.1783

Gondek DC, DeVries V, Nowak EC et al (2008) Transplantation survival is maintained by Granzyme B$^+$ regulatory cells and adaptive regulatory T cells. J Immunol 181:4752–4760. https://doi.org/10.4049/jimmunol.181.7.4752

Graca L, Cobbold SP, Waldmann H (2002) Identification of regulatory T cells in tolerated allografts. J Exp Med 195:1641–1646. https://doi.org/10.1084/jem.20012097

Grossman WJ, Verbsky JW, Barchet W et al (2004) Human T regulatory cells can use the perforin pathway to cause autologous target cell death. Immunity 21:589–601. https://doi.org/10.1016/j.immuni.2004.09.002

Gurkan S, Luan Y, Dhillon N et al (2010) Immune reconstitution following rabbit antithymocyte globulin. Am J Transplant 10:2132–2141. https://doi.org/10.1111/j.1600-6143.2010.03210.x

Hara M, Kingsley CI, Niimi M et al (2001) IL-10 is required for regulatory T cells to mediate tolerance to alloantigens in vivo. J Immunol 166:3789–3796. https://doi.org/10.4049/jimmunol.166.6.3789

Hasegawa H, Matsumoto T (2018) Mechanisms of tolerance induction by dendritic cells in vivo. Front Immunol 9. https://doi.org/10.3389/fimmu.2018.00350

Hermann-Kleiter N, Baier G (2010) NFAT pulls the strings during CD4+ T helper cell effector functions. Blood 115:2989–2997. https://doi.org/10.1182/blood-2009-10-233585

Hippen KL, Merkel SC, Schirm DK et al (2011a) Generation and large-scale expansion of human inducible regulatory T cells that suppress graft-versus-host disease. Am J Transplant 11:1148–1157. https://doi.org/10.1111/j.1600-6143.2011.03558.x

Hippen KL, Merkel SC, Schirm DK et al (2011b) Massive ex vivo expansion of human natural regulatory T cells (T(regs)) with minimal loss of in vivo functional activity. Sci Transl Med 3:83ra41. https://doi.org/10.1126/scitranslmed.3001809

Hippen KL, Riley JL, June CH, Blazar BR (2011c) Clinical perspectives for regulatory T cells in transplantation tolerance. Semin Immunol 23:462–468. https://doi.org/10.1016/j.smim.2011.07.008

Hoffmann P, Eder R, Kunz-Schughart LA et al (2004) Large-scale in vitro expansion of polyclonal human CD4(+)CD25high regulatory T cells. Blood 104:895–903. https://doi.org/10.1182/blood-2004-01-0086

Hoffmann P, Boeld TJ, Eder R et al (2006a) Isolation of CD4+CD25+ regulatory T cells for clinical trials. Biol Blood Marrow Transplant 12:267–274. https://doi.org/10.1016/j.bbmt.2006.01.005

Hoffmann P, Eder R, Boeld TJ et al (2006b) Only the CD45RA+ subpopulation of CD4+CD25high T cells gives rise to homogeneous regulatory T-cell lines upon in vitro expansion. Blood 108:4260–4267. https://doi.org/10.1182/blood-2006-06-027409

Hoffmann P, Boeld TJ, Eder R et al (2009) Loss of FOXP3 expression in natural human CD4+CD25+ regulatory T cells upon repetitive in vitro stimulation. Eur J Immunol 39:1088–1097. https://doi.org/10.1002/eji.200838904

van den Hoogen MWF, Kamburova EG, Baas MC et al (2015) Rituximab as induction therapy after renal transplantation: a randomized, double-blind, placebo-controlled study of efficacy and safety. Am J Transplant 15:407–416. https://doi.org/10.1111/ajt.13052

Hori S, Nomura T, Sakaguchi S (2003) Control of regulatory T cell development by the transcription factor Foxp3. Science 299:1057–1061. https://doi.org/10.1126/science.1079490

Issa F, Hester J, Goto R et al (2010) Ex vivo-expanded human regulatory T cells prevent the rejection of skin allografts in a humanised mouse model. Transplantation 90:1321–1327. https://doi.org/10.1097/TP.0b013e3181ff8772

Jb C, C S, A V et al (2015) Developing in vitro expanded CD45RA+ regulatory T cells as an adoptive cell therapy for Crohn's disease. Gut 65:584–594. https://doi.org/10.1136/gutjnl-2014-306919

Joffre O, Santolaria T, Calise D et al (2008) Prevention of acute and chronic allograft rejection with CD4+CD25+Foxp3+ regulatory T lymphocytes. Nat Med 14:88–92. https://doi.org/10.1038/nm1688

Jonuleit H, Schmitt E, Kakirman H et al (2002) Infectious tolerance: human CD25(+) regulatory T cells convey suppressor activity to conventional CD4(+) T helper cells. J Exp Med 196:255–260. https://doi.org/10.1084/jem.20020394

Josefowicz SZ, Lu L-F, Rudensky AY (2012) Regulatory T cells: mechanisms of differentiation and function. Annu Rev Immunol 30:531–564. https://doi.org/10.1146/annurev.immunol.25.022106.141623

Kingsley CI, Karim M, Bushell AR, Wood KJ (2002) CD25+CD4+ regulatory T cells prevent graft rejection: CTLA-4- and IL-10-dependent immunoregulation of alloresponses. J Immunol 168:1080–1086. https://doi.org/10.4049/jimmunol.168.3.1080

Kwun J, Knechtle SJ (2009) Overcoming chronic rejection-can it B? Transplantation 88:955–961. https://doi.org/10.1097/TP.0b013e3181b96646

Landwehr-Kenzel S, Issa F, Luu S-H et al (2014) Novel GMP-compatible protocol employing an allogeneic B cell bank for clonal expansion of allospecific natural regulatory T cells. Am J Transplant 14:594–606. https://doi.org/10.1111/ajt.12629

Li Y, Koshiba T, Yoshizawa A et al (2004) Analyses of peripheral blood mononuclear cells in operational tolerance after pediatric living donor liver transplantation. Am J Transplant 4:2118–2125. https://doi.org/10.1111/j.1600-6143.2004.00611.x

Li W, Kuhr CS, Zheng XX et al (2008) New insights into mechanisms of spontaneous liver transplant tolerance: the role of Foxp3-expressing CD25+CD4+ regulatory T cells. Am J Transplant 8:1639–1651. https://doi.org/10.1111/j.1600-6143.2008.02300.x

Lin S-J, Lu C-H, Yan D-C et al (2014) Expansion of regulatory T cells from umbilical cord blood and adult peripheral blood CD4(+)CD25 (+) T cells. Immunol Res 60:105–111. https://doi.org/10.1007/s12026-014-8488-1

Liu W, Putnam AL, Xu-yu Z et al (2006) CD127 expression inversely correlates with FoxP3 and suppressive function of human CD4+ T reg cells. J Exp Med 203:1701–1711. https://doi.org/10.1084/jem.20060772

Long E, Wood KJ (2009) Regulatory T cells in transplantation: transferring mouse studies to the clinic. Transplantation 88:1050–1056. https://doi.org/10.1097/TP.0b013e3181bb7913

Ma A, Qi S, Song L et al (2011) Adoptive transfer of CD4+CD25+ regulatory cells combined with low-dose sirolimus and anti-thymocyte globulin delays acute rejection of renal allografts in Cynomolgus monkeys. Int Immunopharmacol 11:618–629. https://doi.org/10.1016/j.intimp.2010.11.001

MacDonald KG, Hoeppli RE, Huang Q et al (2016) Alloantigen-specific regulatory T cells generated with a chimeric antigen receptor. J Clin Invest 126:1413–1424. https://doi.org/10.1172/JCI82771

Macian F (2005) NFAT proteins: key regulators of T-cell development and function. Nat Rev Immunol 5:472–484. https://doi.org/10.1038/nri1632

Marín E, Cuturi MC, Moreau A (2018) Tolerogenic dendritic cells in solid organ transplantation: where do we stand? Front Immunol 9. https://doi.org/10.3389/fimmu.2018.00274

Maude SL, Frey N, Shaw PA et al (2014) Chimeric antigen receptor T cells for sustained remissions in leukemia. N Engl J Med 371:1507–1517. https://doi.org/10.1056/NEJMoa1407222

Min W-P, Zhou D, Ichim TE et al (2003) Inhibitory feedback loop between tolerogenic dendritic cells and regulatory T cells in transplant tolerance. J Immunol 170:1304–1312. https://doi.org/10.4049/jimmunol.170.3.1304

Miroux C, Morales O, Ghazal K et al (2012) In vitro effects of cyclosporine A and tacrolimus on regulatory T-cell proliferation and function. Transplantation 94:123–131. https://doi.org/10.1097/TP.0b013e3182590d8f

Miyara M, Yoshioka Y, Kitoh A et al (2009) Functional delineation and differentiation dynamics of human CD4+ T cells expressing the FoxP3 transcription factor. Immunity 30:899–911. https://doi.org/10.1016/j.immuni.2009.03.019

Nadig SN, Wieckiewicz J, Wu DC et al (2010) In vivo prevention of transplant arteriosclerosis by ex vivo-expanded human regulatory T cells. Nat Med 16:809–813. https://doi.org/10.1038/nm.2154

Nishimura E, Sakihama T, Setoguchi R et al (2004) Induction of antigen-specific immunologic tolerance by in vivo and in vitro antigen-specific expansion of naturally arising Foxp3+CD25+CD4+ regulatory T cells. Int Immunol 16:1189–1201. https://doi.org/10.1093/intimm/dxh122

Noyan F, Zimmermann K, Hardtke-Wolenski M et al (2017) Prevention of allograft rejection by use of regulatory T cells with an MHC-specific chimeric antigen receptor. Am J Transplant 17:917–930. https://doi.org/10.1111/ajt.14175

Ochando J, Ordikhani F, Jordan S et al (2019) Tolerogenic dendritic cells in organ transplantation. Transpl Int 33:113. https://doi.org/10.1111/tri.13504

Ojo AO, Held PJ, Port FK et al (2003) Chronic renal failure after transplantation of a nonrenal organ. N Engl J Med 349:931–940. https://doi.org/10.1056/NEJMoa021744

Okada M, Kanamori M, Someya K et al (2017) Stabilization of Foxp3 expression by CRISPR-dCas9-based epigenome editing in mouse primary T cells. Epigenetics Chromatin 10:24. https://doi.org/10.1186/s13072-017-0129-1

Papp G, Boros P, Nakken B et al (2017) Regulatory immune cells and functions in autoimmunity and transplantation immunology. Autoimmun Rev 16:435–444. https://doi.org/10.1016/j.autrev.2017.03.011

Polansky JK, Kretschmer K, Freyer J et al (2008) DNA methylation controls Foxp3 gene expression. Eur J Immunol 38:1654–1663. https://doi.org/10.1002/eji.200838105

Porter DL, Levine BL, Kalos M et al (2011) Chimeric antigen receptor-modified T cells in chronic lymphoid leukemia. N Engl J Med 365:725–733. https://doi.org/10.1056/NEJMoa1103849

Putnam AL, Brusko TM, Lee MR et al (2009) Expansion of human regulatory T-cells from patients with type 1 diabetes. Diabetes 58:652–662. https://doi.org/10.2337/db08-1168

Putnam AL, Safinia N, Medvec A et al (2013) Clinical grade manufacturing of human alloantigen-reactive regulatory T cells for use in transplantation. Am J Transplant 13:3010–3020. https://doi.org/10.1111/ajt.12433

Qureshi OS, Zheng Y, Nakamura K et al (2011) Trans-endocytosis of CD80 and CD86: a molecular basis for the cell-extrinsic function of CTLA-4. Science 332:600–603. https://doi.org/10.1126/science.1202947

Raffin C, Vo LT, Bluestone JA (2020) Treg cell-based therapies: challenges and perspectives. Nat Rev Immunol 20:158–172. https://doi.org/10.1038/s41577-019-0232-6

Riley JL, June CH, Blazar BR (2009) Human T regulatory cell therapy: take a billion or so and call me in the morning. Immunity 30:656–665. https://doi.org/10.1016/j.immuni.2009.04.006

Roncarolo MG, Gregori S, Bacchetta R, Battaglia M (2014) Tr1 cells and the counter-regulation of immunity: natural mechanisms and therapeutic applications. Curr Top Microbiol Immunol 380:39–68. https://doi.org/10.1007/978-3-662-43492-5_3

Rosen SJ, Harris PE, Hardy MA (2018) State of the art: role of the dendritic cell in induction of allograft tolerance. Transplantation 102:1603–1613. https://doi.org/10.1097/TP.0000000000002239

Sakaguchi S, Sakaguchi N, Asano M et al (1995) Immunologic self-tolerance maintained by activated T cells expressing IL-2 receptor alpha-chains (CD25). Breakdown of a single mechanism of self-tolerance causes various autoimmune diseases. J Immunol 155:1151–1164

Sakaguchi S, Wing K, Onishi Y et al (2009) Regulatory T cells: how do they suppress immune responses? Int Immunol 21:1105–1111. https://doi.org/10.1093/intimm/dxp095

Sakaguchi S, Miyara M, Costantino CM, Hafler DA (2010) FOXP3+ regulatory T cells in the human immune system. Nat Rev Immunol 10:490–500. https://doi.org/10.1038/nri2785

San Segundo D, Fábrega E, López-Hoyos M, Pons F (2007) Reduced numbers of blood natural regulatory T cells in stable liver transplant recipients with high levels of calcineurin inhibitors. Transplant Proc 39:2290–2292. https://doi.org/10.1016/j.transproceed.2007.07.076

Schmidt A, Oberle N, Krammer PH (2012) Molecular mechanisms of Treg-mediated T cell suppression. Front Immunol 3. https://doi.org/10.3389/fimmu.2012.00051

Scottà C, Esposito M, Fazekasova H et al (2013) Differential effects of rapamycin and retinoic acid on expansion, stability and suppressive qualities of human CD4(+)CD25(+)FOXP3(+) T regulatory cell subpopulations. Haematologica 98:1291–1299. https://doi.org/10.3324/haematol.2012.074088

Seay HR, Putnam AL, Cserny J et al (2017) Expansion of human Tregs from cryopreserved umbilical cord blood for GMP-compliant autologous adoptive cell transfer therapy. Mol Ther Methods Clin Dev 4:178–191. https://doi.org/10.1016/j.omtm.2016.12.003

Seddiki N, Santner-Nanan B, Martinson J et al (2006) Expression of interleukin (IL)-2 and IL-7 receptors discriminates between human regulatory and activated T cells. J Exp Med 203:1693–1700. https://doi.org/10.1084/jem.20060468

Segundo DS, Ruiz JC, Izquierdo M et al (2006) Calcineurin inhibitors, but not rapamycin, reduce percentages of CD4+CD25+FOXP3+ regulatory T cells in renal transplant recipients. Transplantation 82:550–557. https://doi.org/10.1097/01.tp.0000229473.95202.50

Shalev I, Schmelzle M, Robson SC, Levy G (2011) Making sense of regulatory T cell suppressive function. Semin Immunol 23:282–292. https://doi.org/10.1016/j.smim.2011.04.003

Shevach EM, Thornton AM (2014) tTregs, pTregs, and iTregs: similarities and differences. Immunol Rev 259:88–102. https://doi.org/10.1111/imr.12160

Shimony O, Nagler A, Gellman YN et al (2012) Anti-T lymphocyte globulin (ATG) induces generation of regulatory T cells, at least part of them express activated CD44. J Clin Immunol 32:173–188. https://doi.org/10.1007/s10875-011-9599-2

Shultz LD, Ishikawa F, Greiner DL (2007) Humanized mice in translational biomedical research. Nat Rev Immunol 7:118–130. https://doi.org/10.1038/nri2017

Sicard A, Koenig A, Morelon E et al (2015) Cell therapy to induce allograft tolerance: time to switch to plan B? Front Immunol 6. https://doi.org/10.3389/fimmu.2015.00149

Stephens LA, Malpass KH, Anderton SM (2009) Curing CNS autoimmune disease with myelin-reactive Foxp3 + Treg. Eur J Immunol 39:1108–1117. https://doi.org/10.1002/eji.200839073

Strauss L, Whiteside TL, Knights A et al (2007) Selective survival of naturally occurring human CD4+CD25+Foxp3+ regulatory T cells cultured with rapamycin. J Immunol 178:320–329. https://doi.org/10.4049/jimmunol.178.1.320

Streitz M, Miloud T, Kapinsky M et al (2013) Standardization of whole blood immune phenotype monitoring for clinical trials: panels and methods from the ONE study. Transplant Res 2:17. https://doi.org/10.1186/2047-1440-2-17

Tang Q, Lee K (2012) Regulatory T-cell therapy for transplantation: how many cells do we need? Curr Opin Organ Transplant 17:349–354. https://doi.org/10.1097/MOT.0b013e328355a992

Tang Q, Henriksen KJ, Bi M et al (2004) In vitro-expanded antigen-specific regulatory T cells suppress autoimmune diabetes. J Exp Med 199:1455–1465. https://doi.org/10.1084/jem.20040139

Tarbell KV, Petit L, Zuo X et al (2007) Dendritic cell–expanded, islet-specific CD4+ CD25+ CD62L+ regulatory T cells restore normoglycemia in diabetic NOD mice. J Exp Med 204:191–201. https://doi.org/10.1084/jem.20061631

Tey S-K (2014) Adoptive T-cell therapy: adverse events and safety switches. Clin Transl Immunol 3:e17. https://doi.org/10.1038/cti.2014.11

Thomson AW, Turnquist HR, Raimondi G (2009) Immunoregulatory functions of mTOR inhibition. Nat Rev Immunol 9:324–337. https://doi.org/10.1038/nri2546

Thomson AW, Metes DM, Ezzelarab MB, Raïch-Regué D (2019) Regulatory dendritic cells for human organ transplantation. Transplant Rev 33:130–136. https://doi.org/10.1016/j.trre.2019.05.001

Thornton AM, Shevach EM (2000) Suppressor effector function of CD4+CD25+ immunoregulatory T cells is antigen nonspecific. J Immunol 164:183–190. https://doi.org/10.4049/jimmunol.164.1.183

Todo S, Yamashita K, Goto R et al (2016) A pilot study of operational tolerance with a regulatory T-cell-based cell therapy in living donor liver transplantation. Hepatology 64:632–643. https://doi.org/10.1002/hep.28459

Tresoldi E, Dell'Albani I, Stabilini A et al (2011) Stability of human rapamycin-expanded CD4+CD25+ T regulatory cells. Haematologica 96:1357–1365. https://doi.org/10.3324/haematol.2011.041483

Trzonkowski P, Bieniaszewska M, Juścińska J et al (2009) First-in-man clinical results of the treatment of patients with graft versus host disease with human ex vivo expanded CD4+CD25+CD127- T regulatory cells. Clin Immunol 133:22–26. https://doi.org/10.1016/j.clim.2009.06.001

Tsang JY-S, Tanriver Y, Jiang S et al (2008) Conferring indirect allospecificity on CD4+CD25+ Tregs by TCR gene transfer favors transplantation tolerance in mice. J Clin Invest 118:3619–3628. https://doi.org/10.1172/JCI33185

Valenzuela NM, Reed EF (2017) Antibody-mediated rejection across solid organ transplants: manifestations, mechanisms, and therapies. J Clin Invest 127:2492–2504. https://doi.org/10.1172/JCI90597

Venken K, Thewissen M, Hellings N et al (2007) A CFSE based assay for measuring CD4+CD25+ regulatory T cell mediated suppression of auto-antigen specific and polyclonal T cell responses. J Immunol Methods 322:1–11. https://doi.org/10.1016/j.jim.2007.01.025

Watanabe T, Masuyama J, Sohma Y et al (2006) CD52 is a novel costimulatory molecule for induction of CD4+ regulatory T cells. Clin Immunol 120:247–259. https://doi.org/10.1016/j.clim.2006.05.006

Watt KDS, Pedersen RA, Kremers WK et al (2010) Evolution of causes and risk factors for mortality post-liver transplant: results of the NIDDK long-term follow-up study. Am J Transplant 10:1420–1427. https://doi.org/10.1111/j.1600-6143.2010.03126.x

Wing K, Sakaguchi S (2010) Regulatory T cells exert checks and balances on self tolerance and autoimmunity. Nat Immunol 11:7–13. https://doi.org/10.1038/ni.1818

Wood KJ, Goto R (2012) Mechanisms of rejection: current perspectives. Transplantation 93:1–10. https://doi.org/10.1097/TP.0b013e31823cab44

Wood KJ, Bushell A, Hester J (2012) Regulatory immune cells in transplantation. Nat Rev Immunol 12:417–430. https://doi.org/10.1038/nri3227

Wu DC, Hester J, Nadig SN et al (2013) Ex vivo expanded human regulatory T cells can prolong survival of a human islet allograft in a humanized mouse model. Transplantation 96:707–716. https://doi.org/10.1097/TP.0b013e31829fa271

Xia G, He J, Leventhal JR (2008) Ex vivo-expanded natural CD4+CD25+ regulatory T cells synergize with host T-cell depletion to promote long-term survival of allografts. Am J Transplant 8:298–306. https://doi.org/10.1111/j.1600-6143.2007.02088.x

Zeiser R, Leveson-Gower DB, Zambricki EA et al (2008) Differential impact of mammalian target of rapamycin inhibition on CD4+CD25+Foxp3+ regulatory T cells compared with conventional CD4+ T cells. Blood 111:453–462. https://doi.org/10.1182/blood-2007-06-094482

Zhou P, Borojevic R, Streutker C et al (2004) Expression of dual TCR on DO11.10 T cells allows for ovalbumin-induced oral tolerance to prevent T cell-mediated colitis directed against unrelated enteric bacterial antigens. J Immunol 172:1515–1523. https://doi.org/10.4049/jimmunol.172.3.1515

The manufacturer's authorised representative in the EU is Springer Nature Customer Service Centre GmbH, Europaplatz 3, 69115 Heidelberg, Germany. If you have any concerns regarding our products, please contact ProductSafety@springernature.com

Printed and bound by CPI Group (UK) Ltd, Croydon, CR0 4YY
30/09/2026
02225317-0020